Frontier Technology for Water Treatment and Pollutant Removal

In Focus – Special Book Series

Frontier Technology for Water Treatment and Pollutant Removal

Editors

Puangrat Kajitvichyanukul, Charongpun Musikavong and Tanapon Phenrat

IWA PUBLISHING

Published by **IWA Publishing**
Alliance House
12 Caxton Street
London SW1H 0QS, UK
Telephone: +44 (0)20 7654 5500
Fax: +44 (0)20 7654 5555
Email: publications@iwap.co.uk
Web: www.iwapublishing.com

First published 2019

British Library Cataloguing in Publication Data
A CIP catalogue record for this book is available from the British Library

ISBN: 9781789061239 (paperback)
ISBN: 9781789061246 (eBook)
ISBN: 9781789061253 (ePUB)

Contents

Editorial: Frontier technology for water treatment and pollutant removal is key for securing the present, correcting the past, and sustaining the future

Presently, the World Health Organization (WHO 2018) reported that 844 million people worldwide cannot access adequate basic drinking-water service. Consequently, the UN Environment's (2019) sixth Global Environment Outlook report revealed that inaccessibility of clean water causes 1.4 million premature deaths annually due to pathogens and toxic chemicals from unsustainable waste and wastewater management. The same report also pointed out that over 80% of the world's wastewater and over 95% in some least developed countries is currently released into the environment without treatment, causing both surface water and groundwater contamination. Thus, appropriate treatment of wastewater before being discharged to the environment is essentially the key to securing public health and quality of life.

Moreover, water decontamination is needed for correcting unintended consequences of past improper practices. As of 2018, 1,566 contaminated sites have been listed on the National Priority List (NPL) in the USA, and only 381 have been sufficiently restored to meet acceptable risk and have been announced for site deletion (US EPA 2018b). In addition, in 2011, approximately 1.2 million potentially contaminated sites and 130,000 contaminated sites were identified in the European Economic Area (European Environment Agency 2014). The costs to remediate these contaminated sites will be large. For example, the US EPA (2004) estimated that, from 2004 to 2033, the cost for remediation of US contaminated sites may be as high as USD 209 billion for as many as 294,000 sites (US EPA 2004). Similarly, for the EU, annual national expenditures for managing contaminated sites are, on average, EUR 10.7 per capita (an average of 0.041% of the national gross domestic product).

However, remediation yields benefits to society from the redevelopment of contaminated sites and cost savings from avoiding health risks from hazardous substance exposure (US EPA 2011). For example, the annual national benefit from redevelopment at Superfund sites was estimated to be USD 54.8 billion for 2017 alone (US EPA 2018a). Finally, for the sustainable future, water reuse is the key for transferring from a linear economy to a circular economy (Voulvoulis 2018). Thus, frontier technology in water treatment and pollutant removal is needed not only for maximizing water reuse but also for rapid detection of contaminants in the recycled water, which requires special care.

For all the reasons above, unsurprisingly, the UN (2018) announced the years 2018 to 2028 as the 'International Decade for Action – Water for Sustainable Development'. To realize this mission, innovative and frontier technologies for water treatment and pollutant removal are important components. This special issue aims to serve as a platform for updating the scientific community with recent progress in this area. We called for submissions from both general authors and presenters of the third Regional IWA Diffuse Pollution Conference in Thailand (19–22 November 2018). From 79 submissions, 25 were accepted, covering frontier technologies in analytical technique, physicochemical treatment, chemical treatment, and biological treatment.

For frontier technologies in analytical techniques, this special issue covers various novel approaches from instrumental-scale to field-scale assessments:

(1) Instrumentally adjusting the atomization and delay times of flame atomic absorption spectrophotometry for reliable copper analysis (Supong & Usapein 2019).
(2) Laboratory-scale examination of malachite green and azure blue for colorimetric analysis of selenium using UV-Vis (Mohapatra & Kirpalani 2019).
(3) Laboratory-scale evaluation of fluorescence excitation–emission matrix spectroscopy (EEMS) to assess the total trihalomethane formation potentials (TTHMFPs) (Jutaporn *et al.* 2019).
(4) Using multivariate analysis with bacteriophages of *Enterococcus faecalis* as sewage-specific fecal indicators to assess coastal water quality (Kongprajug *et al.* 2019).
(5) Calibration and field-scale demonstration of using a molecularly imprinted membrane-passive sampler for polychlorinated biphenyl monitoring (Mkhize *et al.* 2019).

These technologies can serve as a research prototype for assessing water quality parameters and various contaminants including metals, oxyanions, chlorinated organics, and disinfection by-products.

For frontier technologies in physiochemical treatment, various kinds of novel sorbents including triamine-bearing activated rice husk ash (Phan *et al.* 2019), porous pig bone char (Wongrueng *et al.* 2019), modified coconut fiber biochar (Binh & Kajitvichyanukul 2019), and brewery sludge (Kulkarni *et al.* 2019) were studied for broad ranges of

doi: 10.2166/wst.2019.138

contaminants, including DBP precursors, metals, nitrate, pesticides. Moreover, the potential of combining pretreatment techniques including coagulation with ferric chloride and adsorption with activated carbon with thin-film composite-forward osmosis (TFC-FO) filtration for ink printing wastewater treatment was evaluated in terms of treatment and biofouling reduction efficacies (Maknakorn *et al.* 2019). Finally, terracotta aggregate material was evaluated as a low-cost filter for organic matter and fecal coliform (FC) removal at the laboratory scale (Jamwal *et al.* 2019).

For frontier technologies in chemical treatment, various novel catalysts were synthesized and evaluated for their efficacy for pollutant removal. These catalysts are (1) green TiO_2 nanoparticles synthesized from *Coffea arabica* L. in the photocatalytic process (Phuinthiang & Kajitvichyanukul 2019), (2) Co_3O_4/TiO_2 hetero-structure (Chowdhury *et al.* 2019), and (3) TiO_2 coated on PVDF membrane (Penboon *et al.* 2019). Moreover, advances in UV-based treatment techniques are presented. These include (1) using the $UV/S_2O_8^{2-}$ photo-oxidation process in the presence of Al_2O_3 to treat dexamethasone (Shookohi *et al.* 2019), (2) using UV-based activation of peroxymonosulfate and persulfate undergoing copper-redox catalyzing reactions for degradation of tetracycline (Hu *et al.* 2019), and (3) using UV photolysis and UV/chloramine for removal of natural organic matter accountable for disinfection byproduct formation (Hirun-Utok & Phattarapattamawong 2019). In addition, electrochemical oxidation, including a solution plasma process catalyzed by incidentally co-generated copper nanoparticles (Prasertsung *et al.* 2019) and electro-oxidation using boron-doped diamond (Agustina *et al.* 2019), were thoroughly examined.

For frontier technologies in biological treatment, various alternatives for nutrient treatment technology are presented, such as the following:

(1) Coupled anaerobic ammonium oxidation and hydrogenotrophic denitrification for simultaneous NH_4-N and NO_3-N removal (Kamei *et al.* 2019).
(2) Characterization of ammonia-oxidizing bacterial consortia from marine and brackish water (Preena *et al.* 2019).
(3) Use of three-cell entrapment materials including barium alginate (BA), polyvinyl alcohol (PVA), and a mixture of polyvinyl alcohol and barium alginate (PVA-BA) to protect ammonia-oxidizing bacteria from silver nanoparticle toxicity (Giao *et al.* 2019).
(4) Application of pretreatment basins of low impact development (LID) technologies to nutrient reduction (both N and P) (Pascale *et al.* 2019).

Moreover, magnetic nanoparticles were studied for their capability to increase the aggregation and settleability of the sludge. Finally, enzymatic hydrolysis with a commercial lipase was used to treat floatable fatty wastes from dairy and meat food-processing industries.

Finally, the guest editors hope that the findings and discussions included in this special issue will contribute to frontier water treatment and pollutant removal technology for securing the present, correcting the past, and sustaining the future.

Guest Editors
Puangrat Kajitvichyanukul IWA
Naresuan University, Thailand

Charongpun Musikavong
Prince of Songkla University, Thailand

Tanapon Phenrat IWA
Naresuan University, Thailand

REFERENCES

Agustina, F., Bagastyo, A. Y. & Nurhayati, E. 2019 Electro-oxidation of landfill leachate using boron-doped diamond: role of current density, pH and ions. *Water Sci. Technol.* **79** (5), 921–928.

Binh, Q. A. & Kajitvichyanukul, P. 2019 Adsorption mechanism of dichlorvos onto coconut fibre biochar: the significant dependence of H-bonding and the pore-filling mechanism. *Water Sci. Technol.* **79** (5), 866–876.

Chowdhury, M., Kapinga, S., Cummings, F. & Fester, V. 2019 Co_3O_4/TiO_2 hetero-structure for methyl orange dye degradation. *Water Sci. Technol.* **79** (5), 947–957.

European Environment Agency 2014 *Progress in Management of Contaminated Sites*. Agency EE, Copenhagen, Denmark.

Giao, N. T., Limpiyakorn, T., Thuptimdang, P., Ratpukdi, T. & Siripattanakul-Ratpukdi, S. 2019 Reduction of silver nanoparticle toxicity affecting ammonia oxidation using cell entrapment technique. *Water Sci. Technol.* **79** (5), 1007–1016.

Hirun-Utok, C. & Phattarapattamawong, S. 2019 Degradation and transformation of natural organic matter accountable for disinfection byproduct formations by UV photolysis and UV/chlor(am)ine. *Water Sci. Technol.* **79** (5), 929–937.

Hu, J., Zhang, J., Wang, Q., Ye, Q., Xu, H., Zhou, G. & Lu, J. 2019 Efficient degradation of tetracycline by ultraviolet-based activation of peroxymonosulfate and persulfate. *Water Sci. Technol.* **79** (5), 911–920.

Jamwal, P., Phillips, D. & Karlsrud, K. 2019 Assessing local materials for the treatment of wastewater in open drains. *Water Sci. Technol.* **79** (5), 895–904.

Jutaporn, P., Laolertworakul, W., Armstrong, M. D. & Coronell, O. 2019 Fluorescence spectroscopy for assessing trihalomethane precursors removal by MIEX resin. *Water Sci. Technol.* **79** (5), 820–832.

Kamei, T., Eamrat, R., Shinoda, K., Tanaka, Y. & Kazama, F. 2019 Coupled anaerobic ammonium oxidation and hydrogenotrophic denitrification for simultaneous NH_4-N and NO_3-N removal. *Water Sci. Technol.* **79** (5), 975–984.

Kongprajug, A., Booncharoen, N., Jantakee, K., Chyerochana, N., Mongkolsuk, S. & Sirikanchana, K. 2019 Sewage-specific enterococcal bacteriophages and multiple water quality parameters for coastal water quality assessment. *Water Sci. Technol.* **79** (5), 799–807.

Kulkarni, R. M., Shetty, K. V. & Srinikethan, G. 2019 Kinetic and equilibrium modeling of biosorption of nickel (II) and cadmium (II) on brewery sludge. *Water Sci. Technol.* **79** (5), 888–894.

Maknakorn, W., Jutaporn, P. & Khongnakorn, W. 2019 Coagulation and adsorption as pretreatments of thin-film composite–forward osmosis (TFC-FO) for ink printing wastewater treatment. *Water Sci. Technol.* **79** (5), 877–887.

Mkhize, D. S., Quinn, L. P., Nyoni, H., Mamba, B. B. & Msagati, T. A. M. 2019 Calibration and field application of a molecularly imprinted membrane-passive sampler for the sampling of indicator polychlorinated biphenyls in selected aquatic environments of South Africa. *Water Sci. Technol.* **79** (5), 808 819.

Mohapatra, D. P. & Kirpalani, D. M. 2019 Selenium in wastewater: fast analysis method development and advanced oxidation treatment applications. *Water Sci. Technol.* **79** (5), 842–849.

Pascale, N. C., Chastinet, J. J., Bila, D. M., Sant`Anna Jr, G. L., Quitério, S. L. & Vendramel, S. M. R. 2019 Enzymatic hydrolysis of floatable fatty wastes from dairy and meat food-processing industries and further anaerobic digestion. *Water Sci. Technol.* **79** (5), 985–992.

Penboon, L., Khrueakham, A. & Sairiam, S. 2019 Tio$_2$ coated on PVDF membrane for dye wastewater treatment by a photocatalytic membrane. *Water Sci. Technol.* **79** (5), 958–966.

Phan, P. T., Nguyen, T. T., Nguyen, N. H. & Padungthon, S. 2019 Triamine-bearing activated rice husk ash as an advanced functional material for nitrate removal from aqueous solution. *Water Sci. Technol.* **79** (5), 850–856.

Phuinthiang, P. & Kajitvichyanukul, P. 2019 Degradation of paraquat from contaminated water using green TiO_2 nanoparticles synthesized from *Coffea arabica* L. in photocatalytic process. *Water Sci. Technol.* **79** (5), 905–910.

Prasertsung, I., Kaewcharoen, S., Kunpinit, K., Yaowarayt, W., Saito, N. & Phenrat, T. 2019 Enhanced degradation of methylene blue by a solution plasma process catalyzed by incidentally co-generated copper nanoparticles. *Water Sci. Technol.* **79** (5), 967–974.

Preena, P. G., Achuthan, C., Rejish Kumar, V. J., Boobal, R., Deepa, G. D., Puthumana, J., Poulose, S., Surekhamol, I. S. & Bright Singh, I. S. 2019 Community composition of marine and brackish water ammonia-oxidizing consortia developed for aquaculture application. *Water Sci. Technol.* **79** (5), 1017–1028.

Shookohi, R., Faraji, H., Arabkohsar, A., Salari, M. & Mahmoudi, M. M. 2019 The efficiency of UV/$S_2O_8^{2-}$ photo-oxidation process in the presence of Al_2O_3 for the removal of dexamethasone from aqueous solution: kinetic studies. *Water Sci. Technol.* **79** (5), 938–946.

Supong, K. & Usapein, P. 2019 Reliable determination of copper complex ions in synthetic wastewater using flame atomic absorption spectrophotometry. *Water Sci. Technol.* **79** (5), 833–841.

UN Environment 2019 *Global Environmental Outlook-GEO-6: Healthy Planet, Healthy People.* Press CU, Nairobi.

United Nations 2018 International Decade for Action on Water for Sustainable Development, 2018–2028. https://www.un.org/en/events/waterdecade/.

US EPA 2004 *Cleaning Up the Nation's Waste Sites: Markets and Technology Trends.* Office of Solid Waste and Emergency Response, US EPA, Washington, DC.

US EPA 2011 *Beneficial Effects of the Superfund Program.* Office of Superfund Remediation and Technology Innovation, Washington, DC.

US EPA 2018a *Redevelopment Economics at Superfund Sites.* https://www.epa.gov/superfund-redevelopment-initiative/redevelopment-economics-superfund-sites (accessed 23 February 2018).

US EPA 2018b *Superfund: National Priorities List (NPL).* https://www.epa.gov/superfund/superfund-national-priorities-list-npl (accessed 22 February 2018).

Voulvoulis, N. 2018 Water reuse from a circular economy perspective and potential risks from an unregulated approach. *Curr. Opin. Env. Sci. Health* **2**, 32–45.

Wongrueng, A., Rakruam, P., Siri, A. & Siyasukh, A. 2019 Synthesis of porous pig bone char as adsorbent for removal of DBP precursors from surface water. *Water Sci. Technol.* **79** (5), 857–865.

World Health Organization 2018 Drinking-Water. In: *Fact Sheets*, https://www.who.int/news-room/fact-sheets/detail/drinking-water.

Sewage-specific enterococcal bacteriophages and multiple water quality parameters for coastal water quality assessment

Akechai Kongprajug, Namfon Booncharoen, Kanyaluck Jantakee, Natcha Chyerochana, Skorn Mongkolsuk and Kwanrawee Sirikanchana

ABSTRACT

Coastal water quality is deteriorating worldwide. Water quality monitoring is therefore essential for public health risk evaluation and the management of water bodies. This study investigated the feasibility of using bacteriophages of *Enterococcus faecalis* as sewage-specific faecal indicators, together with physicochemical (dissolved oxygen, pH, temperature and total suspended solids) and biological parameters, to assess coastal water quality using multivariate analysis incorporating non-detects. The principal component and cluster analyses demonstrated that coastal water quality was mostly influenced by biological parameters, including *Escherichia coli* and total coliforms, which were found in all 31 sampling sites, and enterococci, which was found in all but two sampling sites. The enterococcal bacteriophages AIM06 and SR14 were detected in 17 and 18 samples at concentrations up to 1,815 and 2,790 PFU/100 mL, respectively. Both bacteriophages co-presented in approximately 80% of phage-positive samples, and the concentrations at each site were not significantly different. Overall, either bacteriophage could be used to differentiate high- and low-level coastal water pollution, as grouped by cluster analysis. This study is the first to investigate the suitability of sewage-specific bacteriophages of *E. faecalis* for monitoring coastal water quality and emphasises the importance of a multivariate analysis with non-detects to facilitate coastal water quality monitoring and management.

Key words | bacteriophage of enterococci, coastal water quality, microbial source tracking, multivariate analysis, non-detects, sewage pollution

Akechai Kongprajug
Kanyaluck Jantakee
Natcha Chyerochana
Skorn Mongkolsuk
Kwanrawee Sirikanchana (corresponding author)
Research Laboratory of Biotechnology,
Chulabhorn Research Institute,
54 Kamphaeng Phet 6 Road, Talat Bang Khen,
Lak Si, Bangkok 10210,
Thailand
E-mail: *kwanrawee@cri.or.th*

Namfon Booncharoen
Program in Applied Biological Sciences:
Environmental Health,
Chulabhorn Graduate Institute, Chulabhorn Royal Academy,
Lak Si, Bangkok 10210,
Thailand

Skorn Mongkolsuk
Kwanrawee Sirikanchana
Center of Excellence on Environmental Health and Toxicology,
CHE, Ministry of Education,
Rama VI Road, Bangkok 10400,
Thailand

INTRODUCTION

Coastal water pollution has become a significant problem in both developed and developing countries. Water quality deterioration can affect coastal water use, such as coastal aquaculture, fisheries, recreation and tourism. Diffuse sources from different land use adjacent to coastal areas can aggravate water quality deterioration already impacted by point sources (Molina *et al.* 2014; Yau *et al.* 2014). Monitoring coastal water quality can not only assess the risk to public health, but also identify the inefficient treatment of wastewater from communities or agriculture (Oun *et al.* 2017). Because a range of pollutants can contaminate receiving water bodies, water quality standards in many countries include multiple water quality parameters, including physicochemical and biological parameters (National Environment Board 1994; US EPA 2012). However, the assessment of physicochemical parameters is often insufficient for tracking the exact sources of pollution. Multivariate analysis has been proven appropriate for retrieving information about and associations among water quality parameters (Wangkahad *et al.* 2017). However, environmental monitoring frequently involves negative detections, meaning that the data are below the detection limits of the methods or instruments used: so-called non-detects, or censored data. Importantly, substitution of negative detections with zero or with the detection limits was reported to cause significant biases in subsequent data analysis; thus, this practice should be discouraged (Helsel 2012). An alternative analytical method of datasets containing non-detects is nonparametric survival

doi: 10.2166/wst.2018.460

analysis, which processes the data based on the relative positions of the data (Helsel 2012; Wangkahad *et al.* 2017).

In addition to groups of physicochemical and biological parameters that reflect water quality, another set of biological parameters could differentiate pollution sources. Microbial source tracking (MST) is a frontier technology that utilises intestinal microorganisms specific to each animal type (e.g., human, pig, cattle, fowl) to track sources of faecal contamination in polluted water. MST indicators provide information about pollution sources that could facilitate the management and restoration of polluted water bodies (Teaf *et al.* 2018). Bacteriophages of enterococci are bacterial viruses that specifically infect their enterococcus bacteria. They have been successfully used to track human sewage pollution in many countries, such as the UK (Purnell *et al.* 2018, 2011), Puerto Rico (Santiago-Rodríguez *et al.* 2010), and Thailand (Wangkahad *et al.* 2017; Booncharoen *et al.* 2018). The detection technique is simple and cost-effective, requiring only the culturing of the bacteriophage using a double-layer agar overlay (Wangkahad *et al.* 2015). Sewage-specific enterococcal bacteriophages AIM06 and SR14 isolated in Thailand are highly specific and sensitive, and have performed well as MST indicators in freshwater (Wangkahad *et al.* 2017); however, information about their performance in seawater is limited and, therefore, urgently needed. The objective of this study was to evaluate the performance of sewage-specific bacteriophages of enterococci for assessing coastal water quality using multivariate analysis incorporating non-detects, which used nonparametric survival analysis procedure.

MATERIALS AND METHODS

Water sample collection and quality control

A total of 31 sampling stations were located along the upper part of the Gulf of Thailand (Figure 1 and Table S1, available with the online version of this paper), from Trat province to Prachuab Kirikan province, matching the sampling stations of Thailand's Pollution Control Department. From June to July 2015, at low tide, 3 L water samples were collected at

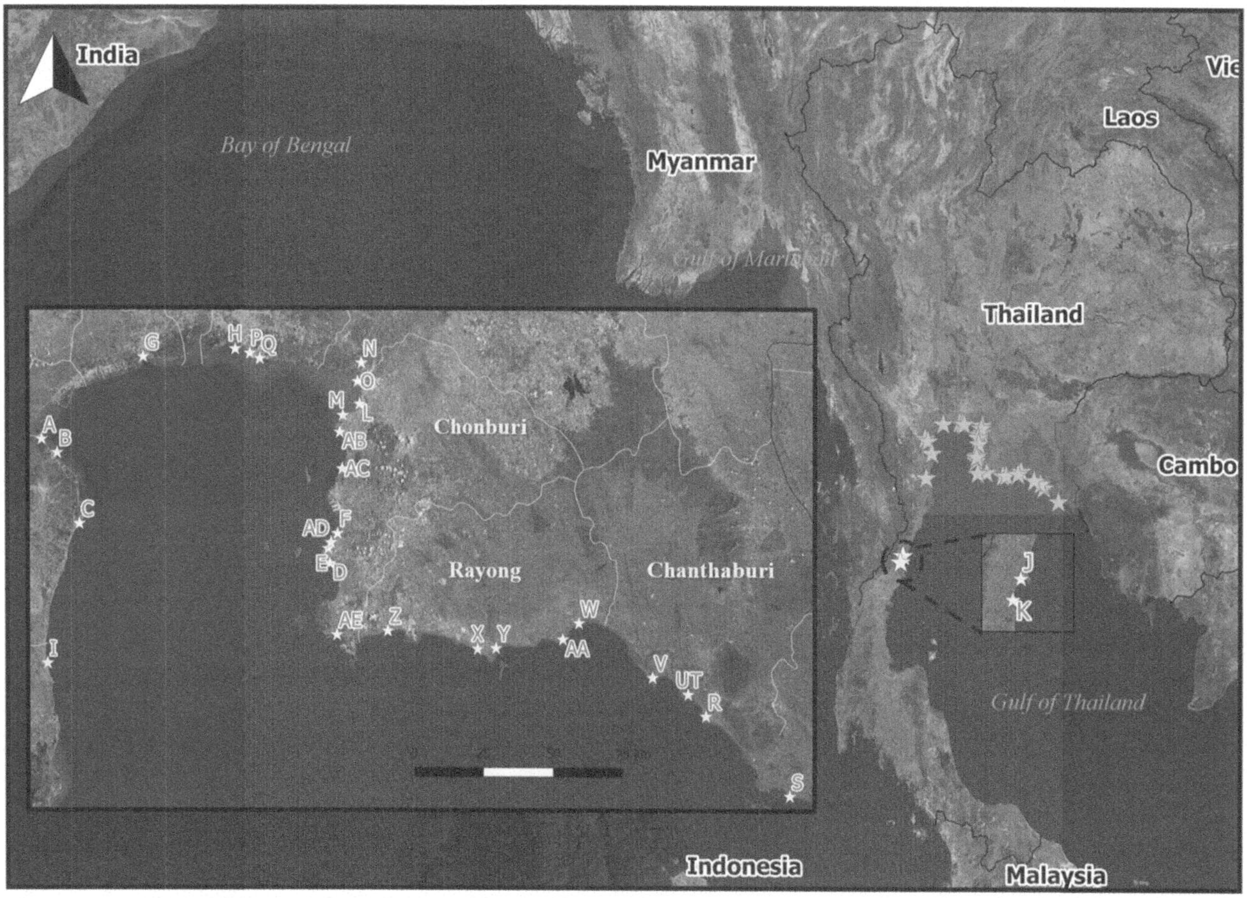

Figure 1 | Locations of coastal water sampling sites along the upper part of the Gulf of Thailand.

30 cm below the sea surface for biological parameters and 1 L was collected at the mid-depth of the water column for total suspended solids (TSS). The samples were stored on ice in sterile bottles before transportation back to the laboratory within 1 day. Dissolved oxygen (DO), pH, and water temperature were measured on site using YSI Pro2030 and YSI60 devices (YSI, Yellow Springs, OH, USA), according to the manufacturer's instructions. Reproducibility of sample preparation and analysis was assessed using field duplicates at two sites, representing low- and high-level microbial pollution (i.e., sites AA and AE, respectively). Three field blank and three laboratory blank samples were also processed. Field blanks were prepared by transporting autoclaved reverse osmosis (RO) water to the field site and then following a protocol similar to that used for the sample collection. Laboratory blanks were prepared by following a protocol similar to that used for processing the sample at the laboratory with autoclaved RO water.

Physicochemical and biological parameters

TSS was analysed using the drying method at 103–105 °C, according to standard protocols (American Public Health Association *et al.* 2017a). Total coliforms and *Escherichia coli* were simultaneously measured using a membrane filtration method with MI medium, containing 4-methylumbelliferyl-β-d-galactopyranoside (MUGal) and indoxyl-β-d-glucuronide (IBDG) (US EPA 2002a). Enterococci were detected with a membrane filtration method using membrane-*Enterococcus* Indoxyl-Beta-D-Glucoside Agar (mEI) (US EPA 2002b). Plates were incubated at 37 °C for up to 24 hours, and the number of colony-forming units (CFU) was recorded separately.

Sewage-specific bacteriophages

Two litres of water sample were centrifuged at 4,000 *g* for 20 min to separate the supernatant and suspended solids. The supernatant was subsequently filtered with a 0.45 μm pore size polyvinylidene fluoride (PVDF) membrane filter to remove bacteria and suspended solids without removing the bacteriophage (Booncharoen *et al.* 2018). One litre each of the resulting filtrate was filtered with a 0.45 μm cellulose nitrate filter (CNF) to retain the bacteriophage on the membrane. Each membrane filter was placed face down on the bacterial lawn of *Enterococcus faecalis* strain AIM06 or strain SR14 (DSM100702 or DSM100701) in tryptic soy agar (Wangkahad *et al.* 2017, 2015). The plates were incubated at 37 °C for 24 hours before the membrane was removed and plaques were counted. Additionally, the sediment pellet resulting from the previous centrifuging step was eluted with 50 ml of the eluent solution (American Public Health Association *et al.* 2017b). The suspension was centrifuged at 4,000 *g* for 20 min, and the pellet was discarded. Each 8–12 mL sample of the supernatant was filtered with a 0.45 μm CNF membrane to retain the bacteriophage on the membrane. Following this protocol, the membrane filter was plated and plaques were counted. The number of plaque-forming units (PFU) for each water sample was the sum of the number from the supernatant portion and the number from the eluate of the sediment portion. To measure the recovery percentage of this bacteriophage detection method, two 1 L representative autoclaved water samples (Sites AF and AG as representatives of high and low bacterial contamination, respectively; Tables S1 and S2, available online) were spiked with 20,000 PFUs of bacteriophages A1 and S4 (Booncharoen *et al.* 2018), representative of bacteriophages infecting *E. faecalis* strains AIM06 and SR14, respectively. The resulting bacteriophages were measured after the water samples were processed as described above. The recovery percentage was calculated as the fraction of bacteriophages recovered.

Statistical analysis

Data analysis of nine parameters, including DO, pH, temperature, TSS, total coliforms, *E. coli*, enterococci, the AIM06 phage and the SR14 phage, was conducted for the 31 samples. The datasets for TSS, enterococci, the AIM06 phage and the SR14 phage contained values lower than the detection limits, or non-detects, requiring statistics for censored data (Helsel 2012). Statistical analysis was performed in the R program (R Core Team 2017). The descriptive statistics for the datasets containing non-detects were calculated using Kaplan-Meier estimates. Normality was tested using the maximum likelihood estimation method, using an adjusted Anderson-Darling value for datasets with non-detects and the Shapiro-Wilk test for those without non-detects. The Mann-Whitney test and Student's t-test were used to compare the two non-normal and normal datasets, respectively. Paired comparisons were conducted for datasets containing non-detects using the generalized Wilcoxon test. All correlation analyses were performed using the non-parametric Kendall's tau analysis using the psych package in R. Multivariate analysis was achieved by first assigning U-score ranks to both censored and uncensored data. Then, a principal component analysis (PCA) was conducted to explore the patterns of multiple variables, and a cluster analysis was performed using Euclidean distance

Table 1 | Descriptive statistics of physicochemical and biological parameters

	Parameters without non-detects					Parameters with non-detects			
	DO (mg/L)	pH	Temperature (°C)	*E. coli* (CFU/100 mL)	Total coliforms (CFU/100 mL)	TSS (mg/L)	Enterococci (CFU/100 mL)	AIM06 phage[a] (PFU/100 mL)	SR14 phage[a] (PFU/100 mL)
Number of values (n)	31	31	31	31	31	31	31	31	31
Normality test[b]	0.033	0.265	0.057	<0.0001	<0.0001	6.909	7.514	7.078	7.316
Passed normality test	No	Yes	Yes	No	No	No	No	No	No
Minimum	1.29	6.87	28.9	50	200	<2.5	<1	<2.6	<3.4
25th percentile	3.41	7.25	29.4	760	2,540	25.0	56	<2.6	<3.4
Median	6.07	7.68	31.2	1,840	5,700	62.0	140	<2.6	<3.4
75th percentile	6.70	7.96	32.0	5,900	49,200	94.0	920	10	10
Maximum	8.81	8.25	34.9	>50,000	>50,000	1,456.0	15,750	1,815	2,790
Mean	5.19	7.64	31.0	10,006	17,041	163.2	1,062	107	143
SD	2.16	0.39	1.6	17,927	20,267	340.0	3,024	384	550

[a]Concentration after multiplying by recovery percentages.
[b]Normality test for datasets without non-detects using the Shapiro-Wilk test (*p*-value) and for datasets with non-detects using maximum likelihood estimation methods (the adjusted Anderson-Darling value).

measures with a k-means algorithm to cluster the group of sampling sites, using the FactoMineR and factoextra packages in R. The optimal number of clusters estimation was computed using the NbClust package in R.

RESULTS AND DISCUSSION

Physicochemical and biological water quality parameters

A total of nine water quality parameters were monitored at 31 sampling sites along the coastal area of Thailand (Table 1). Low DO concentrations (below 4.0 mg/L) were found mostly at sites near estuaries and open markets for fresh produce, meat and seafood. *E. coli* and total coliforms were found at more than 50,000 CFU/100 mL at sites near an estuary, recreational beaches, an open market, and piers, while the highest enterococci (5,750 CFU/100 mL), AIM06 phage and SR14 phage concentrations (1,815 and 2,790 PFU/100 mL, respectively) were detected at a pier. TSS and enterococci datasets contained one and two data points with negative test results, respectively. AIM06 and SR14 phages were detected in 55% (17/31) and 58% (18/31) of the total samples. Both bacteriophages were simultaneously detected in 14 samples, constituting

Table 2 | Reproducibility of analysis calculated from field duplicates

Site ID/statistics	TSS (mg/L)	*E. coli* (CFU/100 mL)	Total coliforms (CFU/100 mL)	Enterococci (CFU/100 mL)	AIM06 phage (PFU/100 mL)	SR14 phage (PFU/100 mL)
AA1	23.0	2,560	4,600	30	2.6	3.4
AA2	29.0	2,480	3,200	32	5.1	3.4
Mean	26.0	2,520	3,900	31	3.8	3.4
SD	4.2	56.6	990.0	1.4	1.8	0.0
%CV	16.3	2.2	25.4	4.6	47.1	0.0
AE1	<2.5	>50,000	>50,000	18,800	938.8	1,279.5
AE2	<2.5	>50,000	>50,000	12,700	1,377.6	1,319.9
Mean	NA[a]	NA	NA	15,750	1,158.2	1,299.7
SD	NA	NA	NA	4,313.4	310.3	28.6
%CV	NA	NA	NA	27.4	26.8	2.2

[a]NA: not applicable.

78–82% of the phage-positive samples. The abundance of AIM06 and SR14 phages in seawater in this study is comparable to their previously reported concentrations in human sewage and freshwater at 2–4 and 1–3 orders of magnitude, respectively (Wangkahad *et al.* 2017). Both bacteriophages in treated wastewater effluents were not detectable, i.e., below a detection limit of 33 PFU/100 mL (Wangkahad *et al.* 2017), implying that the high abundance of bacteriophages in seawater as shown in this study could have derived from untreated sewage.

Quality control and bacteriophage recovery

The reproducibility of the laboratory measurement was investigated using field duplicates of two sampling sites for TSS and microbiological parameters (Table 2). The coefficient of variation (CV) showed values mostly up to 27.4%, with the exception of the AIM06 phage at site AA, for which the CV was 47.1% due to statistical variations in low PFU numbers. Moreover, the three field blanks and three laboratory blanks showed no false detection of any microbial or physicochemical parameters. Furthermore, the recovery percentages for the bacteriophage detection method were 3.92 ± 0.08% and 2.97 ± 0.31% for the AIM06 and SR14 phages, respectively. Although the recovery ratios were lower than those of other methods, this approach required less time and no elution step (Contreras-Coll *et al.* 2002; Méndez *et al.* 2004). The calculated recoveries demonstrated low variation between water samples with high and low bacterial contamination (Sites AF and AG; Table S2). Therefore, the average recovery percentages were used to calculate the actual concentrations of both bacteriophages in all water samples.

Correlations among physicochemical and biological parameters

Among the physicochemical parameters, DO and pH were moderately correlated ($\tau = 0.48$; $p < 0.01$) (Figure 2(a)). Among the biological parameters, *E. coli* showed significant correlations ($p < 0.01$) with total coliforms and enterococci ($\tau = 0.58$ and 0.47, respectively) (Figure 2(b)). Such correlations among bacterial indicators have also been reported in freshwater (Sirikanchana *et al.* 2014; Wangkahad *et al.* 2017, 2015). Moreover, the present study showed a strong correlation between the AIM06 and SR14 phages in seawater ($\tau = 0.60$; $p < 0.001$). This result agrees with previous findings for polluted freshwater, and both bacteriophages have been suggested for use as water quality parameters in such water (Wangkahad *et al.* 2017). In this study, the nonparametric paired Prentice-Wilcoxon test indicated no significant difference between the abundance of AIM06 and SR14 phages in all seawater samples ($p > 0.1$). Consequently, it is suggested that one of the two bacteriophages could be selected to monitor coastal water quality.

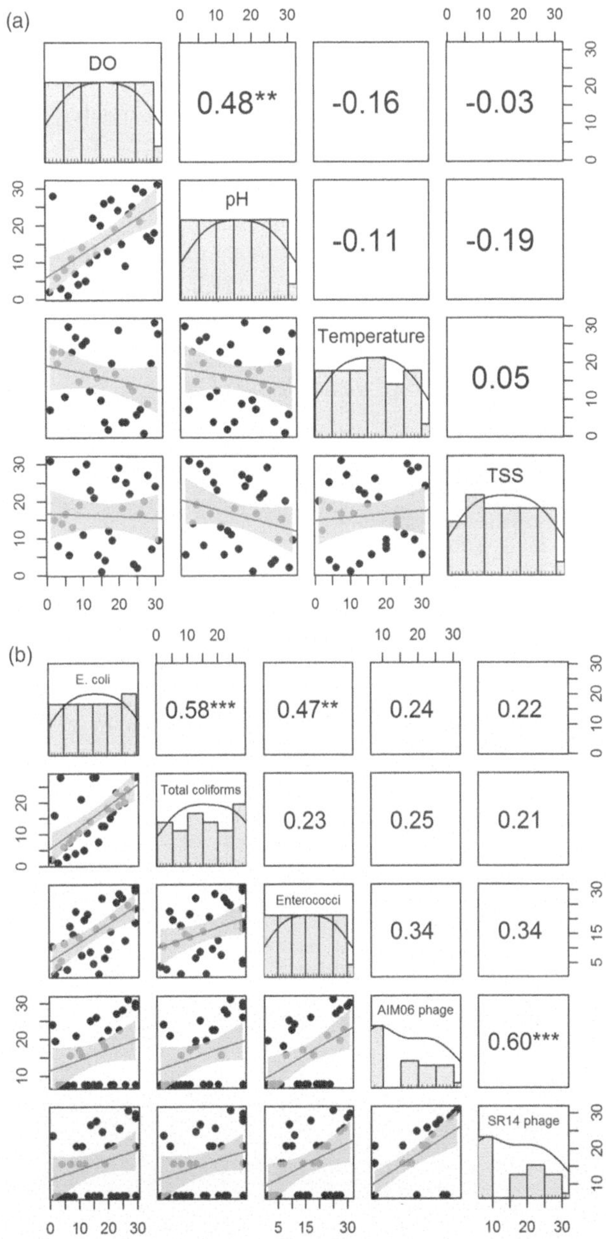

Figure 2 | Kendall's tau correlation coefficient analysis among physicochemical parameters (a) and biological parameters (b). The asterisks indicate significance level: *** ($p < 0.001$), ** ($p < 0.01$) and * ($p < 0.05$). The histogram demonstrates the kernel density estimation and rug plot. The scatter plot with the fitted line and confidence intervals displays the strength of the relationship in the linear fit.

Multivariate and cluster analyses

PCA was performed to convert a set of possibly correlated variables into a set of values of linearly uncorrelated variables called principal components (PCs). The first four PCs showed the eigenvalues of 3.0551, 1.787, 1.2557 and 1.0542, respectively. An eigenvalue of more than 1 indicates that a PC accounts for more variance of all the variables than accounted for by one of the original variables. The cumulative percentage of variance of the first two PCs was 53.8%. The first PC displayed the highest variance of 33.9%, while the second PC showed 19.9% of the variance (Figure S1, available with the online version of this paper). The variables that heavily contributed to the first PC (>11.1% contribution) were enterococci, *E. coli*, AIM06 phage and total coliforms (Figure 3(a)). The variables that heavily contributed to the second PC were temperature, DO, pH and the SR14 phage (Figure 3(b)). In addition, the

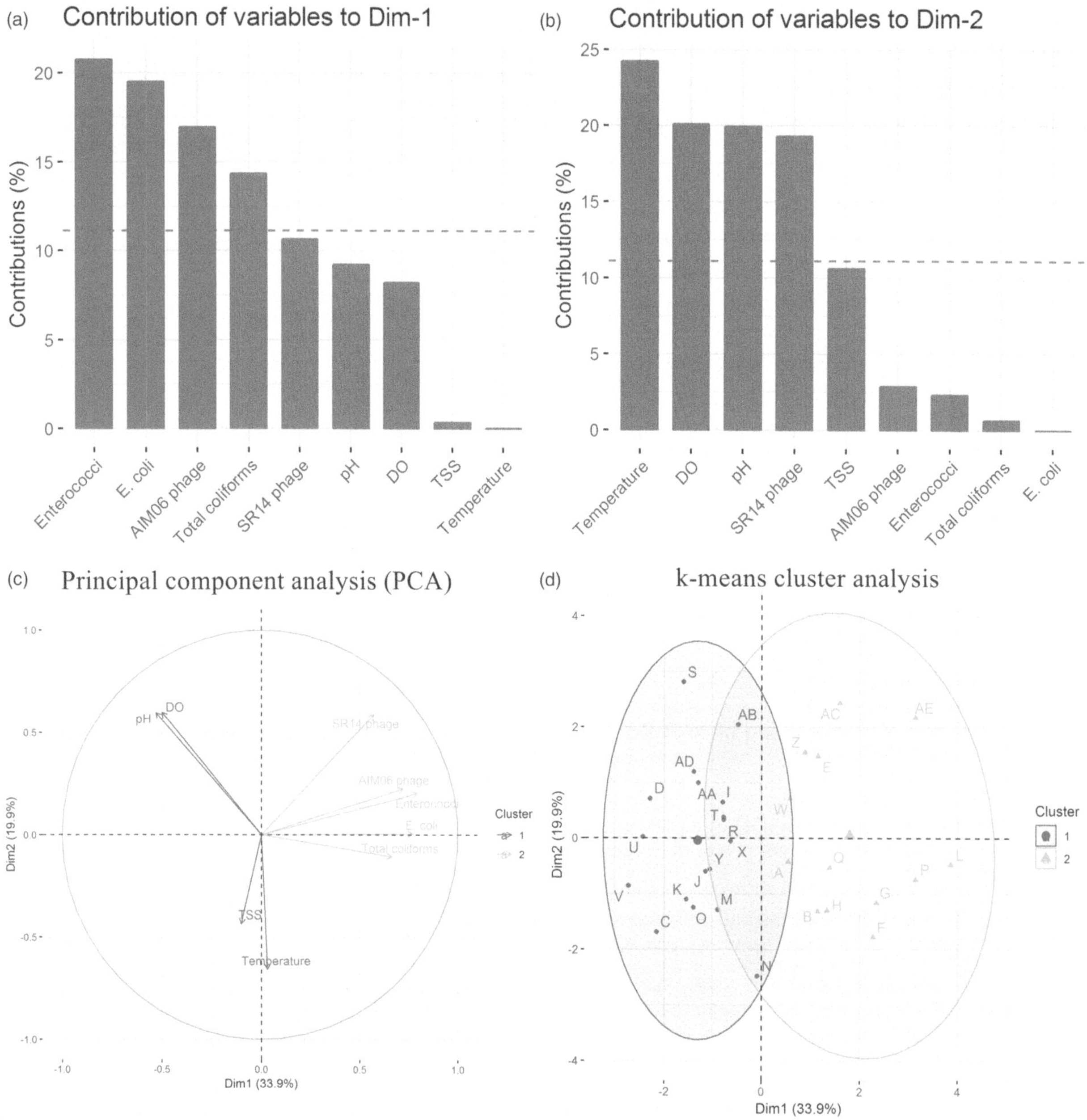

Figure 3 | Multivariate and cluster analyses. The contribution of the variables was extracted to the first PC (a) and the second PC (b). A PCA (c) and cluster analysis of water sampling sites (d) were constructed using Euclidean distance measures with a k-mean algorithm.

variables were classified into separate groups using the k-means clustering algorithm. The optimal number of clusters estimation revealed that two groups were optimal for this dataset (Figure S2, available online). Two groups were clearly classified into physicochemical and biological clusters (Figure 3(c)). The results indicated that both the AIM06 and SR14 phages were associated with the other biological indicators. Furthermore, the cluster analysis could classify the water samples into two groups, comprising 13 sites for the highly contaminated group and 18 sites for the low-contamination group (Figure 3(d)). Cluster centroids indicated that the categorisation of different levels of contamination was mainly affected by the biological parameters, including bacteria and bacteriophage indicators (Table S3, available online).

Comparison of water quality parameters between water samples with high and low levels of contamination

The physicochemical and biological parameters of the two groups with high- and low-level contamination were further analysed (Table S4, available online). DO and pH in the low-contamination group were significantly higher than in the highly contaminated group ($p = 0.0039$ for pH and $p = 0.0156$ for DO) (Figure 4(a) and 4(b)). Temperature and TSS were not significantly different between the two groups, implying that these parameters might not be major parameters indicating coastal water quality. On the other hand, all biological parameters were major parameters influencing coastal water quality. The highly contaminated group had a median *E. coli* concentration of 5,900 CFU/100 mL, which was significantly higher than that of the low-contamination group (1,035 CFU/100 mL; $p = 0.0003$) (Figure 4(c)). A similar trend was observed for total coliforms, for which the highly contaminated group had a median concentration of 24,000 CFU/100 mL, compared to 3,400 CFU/100 mL for the low-contamination group ($p = 0.0048$) (Figure 4(d)).

There were also statistical differences between the groups with high and low levels of contamination for parameters containing non-detected data, comprising enterococci, the AIM06 phage and the SR14 phage. The results showed that the median concentration of enterococci in the highly contaminated group was 1,080 CFU/100 mL, whereas that in the low-contamination group was 81 CFU/100 mL ($p < 0.001$) (Figure 5(a)). The median concentrations of the AIM06 phage were 20.4 PFU/100 mL and below the detection limit (<2.6 PFU/100 mL) in the high- and low-contamination groups, respectively ($p < 0.001$) (Figure 5(b)). The median concentrations of SR14 were 13.5 PFU/100 mL and at the detection limit level (3.4 PFU/100 mL) in the groups with high and low levels of contamination,

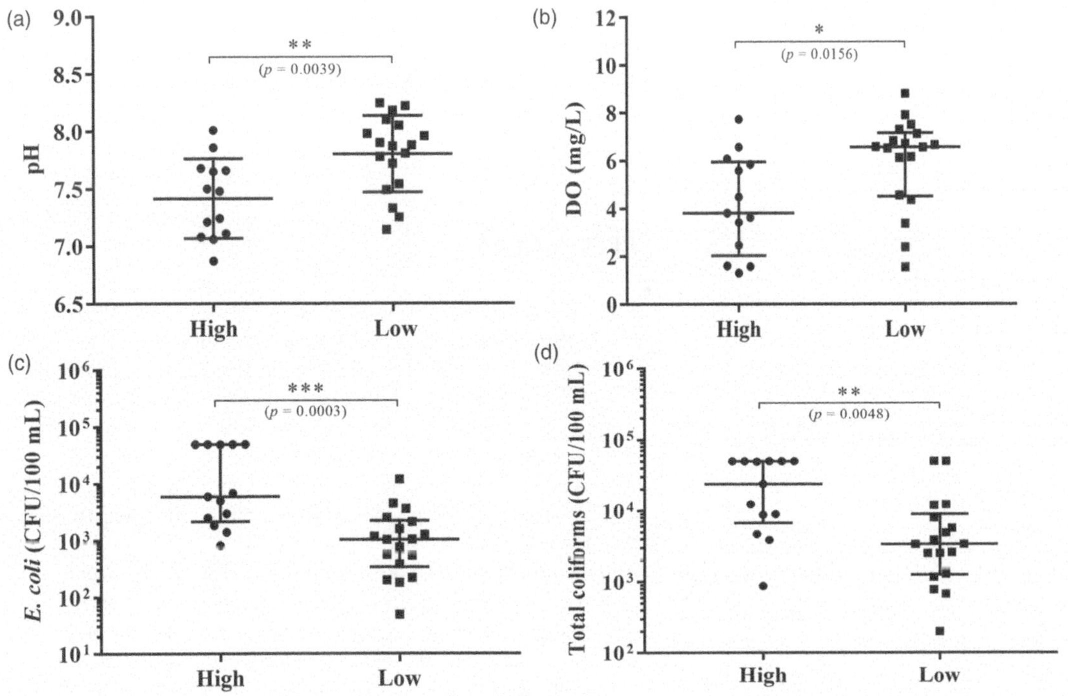

Figure 4 | Scatter plot of parameters without non-detects. The parametric unpaired t-test of pH was represented using mean + SD (a). A non-parametric Mann-Whitney test was used to analyse data on DO (b), *E. coli* (c) and total coliforms (d) and represent them as the median with the interquartile range.

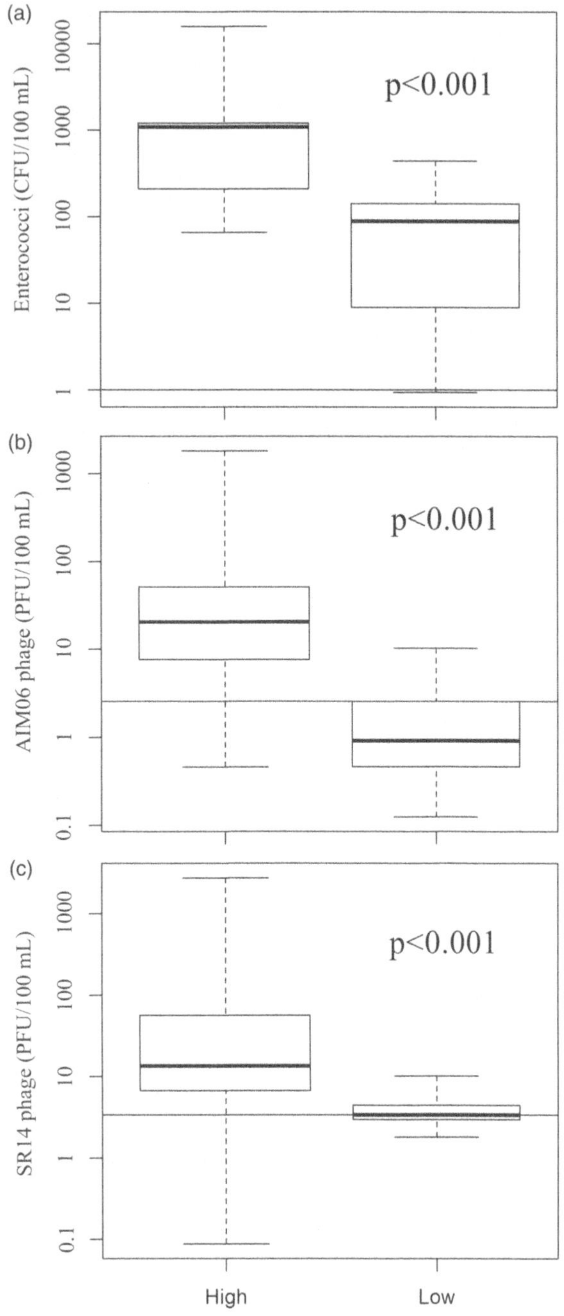

Figure 5 | Box plots of data containing non-detects for enterococci (a), the AIM06 phage (b) and the SR14 phage (c). Three percentiles (25th, 50th and 75th) define the central box, with the maximum and minimum values as whiskers. The horizontal line indicates the detection limit.

respectively ($p = 0.001$) (Figure 5(c)). Although bacteriophages of enterococci have not been monitored in seawater in other geographical regions, sewage-specific enterococcal bacteriophages ENT-49 and ENT-55 were detected at freshwater beaches and harbours at 1–3 orders of magnitude (Vijayavel *et al.* 2014), similar to those found for AIM06 and SR14 phages in this study. The results from this study indicated that both bacteriophages, as well as other bacterial indicators, could be used to assess levels of pollution in coastal water. Because many faecal sources can contribute to deteriorating coastal water quality, the use of microbial source tracking markers to specify human-sewage-specific pollution is, therefore, very beneficial for the management of polluted water bodies (US EPA 2011).

CONCLUSIONS

The sewage-specific AIM06 and SR14 phages were evaluated for their suitability as coastal water quality indicators. Both bacteriophages were present at similar concentrations and were co-present in approximately 80% of the phage-positive samples. The results of this study indicate that microbial contamination is the predominant impact affecting coastal waters along the Gulf of Thailand. This study demonstrated that the sewage-specific AIM06 and SR14 phages were associated with total coliforms, *E. coli* and enterococci. Both bacteriophages could be used to differentiate coastal water with high levels of contamination from coastal water with low levels of contamination.

ACKNOWLEDGEMENTS

This research was financially supported by the Chulabhorn Research Institute. Miss Sirintip Ratthakit is gratefully acknowledged for her assistance in sampling and laboratory analysis.

REFERENCES

American Public Health Association, American Water Works Association, Water Environment Federation 2017a 2540 D. Total Suspended Solids Dried at 103–105°C. In: *Standard Methods for the Examination of Water and Wastewater* (A. D. Eaton, L. S. Clesceri, E. W. Rice & A. E. Greenberg, eds). American Public Health Association, American Water Works Association, Water Environment Federation, Washington, DC, pp. 4–5. Online Edition. Accessed 16 January 2017.

American Public Health Association, American Water Works Association, Water Environment Federation 2017b 9510 F. Recovery of Viruses from Suspended Solids in Water and Wastewater. In: *Standard Methods for the Examination of Water and Wastewater* (A. D. Eaton, L. S. Clesceri, E. W. Rice & A. E. Greenberg, eds). American Public Health Association, American Water Works Association, Water

Environment Federation, Washington, DC, pp. 13–14. Online Edition. Accessed 16 January 2017.

Booncharoen, N., Mongkolsuk, S. & Sirikanchana, K. 2018 Comparative persistence of human sewage-specific enterococcal bacteriophages in freshwater and seawater. *Appl. Microbiol. Biotechnol.* **102**, 6235–6246.

Contreras-Coll, N., Lucena, F., Mooijman, K., Havelaar, A., Pierzo, V., Boque, M., Gawler, A., Holler, C., Lambiri, M., Mirolo, G., Moreno, B., Niemi, M., Sommer, R., Valentin, B., Wiedenmann, A., Young, V. & Jofre, J. 2002 Occurrence and levels of indicator bacteriophages in bathing waters throughout Europe. *Water Res.* **36**, 4963–4974. doi:10.1016/S0043-1354(02)00229-4.

Helsel, D. R. 2012 *Statistics for Censored Environmental Data Using Minitab and R*, 2nd edn. John Wiley & Sons, Inc., Hoboken, New Jersey, USA.

Méndez, J., Audicana, A., Isern, A., Llaneza, J., Moreno, B., Tarancón, M. L., Jofre, J. & Lucena, F. 2004 Standardised evaluation of the performance of a simple membrane filtration-elution method to concentrate bacteriophages from drinking water. *J. Virol. Methods* **117**, 19–25. doi:10.1016/j.jviromet.2003.11.013.

Molina, M., Hunter, S., Cyterski, M., Peed, L. A., Kelty, C. A., Sivaganesan, M., Mooney, T., Prieto, L. & Shanks, O. C. 2014 Factors affecting the presence of human-associated and fecal indicator real-time quantitative PCR genetic markers in urban-impacted recreational beaches. *Water Res.* **64**, 196–208. doi:10.1016/j.watres.2014.06.036.

National Environment Board 1994 Notification of the National Environmental Board: Surface Water Quality Standards, No. 8, B.E. 2537 (1994), issued under the Enhancement and Conservation of National Environmental Quality Act B.E.2535 (1992), The Royal Government Gazette.

Oun, A., Yin, Z., Munir, M. & Xagoraraki, I. 2017 Microbial pollution characterization of water and sediment at two beaches in Saginaw Bay, Michigan. *J. Great Lakes Res.* **43** (3), 64–72. doi:10.1016/j.jglr.2017.01.014.

Purnell, S. E., Ebdon, J. E. & Taylor, H. D. 2011 Bacteriophage lysis of *Enterococcus* host strains: a tool for microbial source tracking? *Environ. Sci. Technol.* **45**, 10699–10705. doi:10.1021/es202141x.

Purnell, S. E., Ebdon, J. E., Wilkins, H. & Taylor, H. D. 2018 Human-specific phages infecting *Enterococcus* host strain MW47: are they reliable microbial source tracking markers? *J. Appl. Microbiol.* doi:10.1111/jam.13700.

R Core Team 2017 *R: A Language and Environment for Statistical Computing*. R Found. Stat. Comput., Vienna, Austria. URL https//www.R-project.org/.

Santiago-Rodríguez, T. M., Dávila, C., González, J., Bonilla, N., Marcos, P., Urdaneta, M., Cadete, M., Monteiro, S., Santos, R., Domingo, J. S. & Toranzos, G. A. 2010 Characterization of *Enterococcus faecalis*-infecting phages (enterophages) as markers of human fecal pollution in recreational waters. *Water Res.* **44**, 4716–4725. doi:10.1016/j.watres.2010.07.078.

Sirikanchana, K., Wangkahad, B. & Mongkolsuk, S. 2014 The capability of non-native strains of *Bacteroides* bacteria to detect bacteriophages as faecal indicators in a tropical area. *J. Appl. Microbiol.* **117**, 1820–1829. doi:10.1111/jam.12646.

Teaf, C. M., Flores, D., Garber, M. & Harwood, V. J. 2018 Toward forensic uses of microbial source tracking. *Microbiol. Spectr.* **6**, 1–14. EMF-0014-2017. doi:10.1128/microbiolspec.EMF-0014-2017. Correspondence.

US EPA 2002a Method 1604: Total Coliforms and Escherichia coli in Water by Membrane Filtration Using a Simultaneous Detection Technique (MI Medium). EPA 821–R-02-024.

US EPA 2002b Method 1600: Enterococci in Water by Membrane Filtration Using membrane-*Enterococcus* Indoxyl-Beta-D-Glucoside Agar (mEI). EPA 821–R-02-022.

US EPA 2011 *Using Microbial Source Tracking to Support TMDL Development and Implementation* 1–64.

US EPA 2012 Recreational Water Quality Criteria. OFFICE OF WATER 820-F-12-058.

Vijayavel, K., Byappanahalli, M. N., Ebdon, J., Taylor, H., Whitman, R. L. & Kashian, D. R. 2014 *Enterococcus* phages as potential tool for identifying sewage inputs in the Great Lakes region. *J. Great Lakes Res.* **40** (4), 989–993.

Wangkahad, B., Bosup, S., Mongkolsuk, S. & Sirikanchana, K. 2015 Occurrence of bacteriophages infecting *Aeromonas*, *Enterobacter*, and *Klebsiella* in water and association with contamination sources in Thailand. *J. Water Health* **13**, 613–624. doi:10.2166/hydro.2013.124.

Wangkahad, B., Mongkolsuk, S. & Sirikanchana, K. 2017 Integrated multivariate analysis with nondetects for the development of human sewage source-tracking tools using bacteriophages of *Enterococcus faecalis*. *Environ. Sci. Technol.* **51**, 2235–2245. doi:10.1021/acs.est.6b04714.

Yau, V. M., Schiff, K. C., Arnold, B. F., Griffith, J. F., Gruber, J. S., Wright, C. C., Wade, T. J., Burns, S., Hayes, J. M., McGee, C., Gold, M., Cao, Y., Boehm, A. B., Weisberg, S. B. & Colford, J. M. 2014 Effect of submarine groundwater discharge on bacterial indicators and swimmer health at Avalon Beach, CA, USA. *Water Res.* **59**, 23–36. doi:10.1016/j.watres.2014.03.050.

First received 30 July 2018; accepted in revised form 22 October 2018. Available online 31 October 2018

Calibration and field application of a molecularly imprinted membrane-passive sampler for the sampling of indicator polychlorinated biphenyls in selected aquatic environments of South Africa

Dennis S. Mkhize, Laura P. Quinn, Hlengilizwe Nyoni, Bhekie B. Mamba and Titus A. M. Msagati

ABSTRACT

A passive sampling device, based on molecularly imprinted membranes (MIM), was fabricated and optimised for sampling polychlorinated biphenyls (PCBs) in aquatic ecosystems. The newly-developed passive sampler was subjected to *in-situ* calibration studies to determine PCB sampling rates under various conditions of water turbulence and temperature. This was carried out by exposing the passive samplers to water spiked with PCBs in a continuous-flow exposure setup. The samplers were preloaded with known concentrations of performance reference compounds (PRCs) prior to exposure. Sampling rates of seven indicator PCBs' congeners (PCBs 28, 52, 101, 118, 138, 153 and 180) ranged between 15.3 and 95.6 L/d for the different environmental conditions investigated. To determine the field suitability, the samplers were preloaded with PRCs and deployed for 10 days at the Roodeplaat and Hartbeespoort Dams, in South Africa. Water samples were taken at the end of the deployment period to compare the spot-and-grab samples to the developed samplers. PCBs 28, 101 and 138 were detected in the samplers deployed at Hartbeespoort Dam. The samplers deployed at the Roodeplaat Dam had quantifiable amounts of PCBs 28, 52, 101, 138 and 180 (0.047–0.828 ng mL^{-1} d^{-1}). The sampler enhanced the detectability of PCB 52 and PCB 180, which were not detected in water samples. The field suitability trials indicated that the developed sampler could successfully be used for PCB monitoring. The sampler enhanced the detection of PCBs that would otherwise be too low to detect in samples collected through the traditional spot-and-grab sampling technique.

Key words | molecularly imprinted membranes, passive sampling, performance reference compounds, polychlorinated biphenyls, time-weighted average concentrations

Dennis S. Mkhize
Laura P. Quinn
National Metrology Institute of South Africa,
Private Bag X34, Lynnwood Ridge, 0040, Pretoria,
South Africa

Dennis S. Mkhize
Department of Applied Chemistry,
Faculty of Science,
University of Johannesburg,
Doornfontein Campus, P.O. Box 17011,
Johannesburg, 2028,
South Africa

Hlengilizwe Nyoni (corresponding author)
Bhekie B. Mamba
Titus A. M. Msagati
College of Science Engineering and Technology,
Nanotechnology and Water Sustainability
Research Unit,
University of South Africa,
UNISA Science Campus, P.O. Box 392 UNISA 0003,
Florida, 1709 Johannesburg,
South Africa
E-mail: *nyonih@unisa.ac.za*

INTRODUCTION

Polychlorinated biphenyls (PCBs) are ubiquitous persistent organic pollutants (POPs) that have attracted the attention of scientists and policy makers alike, due to their toxicity and persistence in the environment (Sun *et al.* 2007; Nieuwoudt *et al.* 2009). They are carcinogenic, neurogenic and endocrine disruptors that affect the reproductive and immune system as well as the functioning of the thyroid hormones (Battershill 1994; Lind *et al.* 2000; Lundberg *et al.* 2006). Therefore, their presence in the environment is of great concern. Out of 209 possible congeners, seven PCBs have been identified as indicator PCBs (PCBs 28, 52, 101, 118, 138, 153 and 180) due to their prevalence within the environment (McFarland & Clarke 1989; Hamers *et al.* 2011).

One of the greatest challenges in PCB sampling is obtaining reliable and accurate data. This could be due to the low concentrations of PCBs in aquatic environments and the intrinsic variations of environmental conditions with time (Vrana *et al.* 2006a, 2006b). Traditional sampling techniques such as spot sampling often require the collection of large volumes of samples and a higher sampling frequency (Bopp *et al.* 2005, 2007; Popp *et al.* 2005). This can increase the turnaround time and cost for analysis, as

doi: 10.2166/wst.2019.044

multiple concentration steps and more intricate sample clean-up steps are required (Bopp *et al.* 2005, 2007; Popp *et al.* 2005). Alternatively, passive sampling devices, such as semi-permeable membrane devices (SPMDs) and the classical MESCO, can be used to augment the shortcomings of these traditional sampling techniques (Kim *et al.* 2014; Alvarez 2010). However, these passive samplers are designed to sample a variety of non-polar pollutants (Kim *et al.* 2014; Piccardo *et al.* 2010), which can result in poor PCB enrichment occurring in low concentrations. Furthermore, recovery of the sequestered analytes from these samplers, such as the SPMD, can result in complex matrices and increased interferences (Esteve-Turrillas *et al.* 2008; Leonard *et al.* 2002). This can result in more challenges in the detection and analysis of PCBs, which often occur at ultra-trace levels in aquatic environments. Therefore, a sampler with high selectivity is important for accurate and reliable sampling of PCBs. In addition, a sampler that can simplify the extraction, clean-up and the overall analysis of PCBs is required.

To augment the shortcomings of these passive sampling techniques, a novel passive sampling device was designed based on our previously synthesised molecularly imprinted membranes (MIMs) (Mkhize *et al.* 2017). The developed sampler was optimised for sampling PCBs in aquatic environments. The molecular imprinting was aimed at enhancing PCB selectivity and enrichment, and this is crucial as PCBs usually occur at ultra-trace levels in aquatic environments (Sun *et al.* 2007). MIMs were produced from molecularly imprinted polymers (MIPs), which were previously obtained from the polymerisation of a mixture of cross-linkers, porogenic solvent and functional monomers in the presence of a template and radical initiator under heat (Cormack & Elorza 2004; Saad *et al.* 2015; Mkhize *et al.* 2017). In this work, we report the calibration and field application of the newly-developed passive sampler for the sampling of indicator PCBs in selected aquatic environments.

THEORY OF SAMPLER AND CALCULATIONS

The accumulation of target analytes by passive samplers usually follows linear, curvilinear or equilibrium phases over the deployment period. When uptake occurs over the linear phase, analyte concentrations in water (C_w) can be calculated as follows:

$$C_w = \frac{N}{R_s t} \tag{1}$$

where C_w is the TWA concentration (ng/L) of the PCBs in water during the exposure time, N is the PCB concentration (ng) in the sampler accumulated over the exposure or deployment period, R_s is the sampling rate (usually in L/d) and t (days) is the exposure or sampling period.

Regression models to estimate a chemical's site-specific sampling rate and TWA concentrations have been developed, and these are based on octanol-partition coefficients (log K_{ow} values) of the analyte, the desorption rate constant, k_e (d^{-1}) of the PRC and the sampler-water partition coefficient (K_{sw}). Isotropic exchange takes place when the uptake and offload kinetics are affected by the mass transfer law, and obey first-order kinetics and the half-lives measured during loss and uptake are approximately identical (Huckins *et al.* 2002). This usually takes place between chemicals and their deuterated and/ or ^{13}C-labelled counterparts (Huckins *et al.* 2002). Therefore, under isotropic exchange conditions, the desorption rate constant k_e of a PRC can be determined according to the following equation (Huckins *et al.* 2002):

$$k_{e-PRC} = \ln \frac{[(N_t|N_0)]}{t} \tag{2}$$

where t is the exposure time for the passive sampler, N_0 and N_t are PRC concentrations in the sampler at time 0 and t, respectively. According to Equation (3), the log K_{sw} is determined from linear regression models using the PRC's log K_{ow}.

$$\log K_{sw} = a_0 + 2.3211 \log K_{ow} - 0.1618 (\log K_{ow})^2 \tag{3}$$

where a_0 is the intercept, which is known to be −2.61 for PCBs. When the sampler volume, V (mL), is known, the R_s-$_{PRC}$ can be determined using K_{sw} and k_e values (already known from the above equations). The equation to calculate R_s-$_{PRC}$ is as follows:

$$R_{s-PRC} = V_s K_{sw} k_e \tag{4}$$

TWA concentrations can only be determined once the sampling rates of target analytes are known. The following equation can be used to estimate the sampling rates:

$$R_{s,i} = R_{s,PRC} \frac{\propto_i}{\propto_{PRC}} \tag{5}$$

where α_i is the compound-specific factor on the sampling rate that can be interpolated with a third-order polynomial obtained from the analysis of experimental data on the

PCBs sampling rates:

$$\log \alpha_i = 0.0130\,(\log K_{ow})^3 - 0.3173\,(\log K_{ow})^2 + 2.244\,(\log K_{ow}) \quad (6)$$

Once PRC-derived sampling rates are available, Equation (6) can be used to determine the sampling rates of other analytes, as long as the log K_{ow} values are known. Furthermore, TWA concentrations can be calculated according to Equation (7), once the sampling rates have been determined:

$$C_w = \frac{N}{V_s K_{sw}\left(1 - e^{\left(\frac{-R_s t}{V_s K_{sw}}\right)}\right)} \quad (7)$$

However, it is well-known that sampling rates are affected by a variety of environmental conditions during deployment, and that they are sampling-site specific (Huckins *et al.* 2002). Furthermore, the dissipation rate constants of PRCs are affected by the same factors as sampling rates, under isotropic exchange conditions. Therefore, the effects of environmental factors on sampling rates are corrected by comparing PRC dissipation rates obtained under controlled laboratory conditions ($k_{e\text{-}PRC,cal}$), with those obtained during field deployments ($k_{e\text{-}PRC,in\text{-}situ}$). This is demonstrated by Equation (8), as follows:

$$R_{s-corrected} = R_{s-cal}\, x \left(\frac{k_{e-PRC,\,in-situ}}{k_{e-PRC,\,cal}}\right) \quad (8)$$

The field PRC dissipations rates are determined by comparing the amount of PRCs loaded into the sampler before deployment with the amount remaining after deployment. This mathematical equation (Equation (8)) accounts for environmental conditions and gives a more accurate account for sampling-site specific sampling rates ($R_{s\text{-}corrected}$), thus resulting in a more correct estimation of TWA concentrations.

EXPERIMENTAL

Chemicals and materials

Polysulfone (PSf, 97%), ethylene glycol dimethacrylate (EDGMA, 98%), methacrylic acid (MAA, 97%) and 4-phenylphenol were purchased from Sigma Aldrich (St Louis, USA). The radical initiator, azobisisobutyronitrile (AIBN, 97%), was obtained from Fluka Chemicals (St Louis, USA). *N*-methyl-2-pyrrolidone (NMP, 98%) was obtained from DLD Scientific (Johannesburg, South Africa). All reagents were used without further purification. The deuterated polyaromatic hydrocarbons (D-PAHs) – phenanthrene-D_{10} (98%), pyrene-D_{10} (98%) and chrysene-D_{12} (98%) were purchased from Sigma Aldrich (St Louis, USA). The selected native PCBs used in this study represented dioxin-like and the seven indicator PCBs, which are the most toxic and most commonly occurring in the environment, respectively (Faroon *et al.* 2003). The native PCBs: PCB 105 (98.8%), PCB 126 (98.8%), PCB 153 (98.8%) and PCB 183 (98.8%), were purchased from Cerilliant Corporation (Texas, USA). Whilst PCB 28 (98.6%), PCB 52 (98.4%), PCB 101 (98.4%), PCB 118 (98.4%), PCB 138 (98.5%), PCB 180 (98.7%) and the nine multi-congener mix solutions (C-CS-01 to C-CS-09) at 10 µg/mL in iso-octane were obtained from AcuStandard Inc. (Connecticut, USA). The $^{13}C_{12}$-labelled mono-ortho PCB mix, used as internal standard ($^{13}C_{12}$-PCBs 114, 118, 123, 156, 157, 167 and 189), and the PRCs, $^{13}C_{12}$-PCB 28, 101, 105 and 153, were purchased from Cambridge Isotope Laboratories, Inc. (Massachusetts, USA). A summary of the physicochemical properties of the different PCBs used, including the log K_{ow} values was obtained from Mackay and co-workers (Mackay *et al.* 1992a, 1992b). MIMs were prepared as described by Mkhize *et al.* (2017).

Sampler design and configuration

The newly-fabricated passive sampler was based on a MIM and was optimised to sample PCBs at high sampling rate. The sampler consists of a MIM with an effective area of approximately 30.3 cm^2, sandwiched between two metal washers. The passive sampler is based on a single-layer system, where the receiving phase is embedded within the membrane. This sampler eliminates any additional diffusive layers, thus increasing sampling rates through a more direct exposure between PCBs in water and receiving phase.

Passive sampler preparation and PRC loading

To determine off-load kinetics, MIMs were cut into disks with an effect area of 30.3 cm^2 and spiked with PRCs prior to exposure and calibration. $^{13}C_{12}$-PCBs and D-PAHs were used as PRCs. An aqueous solution of PRCs was prepared by adding 500 μL of methanolic solution (1,000 ng/mL composite containing $^{13}C_{12}$-PCBs 28, 101, 105, 153, phenanthrene-D_{10}, pyrene-D_{10} and chrysene-D_{12}) to water (50 mL). MIMs were equilibrated overnight in the PRC-enriched water in order to load them with PRCs. The amount of PRCs remaining in the water after equilibration was used to determine the amount loaded into the MIMs. Quantification was done through comprehensive two-dimensional gas chromatography, coupled to time-of-flight mass spectrometry (GC × GC-TOFMS). Samplers were assembled and stored at 4 °C until required. To evaluate the presence of any PRCs in the water used, a negative blank was prepared by exposing MIMs to non-enriched water for 24 hours, followed by GC × GC-TOFMS analysis (see Quantification and instrumental analysis section) of the water after exposure. To determine possible losses during PRC-loading, PRC-enriched water was left to equilibrate for 24 hours in the absence of MIMs.

Passive sampler calibration

Calibration experiments were carried out in a constant concentration flow-through exposure system, consisting of a 40 L assimilation tank equipped with an overhead stirrer (Figure 1(a) and 1(b)) and an overflow to waste. The flow-through set-up allowed experiments to be carried out

(a)

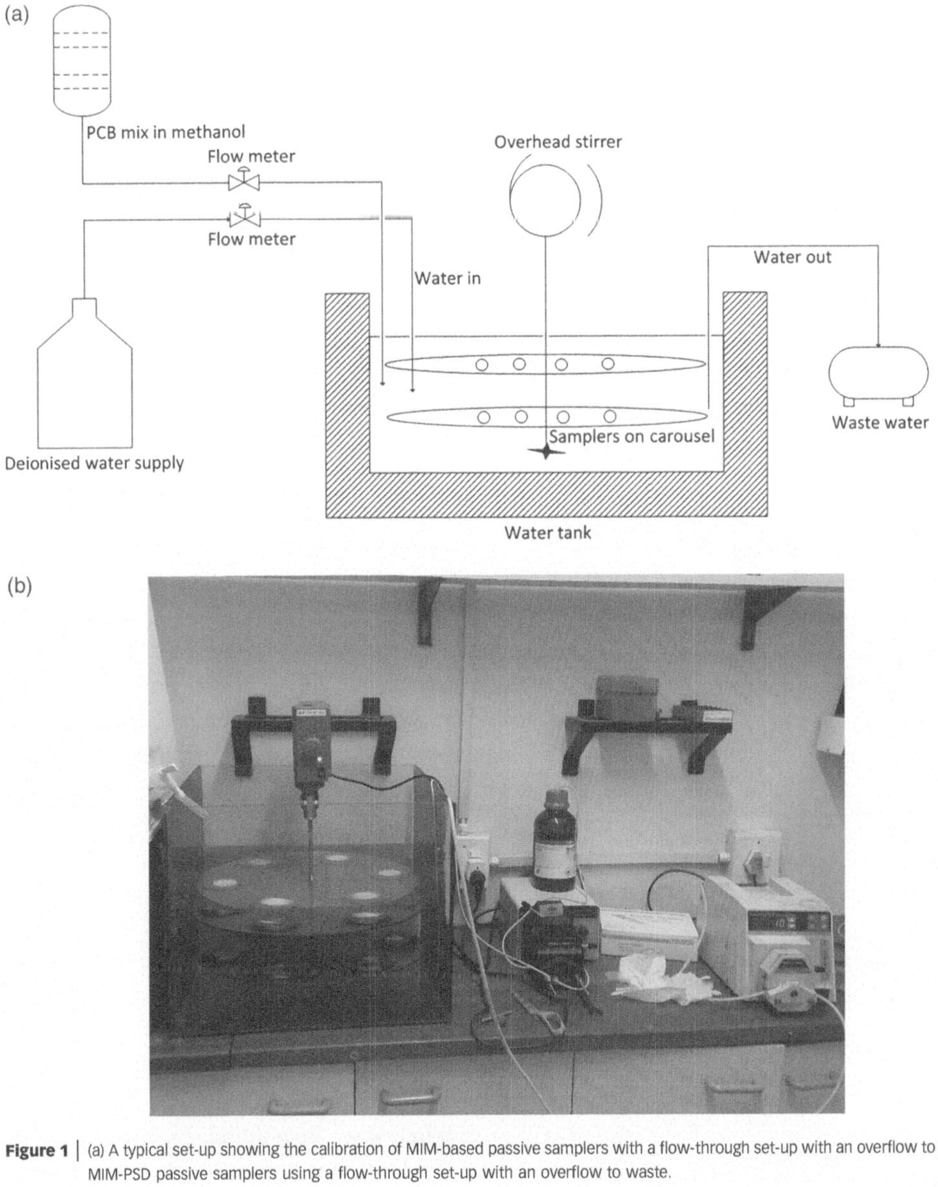

(b)

Figure 1 | (a) A typical set-up showing the calibration of MIM-based passive samplers with a flow-through set-up with an overflow to waste. (b) A typical set-up showing the calibration of MIM-PSD passive samplers using a flow-through set-up with an overflow to waste.

under controlled condition of temperature and PCB concentration, whilst the overhead stirrer was used to investigate effects of water velocity on PCB uptake rates of the sampler.

The experiments were conducted with 24 samplers in a temperature-controlled dark room to minimise photolytic degradation of PRCs. The samplers were mounted on four carousel holders, each capable of housing six passive samplers at a time. Two separate peristaltic pumps were used to introduce water and methanolic solutions of PCBs (PCBs 28, 105, 126, 153 and 183) into the simulation tank at known, controlled flow rates. The feed water was introduced at a flow rate of 2 L/h, whilst the methanol solutions of PCBs were introduced at a flow rate of 100 μL/min (approximately 60 ng/L). The wastewater was pumped out of the simulation tank at the same rate (2 L/h) in order to maintain a constant amount of water. The concentration of PCBs in the simulation tank was maintained in the range of 10–50 ng/L throughout the duration of the experiments. In addition, methanol concentrations in the tank were kept below 1% (v/v) at any given time. PCBs in the simulation tank were allowed to equilibrate for 24 hours prior to exposure of the samplers. The first set of experiments was carried out over a course of 10 days under gentle stirring conditions at room temperature (Table 1).

For statistical purposes, three samplers were removed at set intervals (1, 2, 4, 5, 7, 8, 9 and 10 days) and analysed to determine the uptake and loss of PCBs and PRCs, respectively. The PRC loss data were used to elucidate the sampling rates (R_s) of PCBs. Removed samplers were replaced by empty metal washers to maintain hydrodynamic conditions in the water tank. In addition, three water samples (500 mL) were removed and analysed for PCBs each time a sampler was removed. The water samples were analysed using an ultrasonic extraction method.

Table 1 | A summary of experiments carried out during calibration of the MIM-based passive sampler for different environmental conditions

Experiments	Temperature (°C)	Stirring speed (cycles/min)	Concentration (ng/mL)	Exposure period (days)
1	18	40	50	10
2	21	40	50	10
3	26	40	50	10
4	21	0	50	10
5	21	70	50	10

To determine the effects of hydrodynamic conditions on PCB sampling rates, experiments were also carried out under quiescent conditions (no stirring) at room temperature. The contents of the tank were agitated regularly to prevent a concentration gradient from forming. The samplers were removed from the simulation tank and processed to determine the amount of PRCs remaining after exposure and the amounts of PCBs accumulated (Vrana *et al.* 2006b). To investigate the different parameters (hydrodynamic conditions and temperature) on PCB sampling, a total of five experiments (Table 1) were conducted, which were based on fixing one variable whilst altering the other. The effects of hydrodynamic conditions on sampling rates were investigated by conducting two experiments at different speeds (0 and 70 cycles/min), whilst maintaining a constant temperature (21 °C). The nominal PCB concentration was maintained throughout all the investigations.

Field deployment

For evaluation of the performance for monitoring PCBs under field conditions, PRC-loaded samplers ($n = 3$ at each deployment point) were deployed at the Roodeplaat and Hartbeespoort Dams. The Roodeplaat Dam is located in Pretoria, in the Gauteng province, South Africa (25.62083°S 28.37138°E), whilst Hartbeespoort Dam is located in the North West Province (25.62083°S 28.37138°E), 35 km west of Pretoria. The dams are renowned for poor water quality, resulting from domestic and industrial effluents from surrounding areas (Amdany *et al.* 2014), and were therefore selected as deployment sites for evaluation of the newly-developed MIM-based sampler. Three samplers were deployed at two different harbour points in each dam, denoted Harbour 1 and Harbour 2. Samplers were kept at approximately 4 °C during transportation to the deployment site. Field blanks were taken to each site during deployment and retrieval to account for any possible contamination during transportation and handling (Alvarez 2010). Samplers were immersed approximately 2 m below the surface of the water and secured using a rope. The water temperature during the deployment period ranged between 18 and 21 °C, and no rainfall was recorded over the course of the deployment period. After 10 days of exposure, samplers were retrieved, sealed in foil pre-cleaned with triplicate rinses of acetone and hexane, labelled and stored at approximately 4 °C. In addition, 1 L water samples were collected (glass bottles) in triplicates after deployment at each harbour point. PRC concentrations at time zero and accumulated PCBs, after deployment were determined.

Sampler retrieval and processing

After retrieval, MIMs were extracted overnight in hexane *via* Soxhlet extraction. Prior to extraction, internal standard mix was gravimetrically added to the samplers. The resulting extracts were purified using dual layer Florisil/silica SPE cartridges and reconstituted in 100 μL of toluene prior to GC × GC-TOFMS analysis (see Quantification and instrumental analysis section). This was to determine accumulated PCB amounts and PRCs lost during deployment. All three samplers and field blanks from each harbour point were analysed. The final concentration of the extract contained approximately 625 ng/mL of internal standard mix.

Water sample processing

Water samples from the sampler calibration system were extracted in triplicate using an ultrasonic and liquid-liquid extraction (LLE) with DCM/hexane (3:1%v/v) mixture. The extracts were purified using dual layer Florisil/silica SPE cartridges and reconstituted in 100 μL of toluene prior to GC × GC-TOFMS analysis. Due to larger volumes, field water samples were extracted in an automated extraction system (Horizon™ SPE-DEX 4790 Automated Extractor System) using C_{18} cartridges and eluting with 1:1 ethyl acetate/DCM mixtures. Prior to extraction, internal standard mix was gravimetrically added to the field water samples. The spiked water samples were homogenised and allowed to equilibrate overnight at 4 °C prior to extraction. The final concentration of the extract contained approximately 625 ng/mL of internal standard mix.

Quantification and instrumental analysis

GC × GC-TOFMS analysis was performed on a LECO Pegasus IV system (LECO Corporation, St Joseph, MI, USA). The autosampler, GC and MS methods were optimised based on the method developed by Focant *et al.* (2004) using a Rxi®-5 Sil MS (30 m, 0.25 mm ID, 0.25 μm df) primary column and an Rxi®-17SilMS (1.5 m, 0.25 mm ID, 0.25 μm df) secondary column. The method used was optimised for all 209 PCB congeners. The multi-ramp temperature program used can be summarised as follows: 80 °C for 2 min, then to 150 °C at 25 °C/min, then to 200 °C at 1.5 °C/min, then to 290 °C at 5 °C/min with a 10 min hold. A 20 °C temperature offset between the primary and secondary oven and a 5 s modulation period. The helium flow through the column was 1.5 mL/min with corrected constant flow via pressure ramps. A 1 μL injection was used with an inlet temperature of 250 °C. Data processing and generation of contour plots was achieved through the Leco ChromaTOF™ software. For each analyte, two mass ions were used, one for confirmation and one for quantification. Quantification was based on a seven-point matrix-matched calibration for both passive samplers and water samples. Absolute recoveries were for both passive samplers and water samples.

Quality control

Quality control measures included the use of field blanks during deployment and retrieval of samplers. Furthermore, absolute recovery samples for the passive samplers and collected water samples and solvent blanks in-between GC runs were used during quantification. The use of matrix-matched calibrations was also used as a quality control measure.

RESULTS AND DISCUSSION

Calibration of passive samplers

Passive samplers were spiked with known concentrations of PRCs and exposed to a constant concentration of PCBs in a continuous-flow simulation tank. Preliminary experiments showed that the sampler reached after 10 days of exposure. The amount of PCBs accumulated and PRCs dissipated from the sampler showed a characteristic and satisfactory (correlation between 0.8 and 1) linear relationship with exposure time. The PRC dissipation curves (Figure 2) show a clear decrease in PRC concentrations as a function of time, which occurs under isotropic conditions (Huckins *et al.* 2002).

A summary of the sampler calibration data (supplementary data, Table S1, available with the online version of this paper) shows that the greatest PRC loss was observed for $^{13}C_{12}$-PCB 28, followed by $^{13}C_{12}$-PCBs 101, 105 and 153.

The labelled PCBs, pyrene and chrysene had dissipation rates between 20 and 80%, and were therefore suitable candidates for estimation of sampling rates for the seven indicator PCBs (Alvarez 2010). Phenanthrene completely dissipated from the MIM-based sampler and thus was not used to estimate sampling rates. To minimise variability associated with the use of a single PRC and to ensure more accurate measurements, sampling rates were estimated from more than one

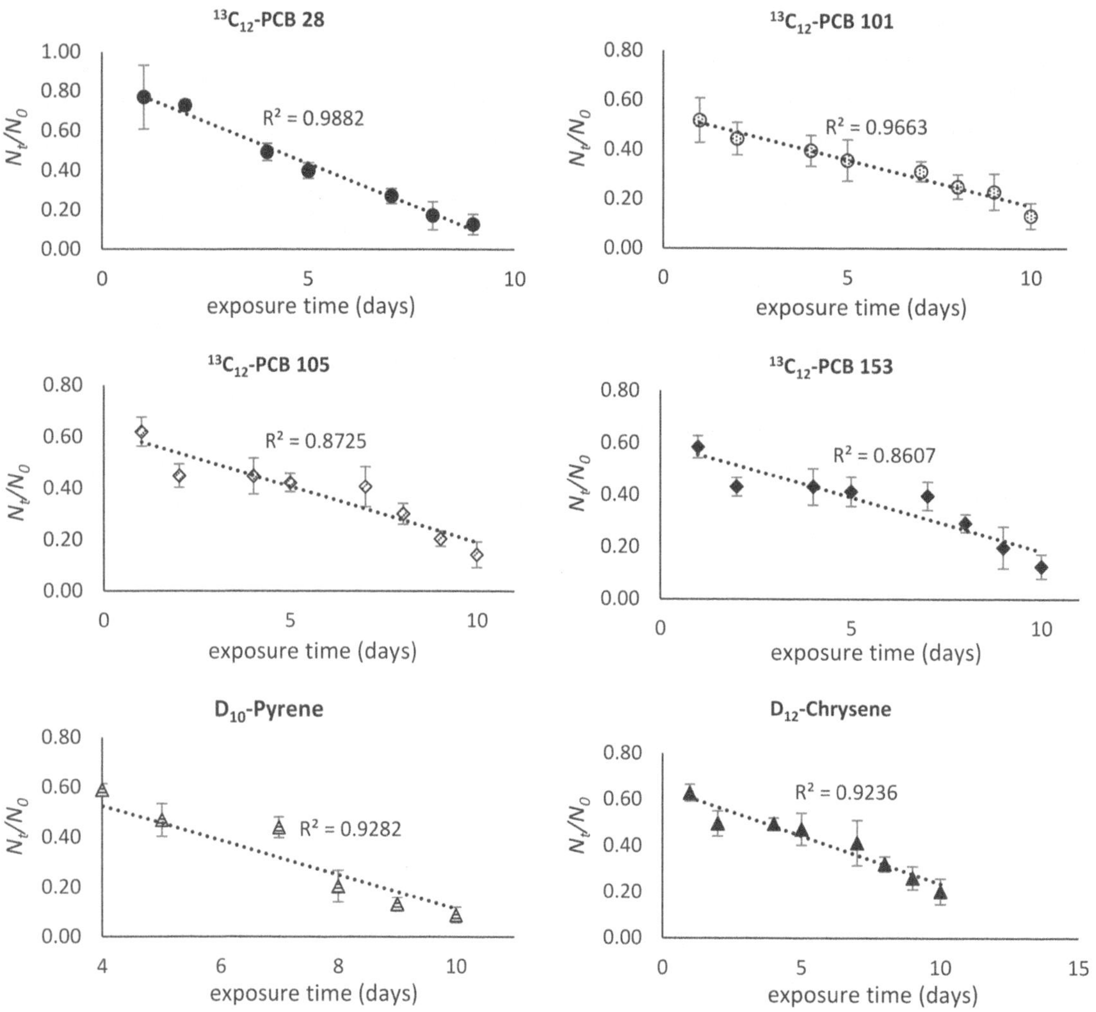

Figure 2 | PRC offload curves from the MIM-based sampler over a ten-day exposure period (18 °C, 40 cycles/min). The results show an average over three separate measurements.

PRC (Alvarez 2010). PCB sampling rates for the different congeners ranged between 20.7 and 71.3 L/d over the three different temperatures investigated. Whilst those obtained under different hydrodynamic conditions ranged between 15.3 and 95.6 L/d for the different congeners. Sampling rates are largely affected by the physicochemical properties of analytes under investigation. The less chlorinated and moderately hydrophobic PCBs had higher sampling, whilst the more chlorinated congeners had the lowest sampling rates. This is consistent with observations from Estoppey *et al.* (2014) in which silicone rubber passive samplers were used for PCB uptake from water. The highest sampling rates were recorded for PCB 28, followed by PCBs 52, 101, 118, 138, 153 and 180. The decrease in sampling rates with higher log K_{ow} values is believed to be due to a decrease in diffusion coefficients and mass transfer kinetics resulting from an increased molecular size (Huckins *et al.* 2002).

Verification of isotropic exchange kinetics

Isotropic exchange kinetics apply when target analytes diffuse from the bulk solution and into the sampler receiving phase, followed by subsequent dissipation of its PRC or labelled analogue from the receiving phase (Huckins *et al.* 2002). Since analyte diffusion and PRC dissipation are governed by the same mass transfer kinetics when measured under the same conditions, PRCs can be used to correct for the effects of environmental conditions during sampling (Vrana *et al.* 2006a, 2006b). When dissipation (k_e) and accumulation rates (R_S) are measured under the same conditions, it is assumed that isotropic exchange kinetics apply, and the PRCs can be used to correct for the effects of environmental conditions during sampling.

Application of the MIM-based sampler under different conditions of water velocities and temperature using seven different PRCs showed sufficient correlation between PCB sampling and PRC dissipation rates (Figure 3). This also demonstrated that PCB uptake and PRC dissipation remained in the linear phase under the different sampler calibration conditions investigated.

This data set represents the sampling and dissipation rates associated with each of the five experiments conducted. The results show the average over three separate measurements. In fact, a correlation was observed for a wider scope of PCB and PRC analogue combinations. This was crucial as it allows for the estimation of sampling rates for PCBs without the use of their carbon-13 labelled PRC analogues. The estimation of sampling rates for PCBs 52, 118, 138 and 180 was determined without the use of their carbon-13 labelled analogues as PRCs. The four $^{13}C_{12}$-labelled PCBs ($^{13}C_{12}$-PCBs 28, 101, 105 and 153) used as PRCs in this study, had sufficient correlation with PCBs 52, 118, 138 and 180 (0.947, 0.987, 0.997 and 0.850) to be used as PRCs.

Effects of temperature and hydrodynamic conditions on sampling rates

The effects of temperature and hydrodynamic conditions on sampling rates were investigated as highlighted in Table 1, and the effects are illustrated in Figure 4. Experiments performed at 18 °C resulted in the lowest sampling rates for all PCBs investigated. Increasing the temperature from 18 to 21 °C resulted in an increase in sampling rates by a factor 1.37. A further increase from 21 to 26 °C had a less notable effect, as the sampling rates only increased by a factor of 1.16 for the different PCBs investigated. Furthermore, increasing the temperature from 16 to 26 °C led to an increase in PCB sampling rates by a factor of 1.60, which was more notable. Increasing temperature results in increased mass transfer kinetics of PCBs towards the sampler receiving phase, thus increasing sampling rates. Increasing temperature also resulted in increased dissipation rate constants (k_e) of PRCs, which further demonstrates the existence of isotropic exchange kinetics between PCBs and PRCs.

Increasing stirring speeds from 0 to 40 cycles/min and 40 to 70 cycles/min increased the sampling rates by a factor of 1.86 and 1.56, respectively. The lowest sampling rates were obtained under static conditions (Figure 4). Increasing stirring speeds from 40 to 70 cycles/min had a less notable effect compared to increasing from 0 to 40 cycles/min. It is likely that the sampler approaches equilibrium with increasing stirring speeds, which could have led to minimal effects on the sampling rates of PCBs. In general, increasing stirring speeds increases mass transfer kinetics, favouring the diffusion of PCBs from the bulk

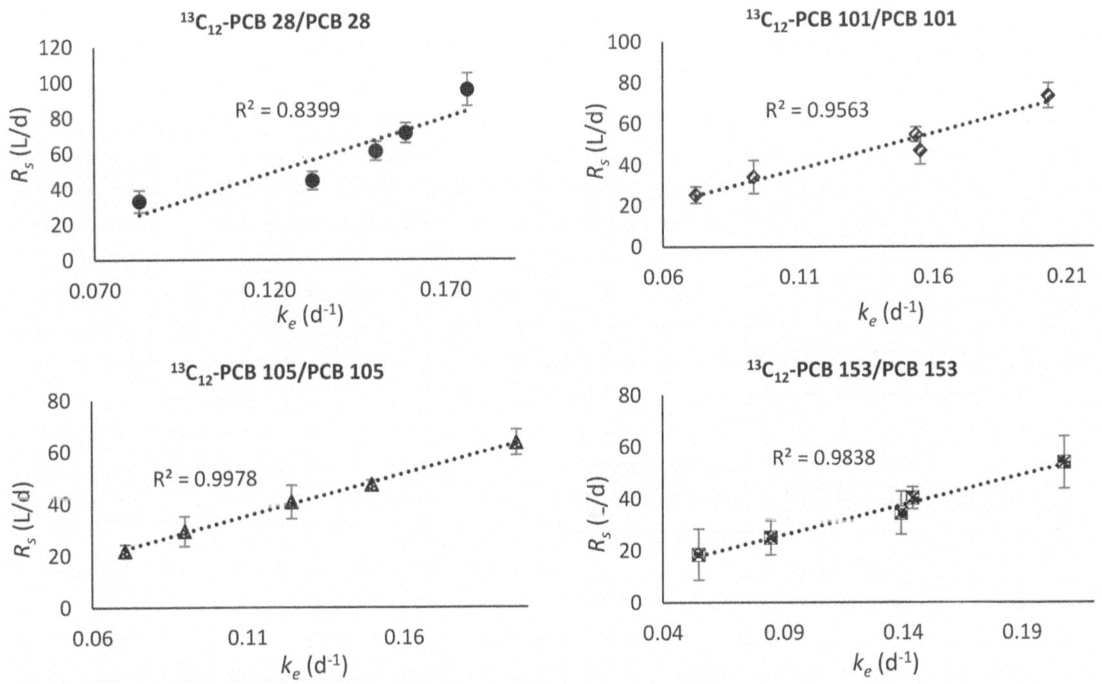

Figure 3 | The verification of isotropic exchange kinetics through correlation between sampling rates (R_s) of PCBs and the dissipation rate constants (k_e), their PRC analogues.

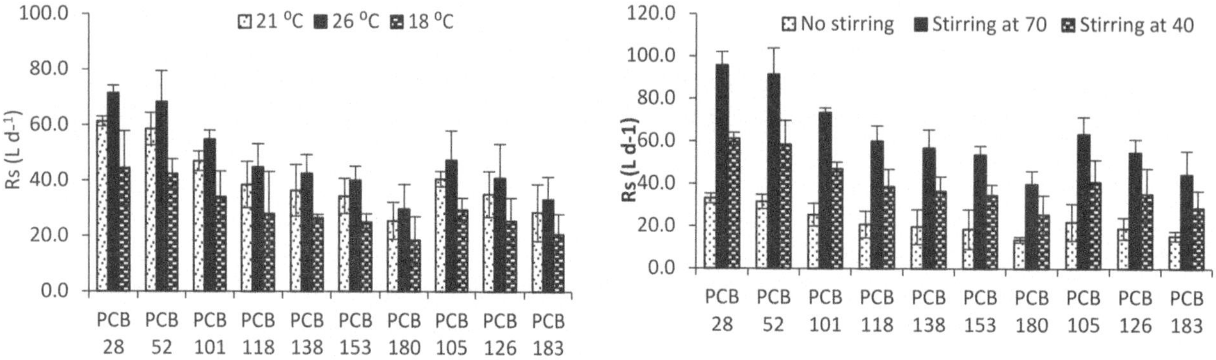

Figure 4 | The effect of temperature (at 40 cycles/min) and hydrodynamic conditions (at 18 °C) on PCB sampling rates. An average of three measurements for each experiment were recorded.

aqueous solution and into the receiving phase of the sampler, thus increasing sampling rates at higher stirring speeds. The high sampling rates suggest suitability of the sampler for PCB sampling. High sampling rates are crucial for PCB sampling, due to their ultra-trace levels in water (Vrana *et al.* 2006a, 2006b).

Deployment of the passive samplers

The newly-developed sampler was deployed at the Hartbeespoort and Roodeplaat Dams to evaluate performance and to determine TWA concentrations of the seven indicator PCBs. Prior to deployment, the samplers were spiked with six PRCs at known concentrations, covering a log K_{ow} range of 4.46 to 6.92. PCBs and PRCs in deployed samplers and water samples collected were quantified through isotope dilution. Concentrations were adjusted according to the PCB recoveries obtained. The absolute recoveries, limit of detection (LOD), limit of quantification (LOQ), TWA concentrations and %RSD of the samples analysed are summarised in Table 2.

Roodeplaat Dam

Over the ten-day sampling period, the samplers had sequestered PCBs 28, 52, 101, 138 and 180 in varying concentrations at the different sampling points. Quantifiable concentrations of PCBs 28, 101 and 138 were found in the water samples, with PCB 28 having the highest freely dissolved concentration. PCB 28 is characterised by a lower log K_{ow} (5.67) (Mackay *et al.* 1992a, 1992b). Therefore, its freely dissolved concentration in water tends to be higher than its more chlorinated counterparts. There was no detectable amount of PCB 52, 118, 153 and 180 in the water samples. These more chlorinated congeners tend to occur in soils and sediments, as opposed to occurring in their freely dissolved form (Agency for Toxic Substances and Disease Registry 2000), thus accounting for their lower detectability in water samples. Using laboratory-derived sampling rates, TWA concentrations were determined from the amounts of PCBs accumulated by the samplers during deployment. Field blanks taken to the Roodeplaat Dam during deployment and retrieval of the samplers had ultra-trace levels of PCB 52 and 101, whilst PCBs 28, 118, 138, 153 and 180 were not detected. Concentrations accumulated in the field blanks were subtracted from amounts in the deployed samplers, which gives more accurate TWA concentrations in the dam.

A comparison of PRC amounts in the samplers before (N_0) and after deployment (N_t) was used to determine dissipation rate constants (k_e), as demonstrated by Equation (2). PRC dissipation rates are sampling-site dependent, due to an intrinsic variation of conditions between aquatic environments. These variations are accounted for by comparing site-specific dissipation rate constants to those obtained under controlled laboratory conditions, as demonstrated by Equation (8). This correction factor gives a more accurate calculation of TWA concentrations. PCB 28 had the highest TWA concentrations (Table 2) across the different samplers in both harbour points. This was followed by PCBs 180, 101, 138 and 52, whilst PCBs 118 and 153 were not detected. The higher concentration of PCB 28 in the samplers could be attributed to a greater abundance in water or a higher sampling rate associated with this congener, or a combination of both factors.

It was interesting to note a presence of PCB 52 and PCB180 in the samplers, which was not detected in the water samples. The lower detectability of PCB 180 in water samples could be attributed to higher log K_{ow} values (Mackay *et al.* 1992a). It has been well established that

Table 2 | Summary of results obtained from MIM-based samplers deployed at the Roodeplaat and Hartbeespoort Dams

						Roodeplaat TWA concentration				Roodeplaat water sample concentration				Hartbeespoort TWA concentration			
PCBs	LODs (ng/mL)	LOQs (ng/mL)	R^2	Mean recoveries	(%RSD)	Harbour 1 (ng mL d^{-1})	Harbour 1 (%RSD)	Harbour 2 (ng mL d^{-1})	Harbour 2 (%RSD)	Harbour 1 (ng mL^{-1})	Harbour 1 (%RSD)	Harbour 2 (ng mL^{-1})	Harbour 2 (%RSD)	Harbour 1 (ng mL d^{-1})	Harbour 1 (%RSD)	Harbour 2 (ng mL d^{-1})	Harbour 2 (%RSD)
28	0.353	1.18	0.966	81.3	2.70	0.828	112	1.52	36.0	4.51	173	0.0051	170	0.754	18.2	0.336	29.8
52	0.307	1.02	0.979	75.5	20.9	0.047	79.1	0.0569	37.7	ND	ND	ND	ND	ND	ND	ND	ND
101	0.306	1.02	0.979	103	20.4	0.310	47.7	0.347	33.9	4.42	173	4.75	4.51	ND	ND	0.0599	70.5
118	0.182	0.606	0.990	110	3.84	ND	ND	N.D.	N.D.	ND	ND	ND	ND	ND	ND	ND	ND
138	0.150	0.500	0.994	121	35.4	0.291	62.6	0.578	13.9	2.83	173	ND	ND	0.0530	56.6	0.0157	70.5
153	0.0984	0.328	0.998	84.1	35.9	ND	ND	ND	ND	ND	ND	ND	ND	ND	ND	ND	ND
180	0.111	0.369	0.997	95.9	16.0	0.436	53.5	ND	ND	ND	ND	ND	ND	ND	ND	ND	ND

ND = not detected.

highly chlorinated PCBs tend to occur in higher concentrations in soils and sediments, as opposed to water, due to their higher log K_{ow} values (Faroon *et al.* 2003). The detection of PCB 52 and PCB 180 in the samplers highlights their presence in water. However, they were not present in concentrations that are easily detectable through spot-and-grab sampling. This demonstrates the enrichment capability of the MIM-based sampler, and its ability to enhance the detection limits of PCBs that would normally occur at ultra-trace levels in aquatic environments. The individual PCBs detected were combined to obtain a sum of the concentrations in the harbour points (1.9124–2.5021 ng mL^{-1} d^{-1}). It is evident that the newly-developed MIM-based sampler enhances the detectability of PCBs occurring in low concentrations in water, in considerably shorter deployment times.

The concentrations detected through spot sampling appear to be highly variable (Table 2), and this could be due to the hydrodynamic properties of the dam. Furthermore, an unrealistically large number of water samples would have to be collected and analysed to get TWA concentrations from spot sampling. Therefore, concentrations obtained from the collected water samples can only be used as an indicator for the presence of PCBs in aquatic systems. The PCB concentrations detected through the MIM-based samplers also showed a great degree of variation, with %RSD ranging between 13.9 and 112.3% (Table 2). This could be attributed to a lack of uniformity across the different membranes in the samplers during MIM fabrication. However, the newly-developed sampler gives a good indication of the presence of PCBs in aquatic environments, and this is achieved through minimal sample preparation, as opposed to spot sampling.

Hartbeespoort Dam

The amount of PRCs before and after deployment were compared and used to correct for sampling-site specific conditions using Equations (2) and (8). Field blanks taken to the deployment site during deployment and retrieval of the samplers showed no quantifiable amounts of any of the seven indicator PCBs. The samplers sequestered PCBs 28, 101 and 138 in varying concentrations over the deployment period, and from these accumulated amounts, TWA concentrations were determined. PCB 28 showed the highest TWA concentration (Table 2), followed by PCBs 101 and 138. There were no quantifiable amounts of PCBs 52, 118, 153 and 180 in the samplers after deployment. PCB 28 appears to have the greatest detectability, which could be attributed to higher sampling rates and a higher freely dissolved concentration in water, or a combination of both factors. Since PCBs 101 and 138 have much higher log K_{ow} values, they tend to partition away from the water and into soils and sediments (Faroon *et al.* 2003). In addition to lower concentrations in water, they have lower sampling rates, which could account for the lower concentrations detected.

The TWA concentrations determined are slightly higher than those obtained by Amdany and co-workers using SPMDs over a 14 day deployment period (Amdany *et al.* 2014). For a more direct comparison, TWA concentrations were compared to those by Amdany and co-workers over the winter period, since the MIM-based sampler were also deployed in winter. They found PCB 28 (0.019 ng/mL) in the highest concentration, followed by PCBs 52 (0.006 ng/mL), 138 (0.003 ng/mL) and 101 (0.003 ng/mL), which is consistent with the findings using the newly-developed MIM sampler. However, they also found PCBs 153, 118 and 180 in varying concentrations, which were not detected using the MIM-based sampler. The variation in concentrations between the MIM-based sampler and those determined by Amdany and co-workers could be due to numerous factors. TWA concentrations could vary from time to time due to hydrodynamic conditions and weather patterns. The samplers were deployed over different times, which could account for the slight variation in concentrations.

CONCLUSION

The study provided the development and evaluation of a new MIM-based passive sampler for sampling PCBs, based on the concept of molecular imprinting. The study also provided calibration data for the newly-developed sampler towards different environmental conditions using PRCs. PCBs were sampled at considerably high sampling rates, which was necessary due to their occurrence at ultra-trace levels in aquatic environments. The sampler was deployed at two dams, the Roodeplaat and Hartbeespoort Dams, to evaluate PCB sampling performance. PRC dissipation rates of the deployed samplers were compared to those obtained under controlled laboratory conditions in order to correct for sampling-site specific variables, and to allow for more accurate estimations of TWA concentrations of PCBs. The newly-developed sampler enhanced the detectability of certain PCB congeners present at ultra-trace levels in aquatic environments, which ultimately minimised the sample preparation required to detect such levels. Furthermore, the sampler provided simplified extraction of sequestered PCBs, with minimal solvents required. The shorter deployment times (10 days) associated with the sampler also eliminates the need for collecting large volumes of water samples frequently. This could minimise costs and the turnaround time for analysis. The MIM-based sampler demonstrated the existence of isotropic exchange between PCBs with different PRC analogues. However, more efforts are still required in the sampler fabrication to achieve less variation in the concentrations obtained.

ACKNOWLEDGEMENTS

The authors would like to thank the Department of Trade and Industry (DTi) for funding the research and the National Research Foundation (NRF) for additional funding.

REFERENCES

Agency for Toxic Substances and Disease Registry 2000 *Toxicological Profile for Polychlorinated Biphenyls (PCBs)*. US Department of Health and Human Resources, Agency for Toxic Substances and Disease Registry, Atlanta, GA, 1–948.

Alvarez, D. A. 2010 Guidelines for the Use of the Semipermeable Membrane Device (SPMD) and the Polar Organic Chemical Integrative Sampler (POCIS) in Environmental Monitoring Studies, chapter 4 (section D), Book 1, Collection of Water Data by Direct Measurement, US Geological Survey (USGS), Reston, VA, pp. 1–28.

Amdany, R., Chimuka, L., Cukrowska, E., Kukučka, P., Kohoutek, J. & Vrana, B. 2014 Investigating the temporal trends in PAH, PCB and OCP concentrations in Hartbeespoort Dam, South Africa, using semipermeable membrane devices (SPMDs). *Water SA* **40**, 425–436.

Battershill, J.M. 1994 Review of the safety assessment of polychlorinated biphenyls (PCBs) with particular reference to reproductive toxicity. *Human & Experimental Toxicology* **13**, 581–597.

Bopp, S., Weiß, H. & Schirmer, K. 2005 Time-integrated monitoring of polycyclic aromatic hydrocarbons (PAHs) in groundwater using the ceramic dosimeter passive sampling device. *Journal of Chromatography A* **1072**, 137–147.

Bopp, S. K., McLachlan, M. S. & Schirmer, K. 2007 Passive sampler for combined chemical and toxicological long-term monitoring of groundwater: the ceramic toximeter. *Environmental Science & Technology* **41**, 6868–6876.

Cormack, P. A. & Elorza, A. Z. 2004 Molecularly imprinted polymers: synthesis and characterisation. *Journal of Chromatography B* **804**, 173–182.

Esteve-Turrillas, F. A., Yusà, V., Pastor, A. & de la Guardia, M. 2008 New perspectives in the use of semipermeable membrane devices as passive samplers. *Talanta* **74**, 443–457.

Estoppey, N., Schopfer, A., Omlin, J., Esseiva, P., Vermeirssen, E. L. M., Delémont, O. & De Alencastro, L. F. 2014 Effect of water velocity on the uptake of polychlorinated biphenyls (PCBs) by silicone rubber (SR) and low-density polyethylene (LDPE) passive samplers: an assessment of the efficiency of performance reference compounds (PRCs) in river-like flow condition. *Sci. Total Environ.* **499**, 319–326.

Faroon, O. M., Keith, L. S., Smith-Simon, C. & De Rosa, C. T. 2003 Concise International Chemical Assessment Document 55. Polychlorinated biphenyls: human health aspects. IPCS Concise Int. Chem. Assess. World Health Organization, Geneva. http://www.who.int/pcs/pubs/pub_cicad.htm. Accessed 2 May 2016.

Focant, J. F., Sjödin, A. & Patterson, D. G. 2004 Improved separation of the 209 polychlorinated biphenyl congeners using comprehensive two-dimensional gas chromatography-time-of-flight mass spectrometry. *J. Chromatogr. A* **1040**, 227–238.

Hamers, T., Kamstra, J. H., Cenijn, P. H., Pencikova, K., Palkova, L., Simeckova, P., Vondracek, J., Andersson, P. L., Stenberg, M. & Machala, M. 2011 *In vitro* toxicity profiling of ultrapure non-dioxin-like polychlorinated biphenyl congeners and their relative toxic contribution to PCB mixtures in humans. *Toxicol. Sci.* **121**, 88–100.

Huckins, J. N., Petty, J. D., Lebo, J. A., Almeida, F. V., Booij, K., Alvarez, D. A., Cranor, W. L., Clark, R. C. & Mogensen, B. B. 2002 Development of the permeability/performance reference compound approach for *in situ* calibration of semipermeable membrane devices. *Environ. Sci. Technol.* **36**, 85–91.

Kim, U. J., Kim, H. Y., Alvarez, D., Lee, I. S. & Oh, J. E. 2014 Using SPMDs for monitoring hydrophobic organic compounds in urban river water in Korea compared with using conventional water grab samples. *Science of the Total Environment* **470**, 1537–1544.

Leonard, A. W., Hyne, R. V. & Pablo, F. 2002 Trimethylpentane-containing passive samplers for predicting time-integrated concentrations of pesticides in water: Laboratory and field studies. *Environmental Toxicology and Chemistry* **21**, 2591–2599.

Lind, P. M., Larsson, S., Oxlund, H., Håkansson, H., Nyberg, K., Eklund, T. & Örberg, J. 2000 Change of bone tissue composition and impaired bone strength in rats exposed to 3,3′,4,4′,5-pentachlorobiphenyl (PCB126). *Toxicology* **150**, 41–51.

Lundberg, R., Lyche, J. L., Ropstad, E., Aleksandersen, M., Rönn, M., Skaare, J. U., Larsson, S., Örberg, J. & Lind, P. M. 2006 Perinatal exposure to PCB 153, but not PCB 126, alters bone tissue composition in female goat offspring. *Toxicology* **228**, 33–40.

Mackay, D., Shiu, W. Y. & Ma, K. C. L. S. S. 1992a *Illustrated Handbook of Physical Chemical Properties of Environmental Fate of Organic Chemicals*, 2nd edn. Lewis Publishers, Michigan, USA.

Mackay, D., Shiu, W. Y. & Ma, K. C. L. S. S. 1992b *Illustrated Handbook of Physical Chemical Properties of Environmental Fate of Organic Chemicals*, 1st edn. Lewis Publishers, Michigan, USA.

McFarland, V. A. & Clarke, J. U. 1989 Environmental occurrence, abundance, and potential toxicity of polychlorinated biphenyl congeners: considerations for a congener-specific analysis. *Environmental Health Perspectives* **81**, 225–239.

Mkhize, D. S., Nyoni, H., Quinn, L. P., Mamba, B. B. & Msagati, T. A. 2017 Molecularly imprinted membranes (MIMs) for selective removal of polychlorinated biphenyls (PCBs) in environmental waters: fabrication and characterization. *Environ. Sci. Pollut. Res. Int.* **24**, 11694–11707.

Nieuwoudt, C., Quinn, L. P., Pieters, R., Jordaan, I., Visser, M., Kylin, H., Borgen, A. R., Giesy, J. P. & Bouwman, H. 2009 Dioxin-like chemicals in soil and sediment from residential and industrial areas in central South Africa. *Chemosphere* **76**, 774–783.

Piccardo, M.T., Stella, A., Pala, M., Balducci, D. & Valerio, F. 2010 Field use of semipermeable membrane devices (SPMDs) for passive air sampling of polycyclic aromatic hydrocarbons: opportunities and limitations. *Atmospheric Environment* **44**, 1947–1951.

Popp, P., Keil, P., Montero, L. & Rückert, M. 2005 Optimized method for the determination of 25 polychlorinated biphenyls in water samples using stir bar sorptive extraction followed by thermodesorption-gas chromatography/mass spectrometry. *J. Chromatogr. A* **1071**, 155–162.

Saad, E. M., Madbouly, A., Ayoub, N. & El Nashar, R. M. 2015 Preparation and application of molecularly imprinted polymer for isolation of chicoric acid from *Chicorium intybus* L. medicinal plant. *Analytica Chimica Acta* **877**, 80–89.

Sun, P., Basu, I., Blanchard, P., Brice, K. A. & Hites, R. A. 2007 Temporal and spatial trends of atmospheric polychlorinated biphenyl concentrations near the Great Lakes. *Environ. Sci. Technol.* **41**, 1131–1136.

Vrana, B., Mills, G. a., Dominiak, E. & Greenwood, R. 2006a Calibration of the chemcatcher passive sampler for the monitoring of priority organic pollutants in water. *Environ. Pollut.* **142**, 333–343.

Vrana, B., Paschke, A. & Popp, P. 2006b Calibration and field performance of membrane-enclosed sorptive coating for integrative passive sampling of persistent organic pollutants in water. *Environ. Pollut.* **144**, 296–307.

First received 10 August 2018; accepted in revised form 2 December 2018. Available online 25 January 2019

Fluorescence spectroscopy for assessing trihalomethane precursors removal by MIEX resin

P. Jutaporn, W. Laolertworakul, M. D. Armstrong and O. Coronell

ABSTRACT

This study investigated the applicability of fluorescence excitation–emission matrix spectroscopy (EEMS) to assess total trihalomethane formation potentials (TTHMFPs) and the ability of magnetic ion exchange (MIEX®) resin to reduce TTHMFP. We treated a surface water and secondary wastewater effluent with MIEX mimicking full-scale operation by repeatedly exposing the same resin batch to additional feed water, with batches ranging from 500 to 5,000 resin bed volumes. Results showed that MIEX was more effective at removing or reducing ultraviolet absorbance (UVA_{254}), dissolved organic carbon (DOC), and TTHMFP in surface water than in secondary effluent. The greater UVA_{254}, DOC and TTHMFP removal for surface waters was explained by the stronger affinity of MIEX for terrestrial dissolved organic matter (DOM) compared to microbial DOM. Fluorescence EEMS results showed that the ratio between terrestrial and microbial fluorescent signals of DOM was significantly greater in surface water than in secondary effluent. Fluorescence surrogate parameters were strongly correlated with TTHMFP, namely, fluorescence intensity of humic-like peak C ($R^2 = 0.98$, $p < 0.01$), protein-like peak T ($R^2 = 0.96$, $p < 0.01$), and fulvic-like peak A ($R^2 = 0.87$, $p < 0.01$). Correlations between fluorescence surrogate parameters and TTHMFP were substantially stronger than correlations between DOC and TTHMFP. Overall, the results indicate that fluorescent parameters extracted from EEMS data can be used as quick surrogate parameters to monitor TTHMFP for a diverse group of raw and MIEX-treated waters.

Key words | dissolved organic matter (DOM) characterization, excitation–emission matrices (EEMs), fluorescence, MIEX®, trihalomethane formation potentials

P. Jutaporn (corresponding author)
W. Laolertworakul
Research Center for Environmental and Hazardous Substance Management, Department of Environmental Engineering, Faculty of Engineering,
Khon Kaen University,
Khon Kaen, 40002,
Thailand
E-mail: *panitju@kku.ac.th*

P. Jutaporn
Center of Excellence on Hazardous Substance Management,
Bangkok 10330,
Thailand

M. D. Armstrong
O. Coronell
Department of Environmental Sciences and Engineering, Gillings School of Global Public Health,
The University of North Carolina at Chapel Hill,
Chapel Hill, NC 27599,
USA

INTRODUCTION

Disinfection is an essential step to provide a clean and safe water supply. While chemical disinfectants (e.g., free chlorine, monochloramine, and chlorine dioxide) are effective at protecting human health from pathogens, their reactions with dissolved organic matter (DOM) can form harmful disinfection by-products (DBPs). Despite the availability of other chemical disinfectants, free chlorine disinfection remains the most widely used disinfectant in the USA (US EPA 2016). Free chlorine is of particular concern because its reaction with DOM forms trihalomethanes (THMs) preferentially over other DBPs and in higher mass concentration (Singer 1999).

Since the discovery of associations between DBPs and adverse health effects, DBPs have become a focus of study in water treatment. The National Cancer Institute declared chloroform to be carcinogenic in laboratory animals (US EPA 2016) and long term exposure to THMs has been associated with higher risk of bladder cancer (US EPA 2016). In 2001, the US Environmental Protection Agency (EPA) promulgated the Stage 1 Disinfectants and Disinfection Byproducts Rule, which set the maximum contaminant level to 80 μg/L for total trihalomethanes (TTHM) (US EPA 2001). As a DBP precursor, the structure of DOM species plays an important role in DBP formation because aqueous chlorine species tend to react with electron-rich sites (e.g., aromatic rings and amino nitrogen) in organic structures (Harrington

doi: 10.2166/wst.2019.036

et al. 1996). For example, THM formation potential (THMFP) was found to be strongly correlated with indicators of DOM aromaticity (i.e., UVA_{254} (Pifer & Fairey 2014) and specific UV absorbance (SUVA) (Singer 1999)), and phenolic carbon content of hydrophobic natural organic matter (NOM) (Harrington *et al.* 1996). The importance of DOM structure on DBP formation is important because the characteristics of organic matter vary among water types. Specifically, compared to organic matter in natural water, effluent organic matter (EfOM) in treated wastewater is less aromatic and more hydrophilic, and is enriched with nitrogen, refractory substrates, and soluble microbial products (Chen *et al.* 2003). Most EfOM contains high levels of DBP precursors that form a wider range of DBPs than do natural waters when treated with chlorine for pathogen control (Krasner *et al.* 2008).

The identification of DOM as a DBP precursor has had a major influence on the selection, design, and operation of water and wastewater treatment strategies to control DBP formation. Magnetic ion exchange (MIEX®) resin was developed by Ixom Watercare Inc. (Australia) specifically to remove DOM in water treatment plants. MIEX is a strong base anion-exchange resin with magnetic properties and a small resin bead size. The magnetic properties make the resin beads act as weak individual magnets, which enhances particle agglomeration and facilitates rapid particle settling. The small resin beads provide high surface areas that enhance DOM removal kinetics and regeneration capacity (Singer & Bilyk 2002).

The effectiveness of MIEX resin to remove bulk dissolved organic carbon (DOC), including THM precursors, in surface water is well established (Singer & Bilyk 2002; Boyer & Singer 2005). For example, THMFP was reduced by more than 60% in MIEX batch experiments using surface water from nine different water utilities across the USA (Singer & Bilyk 2002). By contrast, there are fewer studies that reported the ability of MIEX resin to remove DOC or THM precursors from synthetic wastewater (Nguyen *et al.* 2011) and biologically treated secondary effluent (Zhang *et al.* 2006). Notably, Gan *et al.* (2013) reported that 8–12 mL/L MIEX resin reduced 45–58% of DOC and 39–62% of THMFP from effluent impacted waters. Further, few studies have focused on the quantifiable characteristics of DOM to explain differences in MIEX resin performance in treating effluent and natural water.

DOM has a broad range of concentrations and compositions; its chemistry is complex and highly variable in water (Cory *et al.* 2010). Because organic molecules primarily contain humic substances, which are aromatic chromophores that absorb both visible and UV light, spectrophotometric analysis can be used to characterize DOM. When excited by UV and blue light, a sub-fraction of DOM emits energy as fluorescent light (Stedmon *et al.* 2003; Stedmon & Bro 2008). 3D-fluorescence excitation–emission matrix (EEM) spectroscopy, a series of emission scans in a range of excitation wavelengths, characterizes DOM with minimal sample pretreatment (Coble 1996). Fluorescence excitation–emission matrix spectroscopy (EEMS) is both qualitative and quantitative, providing information regarding the type, structure, and abundance of functional groups. Fluorescent organic matter is generally classified into two distinct groups: terrestrially derived (humic-like substances) and microbially derived (protein-like substances) (McKnight *et al.* 2001; Leenheer & Croué 2003; Stedmon *et al.* 2003). Quantitative fluorescence parameters (e.g., fluorescence index (FI), peak intensity ratios, and maximum fluorescence intensity in different regions) can be extracted from EEM spectra. For example, FI is the ratio of emission intensity (450 nm/500 nm) at 370 nm excitation (McKnight *et al.* 2001) and can be used to distinguish sources of aquatic DOM. A previous study (McKnight *et al.* 2001) reported that terrestrially derived fulvic acids had a low FI (~1.4) and microbially derived DOM had a high FI (~1.9). Typically, FI was low for DOM with high aromaticity and high for DOM with low aromaticity.

EEMS has been used to track DOM in both drinking water and wastewater treatment plants (Chen *et al.* 2003). Strong correlations between THMFP and humic-like fluorophores in samples from drinking water treatment plants were reported previously (Pifer & Fairey 2014), suggesting the application of fluorescence parameters to assess THMFP. Using EEMS analysis, Jutaporn *et al.* (2016) demonstrated that MIEX resin was more effective at removing terrestrially derived DOM than microbially derived DOM, and Arias-Paic *et al.* (2016) reported that MIEX treatment produced water with predominantly microbial character due to its preferential removal of terrestrially derived DOM. However, the correlations of raw water fluorescence parameters with THMFP and the ability of MIEX resin to reduce THMFP remain under-explored, especially with wastewater effluent and wastewater-effluent-impacted water. Moreover, most of the results in the published literature on MIEX performance were obtained using batch experiments with single use of MIEX resin; batch tests do not mimic MIEX conditions in full-scale applications and might overestimate the removal efficiency of MIEX (Kitis *et al.* 2007) Only a limited number of studies use experimental protocols that mimic full-scale operation.

Accordingly, the objective of this study was to investigate the applicability of fluorescence EEMS analysis to assess: (1) the THMFP of surface water and wastewater effluent and (2) the ability of MIEX resin to remove THM precursors from surface water and wastewater effluent. We examined fluorescence spectra of DOM from surface water and wastewater effluent, before and after a MIEX treatment protocol that replicated full-scale, continuous plant operation.

METHODS

Water samples

We collected a surface water and secondary effluent from Orange County, North Carolina, USA. Surface water samples were collected on 10 October 2013 from a drinking water treatment plant intake that used University Lake as source water. Secondary effluent before UV disinfection was collected on 24 January 2014 from Mason Farm Wastewater Treatment Plant (WWTP), a conventional secondary activated sludge process with nitrification. Water samples (50 L) were collected once from each water source and used for all experiments in this work. Upon arrival at the laboratory, water samples were filtered with 1.2 μm glass fiber filters (Whatman GF/C) to remove suspended materials. All samples were stored at 4 °C in the dark and allowed to reach room temperature (20 ± 2 °C) prior to each experiment.

MIEX treatment

MIEX resin (Ixom Watercare Inc., Australia) was stored wet in a 5% NaCl solution, then rinsed with laboratory-grade deionized water prior to use. The resin was allowed to settle in a graduated cylinder for 15 min before measuring the settled resin volume. The resin contact experiments were performed using a six-paddle stirrer jar test (Phipps and Bird Inc., VA, USA) as described by Singer & Bilyk (2002). The experiment followed the multiple-loading test protocol as advised by the manufacturer. The purpose of the MIEX multiple-loading procedure was to replicate the performance of a full-scale continuously operated process using the same resin multiple times without regeneration. The initial resin dose was 2 mL per litre of water (500 bed volume (BV)).

The procedure required mixing the resin with 2 L of raw water for 15 minutes at 100 rpm, followed by 1 minute of settling. We doubled the settling period recommended by the manufacturer from 30 seconds to 1 minute; we also used a magnet to aid settling and decanting. Treated water was decanted into collection vessels. Another 2 L of raw water was added to the resin in the jars and mixing was repeated. The bed volume was calculated as the ratio between the volume of treated water to MIEX resin (Kitis *et al.* 2007). We used the same quantity of resin for each experiment, meaning BV values are defined by the volume of treated water; larger BV values indicate consecutive resin uses. The resin was put in contact with six consecutive volumes of raw water (i.e., repeated six times using the same resin without regeneration), spanning 500–5,000 BV. One sample treated with fresh resin (500 BV) and five composite samples from 1,000–5,000 BV treatments were collected for each water source. MIEX-treated samples were vacuum filtered with 1.2 μm glass fiber filters (Whatman GF/C) after MIEX treatment, as resin fines could have interfered with sample analysis. Each treatment condition was performed in triplicate and the average value was reported.

DOM measurement and characterization

DOC (mg-C/L) and total dissolved nitrogen (TDN, mg-N/L) were measured using a TOC-V organic carbon analyzer (Shimadzu, Atlanta, GA, USA). UVA_{254} measurements were made using a 1-cm quartz cell (Starna, Australia) and a U-2000 spectrophotometer (Hitachi Instruments Inc., Danbury, CT, USA. SUVA values were calculated as UVA_{254} normalized by DOC. As mentioned above, the effluent water was sampled from Mason Farm WWTP, which has good nitrification capability; thus, we expected that the majority of TDN in the effluent was composed of nitrate from the nitrification process (Gan *et al.* 2013).

EEM spectra of the water samples were acquired using a 1-cm quartz cell in a Fluorolog-321 spectro-fluorometer with a xenon arc lamp as the excitation source and a synapse CCD detector (Horiba JobinYvon, Edison, NJ, USA). Excitation wavelengths of 240–450 nm in 5-nm intervals and emission wavelengths of 320–550 nm in 1-nm intervals were used. The blanks were collected from deionized laboratory-grade water daily on the same day as the sample measurement. Inner-filter correction (McKnight *et al.* 2001) was performed with UV absorption spectra data acquired from a diode array UV-Vis spectrophotometer (Hewlett Packard, Palo Alto, CA, USA) and a 1–5 cm quartz cell for an absorption range of 200–600 nm. All EEMs in this study were collected in ratio mode (signal from the sample normalized to the signal from the reference detector: S/R), which corrects the non-uniform output of the lamp over the excitation range (Cory *et al.* 2010). Sample EEMs were corrected for instrumental bias using the excitation and emission correction factor, then corrected for inner filtering effects, and the blank subtracted. The fluorescence character of organic matter in the two water sources could be represented by three main peaks: Peak A (terrestrial fulvic-like, ex 250 nm/em 450 nm), Peak C (terrestrial humic-like, ex 350 nm/em 450 nm), and Peak T (microbial protein-like, ex 275 nm/em 340 nm) (Coble 1996). The fluorescent intensities were normalized by the area under the water-Raman peak (ex 350 nm/em 368–450 nm), reporting data in Raman units (RU) (Stedmon *et al.* 2003). All samples were analyzed at least in duplicate and average values were reported. EEM correction and data processing were performed using Matlab (MATLAB R2016a, MathWorks) and DOM-Fluor Toolbox (Stedmon & Bro 2008). Statistical analyses (i.e., analysis of variance (ANOVA), paired t-test, and regression analysis) were performed using Stata IC 13.1 (StataCorp LP, College Station, TX, USA).

MIEX treatment resulted in minor pH increases of 0.07–0.12 units for surface water and 0.10–0.32 units for secondary effluent. Previous studies reported that the effect of pH on the fluorescence intensity of humic and fulvic like substances was minor in the pH range of 6–8 (Patel-Sorrentino *et al.* 2002) and that no significant pH effect was observed for tryptophan-like fluorescence (Spencer *et al.* 2007). Thus, EEM spectra measurement in this study was performed at the native pH of the samples.

We chlorinated water samples and evaluated THMFP under uniform formation conditions (UFC) (Summers *et al.* 1996). The UFC correspond to the chlorination conditions that yield 1 mg/L free chlorine residual at pH 8 after 24-hour incubation at 20 °C in the dark. Chlorine residuals were measured by a DPD (*N,N*-diethyl-*p*-phenylenediamine) colorimetric method (Ha *et al.* 2004) using a Hach DR890 portable colorimeter. The samples with chlorine residuals of 1 ± 0.2 mg/L as Cl_2 were selected for THM4 analysis (chloroform (Cl_3CH), bromodichloromethane ($BrCl_2CH$), dibromochloromethane (Br_2ClCH), and bromoform (Br_3CH)). THM4 were extracted with methyl tert-butyl ether and analyzed on a Hewlett-Packard 5890 gas chromatograph with electron capture detector. We followed procedures published by the US EPA (EPA Method 551.1; US EPA 1995) with minor modifications. The following temperature program was used: oven held at 35 °C for 22 min, increased at 10 °C/min to 145 °C and held for 2 min, increased at 20 °C/min to 225 °C and held for 10 min, then increased at 20 °C/min to 260 °C and held for 5 min (Lyon *et al.* 2014). The injection volume was 2 μL (splitless), and after 0.5 minutes, the sample was split at a 1:1 ratio. The injection temperature was set at 117 °C and the detector temperature was 290 °C. All samples were analyzed in replicate.

RESULTS AND DISCUSSION

Raw water characteristics

Water quality parameters of the raw water samples are summarized in Table 1. Both water sources had pH values close to neutral (6.85 and 6.95). The surface water had slightly lower DOC (6.55 mg-C/L) than the secondary effluent (7.17 mg-C/L). The secondary effluent had higher conductivity and TDN concentration, and lower UVA_{254} and SUVA values than the surface water. With similar DOC concentrations, high conductivity in the effluent indicated higher concentration of inorganic materials, which can also influence DBP formation (Krasner *et al.* 2008). The low SUVA of the secondary effluent sample is indicative of the non-aromatic and hydrophilic characters of EfOM, consistent with Krasner *et al.* (2008). Additionally, surface water had a lower FI value compared to effluent of (1.47 and 1.99, respectively), indicating that DOM in surface water was terrestrially derived, while EfOM was from microbial origin (McKnight *et al.* 2001). Therefore, these two source waters provide two distinctly different types of waters with which we can evaluate the applicability of EEMS to assess total THMFP (TTHMFP) and the ability of MIEX resin to reduce TTHMFP. Other important aspects of THM formation such as the effect of nitrate, nitrite, and bromide on THM formation have been well established in previous studies (Boyer & Singer 2005).

Removal of DOC and UVA_{254} and reduction of SUVA

Figure 1 illustrates changes in DOC, UVA_{254}, and SUVA in raw water (0 BV) and samples treated with 500–5,000 BV

Table 1 | Water quality of raw waters

Sample	University Lake	Mason Farm
Source	Surface water	Secondary treated effluent
Turbidity	3.34 NTUs	3.02 NTUs
pH	6.85	6.95
Dissolved organic carbon	6.55 ± 0.03 mg-C/L	7.17 ± 0.09 mg-C/L
Total dissolved nitrogen	0.43 ± 0.02 mg-N/L	9.59 ± 0.12 mg-N/L
UV_{254} absorbance	0.150 ± 0.01 cm^{-1}	0.114 ± 0.01 cm^{-1}
Specific UV absorbance	2.29 ± 0.04 L/mg.m	1.59 ± 0.02 L/mg.m
Conductivity	116 ± 5.4 μS/cm	505 ± 6.1 μS/cm
Fluorescence index	1.47 ± 0.01	1.99 ± 0.01

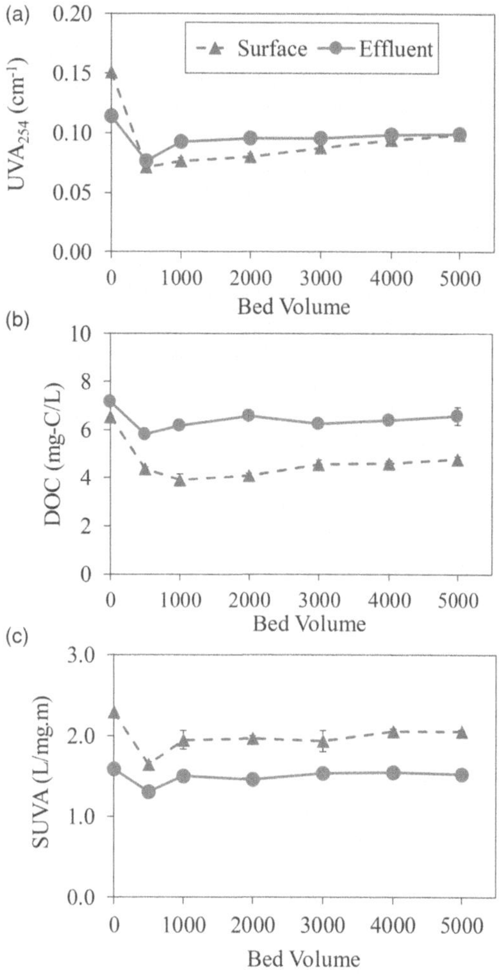

Figure 1 | Quality of raw and MIEX-treated waters measured as (a) UVA_{254}, (b) DOC, and (c) SUVA. MIEX treatment ranged from 500 to 5,000 bed volumes.

of MIEX resin. MIEX treatment removed a substantial amount of UVA_{254} (37–53%) and DOC (28–44%) from the surface water samples, but less (13–33% for UVA_{254} and 8.5–19% for DOC) from the secondary effluent. This result agrees well with previous studies (Gan *et al.* 2013) that reported greater removal of DOC from samples with high SUVA, as the raw water SUVA of secondary effluent was 1.59 L/mg.m (with corresponding lower removals) and that of the surface water was 2.29 L/mg.m (with corresponding higher removals). MIEX treatment preferentially removed UV-absorbing material as indicated by higher removal of UVA_{254} (13–53%, Figure 1(a)) than of DOC (8.5–44%, Figure 1(b)), which resulted in lower SUVA values for the MIEX-treated water (Figure 1(c)). SUVA was reduced by MIEX treatment by 11–28% for surface water and 4.6–18% for secondary effluent. Since high SUVA waters are generally enriched in hydrophobic DOM (Leenheer & Croué 2003), this result is consistent with previous reports that MIEX preferentially removed hydrophobic over hydrophilic DOM (Boyer & Singer 2005).

For this study, the 500–5,000 BV range, equivalent to a resin dose range of 0.2–2 mL/L in a batch operation, covered a typical commercial resin dose of 600 BV, but is relatively low compared to previous batch studies. For example, Gan *et al.* (2013) used 8–12 mL/L MIEX for DBP control and reported 44–82% removal of DOC. With the continuous BV treatment, the low DOC removal reported in this study was as expected.

For both surface water and secondary effluent, increase in BV (i.e., the number of consecutive resin uses) resulted in a decline in removal efficiency of DOM, consistent with a study by Arias-Paic *et al.* (2016) that showed lower BV treatment removed more DOC. Figure 1(b) shows that in this study, for surface water, the 1,000 BV treatment achieved the lowest DOC concentration of 3.91 mg-C/L (40% removal) and the 3,000 BV loading achieved ~4.6 mg-C/L (~31% removal), while larger loadings (>3,000 BV) had marginal differences in DOC removal. A similar trend was observed in secondary effluent, as fresh MIEX resin achieved the greatest removal efficiency at 500 BV (19% removal), and a constant DOC concentration of ~6.0 mg/L (~11% removal) was achieved at larger BVs. In other words, no detectable DOC removal by the resin occurred after a loading of 3,000 BV for surface water and 1,000 BV for secondary effluent. The same trend of resin saturation (i.e., inability to remove additional DOM) was observed for UVA_{254} reductions after 500–1,000 BV (Figure 1(a)). Overall, our results showed that MIEX treatment achieved greater UVA_{254} and DOC removal efficiency in surface water compared to secondary effluent. This greater removal can be explained by the difference in SUVA of these two water sources. DOC removal by MIEX resin was better for high SUVA values of surface water than low SUVA values of secondary effluent, similar to previous studies (Boyer *et al.* 2008; Singer *et al.* 2009). Surface water with high SUVA values has more hydrophobic DOM, whereas secondary effluents with low SUVA values tend to be dominated by hydrophilic acids and neutral DOM. Low DOC removal in waters with low SUVA values occurs because hydrophilic acids contain more carboxylic acid functional groups per milligram of carbon than hydrophobic acids, thereby occupying more sites on the resin (Singer *et al.* 2009) and resulting in low removal.

THMFP formation and its reduction by MIEX

Total THM (TTHM is the sum of the four chloro- and bromo-substituted THM species (Cl_3CH, $BrCL_2CH$, Br_2ClCH, and BR_3CH)). Illustrative effects of MIEX treatment on the formation potential of THM speciation in surface water and secondary effluent is shown in Figure 2(a) and 2(b), respectively. The dominant THM species formed during chlorination of surface water was Cl_3CH, while secondary effluent had higher concentrations of $BrCl_2CH$ and Br_2ClCH. Considering the similarity in DOC concentration (6.55 and 7.17 mg-C/L in surface water and secondary effluent, respectively), TTHMFP was significantly higher in secondary effluent (611 μg/L) than in surface water (256 μg/L). This result suggested the different characteristics of DOM present in surface water and secondary effluent played a major role in formation potentials. NOM in the surface water was rich in humic materials from terrestrial origin, while EfOM was comprised of both refractory substrates and soluble microbial products (SMPs) (Chen *et al.* 2003); thus, an aggregate DOC measurement was not an effective predictor of THMFP for water sources containing different DOM matrices (Krasner *et al.* 2008).

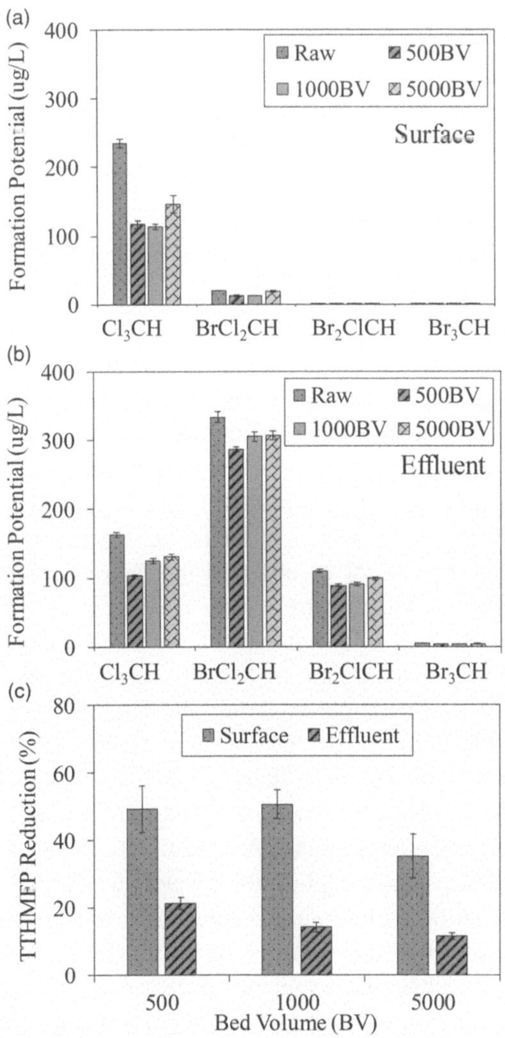

Figure 2 | Impact of selected MIEX loadings on THMFP reduction in (a) surface water, (b) effluent, and (c) TTHMFP reduction by selected MIEX loadings.

TTHMFP reduction by MIEX treatment is shown in Figure 2(c). MIEX treatment achieved greater TTHMFP reduction in surface water than in secondary effluent, which is consistent with the DOC and UVA_{254} results shown in Figure 1. For surface water, 500 BV loading achieved 49% TTHMFP reduction, while increasing the loading from 500 to 1,000 BV did not cause a significant performance change; however, the reduction efficiency decreased substantially to 35% at a loading of 5,000 BV. For secondary effluent, 21% TTHMFP was achieved at a loading of 500 BV, but TTHMFP reduction efficiency dropped to 11% at 5,000 BV. Therefore, differences in DOM composition and source not only affected the speciation of THMs, but also influenced the DOM removal efficiency of MIEX treatments and consequently affected the reduction of TTHMFP.

Fluorescence excitation–emission matrices characterization

Surface water and secondary effluent samples before and after MIEX treatment were characterized using fluorescence excitation–emission matrices spectroscopy (EEMS. Representative EEM spectra obtained for raw and treated water samples are shown in Figure 3. Effects of Rayleigh and Raman scatter were removed for a better representative plot using DOMFluor Toolbox (Stedmon & Bro 2008). Locations of common re-occurring fluorescence components are designated as letter A, C, and T. For the visible region peaks, peak A (ex 250 nm/em 450 nm) in this study had similar positions to those reported for the terrestrial fulvic acids, and peak C (ex 350 nm/em 450 nm) resembled humic-like fluorophore. The position of peak T (ex 275 nm/em 340 nm), which can be seen distinctively in secondary effluent samples, resembled microbially derived protein-like fluorophore (Coble 1996). Therefore, we defined peak A as fulvic-like, peak C as humic-like, and peak T as protein-like DOM.

In all EEM spectra, the highest intensity was observed at peak A and moderate intensity was seen at peak C. Peak T, which is related to SMPs (Chen *et al.* 2003), was

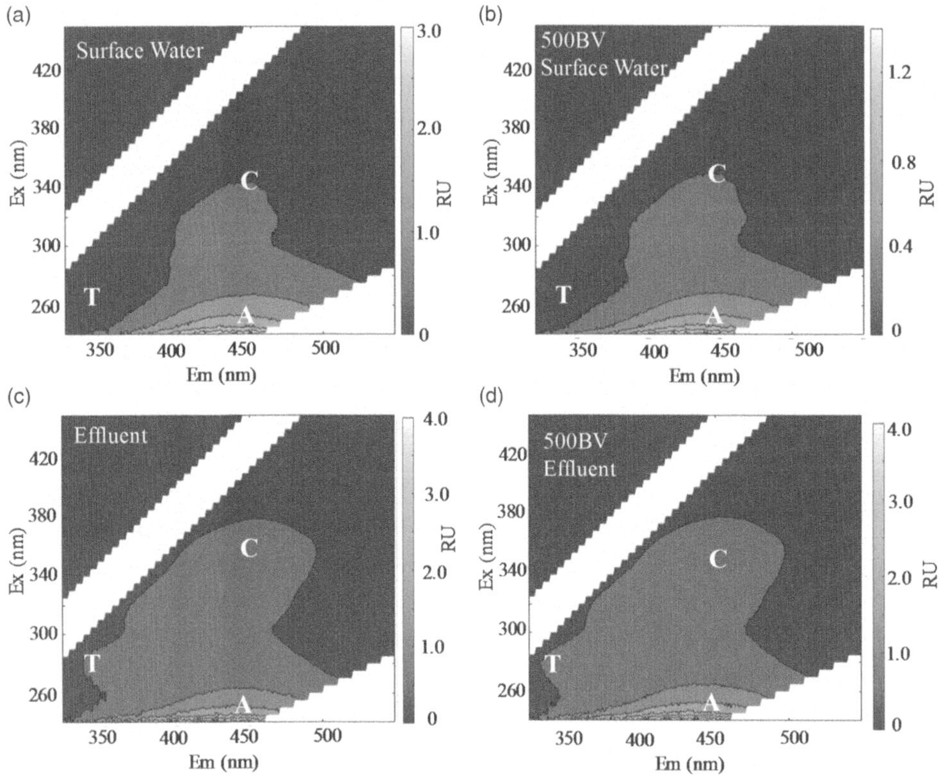

Figure 3 | Fluorescence excitation-emission matrix of (a) surface water, (b) surface water treated with 500 BV MIEX, (c) effluent, and (d) effluent treated with 500 BV MIEX. Positions of the peak A (ex 250 nm/em 450 nm), peak C (ex 350 nm/em 450 nm), and peak T (ex 275 nm/em 340 nm) correspond to fulvic-like, humic-like, and protein-like DOM, respectively. Note that the color axis unit of (a) and (b) are different: maximum intensity was 3.0 RU for raw surface water and 1.4 RU for 500 BV-treated surface water. Please refer to the online version of this paper to see this figure in color: http://dx.doi.org/10.2166/wst.2019.036.

observed more apparently in secondary effluent than in surface water. Comparing the EEM spectra of raw surface water (Figure 3(a)) to 500 BV-treated surface water (Figure 3(b)), loss in fluorescence intensity was observed visually, suggesting effectiveness in fluorescent DOM removal by MIEX treatment. Note that the color scales of Figure 3(a) and 3(b) are different from each other: maximum intensity was 3.0 RU for raw surface water and 1.4 RU for 500 BV-treated water. In the case of secondary effluent (Figure 3(c) and 3(d)), a qualitative difference in the EEM spectra was not observed, suggesting that loss in fluorescence intensity as a result of MIEX treatment was small and that MIEX was not as effective at removing fluorescent DOM from secondary effluent compared to surface water.

To analyze the EEM spectra quantitatively, we extracted fluorescence parameters including maximum fluorescence intensity at peaks A, C and T, as well as peak intensity ratios, FI, and fluorescence signal reduction caused by MIEX treatment. Maximum fluorescence intensity at peaks A, C, and T of untreated surface water and secondary effluent is presented in Figure 4(a). Both water samples had highest intensity at peak A (1.65 and 2.32 RU for surface water and secondary effluent, respectively), followed by moderate intensity at peak C (0.48 and 1.33 RU for surface water and secondary effluent, respectively) and the lowest intensity at peak T (0.26 and 0.77 RU for surface water and secondary effluent, respectively). The quantitative peak intensities confirmed the greater abundance of humic substances compared to protein-like DOM as observed qualitatively from the EEM spectra shown in Figure 3. This result is consistent with literature findings for most natural waters (Stedmon *et al.* 2003) and secondary effluent (Henderson *et al.* 2011). The fluorescence spectra observed for the surface water samples were similar to samples collected from the same treatment plant in 2013 for a separate study (Jutaporn *et al.* 2016).

Figure 4(b) depicts a fluorescence signal ratio of fulvic-like to protein-like DOM peaks (A/T) and fulvic-like to humic-like DOM peaks (A/C) in surface water and secondary effluent. Peak intensity ratio was previously used as a parameter to compare the removals of the related DOM fractions across different treatment processes (Baghoth *et al.* 2011). In this study, surface water had an A/T ratio of 6.4, which was significantly ($p < 0.01$) greater than the A/T ratio of 3.0 of secondary effluent. This result suggests

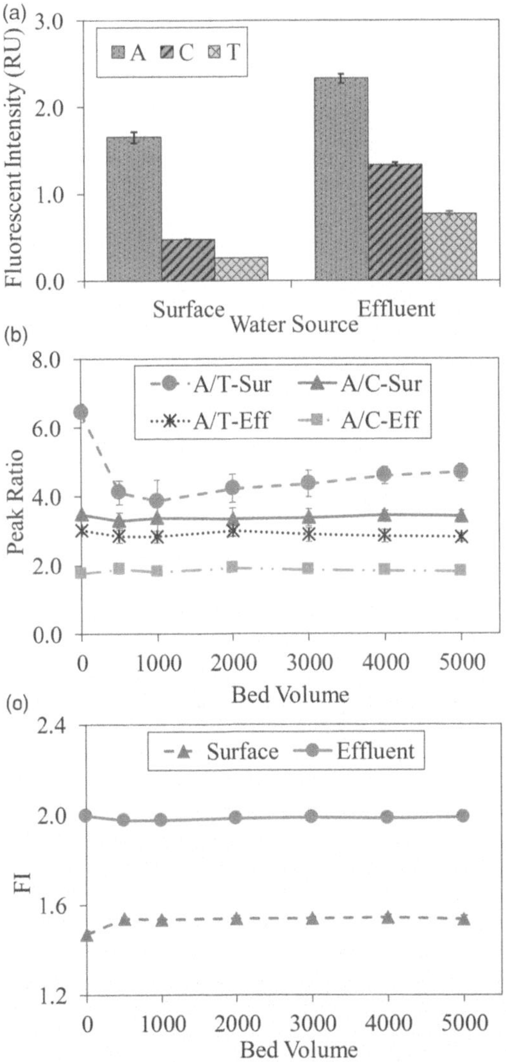

Figure 4 | (a) Fluorescence intensity at peaks A, C, and T of untreated surface water and effluent, (b) ratio of fluorescence signal of the fulvic-like to protein-like DOM peaks (A/T) and fulvic-like to humic-like DOM peaks (A/C), and (c) fluorescence index (FI) with MIEX treatments ranging 500–5,000 BV loadings.

that the two water sources were different in DOM composition and that the surface water was more humic-like, while the secondary effluent had greater abundance of microbial protein-like DOM. After MIEX treatment, the A/T ratio of the surface water decreased substantially from 6.4 to 3.9 at a loading of 1,000 BV, but increased to 4.7 as the BV loading increased from 1,000 BV to 5,000 BV, suggesting the MIEX treatment changed the DOM composition of the surface water. The surface water became less humic after MIEX treatment, as demonstrated by the decrease in A/T ratio compared to the raw surface water, but then humic content increased with continuous resin use. In contrast, the secondary effluent A/T ratios remained constant at ~2.9 across the range of BV treatments (changes were statistically insignificant). Thus, changes in the composition of DOM in the secondary effluent upon MIEX treatment were less apparent. Raw and treated waters had relatively unchanged A/C ratios at ~3.4 for surface water and at ~1.8 for secondary effluent, indicating no significant change in the ratio of fulvic-like to humic-like DOM by MIEX treatment.

FI values were extracted from EEM spectra of all samples and are shown in Figure 4(c). FI is the ratio of emission intensity (450 nm/500 nm) at 370 nm excitation (McKnight *et al.* 2001), and can serve as an index to distinguish sources of aquatic DOM. A previous study (McKnight *et al.* 2001) reported low FI values (~1.4) for terrestrially derived fulvic acids and high FI values (~1.9) for DOM from microbial origin. Typically, FI decreased as aromaticity of DOM increased.

In this study, raw surface water had a low FI value of 1.47 compared to a high FI value of 1.99 of raw secondary effluent (Table 1 and Figure 4(c)), indicating that DOM in surface water was terrestrial and humic, while secondary effluent was enriched with microbial DOM. This result agrees well with the nature of the samples in this study. Surface water was sampled from a lake in North Carolina, in which the DOM sources are dominated by terrestrial plants and soil organic matter. Secondary treated effluent was sampled from an activated sludge wastewater treatment system, where microbial activities are abundant.

With MIEX treatment, the FI value of treated surface water increased, resulting in a maximum FI value of 1.54 at a loading of 2,000 BV. This phenomenon aligns with our observation that MIEX resins remove terrestrially derived organic matter preferentially, resulting in a treated water with predominantly microbial DOM. For raw and treated secondary effluent, FI values remained unchanged at the range of 1.98–1.99, confirming results from the peak ratio analysis that the A/T and A/C ratios generally stayed the same in experiments with secondary effluent. Taken together, the peak ratio and FI results imply that MIEX treatment had a strong affinity for humic-like DOM in surface water, but did not a cause significant change in the DOM composition in secondary effluent. It is also noteworthy that the standard deviations of FI values from replicate experiments were narrower than those of peak ratios or the intensity at different peaks themselves. Therefore, the precision in the FI measurement would allow its use as a reliable parameter to analyze or predict DOM sources from EEM spectra.

To investigate quantitatively the apparent preferential affinity of the MIEX treatments for humic-like DOM,

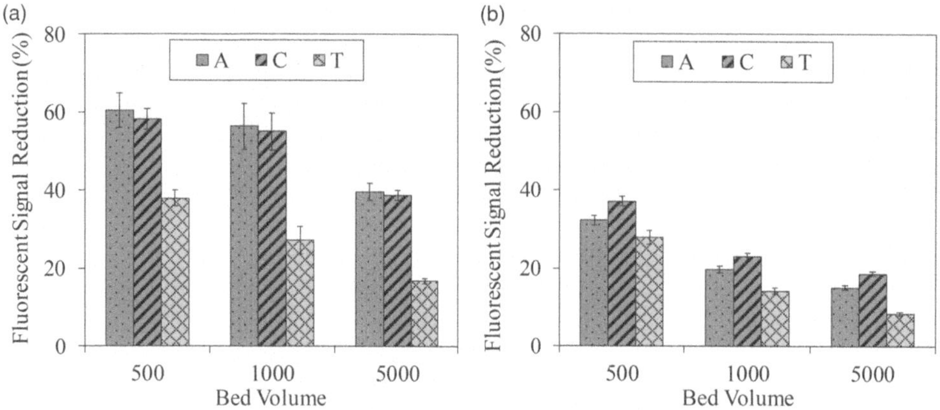

Figure 5 | Performance of fluorescence signal reduction by selected loading of MIEX treatment in (a) surface water and (b) effluent.

fluorescence signals were extracted from the EEM spectra of selected MIEX loadings. The reductions in fluorescence signals are depicted in Figure 5. MIEX treatment achieved greater fluorescence signal reduction with surface water (Figure 5(a)) compared to secondary effluent (Figure 5(b)). One-way ANOVA was performed to ascertain whether there were statistically significant differences in the levels of removal for humic-like and protein-like DOM. For surface water, preferential removal of humic-like materials by MIEX can be seen distinctively, as intensity reduction at peaks A and C was significantly greater ($p < 0.01$) than that at peak T. There was no significant difference ($p = 1.0$) between the fluorescence signal reduction at fulvic-like peak A and humic-like peak C in surface water by MIEX treatment for any loading tested. These results agree well with literature findings for most surface waters and DOM isolates (Arias-Paic *et al.* 2016; Jutaporn *et al.* 2016). In contrast to the surface water and in accordance with the qualitative analysis, preferential fluorescence reduction at peaks A and C was not observed in MIEX treatments with secondary effluent. There was no significant difference ($p = 0.12$) between the reduction of intensity at peak A, C, and T by MIEX treatment. This statistical analysis confirmed that MIEX treatment had preferential removal of humic substances only in surface water.

Correlations between THMFP and surrogate parameters

To investigate relationships between THMFP and water quality parameters, all 42 samples with and without MIEX treatment were included in the linear regression analysis and the results are shown in a correlation matrix in Table 2. Results from triplicate experiments are represented by average values and standard deviation in

Table 2 | Correlation matrix of formation potentials, DOC, UVA_{254}, SUVA, and FI at peaks A, C, and T for all samples with and without MIEX treatment

R^2	Peak A	Peak C	Peak T	FI	UVA_{254}	DOC	SUVA
Cl_3CH-FP	0.10	0.002	0.01	0.12	0.94*	0.24	0.50
$BrCl_2CH$-FP	0.71*	0.92*	0.97*	0.99*	0.00	0.53	0.61
Br_2ClCH-FP	0.71*	0.93*	0.97*	0.98*	0.00	0.53	0.61
Br_3CH-FP	0.76*	0.96*	0.99*	0.95*	0.00	0.58	0.54
TTHMFP	0.87*	0.98*	0.96*	0.90*	0.03	0.72*	0.44
Peak A	1.00	0.90*	0.82*	0.60	0.74*	0.94*	0.01
Peak C		1.00	0.98*	0.85*	0.01	0.74*	0.49
Peak T			1.00	0.92*	0.01	0.64	0.51
FI				1.00	0.02	0.42	0.71*
UVA_{254}					1.00	0.43	0.34
DOC						1.00	0.05
SUVA							1.00

*Correlation is significant at the 0.01 level (two-tailed).

Figure 6. The correlation coefficient (R^2) marked by an asterisk indicates a correlation that is significant at 0.01 confidence level ($p < 0.01$). There were significant correlations observed between formation potentials and UVA_{254}, DOC concentration, fluorescence intensity at peaks A, C, and T, and FI.

The correlations of formation potentials were substantially higher with fluorescence parameters ($R^2 = 0.87–0.98$) than with DOC, UVA_{254}, or with SUVA. The formation potential of chloroform (Cl_3CH-FP), the predominant THM species in the surface water, correlated well only with UVA_{254} ($R^2 = 0.94$, Figure 6(a)). In contrast to earlier findings that UVA_{254} was the most accurate TTHMFP precursor surrogate parameter among natural water sources (river, lake, and groundwater) (Pifer & Fairey 2014), TTHMFP in this study did not correlate with UVA_{254} ($R^2 = 0.03$, $p > 0.01$). A contributing factor to the disparity between our study and Pifer & Fairey (2014) is that the two water sources used in this study, surface water and secondary effluent, contained substantially different DOM matrices as evidenced by the fluorescence parameters and THM formation discussed above. Based on FI values, the DOM in the surface water was from terrestrial sources (FI ~ 1.4), while the DOM in the secondary effluent (FI ~ 1.9) was enriched from microbial sources (McKnight *et al.* 2001). These two water sources also had different predominant THM species: Cl_3CH for surface water and $BrCl_2CH$ for secondary effluent. Even though Cl_3CH correlated well with UVA_{254} ($R^2 = 0.94$, Figure 6(a)), $BrCl_2CH$ did not; therefore, a significant correlation between TTHMFP and UVA_{254} was not found. Accordingly, our results suggested that UVA_{254} is not a good surrogate parameter for TTHMFP for all waters. Furthermore, individual formation potentials of THM species did not correlate with DOC or SUVA (Table 2).

Figure 6 presents some selected regressions from Table 2. The linear fits in Figure 6 correspond to fits to the data from the two water sources combined. Results suggest that UVA_{254} is a good indicator of Cl_3CH-FP ($R^2 = 0.94$), consistent with previous studies (Pifer & Fairey 2014; Yang *et al.* 2015). The correlation between TTHMFP and DOC was significant ($R^2 = 0.72$, Figure 6(b)), but was weaker than the correlations between TTHMFP and fluorescence parameters ($R^2 = 0.87–0.98$, Figure 6(c)). The correlation between TTHMFP and DOC in this study using surface water and effluent water was lower than a previous study using only water from drinking water treatment plants

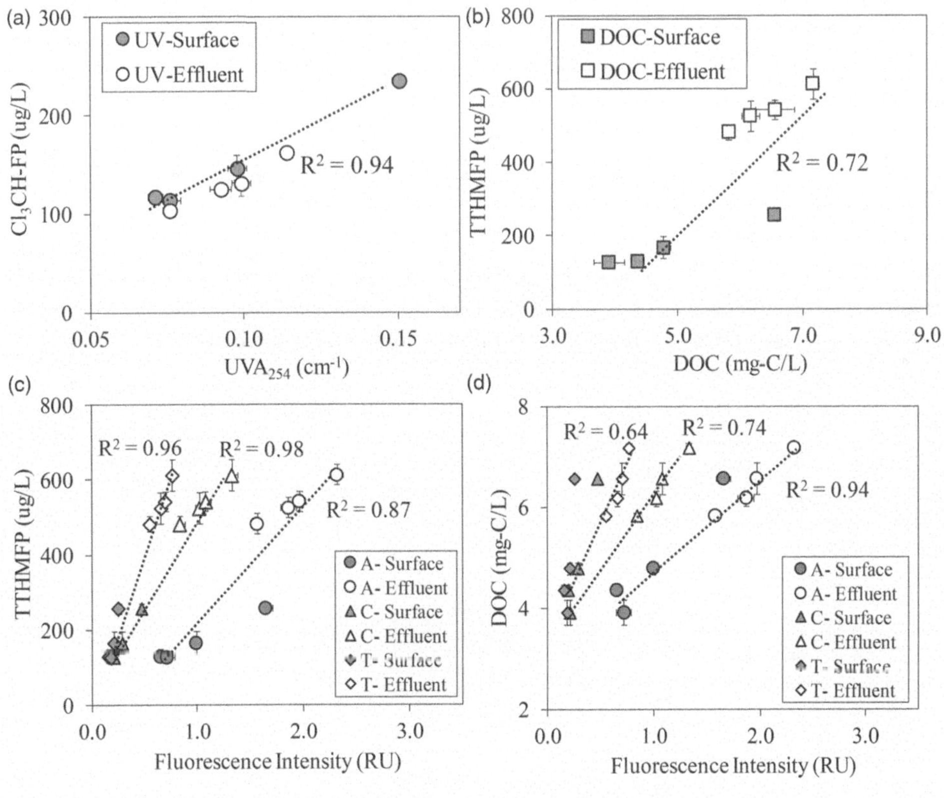

Figure 6 | Regressions describing the relationships between (a) UVA_{254} and formation potential of chloroform (Cl_3CH-FP), (b) DOC and formation potential of TTHM (TTHMFP), (c) fluorescence intensity at peaks A, C, and T and TTHMFP, and (d) DOC and fluorescence intensity at peaks A, C, and T. The regression lines correspond to fits to the data from both water sources combined.

(Yang *et al.* 2015). The surface water and effluent data series in Figure 6(b) were also analyzed separately and the correlation between TTHMFP and DOC was $R^2 = 0.98$ for surface water and $R^2 = 0.97$ for effluent. The difference in correlations of the individual water sample types compared to the aggregate data suggests that DOC can be a good indicator of THMFP for individual types of water sources, but not for a group of water sources with different DOM characteristics.

Of the fluorescence parameters, the intensity of humic-like peak C was most strongly correlated to TTHMFP ($R^2 = 0.98$, Figure 6(c)), suggesting this parameter was the strongest fluorescence-based predictor for TTHMFP. Peak T (protein-like) and peak A (fulvic-like) also correlated with TTHMFP with slightly lower R^2 values of 0.96 and 0.87, respectively. These strong correlations between TTHMFP and fluorescence parameters support a previous finding that fluorescence spectroscopy is a better DOM characterization tool than UV absorbance because it is more sensitive to organic matter than are UV measurements (Leenheer & Croué 2003). A previous study also reported that TTHMFP correlated strongly with humic-like DOM (Yang *et al.* 2015). The results of this study confirm that fluorescence is a useful tool in tracking DBP formation during chlorination of both surface water and secondary treated effluent and that humic-like peak C was the most accurate TTHMFP precursor surrogate parameter.

To better understand differences in DOM composition based on fluorescence surrogates, linear correlations between DOC concentration and fluorescence intensities at peaks A, C, and T were identified (Figure 6(d)). The two terrestrial peaks were good predictors of DOC, with peak A the strongest ($R^2 = 0.94$, $p < 0.01$) and peak C moderate ($R^2 = 0.74$, $p < 0.01$). The correlation between DOC concentration and peak T was the weakest ($R^2 = 0.64$, $p = 0.02$). These correlations suggest that the humic fraction of DOM makes up the majority of DOC, which is consistent with a previous study (Leenheer & Croué 2003) that found humic substances (humic and fulvic acid) typically comprise ~50% of the total DOM composition. While terrestrially and bacterially derived DOM may have contributed to the DOC in surface water, EfOM is comprised of SMPs and refractory materials (Chen *et al.* 2003), so EfOM is derived from different sources than DOM in surface water (Krasner *et al.* 2008). DOC is an aggregate measurement of carbon content and not all fractions of DOC contribute to THM formation equally, which explains the lower correlation coefficients of TTHMFP with DOC. Fluorescence characterization has an advantage over an aggregate DOC measurement because it is more sensitive and selective to DOM fractions and the individual parameters can be used to monitor DOM and TTHMFP for a diverse group of waters.

Correlations in Table 2 should be evaluated together with plots (Figures 2 and 6) of each data set. For example, the correlation between $BrCl_2CH$-FP and peak A was influenced by lower reduction of $BrCl_2CH$-FP (5–37%) compared to those of Cl_3CH-FP (37–50%) (Figure 2(b)). When regression analyses were conducted using both water sources combined as a data set, the R^2 of 0.71 (Table 2) was calculated from the line drawn between two groups of cluster samplings (Figure 7(a)). Even though this correlation was significant ($p < 0.01$), peak A only moderately predicted $BrCl_2CH$-FP for the combined data set. A similar trend was noticed with the other Br-THM species and fluorescence-peak correlations. When we conducted separate correlation analyses for the surface water and effluent water, we discovered the significance of the correlation from the combined data set was driven by the strong

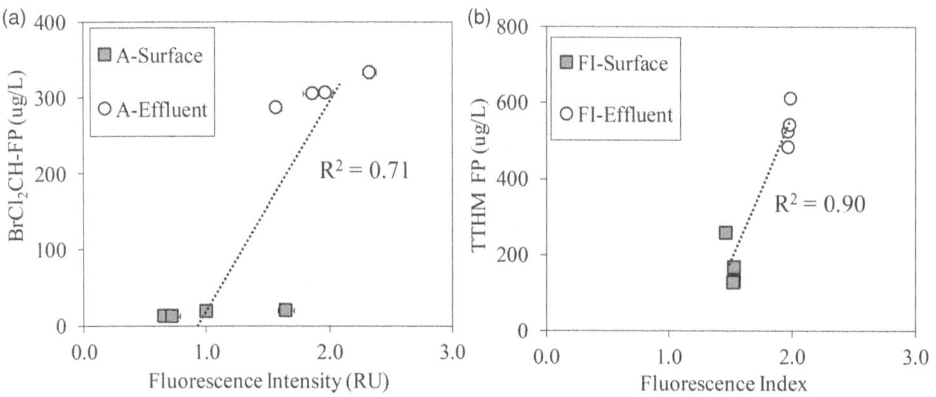

Figure 7 | Regressions describing the relationships between (a) fluorescence intensity at peak A and formation potential of $BrCl_2CH$ and (b) fluorescence index (FI) and TTHMFP. The regression lines correspond to fits to the data from both water sources combined.

correlation between peak A and $BrCl_2CH$-FP for effluent samples ($R^2 = 0.98$). Peak A and $BrCl_2CH$-FP were only moderately correlated for surface water ($R^2 = 0.75$); therefore, peak A may not be an appropriate predictor for Br-THM precursors in all water sources. We performed a similar analysis for the combined data set to calculate a correlation coefficient for TTHMFP and FI values ($R^2 = 0.90$, $p < 0.01$) (Figure 7(b)). Despite the promising correlation, the FI values varied marginally with MIEX treatment (1.47–1.54 for surface water and 1.97–1.99 for effluent), making FI values of each water source clustered together. Thus FI values are poor TTHMFP predictors for two water sources combined, even though they are reliable parameters to analyze or predict DOM sources. A more diverse set of water samples containing a variety of DOM sources is needed to investigate whether FI values could be used to assess TTHMFP.

CONCLUSIONS

We investigated the applicability of fluorescence EEMS to assess TTHMFP and the ability of MIEX resin to reduce TTHMFP. We drew the following conclusions from our study.

- MIEX resin removed DBP precursors more effectively from surface water than from secondary effluent, but also changed the final DOM composition of the treated surface water. We successfully treated surface water with up to 3,000 BV of MIEX resin, which suggests that MIEX resin can be regenerated less frequently than existing practices. Water treatment plants can use this information to help them assess whether MIEX resins are an affordable and appropriate treatment technology for their source water.
- Although UVA_{254} has been used successfully in the past as a surrogate measurement for TTHMFP, our data show it was not a good surrogate parameter for TTHMFP of surface water and secondary effluent combined; therefore, UVA_{254} is not a good surrogate parameter for all water types of different DOM origins.
- Fluorescence parameters, including fluorescence intensity of humic-like peak C, fulvic-like peak A, and protein-like peak T, can potentially be used as surrogate parameters to assess the TTHMP of different raw and MIEX-treated source waters.

Our results demonstrate that fluorescence EEMS can be used as a tool to assess the TTHMFP of raw water sources that contain DOM from different origins and to assess the efficacy of MIEX to reduce TTHMFP.

ACKNOWLEDGEMENT

This work was partially supported by Ixom Watercare Inc. The Royal Thai Office of the Civil Service Commission supported Panitan Jutaporn through a Royal Thai Government Scholarship. M. D. Armstrong acknowledges funding from the US EPA-UNC-CH Cooperative Training Agreement CR-83591401-0. The authors thank Dr Rose Cory and Katherine Harrold for their assistance in fluorescence EEMS analysis, Dr Philip Singer and Ariel Atkinson for their suggestions in THMFP analysis, and Dr Miguel Arias for helpful discussions about MIEX experiments. We also thank our collaborators at the Orange Water and Sewer Authority (Carrboro, NC) who facilitated and helped with water samples collection.

REFERENCES

Arias-Paic, M., Cawley, K. M., Byg, S. & Rosario-Ortiz, F. L. 2016 Enhanced DOC removal using anion and cation ion exchange resins. *Water Research* **88**, 981–989. https://doi.org/10.1016/j.watres.2015.11.019.

Baghoth, S. A., Sharma, S. K. & Amy, G. L. 2011 Tracking natural organic matter (NOM) in a drinking water treatment plant using fluorescence excitation-emission matrices and PARAFAC. *Water Research* **45** (2), 797–809. https://doi.org/10.1016/j.watres.2010.09.005.

Boyer, T. H. & Singer, P. C. 2005 Bench-scale testing of a magnetic ion exchange resin for removal of disinfection by-product precursors. *Water Research* **39** (7), 1265–1276. https://doi.org/10.1016/j.watres.2005.01.002.

Boyer, T. H., Singer, P. C. & Aiken, G. R. 2008 Removal of dissolved organic matter by anion exchange: effect of dissolved organic matter properties. *Environmental Science & Technology* **42** (19), 7431–7437. https://doi.org/10.1021/es800714d.

Chen, W., Westerhoff, P., Leenheer, J. A. & Booksh, K. 2003 Fluorescence excitation-emission matrix regional integration to quantify spectra for dissolved organic matter. *Environmental Science & Technology* **37** (24), 5701–5710.

Coble, P. G. 1996 Characterization of marine and terrestrial DOM in seawater using excitation-emission matrix spectroscopy. *Marine Chemistry* **51** (4), 325–346. https://doi.org/10.1016/0304-4203(95)00062-3.

Cory, R. M., Miller, M. P., McKnight, D. M., Guerard, J. J. & Miller, P. L. 2010 Effect of instrument-specific response on the analysis of fulvic acid fluorescence spectra. *Limnology and Oceanography: Methods* **8**, 67–78. https://doi.org/10.4319/lom.2010.8.67.

Gan, X., Karanfil, T., Kaplan Bekaroglu, S. S. & Shan, J. 2013 The control of N-DBP and C-DBP precursors with MIEX®. *Water*

Research **47**, 1344–1352. https://doi.org/10.1016/j.watres.2012.11.049.

Ha, T.-W., Choo, K.-H. & Choi, S.-J. 2004 Effect of chlorine on adsorption/ultrafiltration treatment for removing natural organic matter in drinking water. *Journal of Colloid and Interface Science* **274**, 587–593. https://doi.org/10.1016/j.jcis.2004.03.010.

Harrington, G. W., Bruchet, A., Rybacki, D. & Singer, P. C. 1996 Characterization of natural organic matter and its reactivity with chlorine. In: *Water Disinfection and Natural Organic Matter* (R. Minear & G. Amy, eds). American Chemical Society, Washington, DC, USA, pp. 138–158. https://doi.org/10.1021/bk-1996-0649.ch010.

Henderson, R. K., Subhi, N., Antony, A., Khan, S. J., Murphy, K. R., Leslie, G. L., Chen, V., Stuetz, R. M. & Le-Clech, P. 2011 Evaluation of effluent organic matter fouling in ultrafiltration treatment using advanced organic characterisation techniques. *Journal of Membrane Science* **382**, 50–59. https://doi.org/10.1016/j.memsci.2011.07.041.

Jutaporn, P., Singer, P. C., Cory, R. M. & Coronell, O. 2016 Minimization of short-term low-pressure membrane fouling using a magnetic ion exchange (MIEX®) resin. *Water Research* **98**, 225–234. https://doi.org/10.1016/j.watres.2016.04.007.

Kitis, M., İlker Harman, B., Yigit, N. O., Beyhan, M., Nguyen, H. & Adams, B. 2007 The removal of natural organic matter from selected Turkish source waters using magnetic ion exchange resin (MIEX®). *Reactive and Functional Polymers* **67**, 1495–1504. https://doi.org/10.1016/j.watres.2016.04.007.

Krasner, S. W., Westerhoff, P., Chen, B., Amy, G. L., Nam, S.-N., Chowdhury, Z. K. & Rittmann, B. E. 2008 *Contribution of Wastewater to DBP Formation*. AWWA Research Foundation, Denver, CO, USA.

Leenheer, J. A. & Croué, J.-P. 2003 Peer reviewed: characterizing aquatic dissolved organic matter, understanding the unknown structures is key to better treatment of drinking water. *Environmental Science & Technology* **1**, 18A–26A. https://doi.org/10.1021/es032333c.

Lyon, B. A., Cory, R. M. & Weinberg, H. S. 2014 Changes in dissolved organic matter fluorescence and disinfection byproduct formation from UV and subsequent chlorination/chloramination. *Journal of Hazardous Materials* **264**, 411–419. https://doi.org/10.1016/j.jhazmat.2013.10.065.

McKnight, D. M., Boyer, E. W., Westerhoff, P. K., Doran, P. T., Kulbe, T. & Andersen, D. T. 2001 Spectrofluorometric characterization of dissolved organic matter for indication of precursor organic material and aromaticity. *Limnology and Oceanography* **46** (1), 38–48. https://doi.org/10.4319/lo.2001.46.1.0038.

Nguyen, T. V., Zhang, R., Vigneswaran, S., Ngo, H. H., Kandasamy, J. & Mathes, P. 2011 Removal of organic matter from effluents by Magnetic Ion Exchange (MIEX®). *Desalination* **276**, 96–102. https://doi.org/10.1016/j.desal.2011.03.028.

Patel-Sorrentino, N., Mounier, S. & Benaim, J. Y. 2002 Excitation-emission fluorescence matrix to study pH influence on organic matter fluorescence in the Amazon basin rivers. *Water Research* **36**, 2571–2581. https://doi.org/10.1016/S0043-1354(01)00469-9.

Pifer, A. D. & Fairey, J. L. 2014 Suitability of organic matter surrogates to predict trihalomethane formation in drinking water sources. *Environmental Engineering Science* **31** (3), 117–126. https://doi.org/10.1089/ees.2013.0247.

Singer, P. C. 1999 Humic substances as precursors for potentially harmful disinfection by-products. *Water Science & Technology* **40** (9), 25–30. https://doi.org/10.1016/S0273-1223(99)00636-8.

Singer, P. C. & Bilyk, K. 2002 Enhanced coagulation using a magnetic ion exchange resin. *Water Research* **36**, 4009–4022. https://doi.org/10.1016/S0043-1354(02)00115-X.

Singer, P., Boyer, T., Holmquist, A. & Morran, J. 2009 Integrated analysis of NOM removal by magnetic ion exchange. *Journal of American Water Works Association* **101** (1), 65–73.

Spencer, R. G. M., Bolton, L. & Baker, A. 2007 Freeze/thaw and pH effects on freshwater dissolved organic matter fluorescence and absorbance properties from a number of UK locations. *Water Research* **41**, 2941–2950. https://doi.org/10.1016/j.watres.2007.04.012.

Stedmon, C. A. & Bro, R. 2008 Characterizing dissolved organic matter fluorescence with parallel factor analysis: a tutorial. *Limnology and Oceanography: Methods* **6**, 572–579. https://doi.org/10.4319/lom.2008.6.572.

Stedmon, C., Markager, S. & Bro, R. 2003 Tracing dissolved organic matter in aquatic environments using a new approach to fluorescence spectroscopy. *Marine Chemistry* **82**, 239–254. https://doi.org/10.1016/S0304-4203(03)00072-0.

Summers, R. S., Hooper, S. M., Shukairy, H. M., Solarik, G. & Owen, D. 1996 Assessing DBP yield: uniform formation conditions. *Journal of American Water Works Association* **88** (6), 80–93. https://doi.org/10.1002/j.1551-8833.1996.tb06573.x.

US EPA 1995 Method 551.1: Determination of chlorination disinfection by-products and chlorinated solvents in drinking water by liquid–liquid extraction and gas chromatography with electron-capture detection. Environmental Monitoring Systems Laboratory, Office of Research and Development, US EPA, Cincinnati, OH, USA.

US EPA 2001 *National Primary Drinking Water Regulations: Stage 1 Disinfectants and Disinfection Byproducts Rule*. https://doi.org/EPA816-F-01-014.

US EPA 2016 *Six-Year Review 3 Technical Support Document for Disinfectants/Disinfection Byproducts Rules*. https://doi.org/EPA-810-R-16-012.

Yang, L., Kim, D., Uzun, H., Karanfil, T. & Hur, J. 2015 Assessing trihalomethanes (THMs) and N-nitrosodimethylamine (NDMA) formation potentials in drinking water treatment plants using fluorescence spectroscopy and parallel factor analysis. *Chemosphere* **121**, 84–91. https://doi.org/10.1016/j.chemosphere.2014.11.033.

Zhang, R., Vigneswaran, S., Ngo, H. H. & Nguyen, H. 2006 Magnetic ion exchange (MIEX®) resin as a pre-treatment to a submerged membrane system in the treatment of biologically treated wastewater. *Desalination* **192**, 296–302.

First received 30 July 2018; accepted in revised form 11 January 2019. Available online 23 January 2019

Reliable determination of copper complex ions in synthetic wastewater using flame atomic absorption spectrophotometry

Kanokporn Supong and Parnuwat Usapein

ABSTRACT

Forms of copper that are highly soluble in aquatic environments are used as chemical reagents in a variety of industries, especially copper complexes. Wastewater containing copper complexes can be difficult to treat and analyse. Normally, flame atomic absorption spectrophotometry (FAAS) is a favourable technique for analysing metal ions in wastewater, but is less reliable for copper analysis owing to the influence of copper oxides. To solve this problem, it is necessary to adjust the atomization and delay times. The objective of this study was to use FAAS to accurately determine the quantity of complexed copper in synthetic wastewater, using increased atomization and delay times. The method showed excellent linearity in the copper standard concentration range of 1–5 mg L^{-1}. The sensitivity of the analysis was 0.023 mg L^{-1}. The percentage of recovery and the relative standard deviation were 85.02% and 0.75%, respectively. The experimental results demonstrated that the estimation of uncertainty from preparation of working standard, repeatability uncertainty, instrument deviation, calibration curve and recovery uncertainty were 8.2×10^{-4}, 4.49×10^{-4}, 7.21×10^{-4}, 21.25×10^{-4}, and 98.19×10^{-4}, respectively. Overall, the results showed the suitability of the FAAS method for determining copper in synthetic wastewater.

Key words | copper, metrological chemistry, reliability, uncertainty, validation method, wastewater

Kanokporn Supong
Parnuwat Usapein (corresponding author)
Rattanakosin College for Sustainable Energy and Environment,
Rajamangala University of Technology Rattanakosin,
Nakhon Pathom 73170,
Thailand
E-mail: *parnuwat.usa@rmutr.ac.th*

INTRODUCTION

In the last decade, there have been major environmental problems in Thailand arising from industrial wastes (Usapein & Chavalparit 2015). One important form of hazardous waste has been wastewater contaminated with copper complexes, which is produced in large quantity by industrial processes such as printed circuit board manufacturing, electronics plating, wire drawing, paint manufacturing, and textile manufacture (Srinivasa Rao *et al.* 2016). Copper complexes are normally quite stable in wastewater, making them difficult to treat. Previous research has indicated that Cu is highly stable and extremely hard to break down for removal from wastewater (Jiang *et al.* 2010). When copper complexes run into water bodies (rivers, lakes, oceans), they can accumulate in aquatic animals. People who frequently eat contaminated food can bioaccumulate copper to levels that are potentially dangerous, with symptoms including vomiting, lethargy, increased blood pressure and respiratory rates, acute haemolytic anaemia, liver damage, neurotoxicity, and neurodegenerative disease. Therefore, means for removing copper from and accurately measuring it in wastewater before discharge are needed.

Inductively coupled plasma (ICP) techniques and flame atomic absorption spectrophotometry (FAAS) are favourable for determining copper ions in wastewater (Bader 2011). FAAS can potentially provide higher sensitivity at lower cost than ICP. However, FAAS analysis of copper complexes in wastewater is interfered with by the formation of oxides, abnormally decreasing their Cu absorbance relative to that of uncomplexed Cu ions and thus worsening the reliability of the technique. Therefore, it is important to adjust the instrument conditions appropriately to ensure accurate measurement (Bader 2011).

Metrological traceability, validation methods, and uncertainty of measurement are the main principles of the field of metrology (Ellison & Williams 2012).

First, the validation method determines the different influences in the whole analytical system that may affect

doi: 10.2166/wst.2019.072

the result. This will confirm that there are no other influences that have not been considered in the analysis. The validation method will confirm the system suitability and linearity. The system suitability test comprehensively considers the instrument, reagents, and analysis. System suitability was calculated from the percentage of relative standard deviation (%RSD). Typically, the %RSD of 10 replicate absorbance measurements will be not more than 7% for each copper ion standard (Ata *et al.* 2015).

Second, the traceability process is based on the accuracy and precision of the trial which resulted in the acceptance of the presented method and the true value. Precision is an index that indicates measurement deviation from the true value and its scatter. Precision is studied by calculating the RSD. Accuracy is studied by calculating the percentage of mean recovery (ISO 2005).

Third, measurement uncertainty represents the dispersion of results obtained by the analysis procedure. Most of the uncertainty sources are linked to specific steps in the analysis procedure, which are subject to random and systematic errors. It is stressed that sample preparation is usually the biggest contributor to measurement uncertainty (Petrov *et al.* 2009). To more clearly understand measurement uncertainty, Santana and colleagues have applied the Ishikawa cause-and-effect diagram to identify the uncertainty for all stages (from sample preparation to analysis). The overall uncertainty includes many sources, for example the preparation of the working standard, instrument deviation, the calibration curve, recovery uncertainty, and repeatability uncertainty (Santana *et al.* 2011).

From previous studies, it is obviously seen that consistent, reliable, and accurate data play a major role in analytical measurement. They can be used to judge the quality, reliability, and consistency of analytical results.

This study focuses on the ionic copper–ammonia complex $Cu(NH_3)_4^{2+}$, a coordination complex consisting of a central copper atom and four ammonia ligands. This complex is very stable, and attempts to analyse its concentration using FAAS conditions suitable for copper ion determination may not be reliable due to incomplete combustion. To solve this problem, three techniques are currently used:

(1) Use of ICP, an expensive technique (Thermo Elemental 2001).
(2) Use of FAAS to analyse copper ions, with the addition of a sample preparation step such as acid digestion to break down the molecular bonds. This step will take at least 1–2 hours and then filtering is required before FAAS analysis. The weakness of this technique is that it often entails loss of copper ions during the sample preparation step (Harvey 2017).
(3) Use of FAAS techniques for analysing copper ion, entailing decomposition of the sample into independent atoms under electric current (e.g. in an electrothermal atomizer or graphite furnace atomizer). However, this technique has long measurement times because of the heating and cooling required of the system (Maryville University 2018). In addition, reproducibility is frequently poor because of the intermittent measurement. This may affect the reliability of the data.

The objective of the present study was to design a reliable method for determining copper concentration in synthetic wastewater using FAAS. We have proposed new conditions to adjust the FAAS instrument for determination of copper occurring in the ionic copper–ammonia complex, namely increases in the atomization time and delay time. Chemical metrology was applied to confirm that the proposed method is reliable and feasible for measurement of copper ions in copper–ammonia complex form.

MATERIALS AND METHODS

Materials and reagents

A standard stock solution of 1,000 mg L^{-1} copper ions was purchased from Merck, Germany. All standard solutions were prepared by diluting the stock solution. Synthetic wastewater was prepared by adding NH_4OH (30% w/w) to a 2 mg L^{-1} standard copper solution to achieve the Cu^{2+}:NH_4OH molar ratio of 1:4, then diluting to 100 mL with deionized water. All solutions were prepared in deionized water.

Apparatus

An atomic absorption spectrophotometer (SpectrAA-200, PerkinElmer) was used to perform all measurements of copper concentration (Figure 1). This machine consists of eight turret lamps emitting in the 190–900 nm spectral region, and has a deuterium-arc background corrector that can accurately correct for combined atomic plus background absorption signals as high as 3 absorbance units. To integrate and process data, the manufacturer-supplied software Winlab32 was used. The spectroscopic conditions were set as follows: bandwidth 2.7/0.8 nm with 1.0 filter factor and deuterium arc background correction. The integration time was set for 7.0 s at 15.0 mA lamp current. The analytes were detected at 324.8 nm.

Figure 1 | Atomic absorption spectrophotometer used in this study.

Experimental setup for reliable determination of complexed copper

Assay method

To set suitable spectroscopic conditions, first a steady baseline was recorded. A standard solution of Cu was injected, and spectrograms were recorded by measuring the absorbance at 324.8 nm. A copper hollow-cathode lamp was used as the radiation source and an air–acetylene flame was used with the air and fuel flowrates of 10 and 2.5 L min^{-1}. According to guidelines from the manufacturer's manual, the Cu concentration of 1.3 mg L^{-1} will give a signal of 0.2 absorbance units. The characteristic concentration was calculated as follows:

$$S_{char} = [0.0044 \times S_{std}]/\overline{Abs} \quad (1)$$

where S_{char} and S_{std} are the characteristic and standard concentrations, respectively (mg L^{-1}), and *Abs* is the measured absorbance.

Validation method

System suitability. The Cu working standard used (100 mg L^{-1} Cu) was prepared from the standard solution (1,000 mg L^{-1} Cu). The working standard was diluted to prepare standard samples of 1–5 mg L^{-1} Cu, which were each aspirated into the flame of an atomic absorption spectrophotometer and measured at 324.8 nm. At least 10 readings were taken for each sample. Table 1 lists the absorbance results for these samples and a synthetic wastewater sample, including %RSDs from 10 readings.

Table 1 | Assessment for system suitability of copper analysis in wastewater using FAAS

Copper standard (mg L^{-1})	Absorbance (mean)	% Relative standard deviation (%RSD)
1	0.161	0.51
2	0.321	0.33
3	0.454	0.29
4	0.592	0.39
5	0.720	0.10
Synthetic wastewater	0.311	0.41

Acceptance criteria: %RSD of copper absorbance not greater than 7.0%.

Linearity. To assess the system suitability, the linearity of the average absorption results from each copper standard was evaluated by plotting the results with a trend line as shown in Figure 2.

Traceability process

Accuracy was assessed from five replicate determinations of 2 mg L^{-1} chelated copper. The solution was spiked with three different copper standard solutions of concentration 1, 2, and 3 mg L^{-1}. The spiked sample solutions were analysed according to the analytical method. The recovery was calculated as follows:

$$\text{Recovery (\%)} = \frac{[S_{spike} - S_{complex\_Cu}]}{S_{spike}} \times 100 \quad (2)$$

where S_{spike} is the calculated copper in the spiked sample (mg L^{-1}) and $S_{complex\_Cu}$ is the copper concentration of the copper complex solution (mg L^{-1}).

Sources of uncertainty

An Ishikawa diagram was prepared to identify the possible causes of uncertainty in the measurement results (Santana *et al.* 2011). The diagram consists of a single hierarchical structure with the objective of identifying the sources of uncertainty associated with the measurement protocol for the metal ions. Five major influences on uncertainty were identified, as shown in the diagram (Figure 3).

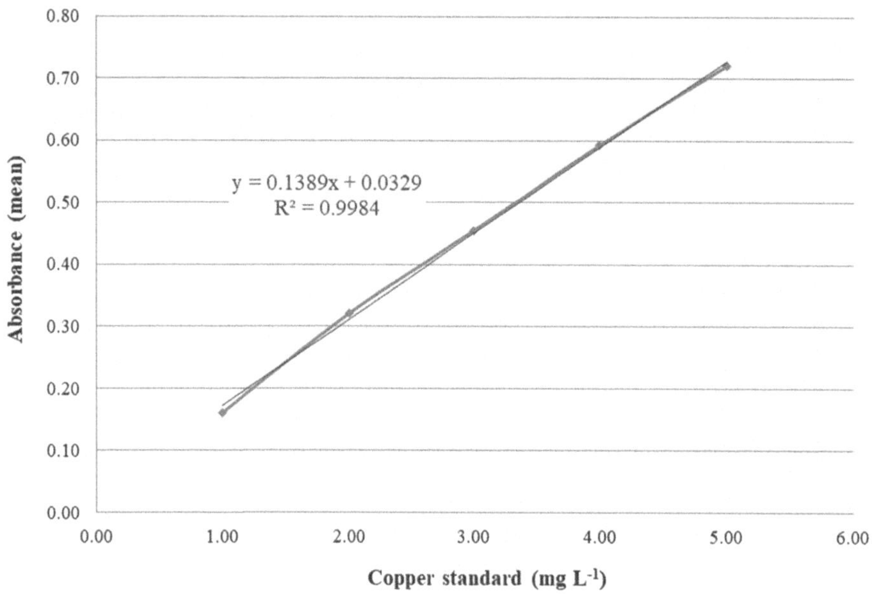

Figure 2 | Linearity verification of FAAS copper measurement.

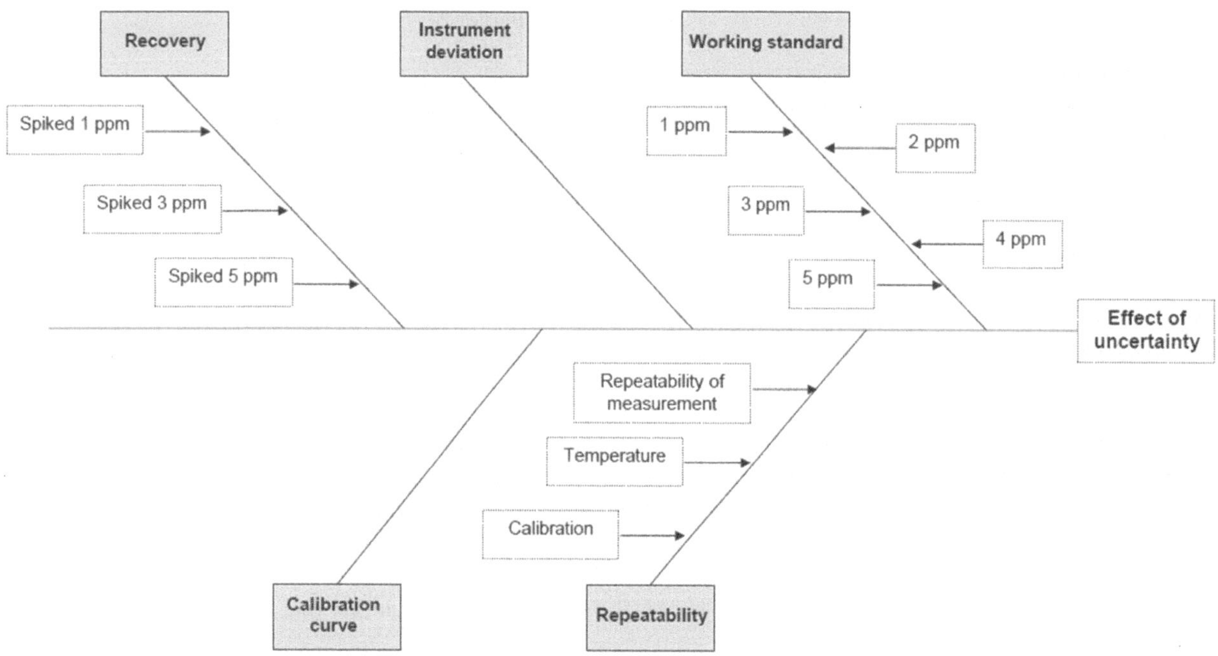

Figure 3 | Ishikawa diagram for uncertainty analysis.

RESULTS AND DISCUSSION

We optimized the FAAS procedure to improve the stability of the assay method. In this study, we hypothesized that the copper complexes can form copper oxides. These oxides are larger than copper ions, and scatter light, reducing their detection and thereby affecting the copper determination. This problem was solved by increasing the atomization time, which was the parameter used in the analysis.

Characteristic result

Given the set conditions (atomization time of 7 s and delay time of 7 s), it is necessary to consider the sensitivity of

Table 2 | Characteristic concentration results

Parameter	Value
Characteristic concentration based on this experiment	0.023 mg L^{-1}
Characteristic concentration based on instrument manual	0.025 mg L^{-1}
Instrument tolerance	8%

copper ion measurement that can be attained in the instrument's available range. To determine this issue, a solution of copper–ammonia complex (1.3 mg L^{-1}) was prepared. As mentioned above, this solution has the expected absorbance value of 0.2; this value was used to calculate the characteristic concentration. From the experimental results, it was found that the characteristic concentration was 0.023 mg L^{-1} (as shown in Table 2). Compared with the optimum sensitivity listed in the instrument manual, there was a difference of about 8% for the measurement of copper metal ions. However, this difference is acceptable because it is no more than 20%, as defined in the instrument manual (PerkinElmer 2011).

System suitability result

Table 1 lists the results for system suitability acquired by aspirating 10 replicates each of copper standard solutions. It appears that the RSD of the experimental data is 0.34%. Normally, if the %RSD from the experiment has a wide range, this may indicate measurement error of the instrument. This can be caused by: (1) excessive sample concentration; (2) highly contaminated matrix solutions; (3) the presence of light-scattering organic compounds; and/or (4) inadequate combustion conditions causing clogging of the FAAS. However, to avoid factors (1)–(3), this study used a standard solution of copper with a light absorption value in accordance with the concentration. Therefore, if the results of %RSD are quite wide, it would likely be caused by factor (4). On the other hand, if the %RSD is rather narrow, then this indicates that the conditions were appropriate to achieve complete combustion. The %RSD obtained in this study was within an acceptable interval, and thus we consider the present method to be sufficiently precise. Thus, the present FAAS method appears suitable for the determination of copper in complexed form in synthetic wastewater.

Linearity results

Calibration curves of copper were established by graphing average absorbances versus the concentrations of each copper standard; the curve was linear throughout the studied range of 1–5 mg L^{-1} copper, with the correlation coefficient of 0.9984 (Figure 2). In this study, the linear least squares fitting technique was applied as the form of linear regression. This verified the linearity of the proposed FAAS method for the determination of copper in the complexed form in synthetic wastewater.

Traceability process results

The recovery results in Table 3 indicate that the concentration obtained from the experiment was close to the true value. The accuracy was assessed from 10 replicate determinations of samples containing 1 and 3 mg L^{-1} of copper. The absolute means obtained were 82.61% and 87.44%, respectively, with RSDs of 1.11% and 0.39%, respectively, the limit of detection (LOD) of 0.0042 mg L^{-1}, and the limit of quantitation (LOQ) of 0.0140 mg L^{-1}. From the result, it was found that the recovery percentage of sample containing 1 mg L^{-1} of copper was less than 85%. This often occurred when a dilute solution was analysed. Although there is no analytical error, it is well known that the recovery percentage is not equal to 100% when the analytical concentration of sample is decreased. Therefore, the absolute mean of recovery percentage (between sample

Table 3 | Copper recoveries using the spiking levels of 1 and 3 mg L^{-1} of copper

Solution	Concentration (mg L^{-1})	Total concentration (mg L^{-1})	Concentration obtained (mg L^{-1})	Concentration recovered (mg L^{-1})	Recovery (%)	RSD (%)	LOD	LOQ
Sample	1.9493	2.9493	2.7754	0.8261	82.61	1.11	0.0042	0.014
Standard added	1							
Sample	1.9493	4.9493	4.5725	2.6232	87.44	0.39		
Standard added	3							

Acceptance criteria: 85% ≤ recovery ≤115%.

containing 1 and 3 mg L^{-1} of copper) was selected with the value of 85.03%.

Measurement sources of uncertainty

Uncertainty in preparation of working standard

The certified standard used in these experiments was quoted as having a concentration and uncertainty of 1,000 ± 0.0005 mg L^{-1} of copper ions. Stock solutions (100 mg L^{-1} of copper ions) were prepared by diluting the certified standard by 10 times. Thus, the uncertainty in the preparation of stock solutions can be calculated by using Equation (3). The results from the experiment showed that the uncertainty of the pipettes was equal to 0.008, 0.02, 0.04, 0.04, and 0.04, respectively, while the uncertainty of the volumetric flask was equal to 0.01. Therefore, the uncertainty of sample preparation was calculated as follows:

$$\frac{\mu_{stdCu100ppm}}{100} = \sqrt{\left(\frac{0.005}{1{,}000}\right)^2 + \left(\frac{0.03}{10}\right)^2 + \left(\frac{0.01}{100}\right)^2} = 0.0300 \quad (3)$$

where $\mu_{stdCu100ppm}$ is the uncertainty of the copper standard solution at 100 ppm.

All the standards used to make the calibration curves were prepared from the stock solution. The total uncertainty in the preparation of the working standards is calculated using Equations (4) and (5):

$$\mu(C_{px}) = \sqrt{\sum_{i=1}^{n} \left(\frac{\mu_{std.Cu.i}}{i}\right)^2} \quad (4)$$

where $\mu(C_{px})$ is the uncertainty of each standard used to build the calibration curve, $\mu_{std.Cu.i}$ is the uncertainty of the copper standard at i ppm, and i is the concentration of copper standard (ppm).

$$\frac{\mu_{std.Cu.i}}{i} = \sqrt{\left(\frac{\mu_{Cert}}{C_{Cert}}\right)^2 + \left(\frac{\mu_{DS}}{C_{DS}}\right)^2 + \left(\frac{\mu_{DW}}{C_{DW}}\right)^2} \quad (5)$$

where μ_{Cert} is the uncertainty of the certified reference given by the producer, C_{Cert} is the concentration of certified reference (ppm), μ_{DS} is the uncertainty of dilution for preparing stock solution, DS is the volume of the volumetric flask for preparing stock solution, μ_{DW} is the uncertainty of diluted stock solution for preparing working standard, DW is the volume of the volumetric flask for preparing working standard.

Table 4 summarizes the uncertainty of the preparation of the working standards by FAAS; the total uncertainty of this preparation was 0.0882.

Table 4 | Summary of uncertainty in the preparation of working standards by FAAS

	Standard (mg L^{-1})					
	1	2	3	4	5	Total
Uncertainty (mg L^{-1})	0.0855	0.0104	0.0133	0.0104	0.0085	0.0882

Repeatability uncertainty

The repeatability uncertainty depends on three major influences: certificate, repeatability, and temperature. The manufacturer uncertainty was calculated assuming a triangular distribution. The repeatability uncertainty of volume was estimated by a series of 10 filling and weighting tests in a 100 mL volumetric flask. The repeatability uncertainty of temperature can be calculated between temperature ranges with the coefficient of volume expansion. Thus, the total of repeatability uncertainty was combined as given in Equation (6):

$$\mu(V_X) = \sqrt{\mu(V_p)^2 + \mu(V_{repe})^2 + \mu(V_T)^2} \quad (6)$$

where $\mu(V_p) = \dfrac{\text{Uncertainty of the glassware}}{\sqrt{6}}$

$$\mu(V_{repe}) = \frac{\text{Standard deviation of the measures}}{\sqrt{\text{Number of measures}}}$$

$$\mu(V_T) = \frac{\text{Variation of temperature} \times \text{expansion coefficient of water} \times \text{volume}}{\sqrt{3}}$$

Table 5 summarizes the repeatability uncertainty results.

Instrument deviation

The uncertainty of instrument deviation was estimated from the calculation of the system suitability result according to

Table 5 | Summary of repeatability uncertainty of the FAAS method

Parameters	Value	Distribution	Relative uncertainty
Certificate ($\mu(V_P)$)	0.1	$\sqrt{6}$	0.0408
Repeatability ($\mu(V_{repe})$)	0.0454	$\sqrt{10}$	0.0144
Temperature ($\mu(V_T)$)	$100\times1\times2.1\times10^{-2}$	$\sqrt{3}$	0.0121
Total $\left(\sqrt{\mu(V_T)^2+\mu(V_P)^2+\mu(V_{repe})^2}\right)$	0.0449	1	0.0449

Equation (7).

$$\mu(Repe) = \text{Max RSD}/\sqrt{n} \tag{7}$$

In the experimental results, the greatest relative standard deviation of the interval was 0.51 and the total number of measurements was 50. Thus, the instrument deviation was calculated to be 0.0721 using Equation (7).

Uncertainty of calibration curve

The uncertainty from the calibration curve was estimated from the linearity calculation as follows:

$$\mu(c_X) = \frac{S}{B_1}\times\sqrt{\frac{1}{p}+\frac{1}{n}+\frac{(c_{X-}\ c_{\text{mean}})^2}{S_{XX}}} \tag{8}$$

p = number of measurements to determine sample
n = number of measurements for the calibration curve
c_X = concentration of sample from the calibration curve
c_{mean} = mean value of the different calibration standards
S = residual standard deviation

$$S_{XX} = \sum_{j=1}^{n}(c_j - c_{mean})^2$$

$$S = \sqrt{\frac{\sum[A_j - (B_0 + B_1\times c_j)]^2}{n-2}} \tag{9}$$

A_j = absorbance of standard obtained from the calibration curve
c_j = concentration of standard obtained from the calibration curve
B_0 = intercept of y axis of the calibration curve
B_1 = slope of the calibration curve
j = number of measurements of standard.

As shown in Figure 2, the calibration curves for copper were plotted as absorbances versus concentrations. The curve was well linear within the studied concentration range of 1–5 mg L^{-1} copper, with the coefficient of determination of 0.998. Linear regression was carried out by means of the least squares method. The experimental results showed that $B_1 = 0.138$ and $B_0 = 0.032$. The synthetic wastewater was measured to have the copper concentration of 2.02 mg L^{-1} by using the calibration curve. Then, S and S_{XX} were calculated, giving $S = 0.0356$ and $S_{XX} = 101.623$. Therefore, the uncertainty calculated by using Equations (8) and (9) was 0.0932.

Recovery uncertainty

The uncertainty from recovery was estimated from the traceability result and calculated as follows:

$$\mu(Rec) = \overline{REC}/\sqrt{s} \tag{10}$$

where $\overline{REC}$ is the recovery average and s is the standard deviation of the recovery.

Accuracy was assessed from 10 replicate determinations of two different solutions containing 1 and 3 mg L^{-1} of copper. The average of recovery uncertainty was found to be 0.8502 and the standard deviation of the recovery was 0.75. Therefore, the recovery uncertainty calculated in this study was 0.9818.

Combined and expanded uncertainty

The combined standard uncertainty and the expanded uncertainty were calculated by using Equations (11) and (12), respectively.

$$\frac{\mu C_s}{C_s} = \sqrt{\left(\frac{(\mu(c_X))}{c_X}\right)^2 + \left(\frac{\mu(Repe)}{100}\right)^2 + \left(\frac{\mu(Rec)}{100}\right)^2 + \left(\frac{\mu(V_X)}{100}\right)^2 + \left(\frac{\mu(C_{px})}{100}\right)^2} \tag{11}$$

$$U = k\mu C_s \tag{12}$$

where k is the coverage factor (in this study, 95% confidence level was selected, and k is equal to 2).

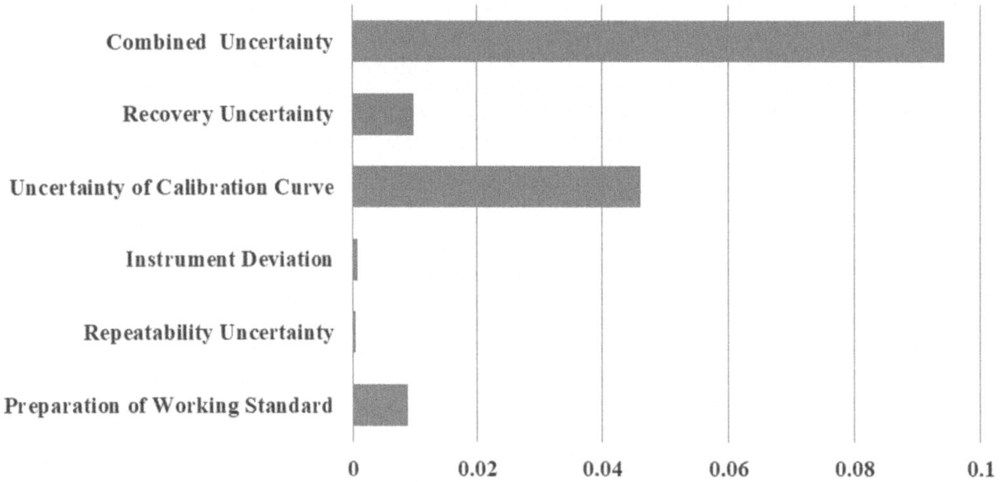

Figure 4 | Estimation of uncertainty from all sources.

In this experiment, the relative standard uncertainty was calculated to be 0.0943, and the expanded uncertainty was 0.1886. The expanded uncertainty was calculated by using the coverage factor of two, giving a confidence level of 95%. The contributions of the different parameters and influence quantities to the measurement's uncertainty are illustrated in Figure 4.

CONCLUSIONS

The FAAS method presented herein is accurate, robust, linear, and precise. The method is suitable for the quantification of Cu in the form of copper complexes in wastewater. The validation method had characteristic concentration of 0.023 mg L^{-1}, system suitability 0.10–0.51, and linearity over the range of 1-5 mg L^{-1}. It seems likely that the procedure is consistent, dependable, and precise. All sources of uncertainty were identified; the source of greatest uncertainty was the calibration curve, followed by recovery uncertainty and then by the preparation of the working standard. This approach is suitable for determination of not only Cu, but also other metals in complexed form in wastewater. Therefore, it should be included in the establishment of procedures to ensure the quality, safety, and efficiency of analysing metals in complexed forms in wastewater.

ACKNOWLEDGEMENTS

The authors are grateful to the Faculty of Science of King Mongkut's Institute of Technology Ladkrabang and to the Rajamangala University of Technology Rattanakosin for their support of this research.

REFERENCES

Ata, S., Hamid, F., Ahmed, M., Wattoo, M. H. S., Tirmizi, S. A. & Wadood, A. 2015 A method optimization study for atomic absorption spectrophotometric determination of total zinc in insulin using direct aspiration technique. *Alexandria Journal of Medicine* **51**, 19–23.

Bader, N. R. 2011 Sample preparation for flame atomic absorption spectroscopy: an overview. *Rasayan Journal of Chemistry* **4** (1), 49–55.

Ellison, S. L. R. & Williams, A. (eds) 2012 *Eurachem/CITAC Guide: Quantifying Uncertainty in Analytical Measurement*, 3rd edn. Eurachem/CITAC, http://www.eurachem.org.

Harvey, D. 2017 *Analytical Chemistry 2.0; 10.4: Atomic Absorption Spectroscopy; Representative Method 10.2: Determination of Cu and Zn in Tissue Samples.* https://chem.libretexts.org/Textbook_Maps/Analytical_Chemistry/Book%3A_Analytical_Chemistry_2.0_(Harvey)/10_Spectroscopic_Methods/10.4%3A_Atomic_Absorption_Spectroscopy (accessed 10 December 2018).

ISO 2005 *ISO/IEC 17025:2005 – General Requirement for the Competence of Testing and Calibration Laboratories*, 2nd edn. International Organization for Standardization, Geneva, Switzerland.

Jiang, S., Qu, J. & Xiong, Y. 2010 Removal of chelated copper from wastewaters by Fe^{2+}-based replacement–precipitation. *Environmental Chemistry Letters* **8** (4), 339–342.

Maryville University 2018 *Atomic Absorption Spectroscopy Learning Module: Sample Atomization.* http://blogs.maryville.edu/aas/atomization-source/ (accessed 10 December 2018).

PerkinElmer 2011 *Atomic Spectroscopy: A Guide to Selecting the Appropriate Technique and System.*

https://www.perkinelmer.com/Content/relatedmaterials/brochures/bro_worldleaderaaicpmsicpms.pdf (accessed 8 June 2018).

Petrov, A., Gentscheva, G., Havezov, I. & Ivanova, E. 2009 Determination of the uncertainty of the flame atomic absorption spectrometer for copper, cobalt, cadmium, and nickel. *Analytical Letters* **42** (16), 2509–2519.

Santana, L., Sarkis, E. S., Azevedo, S., Ulrich, J. C. & Hortellani, M. A. 2011 Estimate of uncertainty of measurement in the determination of Zn, Cu, Hg and Pb in fish muscle. In: *International Nuclear Atlantic Conference INAC 2011, Belo Horizonte, MG, Brazil.*

Srinivasa Rao, V., Ramachandra Murty, C. V., Kalyani, G. & Joga Rao, H. 2016 Process optimization studies on copper adsorption onto 5 Å molecular sieves using response surface methodology. *Research Journal of Pharmaceutical, Biological and Chemical Sciences* **7** (4), 1515–1524.

Thermo Elemental 2001 *AAS, GFAAS, ICP or ICP-MS? Which Technique Should I Use? An Elementary Overview of Elemental Analysis.* http://oliver.chemistry.ucsc.edu/122/Lab5%20Handout.pdf (accessed 5 December 2018).

Usapein, P. & Chavalparit, O. 2015 Evaluating the performance of 3R options to reduce landfill wastes using the 3R indicator (3RI): case study of polyethylene factories in Thailand. *Journal of Material Cycles and Waste Management* **17** (2), 303–311.

First received 16 August 2018; accepted in revised form 10 February 2019. Available online 21 February 2019

Selenium in wastewater: fast analysis method development and advanced oxidation treatment applications

Dipti Prakash Mohapatra and Deepak M. Kirpalani

ABSTRACT

Selenium, a ubiquitous non-metal in nature, is potentially toxic to natural ecosystems due to its bioaccumulation potential. Due to increased monitoring and enforcement of selenium regulations, the need to be able to measure and treat selenium efficiently has taken on an increased importance. The principal aqueous forms of inorganic selenium are selenite (Se(IV)) and selenate (Se(VI)). Selenate, due to its high mobility and lack of affinity to conventional adsorbents, is typically much more difficult to treat and remove. To address both measurement and removal, an analytical method is reported for quantification of selenium in wastewater (WW) using UV-Vis spectrophotometer followed by removal studies using advanced oxidation processes (AOPs). Malachite green and azure blue were selected for colorimetric analysis using UV-Vis. Malachite green indicator showed the best results for analysis. The reported UV-Vis method was applied to establish the effect of AOPs on selenium removal. It was noted that all of the AOP treated samples showed removal of selenium and it was established that the UV-Vis method has a lower limit of detection at 2 mg/L. Further, through this study, it was found that the chemical cavitation yield and selenium removal efficiency peaked at low frequency ultrasound of 40 kHz.

Key words | advanced oxidation processes, analytical method development, selenium, wastewater

Dipti Prakash Mohapatra (corresponding author)
Deepak M. Kirpalani
National Research Council of Canada,
Energy Mining and Environment Portfolio,
1200 Montreal Road, Ottawa, ON, K1A 0R6,
Canada
E-mail: *dipti.mohapatra@nrc-cnrc.gc.ca*

INTRODUCTION

Selenium is an essential trace element needed for all living organisms. Selenium, a non-metal, shares many similar physiological properties with arsenic and chemical behavior with sulfur. It has both chalcophilic (strong affinity with sulfur) and siderophilic (able to form a metallic bond with iron) properties (Nakamaru *et al.* 2005; Vesper *et al.* 2008). Despite its essentiality at low concentrations, selenium is a potential toxic element to natural ecosystems due to its bioaccumulation potential. A recent study by Lemly (2018) on selenium poisoning of fish by coal ash wastewater in Herrington Lake, Kentucky showed a total selenium-induced mortality in excess of 25% per year and further confirmed that coal ash discharges into Herrington Lake are contributing selenium to the lake that impacted aquatic life negatively. Elemental selenium is relatively nontoxic and is considered to be an essential trace element, however, it quickly becomes toxic at higher than recommended levels. Selenium pollution has a wide environmental impact range in aquatic and terrestrial life, ecosystems and even agricultural processes (Chapman *et al.* 2010; El-Ramady *et al.* 2014). Environmental contamination by selenium typically occurs at low concentrations, and the importance in remediating selenium-containing environments has been deemed important. Selenium detected in soil and water which further can enter the food chain through the roots of plants or aquatic organisms, leading to possible long term environmental impacts (Staicu *et al.* 2015). Metallurgical coal mining is considered to be one of the major sources of selenium in wastewater (WW). Selenium source in WW is due to presence of selenium in mine drainage, oil refineries, residues from fossil fuel thermoelectric power plants, metal ores and agricultural drainage water.

Fast determination of selenium in different matrices is an important task in analytical chemistry. The toxic and deficiency levels of selenium for metabolism are rather narrow. There are research works existing in the literature on the development of new analytical methods for the quantitative extraction of selenium from different matrices

doi: 10.2166/wst.2019.010

including wastewater (Roehl *et al.* 2002; Vriens *et al.* 2015). A range of analytical techniques such as inductively coupled plasma mass spectrometry (ICP-MS), inductively coupled plasma atomic emission spectrometry, electrospray ionization mass spectrometry, hydride generation atomic fluorescence spectrometry, liquid chromatography with tandem mass spectrometry, high-performance liquid chromatography–inductively coupled plasma mass spectrometry, and sequential injection analysis–chemiluminescence, are used for the detection of trace and ultra-trace amounts of selenium and its species (Sutton *et al.* 1997; Roehl *et al.* 2002; Reemtsma 2003; LeBlanc *et al.* 2018).

Spectrophotometric techniques are generally preferred for the determination of analytes because of their simplicity and low cost. The determination of analytes at trace levels can be achieved using a suitable complexing agent. The main drawback of the spectrophotometric method is its poor selectivity. The complex matrix components generally interfere to a large extent; therefore, in most cases, the masking of the interfering species is unavoidable. In addition, the elimination of matrix effects and the determination of analytes present at trace and ultra-trace levels are generally performed using different separation and extraction methods. The first part of this research focuses on the development of a quantification method based on UV-Vis spectrophotometer for analysis of selenium in WW.

Physical, chemical and biological treatment technologies have shown potential for removal of selenium. However, limited success was reported in the removal of selenium from water to less than 5 μg/L at any scale (Sandy & DiSante 2010). Biological processes for selenium removal that focus on microbial reduction of selenium are new approaches that apply facultative heterotrophic bacteria to reduce selenite and selenate. In this process, selenate is first reduced to selenite, and subsequently selenite is reduced by sulfide and precipitates from the solution as selenium sulfide (He & Yao 2010; Hageman *et al.* 2017). The latter is bio-reduced to elemental selenium or organic forms of selenium and sulfide. However, biological process for the selenium removal from mining wastewater is limited by influent concentration and higher operational cost (Tan *et al.* 2016). Li *et al.* (2018) reviewed the adsorption properties of various highly porous metal–organic framework (MOF)-based materials for the efficient removal of toxic metal ions. They reported that MOF-based materials can be used as an advanced treatment method for the removal of toxic metal ions including selenium from WW due to various attractive characteristics, such as high adsorption capacity, fast adsorption kinetics, superior selectivity and excellent reusability. Zhao *et al.* (2018) reviewed use of polymer-based nanocomposites for the removal of hazardous metal ions from WW. They reported that polymer-based nanocomposites can be used for the removal of diverse heavy metal ions including selenium from WW as these nanocomposites often present superior physical, chemical and mechanical properties, as well as superior compatibility, as compared with single polymers.

Among the different treatment options for the selenium removal, catalytic oxidation treatment has the advantage of forming elemental selenium nanospheres which exhibit unique optical and spectral properties for various industrial applications, i.e. medical, electrical, and manufacturing processes. Advanced oxidation processes (AOPs) has been widely used for the treatment of WW to enhance chemical reaction rate and facilitate the removal of several pollutants. AOPs such as sonication treatment in liquid media involved the generation, expansion, oscillations, splitting and implosions of numerous tiny gas voids called cavities or bubbles. As a result of implosion of cavitation bubbles, extreme localized temperatures and pressures are generated in the centre of the collapsed bubble. This leads to solute thermolysis as well as the formation of hydroxyl radical and hydrogen peroxide. The formation of hydroxyl radical in medium is responsible for the degradation of polar organic compounds (Kirpalani & McQuinn 2006; Mohapatra *et al.* 2011). Further, among the many AOPs, the Fenton process has been proven to be effective for treating recalcitrant organic compounds by using Fe^{2+} and H_2O_2 to produce hydroxyl (-OH) radicals.

The second part of this study focuses on the development of an optimum advanced oxidation process for the treatment of mining wastewater for removal of selenium. This two pronged study addresses the following specific objectives: (i) fast quantification method development for analysis of selenium in wastewater using UV-Vis spectrophotometry; (ii) to develop an optimum condition of AOPs such as ultrasonication, Fenton's oxidation, ferro-sonication (combination of ultrasonication with Fenton's oxidation) and ozonation processes for selenium removal from wastewater.

MATERIAL AND METHODS

Reagents and chemical solution preparation

All chemicals used were of analytical grade. Malachite green, azure B, Arabic gum, methanol, ferrous sulfate anhydrite were purchased from Sigma Aldrich (Ontario, Canada). High performance liquid chromatography (HPLC) grade water was prepared in the laboratory using a Milli-Q/Milli-RO Millipore system (Milford, MA, USA). Nitric acid, hydro-chloric acid, and hydrogen peroxide were supplied by Fisher scientific (Ontario, Canada).

Arabic gum solution was prepared by adding 10 g of Arabic gum powdered (Sigma Aldrich, Canada) in 100 mL ultrapure Milli-Q water (ISO 3696) followed with continuous stirring due to its limited solubility in water. The mixture was continuously stirred at 60 °C to dissolve the Arabic gum completely. Malachite green solution (indicator) was prepared by dissolving 0.05 g of malachite green (Sigma Aldrich, Canada) in 10 mL of Milli-Q water followed by stirring until crystals are dissolved. A (0.1%) solution of azure B in methanol–water mixture was used. Potassium iodide solution was prepared by adding 10 g of potassium iodide in 100 mL volumetric flask.

Standard stock solution of Se was prepared by adding 1.0 g of Se^0 (elemental) (Sigma Aldrich, Canada) into 50 mL (1 + 1) HNO_3 solution followed by heating the solution at 80 °C. After proper mixing, the solution was cooled down to room temperature, followed by the addition of 5 mL of sulfuric acid and heating the solution until transparency appears. Standard stock solution of Se(VI) was prepared by dissolving 0.2393 g Na_2SeO_4 (Sigma Aldrich, Canada) in Milli-Q water and top it up to the 100 mL mark in a volumetric flask.

Wastewater sample

Control samples and standards for UV-Vis analysis were prepared by introducing reagent grade sodium selenite into double distilled water (conductivity <0.2 µS/cm) with various concentrations. Wastewater samples were collected from a metallurgical coal operations site in British Columbia, Canada. Typical ICP and IC assay for the WW used in this study is presented in Table 1.

Table 1 | Characteristics of the wastewater used for experiments

	Units	Concentration
pH		7.1
NO_3-N	mg/L as N	58–65
Calcium	mg/L	450
Magnesium	mg/L	380–440
Chloride	mg/L	20.5
Sulphate	mg/L	1,750–2,800
TDS	mg/L	3,799
TSS	mg/L	6.8
Alkalinity	mg/L as $CaCO_3$	550–600
Phosphorus	mg/L	0.007
Temperature	°C	5

TDS, total dissolved solids; TSS, total suspended solids.

Procedure for determination of Se(IV) and Se(VI) with azure B

An aliquot of the solution containing 2.0–10.0 µg/mL of selenium was transferred into a series of 10 mL calibrated flasks. A volume of 1 mL of 2% potassium iodide solution followed by 1 mL of 2 M hydrochloric acid was added to it. The mixture was gently shaken until the appearance of yellow colour, indicating the liberation of iodine. Azure B (0.1%, 0.5 mL) was added to the solution and the reaction mixture was shaken for 2 min. The contents were diluted to 10 mL in a standard flask with distilled water. The absorbance of the resulting solution was measured at 644 nm against a reagent blank. A blank was prepared by replacing the selenium solution with distilled water. The absorbance corresponding to the bleached colour, which in turn corresponds to the selenium concentration, was obtained by subtracting the absorbance of the blank solution from that of the test solution. The amount of the selenium present in the volume taken was computed from the calibration graph.

Procedure for determination of Se(IV) and Se(VI) with malachite green

An aliquot of the solution containing 2.0–10.0 µg/mL of selenium was transferred into a series of 10 mL calibrated flasks followed by the addition of 1 mL of 6 mol/L of hydrochloric acid (HCl), 3 mL of 0.05% malachite green, 3 mL of 10% potassium iodide (KI) and 3 mL of 10% Arabic gum solution. Further, the flask was topped with 5 mL of double distilled water up to the 25 mL mark. Analysis was carried out by using UV-Vis spectrometry at wavelength of 670 nm in 1 cm quartz cell to observe the absorbency of Se(VI) after 30 min.

Further for Se(IV) quantification, 5 mL of Se(VI) standard solution transfer into a 10 mL tube followed by addition of 5 mL of 9 M hydrochloric acid. The mixture was then heated with boiling water bath for some time followed by cool down to room temperature. Se(VI) has been reduce to Se(IV) and analysis was carried out by following the procedure used for Se(IV) determination.

Advanced oxidation treatment of wastewater

Sonication experiment

Low frequency (20 kHz and 40 kHz) ultrasonic experiments were carried out by using an ultrasonic homogenizer (Autotune 750 W, Cole-Parmer Instruments, Vernon Hills, IL, USA). A mid frequency (574 kHz) sonochemical processing system was assembled using a broadband transducer (Ultraschalltechnik-Meinhardt GMBH, Germany). The transducer was installed at the bottom of a coolant-jacketed glass column reactor with a diameter of 5 cm and 100 cm in height. The ultrasound energy was supplied by a power amplifier (HM8001-2) through a function generator (HM 8030-5 and HM 8032). Sonication experiments were carried out at different frequency conditions using two different broadband transducers with the same effective diameter. The reactor was supplied with different power input starting from 16.67 to 83.33%. The cooling system was operated to maintain a constant temperature. The experiment was carried out with a sample volume of 200 mL held within a jacked glass cooling column with 30 min ultrasonic time. Thereafter, ultrasonic treatment was carried out at 20, 40 and 574 kHz, referred as US1, US2 and US3, respectively.

Fenton oxidation and ferro-sonication treatment

Wastewater sample of 500 mL was acidified to pH 3 by using 10 N H_2SO_4 and placed in a 1 L flask. Fenton's oxidation process was conducted at pH 3 to avoid iron hydroxide precipitation. To initiate Fenton's oxidation reaction, H_2O_2 and $FeSO_4$ solution were added to the WW samples with three different conditions: 0.2 mM, 0.4 mM, and 0.8 mM H_2O_2 and $[H_2O_2]_0/[Fe^{2+}]_0 = 50$. Fenton's oxidation was carried out at ambient temperature for 60 min using a shaker at 150 rpm. Thereafter, Fenton's oxidation treatment was carried out under 0.2 mM, 0.4 mM, and 0.8 mM H_2O_2 concentration referred to as FO1, FO2 and FO3, respectively.

Further, for ferro-sonication treatment, 500 mL of WW sample was acidified to pH 3 by using 10 N H_2SO_4 and placed in a 1 L flask. Ferro-sonication pre-treatment was carried out by addition of 0.4 mM of $FeSO_4$ solution in WW followed by ultrasonic pre-treatment under three frequency conditions such as 20 kHz, 40 kHz and 574 kHz with 60 min ultrasonic time. Ferro-sonication treatment carried out under ultrasonic frequency 20, 40 and 574 kHz referred as FS1, FS2 and FS3, respectively.

Ozonation treatment

Ozone was produced using dry air by a laboratory ozone generator (Wedeco, Xylem Water Solutions, Herford GmbH, Germany) which was set at different flow rates of 10, 20, and 30 L/h. The ozonation treatment was carried out for 1 h in an ozone reactor of 1 L working volume capacity, where a ceramic fine bubble diffuser located at the bottom of the reactor and the wastewater was thoroughly mixed with magnetic stirrer to provide efficient mixing conditions in order to ensure the homogeneous distribution of the ozone in the liquid phase. The concentration in the gas-stream was measured optically using an ozone-in-off-gas analyzer (BMT 964 OG, BMT Messtechnik GmbH, Stahnsdorf, Germany). Thereafter, ozonation treatment was carried out with flow rates of 10, 20, and 30 L/h, referred to as OZ1, OZ2 and OZ3 respectively.

RESULTS AND DISCUSSION

Method development for selenium analysis in wastewater samples

Selenium (IV) reacts with potassium iodide in the presence of dilute hydrochloric acid to liberate iodine quantitatively. The liberated iodine selectively oxidizes malachite green dye. The green colour of the dye was developed in an acetate buffer medium (pH 4.5–5.5) on heating in a water bath for 5 min. The absorbance of the dye was measured after 20 min of dilution. The malachite green dye showed maximum absorption at 650 nm, and the reagent blank had negligible absorbance at this wavelength (Figure 1).

The experiments were also carried out by using azure B as the colour agent. The reaction of selenium with potassium iodide in acidic medium liberates iodine. The liberated iodine bleaches the violet colour of azure B to colourless leucoform and the absorbance is measured at 644 nm (Figure 2). From Figures 1 and 2, it was observed that both malachite green and azure B were highly selective and sensitive for the spectrophotometric determination of selenium in WW samples. The adherence to Beer's law was studied by measuring the absorbance values of solutions, varying selenium concentration. A straight line graph was obtained by plotting absorbance against concentration of selenium. The results showed that at 644 nm, Beer's law was obeyed up to 2 μg/mL of selenium. The molar absorptivity and Sandell's sensitivity for the coloured system were found to be 0.82×10^5 L/moL/cm and 6.53×10^{-4} μg/cm^2, respectively.

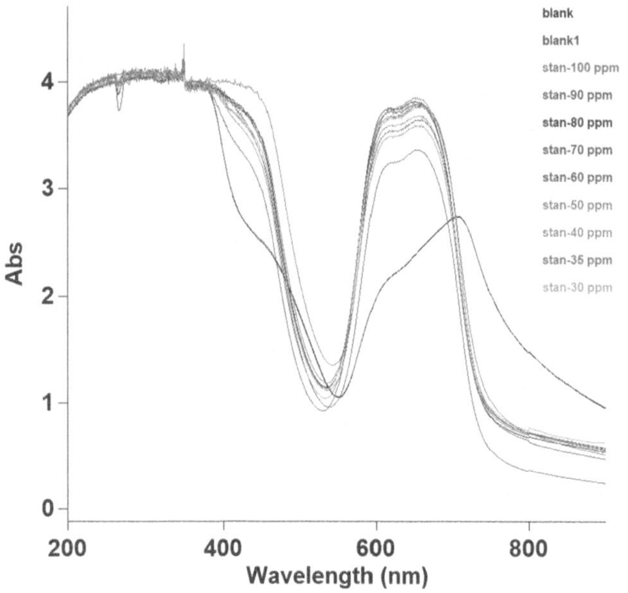

Figure 1 | Absorption spectra of different selenium standard solution with malachite green using UV-Vis spectrophotometry.

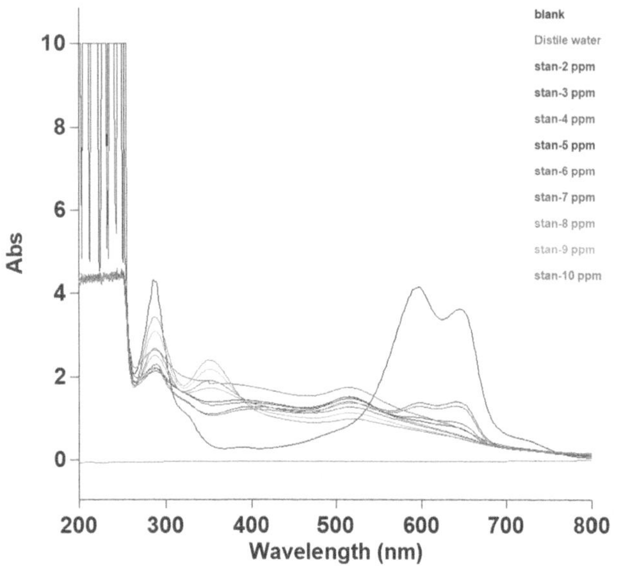

Figure 2 | Absorption spectra of different selenium standard solution with azure B using UV-Vis spectrophotometry.

Krishnaiah *et al.* 2003 studied the use of the spectrophotometric method for determination of selenium in environmental samples. They reported that the statistical analysis of different selenium concentrations showed use of spectrophotometric method produced good precision and accuracy and further obeyed Beer's law for selenium in the concentration range 0.03–3.5 µg/mL (pink colour product) and 0.07–2.5 µg/mL (orange red colour product).

The developed method was applied to the quantitative determination of traces of selenium in mining effluent water. The results obtained by using developed UV-Vis methods were compared with the ICP-MS results. The reagent provides a simple, rapid, sensitive and accurate method for the spectrophotometric determination of selenium. The method needs neither heating for the complete colour development nor extraction into any organic phase. However, it was observed that this developed method worked for selenium determination to a minimum concentration of 2 mg/L. The proposed method has been successfully applied to the determination of traces of selenium in mining effluent water with selenium concentration above 2 mg/L.

Development of an optimum advanced oxidation process for selenium removal from mine wastewater

The selenium concentration was measured by using the developed UV-Vis spectrophotometric method after each treatment. Each analysis was repeated three times. The results were compared with blank solutions of a known concentration of selenium. All the results obtained by UV-Vis were compared with the ICP-MS results for validation of method developed.

The effectiveness of AOPs such as ozonation (O3), ultrasonication (US), Fenton's oxidation (FO) and ferro-sonication (FS) processes was analyzed in terms of reduction of total selenium concentration from WW. Different results obtained during the treatment process is presented in Figure 3.

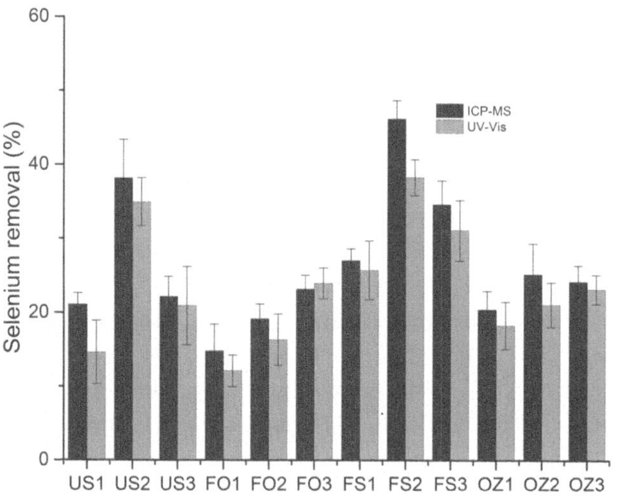

Figure 3 | Selenium removal from wastewater under different AOPs as US1 (20 kHz), US2 (40 kHz), US3 (574 kHz), FO1 (0.2 mM H_2O_2), FO2 (0.4 mM), FO3 (0.8 mM), FS1 (0.4 mM of $FeSO_4$, 20 kHz), FS2 (0.4 mM of $FeSO_4$, 40 kHz), FS3 (0.4 mM of FeSO4, 574 kHz), OZ1 (10 L/h), OZ2 (20 L/h) and OZ3 (30 L/h).

Among all the conditions of the AOPs, ferro-sonication process (combination of ultrasonication and Fenton's oxidation) carried out by adding 0.4 mM $FeSO_4.7H_2O$, ultrasonic frequency 40 kHz, pH 3.0, sample volume 500 mL showed higher removal of selenium (47%) in both UV-Vis spectrophotometry analysis and ICP-MS analysis. Further, ultrasonication treatment carried out with ultrasonic frequency of 40 kHz also showed a comparative selenium removal (38%) from wastewater. However, ozonation and Fenton's oxidation treatment of WW resulted in lower selenium removal as compared to ferro-sonication processes.

The order of efficiency of AOPs for selenium reduction from WW was: FS2 > US2 > FS3 > FS1 > OZ2 > OZ3 > US3 > US1 > FO3 > OZ1 > FO2 > FO, respectively. The reason for higher reduction of selenium concentration during ferro-sonication treatment processes was due to the addition of FeSO4, which enhanced the formation of hydroxyl (OH) radicals from hydrogen peroxide that was produced in the solution during the ultrasonic processes. Ultrasonication induces acoustic cavitation into liquid media, involving the generation, expansion, oscillations, splitting and implosions of numerous tiny gas voids called cavities or commonly referred to as cavitation bubbles. As a result of implosion of cavitation bubbles, extreme temperature and pressure is generated in the centre of the collapsed bubble leading to solute thermolysis, as well as the formation of hydroxyl radical and hydrogen peroxide. The hydroxyl radicals generated during Fenton and ferro-sonication pre-treatment processes have an oxidizing potential of 2.8 V as compared to normal hydrogen electrode and are capable of oxidizing a wide range of metals including selenium from WW. Further, higher selenium reduction from WW in ferro-sonication as compared to Fenton's oxidation processes is believed to be due to the higher efficiency of production of OH radicals.

Further, the reduction in chemical oxygen demand (COD) and ammonia-nitrogen during ozonation (O3), US, FO and FS processes was also observed (Figure 4). Higher reduction of COD (54%) and ammonia-nitrogen (42%) was observed during ozonation treatment with ozone dose 30 L/h, ozonation time 30 min, sample volume 350 mL, pH 7.68. Comparative results for COD and ammonia-nitrogen reduction was also observed during ferro-sonication treatment carried out with the addition of 0.4 mM $FeSO_4.7H_2O$, pH 3 with frequency of 40 kHz, power input 80%, sonication time 1 h, sample volume 500 mL. Further, it was observed that these conditions also favourable for the solubilisation of COD and ammonia-nitrogen removal from mining effluent water. During these pre-treatment processes, organic substances were transferred from solid to the aqueous phase, which resulted in the increase of soluble protein and soluble carbohydrate concentration in the aqueous phase and decrease of COD and SOC in the solid phase. Höfl *et al.* (1997) compared different AOPs for COD removal from WW. They reported that processes based on hydroxyl radicals were less selective but considerably more effective in COD degradation.

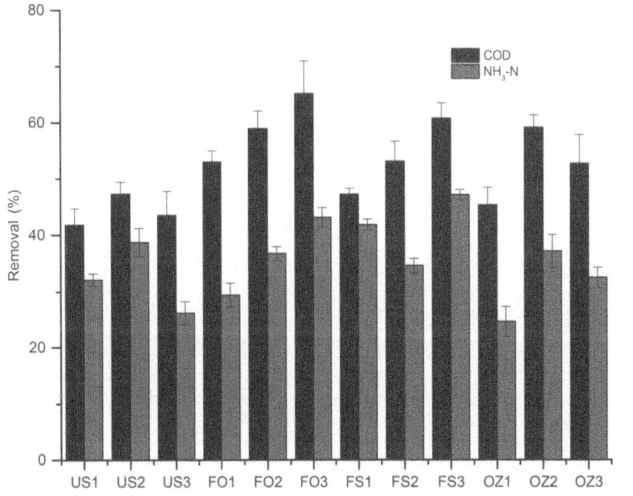

Figure 4 | Reduction in COD and ammonia-nitrogen during different advanced oxidation treatment of wastewater.

Optimization of ferro-sonication treatment for selenium removal from mining wastewater was carried out in a series of different concentration of $FeSO_4.7H_2O$ (0.2 to 1.0 mM), pH (2.5 to 4), shaking for 1 h at 200 rpm with sample volume of 300 mL and ultrasonic frequency of 40 kHz. All the experiments were carried out in effluent water as well as in standard solution (Milli-Q water spiked with a known concentration of selenium). The experimental conditions and percentage of selenium removed is presented in Table 2. Among all the ferro-sonication conditions examined in this study, the $FeSO_4.7H_2O$ concentration of 0.8 mM with pH 2.5 showed the maximum removal efficiency (65% and 53%) of selenium from standard solution and mining wastewater, respectively. However, it was observed that by increasing $FeSO_4.7H_2O$ concentration from 0.2 to 0.8 mM an increase in efficiency of the ferro-sonication treatment with higher selenium removal from standard solution was shown, as well from mining wastewater. Further, optimization of ferro-sonication treatment ultrasonic frequency of 574 kHz for selenium removal from mining wastewater was carried out in a series of different concentrations of $FeSO_4.7H_2O$ (0.2 to 1.0 mM), pH (2.5 to 4), shaking for 1 h at 200 rpm with sample volume of 300 mL. The experimental conditions

Table 2 | Optimization of ferro-sonication treatment under different concentrations of $FeSO_4.7H_2O$ and ultrasonic frequency of 40 kHz for selenium removal (%)

	0.2 mM		0.4 mM		0.6 mM		0.8 mM		1.0 mM	
$FeSO_4$ conc.	**stnd**	**eff**	**stnd**	**eff**	**stnd**	**eff**	**stnd**	**eff**	**stnd**	**eff**
pH 4	29	21	38	32	36	31	41	37	48	42
pH 3.5	36	28	46	27	48	41	46	43	43	37
pH 3	38	23	49	41	57	49	55	46	57	49
pH 2.5	42	33	56	45	55	53	65	53	60	57

Stnd, standard solution; eff, effluent water.

Table 3 | Optimization of ferro-sonication treatment under different concentration of $FeSO_4.7H_2O$ and ultrasonic frequency of 574 kHz for selenium removal (%)

	0.2 mM		0.4 mM		0.6 mM		0.8 mM	
$FeSO_4$ conc.	**stnd**	**eff**	**stnd**	**eff**	**stnd**	**eff**	**stnd**	**eff**
pH 4	47	39	54	48	33	27	51	43
pH 3.5	42	40	61	53	59	45	59	46
pH 3	31	26	51	44	37	28	32	28
pH 2.5	27	24	29	23	63	57	39	35

Stnd, standard solution; eff, effluent water.

and percentage of selenium removed is presented in Table 3. Among all the ferro-sonication conditions tried under 574 kHz ultrasonic frequency, $FeSO_4.7H_2O$ concentration of 0.6 mM with pH 2.5 showed the maximum removal efficiency (63% and 57%) of selenium from standard solution and mining wastewater, respectively. Further, decrease in pH from 4 to 2.5 showed that ferro-sonication treatment with higher $FeSO_4.7H_2O$ concentration and lower pH is a favourable condition for higher removal of selenium from mining wastewater. The pH value has a decisive effect on the oxidation potential of -OH radicals because of the reciprocal relation of the oxidation potential to the pH value. Furthermore, the concentration of inorganic carbon and the hydrolytic speciation of Fe(III) species are strongly affected by the pH value (Mohapatra *et al.* 2013; Pulicharla *et al.* 2017). Higher removal of selenium from mining wastewater during ferro-sonication process was due to the addition of $FeSO_4$, as well as production of H_2O_2 in the medium during ultrasonic treatment, which enhanced the formation of hydroxyl (OH) radicals in the solution.

CONCLUSIONS

The UV-Vis spectrophotometric method development for analysis of selenium in wastewater samples showed that malachite green indicator was best-suited for selenium analysis. The colour indicator provides a simple, rapid, sensitive and accurate method for spectrophotometric quantification of selenium in wastewater samples. UV-Vis produced comparable results as ICP-MS; however, the UV-Vis approach is limited to concentrations above 2 mg/L. Further, among the AOPs tested (ultrasonication, Fenton's oxidation, ferro-sonication and ozonation) for selenium reduction from wastewater samples, ferro-sonication treatment (a combination of ultrasonication and Fenton's oxidation) showed the best method for selenium removal from wastewater. Ferro-sonication treatment process showed higher selenium removal efficiency due to the utilization of iron to generate more OH radicals. Further, through this study, it was found that the ultrasonic treatment carried out at lower frequency (40 kHz) showed higher removal of selenium from effluent water as compared to mid-frequency ultrasound treatment at 574 kHz. This shows that the removal is influenced by higher cavitation yield as the cavitation yield decreases at higher frequencies due to the formation of capillary instabilities.

ACKNOWLEDGEMENT

The authors would like to acknowledge the technical discussion with Canadian industry partners through the Environmental Advances in Mining Program at the National Research Council of Canada in outlining this manuscript.

REFERENCES

Chapman, P. M., Adams, W. J., Brooks, M. L., Delos, C. G., Luoma, S. N., Maher, W. A., Ohlendorf, H. M., Presser, T. S. & Shaw, D. P. (eds) 2010 *Ecological Assessment of Selenium in the Aquatic Environment.* SETAC Press, Pensacola, FL, USA.

El-Ramady, H. R., Domokos-Szabolcsy, E., Abdalla, N. A., Shalaby, T. A., Sztrik, A., Prokisch, J. & Fári, M. 2014 Selenium and nano-selenium in agroecosystems. *Environmental Chemistry Letters* **12** (4), 495–510.

Hageman, S. P. W., Weijden, R. D., Stams, A. J. M., Cappellen, P. & Buisman, C. J. N. 2017 Microbial selenium sulfide reduction for selenium recovery from wastewater. *Journal of Hazardous Materials* **329**, 110–119.

He, Q. & Yao, K. 2010 Microbial reduction of selenium oxyanions by *Anaeromyxobacter dehalogenans*. *Bioresource Technology* **101**, 3760–3764.

Höfl, C., Sigl, G., Specht, O., Wurdack, I. & Wabner, D. 1997 Oxidative degradation of AOX and COD by different advanced oxidation processes: a comparative study with two samples of a pharmaceutical wastewater. *Water Science and Technology* **35**, 257–264.

Kirpalani, D. M. & McQuinn, K. J. 2006 Experimental quantification of cavitation yield revisited: focus on high frequency ultrasound reactors. *Ultrasonic Sonochemistry* **13**, 1–5.

Krishnaiah, L., Suresh, K., Suvardhan, K. & Chiranjeevi, P. 2003 Simple spectrophotometric determination of traces of selenium in environmental samples. In: *Proceedings of the Third International Conference on Environment And Health*, J. Martin, V. Bunch, Madha Suresh & T. Vasantha (eds), Chennai, India, pp. 217–225.

LeBlanc, K. L., Kumkrong, P., Mercier, P. H. J. & Mester, Z. 2018 Selenium analysis in waters. Part 2: Speciation methods. *Science of the Total Environment* **640–641**, 1635–1651.

Lemly, A. D. 2018 Selenium poisoning of fish by coal ash wastewater in Herrington Lake, Kentucky. *Ecotoxicology and Environmental Safety* **150**, 49–53.

Li, J., Wang, X., Zhao, G., Chen, C., Chai, Z., Alsaedi, A., Hayat, T. & Wang, X. 2018 Metal–organic framework-based materials: superior adsorbents for the capture of toxic and radioactive metal ions. *Chemical Society Reviews* **47**, 2322–2356.

Mohapatra, D. P., Brar, S. K., Tyagi, R. D. & Surampalli, R. Y. 2011 Concomitant degradation of bisphenol A during ultrasonication and Fenton oxidation and production of biofertilizer from wastewater sludge. *Ultrasonics Sonochemistry* **18**, 1018–1027.

Mohapatra, D. P., Brar, S. K., Tyagi, R. D., Picard, P. & Surampalli, R. Y. 2013 A comparative study of ultrasonication, Fenton's oxidation and ferro-sonication treatment for degradation of carbamazepine from wastewater and toxicity test by yeast estrogen screen (YES) assay. *Science of the Total Environment* **447**, 280–285.

Nakamaru, Y., Tagami, K. & Uchida, S. 2005 Distribution coefficient of selenium in Japanese agricultural soils. *Chemosphere* **58** (10), 1347–1354.

Pulicharla, R., Brar, S. K., Rouissi, T., Auger, S., Drogui, P., Verma, M. & Surampalli, R. Y. 2017 Degradation of chlortetracycline in wastewater sludge by ultrasonication, Fenton oxidation, and ferro-sonication. *Ultrasonics Sonochemistry* **34**, 332–342.

Reemtsma, T. 2003 Liquid chromatography–mass spectrometry and strategies for trace-level analysis of polar organic pollutants. *Journal of Chromatography A* **1000**, 477–501.

Roehl, R., Slingsby, R., Avdalovic, N. & Jackson, P. E. 2002 Applications of ion chromatograph with electrospray mass spectrometric detection to the determination of environmental contaminants in water. *Journal of Chromatography A* **956**, 245–254.

Sandy, T. & DiSante, C. 2010 *Review of Available Technologies for the Removal of Selenium From Water –Report to North American Metals Council*. CH2M HILL, 1–233.

Staicu, L. C., van Hullebusch, E. D., Lens, P. N. L., Pilon-Smits, E. A. H. & Oturan, M. A. 2015 Electrocoagulation of colloidal biogenic selenium. *Environmental Science and Pollution Research* **22**, 3127–3137.

Sutton, K., Sutton, R. M. C. & Caruso, J. A. 1997 Inductively coupled plasma mass spectrometric detection for chromatography and capillary electrophoresis. *Journal of Chromatography A* **789**, 85–126.

Tan, L. C., Nancharaiah, Y. V., Hullebusch, E. D. & Lens, P. N. L. 2016 Selenium: environmental significance, pollution, and biological treatment technologies. *Biotechnology Advances* **34**, 886–907.

Vesper, D. J., Roy, M. & Rhoads, C. J. 2008 Selenium distribution and mode of occurrence in the Kanawha Formation, southern West Virginia, U.S.A. *International Journal of Coal Geology* **73**, 237–249.

Vriens, B., Mathis, M., Winkel, L. H. E. & Berg, M. 2015 Quantification of volatile-alkylated selenium and sulfur in complex aqueous media using solid-phase micro-extraction. *Journal of Chromatography A* **1407**, 11–20.

Zhao, G., Huang, X., Tang, Z., Huang, Q., Niu, F. & Wang, X. 2018 Polymer-based nanocomposites for heavy metal ions removal from aqueous solution: a review. *Polymer Chemistry* **9**, 3562–3582.

First received 21 August 2018; accepted in revised form 22 December 2018. Available online 9 January 2019

Triamine-bearing activated rice husk ash as an advanced functional material for nitrate removal from aqueous solution

Phuoc Toan Phan, Trung Thanh Nguyen, Nhat Huy Nguyen and Surapol Padungthon

ABSTRACT

In this study, we proposed a simple method for preparing triamine-bearing activated rice husk ash material (TRI-ARHA) as an advanced functional material with high nitrate adsorption capacity and durability. The novel TRI-ARHA was generated via a post-synthesis grafting method and modified by acidification to form ammonium moiety. TRI-ARHA was then characterized by using thermogravimetric analysis, Fourier transform infrared spectroscopy, X-ray diffraction, scanning electron microscopy (SEM), energy-dispersive X-ray spectroscopy and SEM-mapping. The TRI-ARHA had high nitrate adsorption capacity of 163.4 mgNO_3^-/g, which was 2.06 times higher than that of a commercial anion exchange resin (Akualite A420) due to strong interactions of Si, C, and N in the material structure. This functional material with good durability after 10 cycles could be very promising for adsorption of other anions (e.g. sulfate and phosphate) in water and wastewater treatment.

Key words | activated rice husk ash, adsorption, nitrate, triamine, water treatment

Phuoc Toan Phan
Trung Thanh Nguyen (corresponding author)
Faculty of Engineering – Technology – Environment,
An Giang University,
18 Ung Van Khiem street, Dong Xuyen Dist., Long Xuyen City, An Giang Prov.,
Vietnam
E-mail: *ntthanh@agu.edu.vn*

Phuoc Toan Phan
Nhat Huy Nguyen
Faculty of Environment and Natural Resources,
Ho Chi Minh City University of Technology,
VNU-HCM,
268 Ly Thuong Kiet St., Dist. 10, Ho Chi Minh City,
Vietnam

Surapol Padungthon
Department of Environmental Engineering,
Khon Kaen University,
123 Moo 16 Mittraphap Rd., Nai-Muang, Muang District, Khon Kaen 40002,
Thailand

INTRODUCTION

The contamination by nitrogen-containing compounds (e.g. ammonia, nitrate and nitrite) in drinking water, surface water and various types of wastewater is an issue in many countries. The sources of these compounds can be attributed to the improper or incomplete treatment of industrial and domestic wastewater as well as extensive use of nitrogen fertilizers. Literature has reported that high concentration of nitrate could cause human health issues such as blue baby syndrome (methemoglobinemia) (Ward & Brender 2018). For nitrate removal, different technologies have been developed in the past decade, including nitrification, denitrification, adsorption, reverse osmosis, ammonia stripping, chemical coagulation, filtration, and electrodialysis (Tatarchuk *et al.* 2016, 2018a, 2019; Tyagi *et al.* 2018). Recently, adsorption has been widely studied due to its simplicity, sludge-free operation, easiness in handling, and availability of various adsorbents (Bhatnagar & Sillanpää 2011). Among them, adsorbents based on agricultural by-products have attracted attention because of their special features such as biodegradability, biocompatibility, and renewability (Xu *et al.* 2010; Song *et al.* 2016).

In Vietnam, there is a huge amount of rice husk ash (RHA) produced annually from burning of rice husk; however, very little industrial application of RHA has been developed. Therefore, our previous work focused on environmental applications of the activated rice husk ash (ARHA) for removals of organic compounds, arsenic, CO_2, and H_2S (Thanh 2016). Dimethylamine–rice husk has also been applied for nitrate removal with high efficiency; however, this study used many chemicals such as *N,N*-dimethylformamide, epichlorohydrin, and pyridine and treatment at 100 °C for activation of raw rice husk (Katal *et al.* 2012). To the best of our knowledge, there is very limited information on the activation of RHA waste at ambient condition without surfactants or

doi: 10.2166/wst.2018.445

high temperature treatment, and for application for nitrate adsorption.

In this study, the TRI-ARHA was investigated as a novel adsorbent for nitrate removal in the laboratory condition and compared to a commercial exchange resin (Akualite A420, China). This study could demonstrate the potential of TRI-ARHA for effective nitrate removal and the sustainable use of RHA resource in Vietnam.

MATERIALS AND METHODS

Chemicals

Chemicals such as NaOH, HF, HCl – 36 vol.% (made in China); N^1-(3-trimethoxysilylpropyl) dimethylene triamine (named triaminesilane), toluene, pentane (Merck); and DI (deionized) water were used in the adsorbent synthesis protocol. KBr and stock nitrate solution (Merck) with concentration of 1,000 mg/L were used for Fourier transform infrared (Fourier transform infrared (FTIR)) analysis and nitrate adsorption experiments, respectively. In this study, RHA was collected from a brick production company in An Giang province. Anion exchange resin (Akualite A420 with D = 0.5 mm, China) was used as a comparison to the as-synthesized TRI-ARHA (Table 1).

Generation of ARHA and TRI-ARHA

The novel support from RHA with high surface area was generated via a simple HF corrosion process worked at ambient condition without the need for surfactant or thermal treatment (Thanh 2016). In a typical procedure, 20 mg of the dried RHA were added into 80 mL of aqueous HF solution (5 vol.%). After 30 mins of stirring at the ambient condition (i.e. 1 atm and room temperature), the ARHA was obtained after washing and filtering with DI water several times and finally drying in an oven at 100 °C overnight.

Table 1 | Characteristics of anion exchange resin (Akualite A420)

Matrix	Styrene–divinylbenzene copolymer
Functional group	$-N\text{-}(CH_3)_3$
Ionic form	Cl^-
Total capacity	min 1.0 eq/L (Cl^- form)
Particle size	0.5 μm
pH_{PZC}	7.1

The incorporation of amine was achieved via surface grafting, which is simple, effective, and widely applied for loading of amine into mesoporous silica materials (Thanh 2016). A sample of ARHA was loaded into a multi-neck glass flask containing 150 mL of toluene. Once a homogeneous mixture was obtained, 0.3 mL of deionized water per 1 g of ARHA was added and stirred for 30 min. The glass flask was then submerged in a silicon oil bath set at 85 °C using a temperature-controlled stirring hotplate with an external temperature probe. Triaminesilane (3 mL per gram of ARHA) was subsequently added to the mixture and left under stirring for 16 h. The material was filtered and washed with copious amounts of toluene, then pentane. Finally, the recovered solid was dried at 100 °C in a natural convection oven for 1 h and labeled as TRI-ARHA. Before each experiment, TRI-ARHA and Akualite A420 were pretreated by immersing 1 g of each adsorbent into 1 L of 0.1 M HCl solution for 3 h, then filtrated, and dried at ambient condition overnight. The acidified TRI-ARHA was denoted as TRI-ARHA$^+$. The regeneration of the used materials was exactly the same as the pretreatment procedure.

Physicochemical characteristics

The surface chemical property of the samples was carried out using FTIR spectroscopy (Bruker-FTIR with spectrum scanned from 500 to 4,000 cm^{-1}). The thermogravimetric analysis (TGA) pattern of samples was recorded by using a TGA Q500 machine in ambient condition (100 mL/min) at rate of 10 °C/min, and temperature range of 30–1,000 °C. Wide angle X-ray diffraction (XRD) patterns were recorded for sample powder using a D2 Phaser XRD 300 W diffractometer equipped with Cu Kα radiation (λ = 1.5406 Å) at a step size of 0.01° and 30 s step time. The morphology, particle sizes and amine dispersion of RHA, ARHA, and TRI-ARHA were evaluated by scanning electron microscopy (SEM) and SEM-mapping technique. Additionally, the average composition and amine loading of TRI-ARHA were determined by energy-dispersive X-ray spectroscopy (EDX) and TGA performance, respectively. The Brunauer–Emmett–Teller (BET) surface area was determined from N_2 adsorption/desorption isotherms at 77 K (Porous Materials, BET-202A). Before BET measurement, the sample was degassed at 150 °C for 3 h to remove completely the water inside the pores of these oxides. Accordingly, the BET data shown here correspond to the annealed samples.

Batch adsorption experiments

In order to clarify the comparison for nitrate removal capacity of TRI-ARHA and Akualite A420, the initial aqueous nitrate solution was set up at pH ~7, temperature of 30 ± 0.5 °C, and nitrate concentration of 50 $mgNO_3^-$-N/L. For batch adsorption experiments, 0.03 g of the adsorbent (TRI-ARHA or Akualite A420) was added into 50 mL of aqueous nitrate solution and shaken in a thermostatic water-bath shaker operated at 120 rpm. After reaching equilibrium condition (i.e. 10 min as obtained from preliminary study), the adsorbents were isolated by centrifuging at 10,000 rpm, and nitrate concentration in the supernatant was analyzed by UV-visible spectrophotometer method (SPECORD 210 Plus, Analytik Jena, Germany) of sodium salicylate reagent at wavelength of 410 nm. All experiments were replicated three times and the adsorption capacity (defined as amount of nitrate adsorbed per mass unit of adsorbent) was calculated from the following expression:

$$q_e = \frac{C_o - C_e}{m} \times V$$

$$H\% = \frac{C_o - C_t}{C_o} \times 100$$

where q_e (mg/g) is the adsorption capacity and $H\%$ is the removal efficiency. C_o, C_t, and C_e (mg/L) are the initial, time t, and equilibrium nitrate concentrations in the aqueous solution, respectively. V (mL) is the volume of the experimental solution (50 mL) and m (mg) is the adsorbent weight.

RESULTS AND DISCUSSION

Adsorbent characterization

FTIR was used to identify the chemical composition and amine group in the structure of TRI-ARHA and anion exchange resin (Akualite A420). The FTIR spectra of RHA-based materials and Akualite A420 commercial anion exchange resin are shown in Figure 1. For ARHA material, its FTIR spectrum is more simple than that of RHA. The FTIR patterns of ARHA and RHA materials were characterized by oscillations of Si-O-Si (1,100 cm^{-1}), C=C (1,600 cm^{-1}), C=O (1,730 cm^{-1}), and –OH (3,450 cm^{-1}) (Kane *et al.* 2018; Raghuvanshi *et al.* 2018; Tatarchuk *et al.* 2018b). The peak around 3,442 cm^{-1} for strong bands of the -OH stretching vibrations was observed for both anion resin Akualite A420 and TRI-ARHA (silanol groups on ARHA surface). On the lower frequency side band for Akualite A420 resin, some peaks at wavenumbers of 3,018 cm^{-1} and 2,922 cm^{-1} were related to the stretching vibrations of the ring C-H bonds and $-CH_2$ groups of the matrix for cross-linked polystyrene structure (Sowmya & Meenakshi 2013). At 1,601 cm^{-1}, the presence of C-C bonds of the styrene ring was found (Lazar *et al.* 2014). The band at 1,481 cm^{-1} is due to symmetric and asymmetric vibrations of the methyl group of quaternary nitrogen (Wołowicz & Hubicki 2009; Gandhi *et al.* 2011). However, the peaks at the wavenumbers of 1,039 and 1,128 cm^{-1} were identified as the stretching vibrations of the benzene ring from the linked matrix of styrene–divinylbenzene in the resin structure (Lee *et al.* 2003). For TRI-ARHA material, the stretching vibration of amine group was found at wavenumber of 1,480 cm^{-1}, which was also observed for the

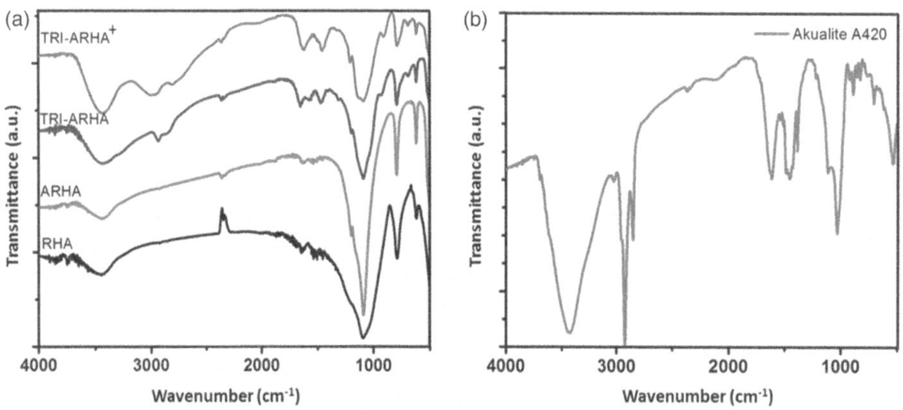

Figure 1 | FTIR spectra of (a) RHA, ARHA, TRI-ARHA, TRI-ARHA+ and (b) Akualite A420 exchange resin.

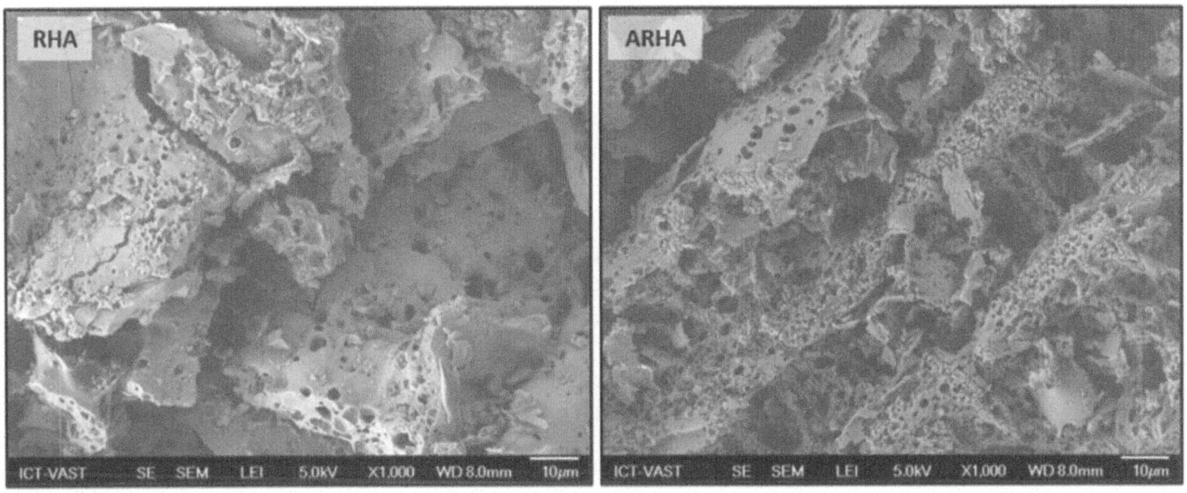

Figure 2 | SEM images of RHA and ARHA.

FTIR spectrum of Akualite A420 exchange resin. However, the FTIR spectrum of TRI-ARHA material showed the presence of vibrations of silica and organic chemicals such as Si-H (650–840 cm^{-1}), Si-O-Si (1,030–1,130 cm^{-1}), C=C (1,650 cm^{-1}), C-H (2,930 cm^{-1}), and –OH (3,420 cm^{-1}) (Ibrahim *et al.* 1980). This was evidence for the presence of amine groups on the porous SiO_2 surface of TRI-ARHA material.

SEM images of RHA and ARHA are shown in Figure 2. These images showed that there were a large number of holes in the ARHA surface and a large number of networks inside the solid structure as a result of the reaction of HF and SiO_2 component of RHA. For TRI-ARHA sample, the techniques of SEM, SEM-EDX, and SEM-mapping were carried out carefully and the results are shown in Figure 3. After amine grafting, TRI-ARHA contains O, C, Si, and N atoms with the weight percentages of 36.4, 33.6, 22.6, and 7.4 wt.%, respectively. These atoms are well dispersed on the TRI-ARHA surface. However, the crystalline structure of TRI-ARHA and

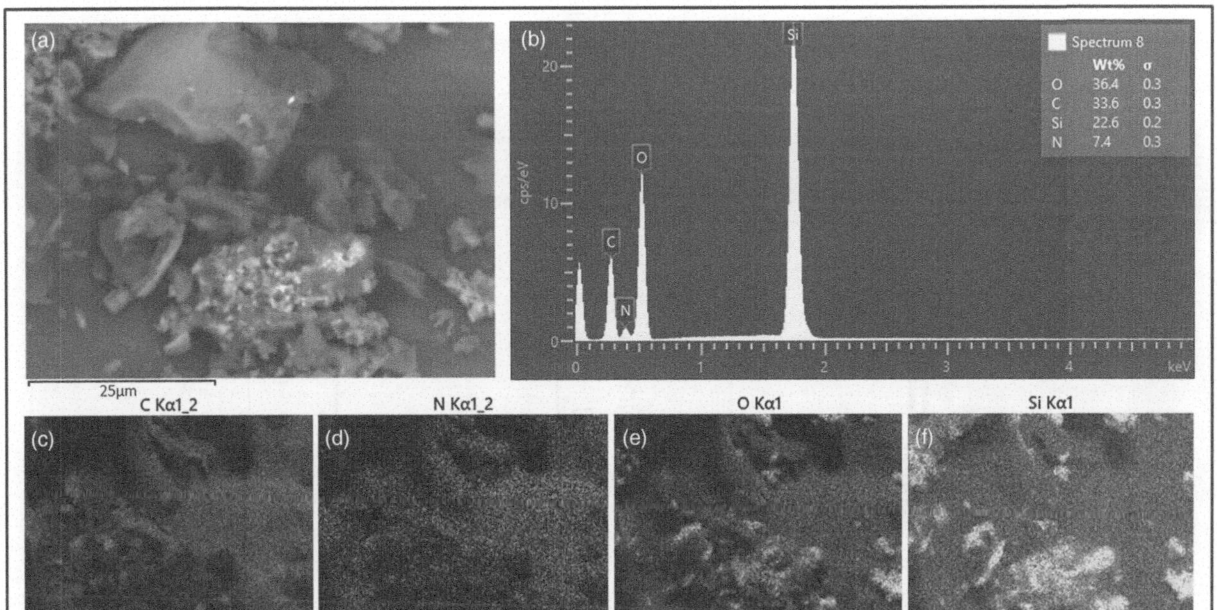

Figure 3 | SEM images (a), EDX (b), C mapping (c), N mapping (d), O mapping (e) and Si mapping (f) of TRI-ARHA.

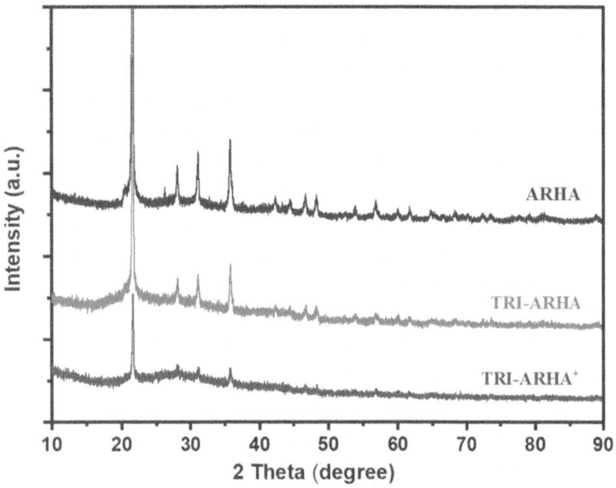

Figure 4 | XRD patterns of ARHA, TRI-ARHA and TRI-ARHA$^+$.

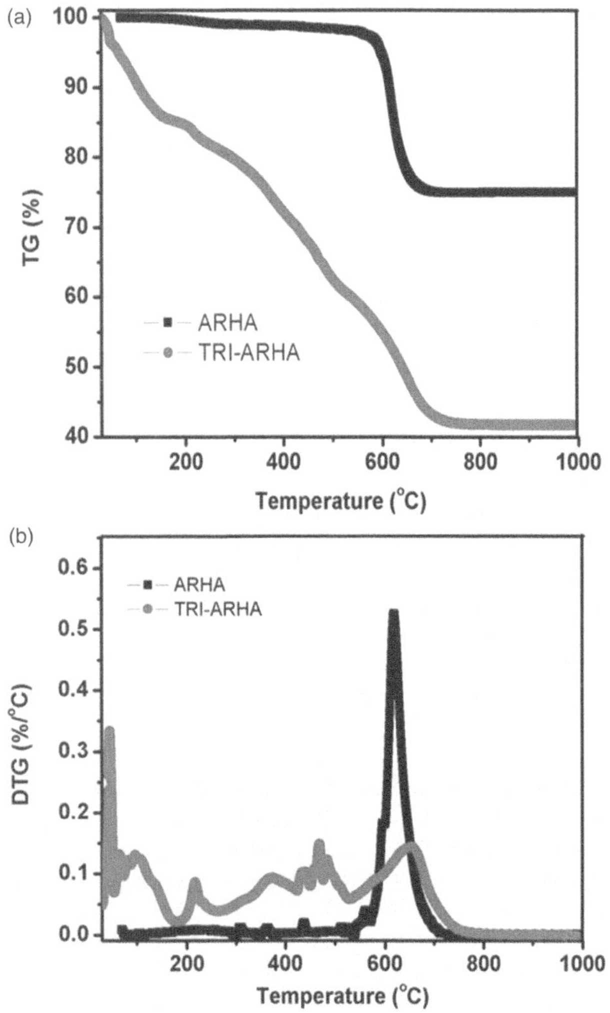

Figure 5 | (a) TGA and (b) differential TGA curves of ARHA and TRI-ARHA.

TRI-ARHA$^+$ (acidified TRI-ARHA) samples did not change as compared to the ARHA sample (Figure 4).

Thermogravimetric analysis results are shown in Figure 5 over a wide temperature range from 30 to 1,000 °C. It was obvious that the mass of sample was reduced with the increased temperature and three mass loss stages can be typically distinguished. The initial period from 30 to 200 °C had a mass loss of ~16%, which can be attributed to the release of moisture and low weight volatile compounds, with an endothermic peak at 50–100 °C. Next, the mass loss of the second stage occurred from 200 to 430 °C. At high pyrolysis temperature of the second stage, the decrease of sample with the increased temperature was mainly due to dehydration of the hydroxyl group. Most of the pyrolysis process occurred in the third stage of activation as the temperature increased from 450 to 620 °C. After this period, the weight loss of sample reached ~56%. During the last stage, various volatile components were released and loose porous char was formed. The second and third stages were used as the baseline amount for the organic moieties of sample (Shahbazi *et al.* 2011) and the amine loadings results were then calculated and are presented in Table 2. The amine loaded on the surface of ARHA was determined as ~23 wt.%.

Table 2 | BET surface area and amine content of ARHA and TRI-ARHA

Adsorbent	BET surface area (m^2/g)	Amine content (%)
ARHA	415	–
TRI-ARHA	402	~23

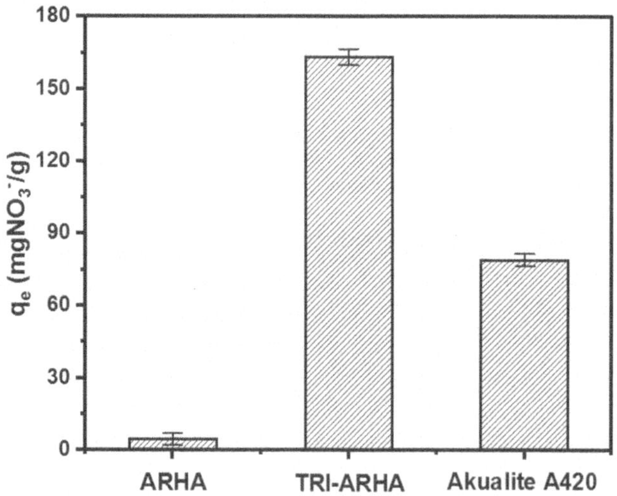

Figure 6 | Nitrate adsorption capacity of ARHA, TRI-ARHA, and Akualite A420.

Table 3 | Comparison of nitrate adsorption capacity of TRI-ARHA with other reported adsorbents

Adsorbent	Adsorption capacity (mg/g)	Conditions	Reference
Triamine-bearing activated rice husk ash	163.4	50 $mgNO_3^-$-N/L, 30 °C, 10 min, pH ~ 7	Present study
Ammonium functionalized mesoporous MCM-48 silica	30.0	300 $mgNO_3$/L, 25 °C, 60 min, pH ~ 5	Hamoudi *et al.* (2007)
Ammonium-functionalized mesoporous SBA-15 silica	55.8	300 $mgNO_3$/L, 25 °C, 60 min, pH ~ 5	Hamoudi *et al.* (2011)
Magnetic amine-crosslinked biopolymer-based corn stalk	58.5	240 $mgNO_3$/L, 25 °C, 4 h, pH ~ 7	Song *et al.* (2016)
Conditioned cross-linked chitosan beads	75.0	250 $mgNO_3$/L, 30 °C, 24 h, pH ~ 5	Chatterjee *et al.* (2009)

Nitrate removal capacity of TRI-ARHA material and its durability

The nitrate removal capacities of ARHA, TRI-ARHA, and Akualite A420 resin materials are illustrated in Figure 6. It is observed that TRI-ARHA material had the highest capacity for nitrate removal, which was 34.8 times and 2.06 times higher than that of ARHA and Akualite A420, respectively. Thus, the order of nitrate adsorption is TRI-ARHA > Akualite A420 > >ARHA. The strong interaction of Si, C, and N in the structure and the high amine-loading amount of TRI-ARHA could account for its superior nitrate adsorption as compared to commercial resin. The comparative study on nitrate adsorption capacity of TRI-ARHA with other reported adsorbent materials are shown in Table 3. It is observed that the synthesized TRI-ARHA material was superior in nitrate adsorption capacity as compared with other reported adsorbents.

The recyclability test for TRI-ARHA and Akualite A420 was conducted and the results are displayed in Figure 7. It can be seen that the nitrate adsorption capacities of these two materials were very stable after 10 cycles. With the high and stable adsorption capacities, TRI-ARHA material would be a very promising material for nitrate removal in advanced water and wastewater treatment.

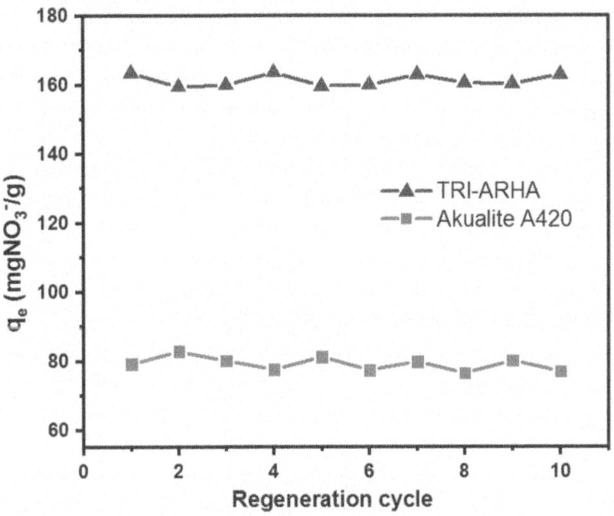

Figure 7 | Durability test of TRI-ARHA, and Akualite A420.

CONCLUSIONS

The TRI-ARHA was successfully prepared using RHA waste without the need for surfactants or high temperature for activation. The TRI-ARHA material showed much higher nitrate adsorption capacity (>160 $mgNO_3^-$/g) as compared to commercial anion exchange resin Akualite A420 (~80 $mgNO_3^-$/g), and also throughout 10 cycles of adsorption–desorption. This study also indicates that there is an effective and sustainable use of RHA since TRI-ARHA could be a very promising and potential adsorbent for removal of nitrate and other anions in advanced water and wastewater treatment. Future works could focus on the kinetics, isotherms, and effects of other anions on the adsorption of nitrate in aqueous solution.

ACKNOWLEDGEMENTS

The authors are grateful for support from Faculty of Engineering – Technology – Environment, An Giang University, Vietnam and Faculty of Environment and Natural Resources, Ho Chi Minh City University of Technology, Vietnam.

REFERENCES

Bhatnagar, A. & Sillanpää, M. 2011 A review of emerging adsorbents for nitrate removal from water. *Chemical Engineering Journal* **168** (2), 493–504.

Chatterjee, S., Lee, D. S., Lee, M. W. & Woo, S. H. 2009 Nitrate removal from aqueous solutions by cross-linked chitosan beads conditioned with sodium bisulfate. *Journal of Hazardous Materials* **166** (1), 508–513.

Gandhi, M. R., Kalaivani, G. & Meenakshia, S. 2011 Sorption of chromate and fluoride onto duolite a 171 anion exchange resin – a comparative study. *Elixir Pollution* **32**, 2034–2040.

Hamoudi, S., Saad, R. & Belkacemi, K. 2007 Adsorptive removal of phosphate and nitrate anions from aqueous solutions using ammonium-functionalized mesoporous silica. *Industrial & Engineering Chemistry Research* **46** (25), 8806–8812.

Hamoudi, S., El-Nemr, A., Bouguerra, M. & Belkacemi, K. 2011 Adsorptive removal of nitrate and phosphate anions from aqueous solutions using functionalised SBA-15: effects of the organic functional group. *The Canadian Journal of Chemical Engineering* **90** (1), 34–40.

Ibrahim, D. M., El-Hemaly, S. A. & Abdel-Kerim, F. M. 1980 Study of rice-husk ash silica by infrared spectroscopy. *Thermochimica Acta* **37** (3), 307–314.

Kane, S. N., Raghuvanshi, S., Satalkar, M., Reddy, V. R., Deshpande, U. P., Tatarchuk, T. R. & Mazaleyrat, F. 2018 Synthesis, characterization and antistructure modeling of Ni nano ferrite. *AIP Conference Proceedings* **1953** (1), 030089.

Katal, R., Baei, M. S., Rahmati, H. T. & Esfandian, H. 2012 Kinetic, isotherm and thermodynamic study of nitrate adsorption from aqueous solution using modified rice husk. *Journal of Industrial and Engineering Chemistry* **18** (1), 295–302.

Lazar, L., Bandrabur, B., Tataru-Fărmuş, R.-E., Drobotă, M., Bulgariu, L. & Gutt, G. 2014 FTIR analysis of ion exchange resins with application in permanent hard water softening. *Environmental Engineering & Management Journal (EEMJ)* **13** (9), 2145–2152.

Lee, B., Bao, L. L., Im, H.-J., Dai, S., Hagaman, E. W. & Lin, J. 2003 Synthesis and characterization of organic–inorganic hybrid mesoporous anion-exchange resins for perrhenate (ReO_4^-) anion adsorption. *Langmuir* **19** (10), 4246–4252.

Raghuvanshi, S., Kane, S. N., Tatarchuk, T. R. & Mazaleyrat, F. 2018 Effect of Zn addition on structural, magnetic properties, antistructural modeling of $Co_{1-x}Zn_xFe_2O_4$ nano ferrite. *AIP Conference Proceedings* **1953** (1), 030055.

Shahbazi, A., Younesi, H. & Badiei, A. 2011 Functionalized SBA-15 mesoporous silica by melamine-based dendrimer amines for adsorptive characteristics of Pb(II), Cu(II) and Cd(II) heavy metal ions in batch and fixed bed column. *Chemical Engineering Journal* **168** (2), 505–518.

Song, W., Gao, B., Xu, X., Wang, F., Xue, N., Sun, S., Song, W. & Jia, R. 2016 Adsorption of nitrate from aqueous solution by magnetic amine-crosslinked biopolymer based corn stalk and its chemical regeneration property. *Journal of Hazardous Materials* **304**, 280–290.

Sowmya, A. & Meenakshi, S. 2013 Removal of nitrate and phosphate anions from aqueous solutions using strong base anion exchange resin. *Desalination and Water Treatment* **51** (37–39), 7145–7156.

Tatarchuk, T., Bououdina, M., Vijaya, J. J. & Kennedy, L. J. 2016 Spinel ferrite nanoparticles: synthesis, crystal structure, properties, and perspective applications. In: *Nanophysics, Nanomaterials, Interface Studies, and Applications: Selected Proceedings of the 4th International Conference Nanotechnology and Nanomaterials (NANO2016), August 24–27, 2016, Lviv, Ukraine*. Springer, Cham, Switzerland, pp. 305–325.

Tatarchuk, T., Peter, A., Al-Najar, B., Vijaya, J. & Bououdina, M. 2018a Photocatalysis: activity of nanomaterials. In: *Nanotechnology in Environmental Science* (C. H. Hussain & A. J. Mishra, eds). Wiley-VCH, Weinheim, Germany, pp. 211–291.

Tatarchuk, T. R., Paliychuk, N. D., Bououdina, M., Al-Najar, B., Pacia, M., Macyk, W. & Shyichuk, A. 2018b Effect of cobalt substitution on structural, elastic, magnetic and optical properties of zinc ferrite nanoparticles. *Journal of Alloys and Compounds* **731**, 1256–1266.

Tatarchuk, T., Bououdina, M., Al-Najar, B. & Bitra, R. B. 2019 Green and ecofriendly materials for the remediation of inorganic and organic pollutants in water. In: *A New Generation Material Graphene: Applications in Water Technology* (M. Naushad, ed.). Springer International Publishing, Cham, Switzerland, pp. 69–110.

Thanh, N. T. 2016 Amine-bearing activated rice husk ash for CO_2 and H_2S gas removals from biogas. *KKU Engineering Journal* **43** (S3), 396–398.

Tyagi, S., Rawtani, D., Khatri, N. & Tharmavaram, M. 2018 Strategies for nitrate removal from aqueous environment using nanotechnology: a review. *Journal of Water Process Engineering* **21**, 84–95.

Ward, M. H. & Brender, J. D. 2018 Drinking water nitrate and human health. In: *Reference Module in Earth Systems and Environmental Sciences*. Elsevier, https://doi.org/10.1016/B978-0-12-409548-9.11245-X.

Wołowicz, A. & Hubicki, Z. 2009 Sorption of palladium(II) complexes onto the styrene–divinylbenzene anion exchange resins. *Chemical Engineering Journal* **152** (1), 72–79.

Xu, X., Gao, B.-Y., Yue, Q.-Y. & Zhong, Q.-Q. 2010 Preparation of agricultural by-product based anion exchanger and its utilization for nitrate and phosphate removal. *Bioresource Technology* **101** (22), 8558–8564.

First received 26 July 2018; accepted in revised form 11 October 2018. Available online 16 October 2018

Synthesis of porous pig bone char as adsorbent for removal of DBP precursors from surface water

Aunnop Wongrueng, Pharkphum Rakruam, Alongorn Siri and Adisak Siyasukh

ABSTRACT

This research study aims to investigate the efficiency of synthesized porous pig bone char (PBC) for reduction of disinfection by-product (DBP) precursors from surface water. Dissolved organic matter (DOM) is commonly present in natural water and acts as a disinfection by-product precursor. Adsorption is one of the promising technologies that is commonly applied for DOM removal. Interestingly, the properties of pig bone are such that it has a surface area and pore volumes that can adsorb DOM. Pig bone was synthesized as porous bone char (PBC). The results show that synthesized PBC at 900 °C (PBC-900 °C) provides a high volume of mesoporous structure. The adsorption process was best fitted with the pseudo-second-order and Freundlich isotherm model. Thus, the mechanisms occurred on the multilayer adsorption of the surface. PBC-900 °C can remove approximately 70–80% of DOM with varying concentrations, from 0.2 g/L to 0.8 g/L. Furthermore, the results of fluorescence excitation–emission (FEEM) showed that humic acids and humic-like substances in water can be removed by using PBC at concentrations higher than 0.4 g/L. From the obtained results, it can be concluded that PBC is an alternative low-cost adsorbent which can be utilized for reduction of DBP precursors from water.

Key words | adsorption, dissolved organic matter, pig bone, porous bone char, surface water, trihalomethane formation potentials (THMFPs)

Aunnop Wongrueng
Pharkphum Rakruam (corresponding author)
Department of Environmental Engineering, Faculty of Engineering,
Chiang Mai University,
Chiang Mai 50200,
Thailand
and
Center of Excellence on Hazardous Substance Management,
Chulalongkorn University,
Bangkok,
Thailand
E-mail: *pharkphum@eng.cmu.ac.th*

Aunnop Wongrueng
Research Program in Hazardous Substance Management in Agricultural Industry,
Center of Excellence on Hazardous Substance Management (HSM),
Bangkok 10330,
Thailand

Alongorn Siri
International Postgraduate Programs in Environmental Management, Graduate School,
Chulalongkorn University,
Bangkok,
Thailand

Adisak Siyasukh
Department of Industrial Chemistry, Faculty of Science,
Chiang Mai University,
Chiang Mai 50200,
Thailand

INTRODUCTION

Natural organic matter (NOM) is a major component in aquatic environments (Sillanpää 2015). NOM can cause many problems in drinking water because of processes such as oxidation, coagulation, disinfection, and adsorption (Jacangelo *et al.* 1995). The important compound of NOM is dissolved organic matter (DOM). DOM is generally present in several water sources (e.g. surface water, lakes, reservoirs, and ground water). High concentrations of DOM may cause severe deterioration in the water quality of drinking water because DOM is a precursor of disinfection by-products (DBPs) (Cowman & Singer 1995). Reactions between DOM and chlorine create DBPs, which include trihalomethanes (THMs) and haloacetic acids (HAAs) (Chellam & Krasner 2001). THMs include chloroform, bromodichloromethane, dibromochloromethane, and bromoform. They are classified as possible carcinogens to humans (USEPA 2004).

Several technologies, for example coagulation, ion exchange, advanced oxidation processes, membrane filtration, electrochemical methods and adsorption, have been used to remove DOM. Adsorption is one of the technologies that is widely used in drinking water purification processes due to its efficiency and convenience. Activated

doi: 10.2166/wst.2018.486

carbon is commonly used as an adsorbent for the removal of DOM because it has high efficiency (Bond *et al.* 2011). Nevertheless, the cost of using activated carbon is still high for developing countries.

In Thailand, animals are slaughtered for meeting people's demand for meat. Large amounts of animal bone become bone waste, which affects the environment. Although cattle farms use animal bone to feed their animals, the amount of animal bone waste still exceeds the demand for feeding. Interestingly, the properties of animal bone are such that it has a surface area and pore volumes that can adsorb various substances. Accordingly, many research studies have utilized animal bone as bone char to adsorb various substances including fluoride (Sawangjang 2016), arsenic (Chen *et al.* 2008), and chromium (Hyder *et al.* 2014). Nevertheless, there are few research studies on DOM removal by animal bone char. The commonly used adsorbents for DOM removal are granular activated carbon (GAC) (Wang *et al.* 2015) and powdered activated carbon (PAC) (Fabris *et al.* 2008). Thus, removal of DOM by animal bone char has not been well investigated.

Pig bone was selected as the adsorbent for the removal of DBP precursors from surface water. Because the properties of pig bone include high surface area (Sawangjang 2016), it can be considered a good characteristic adsorbent. The aim of this study was to synthesize and characterize porous pig bone char. The efficiency of porous pig bone char for removal of DBP precursors was investigated by using the adsorption kinetic and isotherm models.

MATERIALS AND METHODS

Raw surface water

A volume of 60 L of raw water was collected from the Ping River in the rainy season (2016) and used for all the research. The quantity and characteristics of the organic matter in the raw water were investigated by measuring various parameters including dissolved organic carbon (DOC), trihalomethane formation potentials (THMFPs), UV absorbance at wavelength 254 nm (UV254), and fluorescence excitation–emission matrix (FEEM).

Synthesis of porous pig bone char

The porous pig bone char (PBC) used in this study was synthesized from raw pig bone. First, raw pig bone was prepared by boiling the same at 100 °C to remove any fat that remained inside the bone. It was boiled several times until the boiling water become white and transparent, indicating that the colloid was completely released from the pig bone (Rojas-Mayorga *et al.* 2013). After that, it was rinsed with deionized water. Second, the clean pig bone was dried in an incubator at 100 °C for 24 h to eliminate moisture. Next, the dry bone was crushed by a hammer into fragments of 1–2 cm and crushed again by a mortar. Next, it was put into a ball mill for 2 days to produce small sizes of the bone. After that, it was filtered using a sieve at 250 μm, and any moisture was removed by drying in an incubator at 100 °C for 24 h. Finally, the bone char was synthesized by pyrolyzing in a horizontal tube for 2 hours under nitrogen ambient with varied temperature conditions, at 650 °C and 900 °C. The final particle size of pig bone char was lower than 256 μm. The morphology and elements of the synthesized porous pig bone char were investigated by using a scanning electron microscope and energy-dispersive X-ray spectroscopy (EDX). Furthermore, the physicochemical properties of the porous pig bone char were analyzed by various parameters including surface area (BET), point of zero charge (PZC), and surface charge density.

Adsorption study

The adsorption kinetics of the DOC by the porous PBC were studied by varying the concentrations of the PBC at 30, 60, 90, and 120 mg with 150 mL of raw surface water. It was represented at concentrations of 0.2, 0.4, 0.6, and 0.8 g/L, respectively. The experiment was conducted at room temperature and the speed of shaking was 200 revolutions per minute (rpm). Water samples were collected at various times, including 0, 1, 2, 3, 4, 5, 10, 20, 30, 40, 50, 60, 120, 180, 360, 720, and 1,440 min. In addition, the adsorption isotherm was calculated from the results obtained from the adsorption kinetics. The collected water samples were filtered through a 0.45 μm nylon syringe filter for removing the PBC. After that, the samples were analyzed for any remaining DOC.

DBP precursors removal experiment

The amount (concentration) of PBC used for this experiment was based on the results obtained from the study on the adsorption kinetics. PBC was mixed with 150 mL of water sample in an Erlenmeyer flask and shaken at 200 rpm with the temperature controlled at 25 °C until equilibrium time. The water sample was collected and measured for DOM concentration. The DOM concentration was determined by measuring the DOM surrogate parameters, including DOC concentration, UV-254, and THMFPs.

Analytical methods

All the water samples were filtered through GF/F followed by 0.45 μm nylon prior to measurement of DOM surrogate parameters. Milli-Q water (ELGA) was used on every sample for a clean system and blank sample preparation. The DOC concentration in the water samples was measured in accordance with Standard Methods 5310 total organic carbon (TOC) (APHA/AWWA/WEF 2018), AJ-analyzer multi N/C 3100, multiWin 4.09. The UV-254 of the water samples was analyzed in accordance with the Standard Methods 5910B ultraviolet absorption method (APHA/AWWA/WEF 2018) by using Perkin-Elmer Model Lambda 365, UV/Vis spectrophotometer: Lambda 365 with matched quartz cells that provided a path length of 1 mm. The THMFPs measurements were taken according to Standard Methods 5710 A-D (APHA/AWWA/WEF 2018). Phosphate solution was used as the buffer solution before incubation at 25 ± 2 °C in amber bottles with PTFE liners. At the end of the 24 h reaction period, the free chlorine remaining in the water samples should be between 3 mg/L and 5 mg/L. The residual chlorine was measured according to the Standard Methods 4500-Cl G (APHA/AWWA/WEF 2018). The THMs were extracted with MTBE in accordance with EPA551 (USEPA 1995). Agilent 6890 Gas Chromatograph with an electron capture detector was used for measuring the THMs. The FEEM was analyzed using a spectrofluorometer with excitation wavelengths in the range of 220 nm to 600 nm.

RESULTS AND DISCUSSION

Water quality

The average pH of the Ping River was 8.4 and the average turbidity was 51.37 NTU. The concentration of the DOC was 4.492 mg/L. The THMFPs of the Ping River is 1,446.26 μg/L, which indicates that the DOM in the Ping River had greatly reacted with chlorine to form THMs. Among the THM species, chloroform was found to be at the highest concentration (1,344.09 μg/L), which corresponds well with the findings of various research studies (Rodriguez *et al.* 2004; Krutklom 2013). Hence, removal of DOM is necessary before it reacts with chlorine during the disinfection procedure in the drinking water purification process.

Morphology and characteristics of porous pig bone char

The morphology of the PBC which was pyrolyzed at 650 °C (PBC-650 °C) and the morphology of the PBC which was pyrolyzed at 900 °C (PBC-900 °C) are shown in Figure 1. The results of the morphology of both the PBC samples showed that they had smooth surfaces.

The elements of PBC including carbon (C), oxygen (O), phosphorus (P), calcium (Ca), and nitrogen (N) were analyzed, and the results are shown in Table 1. From the results, it was evident that the carbon remaining in PBC-650 °C was higher than that in PBC-900 °C. This can be taken as indicating that pyrolysis at higher temperatures can remove more carbon in pig bone. The carbon element

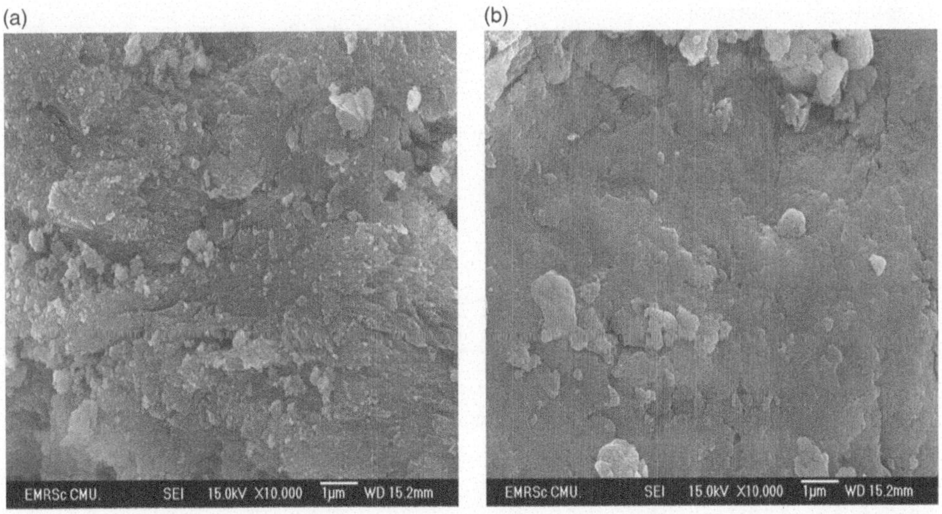

Figure 1 | The morphology of porous pig bone char pyrolyzed at (a) 650°C and (b) 900°C.

Table 1 | Results of energy-dispersive X-ray (EDX) spectroscopy of porous pig bone char at 650 °C

	PBC – 650 °C		PBC – 900 °C	
Elements	**Weight%**	**Atomic%**	**Weight%**	**Atomic%**
C	8.67	15.35	8.05	13.48
O	42.15	56.03	2.18	3.13
P	15.82	10.87	48.06	60.41
Ca	33.37	17.72	13.9	9.03
N	–	–	27.81	13.95
Total	100		100	

in the pig bone represents the organic content in the pig bone char (Leyva-Ramos *et al.* 2010).

Physicochemical characteristics of porous pig bone char adsorbents

The N_2 adsorption–desorption isotherms of both the PBC samples were investigated. Based on the IUPAC classification, both of the PBC samples were type IV and had mesoporous structures, which means large pore volumes and high surface areas, and these were defined as good properties for an adsorbent (Patel *et al.* 2015). The BET surface area and the volume of the mesoporous structure are summarized in Table 2. The results show that PBC-900 °C had lower BET surface area than PBC-650 °C ; however, PBC-900 °C should be utilized as an alternative adsorbent as the volume of its mesoporous structure was greater than that of PBC-650 °C. The mesoporous structure provides a large pore volume and high surface area, which are defined as good properties for an adsorbent, and it exhibits good electrochemical performance (Patel *et al.* 2015; Tian *et al.* 2018). An adsorption process depends on the compatibility of molecular size and pore width of the adsorbates and the adsorbents, respectively; a small molecule prefers the micropore volume (pore width <2 nm in diameter), and a larger molecule favors the mesopore volume (2≤ pore width ≤50 nm). Although the micropore is responsible for increasing the specific surface area, the absorption of large molecules in a liquid medium depends on the porosity of the mesopore range rather than the micropore (Newcombe 1997; Zimmerman *et al.* 2004; Aschermann *et al.* 2018; Siyasukh *et al.* 2018). Moreover, the target pollutant in this study is the DOM, for which the molecular size is normally between 1 and 7 nm on average (Kennedy & Summers 2015). For this reason, the criteria for selecting an absorbent material should be determined by the mesopore volume rather than the surface area or micropore volume. It is, therefore, more suitable to choose PBC-900 °C, whose ratio of mesopore to micropore volume is 7.7, for use as the adsorbent in this study, rather than PBC-600 °C, for which the ratio is 6.0.

Table 2 | Physicochemical characteristics of porous pig bone char

Adsorbent	BET surface area (m²/g)	Vmic_DR (cm³/g)	Vmeso (cm³/g)
PBC-650 °C	100.63	0.0400	0.2400
PBC-900 °C	78.00	0.0323	0.2490

The point of zero charge (PZC) expresses the pH value that affects the surface charge of the adsorbent. The PZC of the porous bone char was found to be approximately in the pH range of 6.5–8.5. If the pH of the solution was above the PZC of the porous pig bone char (>8.5), the surface of the porous pig bone char was considered to be negatively charged. On the other hand, if the pH of the solution was below the PZC of the porous pig bone char (<6.5), the surface of the porous pig bone char was considered to be positively charged. The PZC of the porous pig bone char was similar to that of PAC (poly aluminium chloride), which is generally used as the adsorbent in water treatment processes. Prarat (2011) found that the PZC and the surface charge density values of PAC were 9.5 and in the range of 8–10, respectively.

Adsorption kinetics

Based on the results regarding the morphology and the characteristics of PBC, the PBC synthesized at 900 °C (PBC-900 °C) was used as the adsorbent for the DOM experiment. The results of the adsorption kinetics of the DOC of PBC-900 °C are shown in Figure 2, where q_t represents the amount of DOC adsorbed at time.

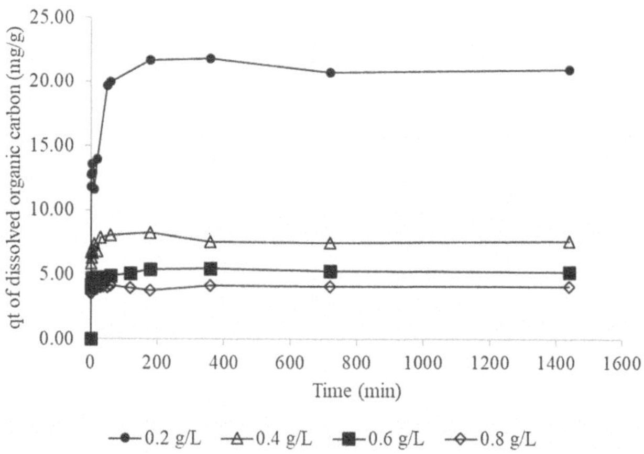

Figure 2 | The adsorption kinetics of DOC at various concentrations of PBC-900 °C.

The results presented in Figure 2 show that the adsorption was fast within the short contact time. After 1 h, the adsorption of the PBC-900 °C adsorbent regularly slowed until it reached the equilibrium. The equilibrium time of adsorbent dosage at 0.2 g/L was 12 hours. For the other absorbent dosages, the equilibrium time was 6 hours. Table 3 shows that the adsorption capacity values of DOC at equilibrium times were 19.97, 8.505, 5.589, and 4.420 mg/g at adsorbent concentrations of 0.2, 0.4, 0.6 and 0.8 g/L, respectively.

The pseudo-first-order and the pseudo-second-order were applied to examine the adsorption kinetics of the DOM onto the porous pig bone char adsorbents, as shown in Table 3. The constant (k) values of the pseudo-first-order were calculated from the slope of each regression line. Besides, the square of the slope divided by the intercept of the regression line was calculated for the constant (k) of the pseudo-second-order kinetic model. In this study, the R^2 (correlation coefficient) value was considered for defining the kinetic model that is the best fit for the adsorbents.

The results showed that the R^2 (the correlation coefficient) value of the pseudo-second-order kinetic model was higher than the R^2 value of the pseudo-first-order model for all conditions. So, it can be concluded that the adsorption kinetics of PBC-900 °C were a fit with the pseudo-second-order model. It presented chemisorption (Wang *et al.* 2007), which is an electrostatic interaction between DOC and PBC-900 °C adsorbents. The porous bone char used in this study has a hydroxyapatite structure $Ca_{10}(PO_4)_6(OH)_2$. It can be used as a host material for ion exchange. So, the DOM, which has an anion charge, can be adsorbed. This finding is similar to those of various research studies that used bone char adsorbents and reported that it was the best fit to the pseudo-second-order kinetic model (Rojas-Mayorga *et al.* 2013; Medellin-Castillo *et al.* 2014; Cazetta *et al.* 2016; Sawangjang 2016).

From Table 3, it is clear that the initial adsorption rate (h) of PBC-900 °C followed the following order: PBC-0.2 g/L > PBC-0.4 g/L > PBC-0.6 g/L > PBC-0.8 g/L. The high h values indicate the fast adsorption process. Thus, the result indicates that the adsorption process involves not only the amount of adsorbent used but also the electrostatic interaction and the interparticle diffusion of adsorbent and DOM (Permrungruang 2013).

The study of the adsorption isotherm of PBC-900 °C involved investigation by varying the concentration (0.2, 0.4, 0.6, and 0.8 g/L). The R^2 values of the Langmuir and Freundlich models of porous bone char at 900 °C were investigated. The adsorption isotherm results of the Langmuir and the Freundlich models are shown in Table 4. The R^2 values of the Langmuir and the Freundlich models were 0.8802 and 0.9575, respectively. Obviously, the R^2 value of the Freundlich isotherm model is higher than the R^2 value of the Langmuir isotherm model. Hence, it can be concluded that PBC-900 °C is better fitted to the Freundlich isotherm model and that the adsorption process occurs via multilayer adsorption on the surface (Jimenez-Reyes & Solache-Rios 2010).

Reduction of DOC, UV-254, and FEEM

The efficiency of PBC-900 °C for DOM reduction was investigated by measuring various parameters including DOC, THMFP, and FEEM. The results of DOC reduction by porous bone char are shown in Figure 3.

The results showed that the DOC concentration was strongly adsorbed on varying the concentration of PBC-900 °C adsorbent as 0.2, 0.4, 0.6, and 0.8 g/L. The increase of the PBC-900 °C provided higher DOC reduction, but it was not significantly different for the different concentrations. The DOC concentration in raw water was 4.492 mg/L. In the adsorption experiment, the DOC that remained after adsorption by PBC-900 °C at 0.2, 0.4, 0.6, and 0.8 g/L was 1.310, 1.090, 0.957, and 0.954 mg/L, respectively. The percentages of DOC reduction of PBC-900 °C at 0.2, 0.4, 0.6, and 0.8 g/L were higher than 70%. The result was compared with DOC reduction by

Table 3 | Kinetic parameters of DOC adsorption on porous pig bone char at 900 °C

		Pseudo-first-order			Pseudo-second-order			
PBC (g/L)	**Qe, exp (mg/g)**	**Qe, cal mg/g**	**K1, min^{-1}**	**R^2**	**Qe, cal mg/g**	**K2, min^{-1}**	**R^2**	**h (mg/g.min)**
0.2	19.97	2.49	−0.0005	0.1227	20.05	0.0475	0.9995	18.94
0.4	8.505	4.05	−0.0003	0.1566	8.64	0.1317	0.9998	9.53
0.6	5.589	0.58	0.0002	0.095	5.31	0.1902	0.9999	5.94
0.8	4.420	0	0.0006	0.0607	4.10	0.2443	0.9999	4.77

Table 4 | Isotherm parameters of DOM adsorption on porous pig bone char at 900 °C

Model	Parameter	PBC
Langmuir	R^2	0.8802
	q_m, mg/g	15.91
	K_L, L/mg	0.4876
Freundlich	R^2	0.9575
	n	2.26
	K_F, L/g	3.6199

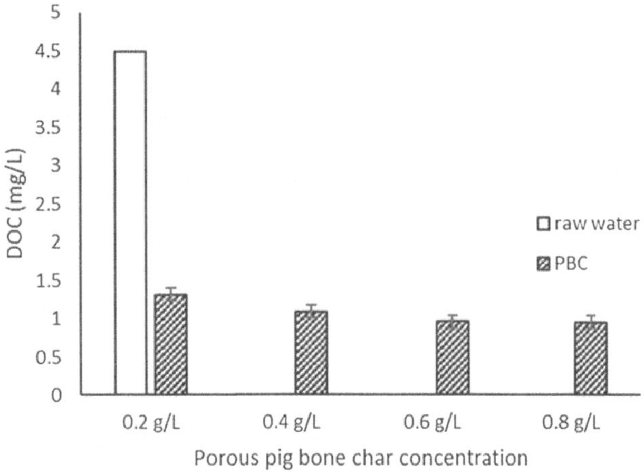

Figure 3 | Reduction of DOC by porous bone char at different concentrations.

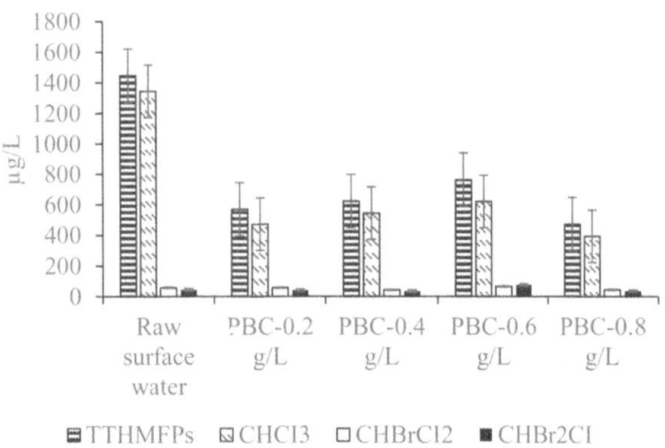

Figure 4 | Reduction of THMFPs by porous bone char at different concentrations.

commercial PAC, studied by Wudthigarn (2015). It was found that the use of synthesized PBC-900 °C provided greater DOC reduction than commercial PAC (43.3%).

The THMFPs in raw surface water and after adsorption by PBC-900 °C are shown in Figure 4. The THMFPs of raw water was high, at 1,446.26 μg/L. After the adsorption experiment, it was observed that the values of THMFPs had decreased to 569.87, 621.73, 762.12, and 473.06 μg/L for PBC-900 °C at 0.2, 0.4, 0.6, and 0.8 g/L, respectively. Chloroform was found in the highest concentration among the THM species. The concentration of chloroform in raw water was 1,344.09 μg/L. It was found to have decreased to 472.23, 543.84, 620.43, and 392.82 μg/L after adsorption by PBC-900 °C at 0.2, 0.4, 0.6, and 0.8 g/L, respectively. The results showed that the THMFPs greatly decreased due to the reduction in the DBPs precursor DOC. The DBPs precursor is an important factor that affects THMFPs (Hua & Reckhow 2007; Bond *et al.* 2009). The efficiency of porous pig bone char to reduce THMFPs was found to be in the range of 47.3% to 67.3%. When compared with the efficiency of commercial PAC, as studied by Wudthigarn (2015), it was found that the efficiency of commercial PAC for THMFPs reduction was 70%.

The characteristics of DOM in raw water and treated water by PBC-900 °C were investigated by using the FEEM technique. The result of the FEEM method using raw water from the Ping River is illustrated in Figure 5. The result showed that two peaks of DOM (A and B) were found in raw water. The results after adsorption by porous bone char at different concentrations are shown in Figure 5. The results presented in Figure 5 show that two peaks of DOM (A and B) were found when PBC-900 °C was used at 0.2 g/L. In contrast, the fluorescent peak was not detected in the other porous bone char concentrations (0.4, 0.6, and 0.8 g/L). Based on the PAC adsorption, the fluorescent peak was not detected after the adsorption process (Wudthigarn 2015).

CONCLUSIONS

From the results of the morphology and the characteristics of synthesizing porous pig bone char, it can be concluded that PBC-900 °C not only provides a mesoporous structure but also gives a higher volume of mesopores than PBC-650 °C. The pH of the PZC of the porous pig bone char was similar to that of the commercial PAC adsorbent. Furthermore, the adsorption process was best fitted with the pseudo-second-order and the adsorption capacity at equilibrium times (19.97 mg/g within 1 h) was found at the concentration of 0.2 g/L of porous pig bone char. In addition, the Freundlich isotherm model was observed to give the highest R^2 values. So, the mechanisms could be concluded as occurring in the form of multilayer adsorption on the surface.

From the DBPs precursor reduction results, the efficiency of the DBPs precursor reduction was found to be in the range of 70–80% with varying concentrations of PBC-900 °C (0.2 g/L to 0.8 g/L). Chloroform was the major

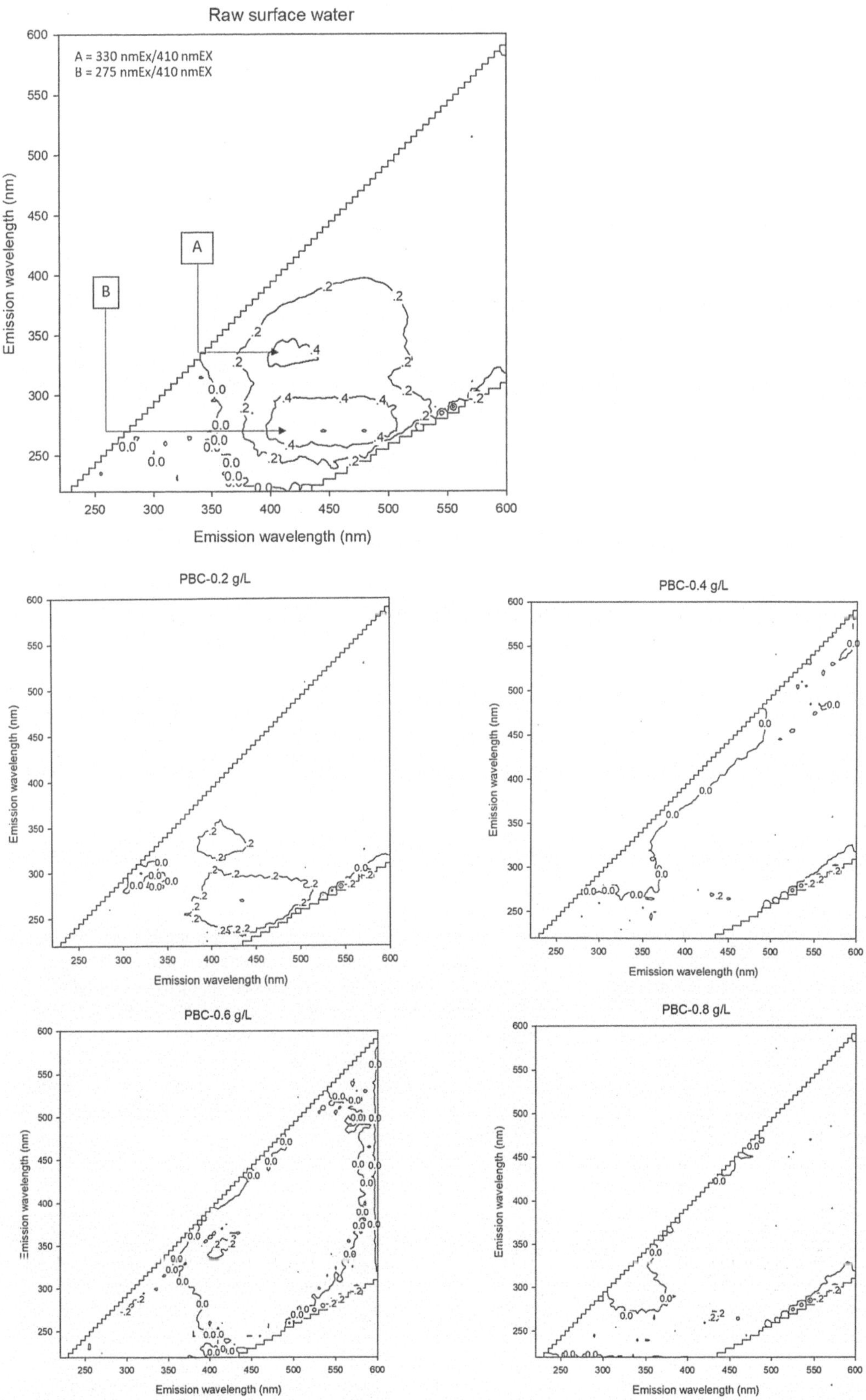

Figure 5 | FEEM of raw water and treated water at different concentrations of PBC-900 °C.

species in the THMFPs, both in raw water and treated water. The THMFPs were found to have significantly decreased by PBC-900 °C adsorption, at 47.3% to 67.3%. From the results obtained from FEEM, it can be concluded that using 0.2 g/L concentration of PBC-900 °C cannot remove the humic acids and the humic-like substances in water. On the other hand, increasing the porous bone char concentrations to 0.4, 0.6, and 0.8 g/L could remove those substances. From the obtained results, it can be concluded that synthesized porous pig bone char is an alternative low-cost adsorbent which can be utilized for reduction of DBP precursors from water.

ACKNOWLEDGEMENTS

The authors would like to thank the Center of Excellence on Hazardous Substance Management (HSM), Chulalongkorn University, Bangkok, Thailand, for the financial support.

REFERENCES

APHA, AWWA & WEF 2018 *Standard Methods for the Examination of Water and Wastewater*, 23rd edn. APHA, Washington, DC, USA.

Aschermann, G., Zietzschmann, F. & Jekel, M. 2018 Influence of dissolved organic matter and activated carbon pore characteristics on organic micropollutant desorption. *Water Research* **133**, 123–131.

Bond, T., Goslan, E. H., Jefferson, B., Roddick, F., Fan, L. & Parsons, S. A. 2009 Chemical and biological oxidation of NOM surrogates and effect on HAA formation. *Water Research* **43** (10), 2615–2622.

Bond, T., Goslan, E. H., Parsons, S. A. & Jefferson, B. 2011 Treatment of disinfection by-product precursors. *Environ. Technol.* **32** (1–2), 1–25.

Cazetta, A. L., Martins, A. C., Pezoti, O., Bedin, K. C., Beltrame, K. K., Asefa, T. & Almeida, V. C. 2016 Synthesis and application of N–S-doped mesoporous carbon obtained from nanocasting method using bone char as heteroatom precursor and template. *Chemical Engineering Journal* **300**, 54–63.

Chen, Y. N., Chai, L. Y. & Shu, Y. D. 2008 Study of arsenic (V) adsorption on bone char from aqueous solution. *Journal of Hazardous Materials* **160** (1), 168–172.

Chellam, S. & Krasner, S. W. 2001 Disinfection byproduct relationships and speciation in chlorinated nanofiltered waters. *Environ. Sci. Technol.* **35** (19), 3988–3999.

Cowman, G. A. & Singer, P. C. 1995 Effect of bromide ion on haloacetic acid speciation resulting from chlorination and chloramination of aquatic humic substances. *Environ. Sci. Technol.* **30** (1), 16–24.

Fabris, R., Chow, C. W. K., Drikas, M. & Eikebrokk, B. 2008 Comparison of NOM character in selected Australian and Norwegian drinking waters. *Water Research* **42** (15), 4188–4196.

Hua, G. & Reckhow, D. A. 2007 Characterization of disinfection byproduct precursors based on hydrophobicity and molecular size. *Environ. Sci. Technol.* **41** (9), 3309–3315.

Hyder, A. H., Begum, S. A. & Egiebor, N. O. 2014 Sorption studies of Cr (VI) from aqueous solution using bio-char as an adsorbent. *Water Sci Technol* **69** (11), 2265–2271.

Jacangelo, J., DeMarco, J., Owen, D. & Randtke, S. 1995 Selected processes for removing NOM: an overview. *Am. Water Works Assoc.* **87** (1), 64–77.

Jimenez-Reyes, M. & Solache-Rios, M. 2010 Sorption behavior of fluoride ions from aqueous solutions by hydroxyapatite. *Journal of Hazardous Materials* **180** (1–3), 297–302.

Kennedy, A. M. & Summers, R. S. 2015 Effect of DOM size on organic micropollutant adsorption by GAC. *Environ. Sci. Technol.* **49** (11), 6617–6624.

Krutklom, N. 2013 *Dissolved Organic Matter Reduction Using Ceramic Membrane Combined with Alum*. Master Thesis, Environmental Engineering, Chiang Mai University, Chiang Mai, Thailand.

Leyva-Ramos, R., Rivera-Utrilla, J., Medellin-Castillo, N. A. & Sanchez-Polo, M. 2010 Kinetic modeling of fluoride adsorption from aqueous solution onto bone char. *Chemical Engineering Journal* **158** (3), 458–467.

Medellin-Castillo, N. A., Leyva-Ramos, R., Padilla-Ortega, E., Ocampo Perez, R., Flores-Cano, J. V. & Berber-Mendoza, M. S. 2014 Adsorption capacity of bone char for removing fluoride from water solution. Role of hydroxyapatite content, adsorption mechanism and competing anions. *Journal of Industrial and Engineering Chemistry* **20** (6), 4014–4021.

Newcombe, G. 1997 Influence of characterised natural organic material on activated carbon adsorption: II. Effect on pore volume distribution and adsorption of 2-methylisoborneol. *Water Research* **31** (5), 1065–1073.

Patel, S., Han, J., Qiu, W. & Gao, W. 2015 Synthesis and characterisation of mesoporous bone char obtained by pyrolysis of animal bones, for environmental application. *Journal of Environmental Chemical Engineering* **3** (4), 2368–2377.

Permrungruang, J. 2013 *Effects of Surface Functional Groups and Natural Organic Matter on Clofibric Acid Adsorption by Mesoporous Silicate SBA-15*. Master Thesis, *Environmental Management (Interdisciplinary Program)*, Chulalongkorn University, Bangkok, Thailand

Prarat, P. 2011 Adsorption characteristics of haloacetonitriles on functionalized silica-based porous materials in aqueous solution. *Journal of Hazardous Materials* **192** (3), 1210–1218.

Rodriguez, M. J., Serodes, J. & Levallois, P. 2004 Behavior of trihalomethanes and haloactic acids in a drinking water distribution system. *Water Research* **38** (20), 4367–4382.

Rojas-Mayorga, C. K., Bonilla-Petriciolet, A., Aguayo-Villarreal, I. A., Hernandez-Montoya, V., Moreno-Virgen, M. R., Tovar-Gomez, R. & Montes-Moran, M. A. 2013 Optimization of pyrolysis conditions and adsorption properties of bone char for fluoride removal from water. *Journal of Analytical and Applied Pyrolysis* **104**, 10–18.

Sawangjang, B. 2016 *Adsorption Mechanism of Bone Chars for Defluoridation of Groundwater.* Master Thesis, Environmental Management, Chulalongkorn University, Bangkok, Thailand.

Sillanpää, M. 2015 *Natural Organic Matter in Water: Characterization and Treatment Methods.* IWA Publishing, London, UK.

Siyasukh, A., Chimupala, Y. & Tonanon, N. 2018 Preparation of magnetic hierarchical porous carbon spheres with graphitic features for high methyl orange adsorption capacity. *Carbon* **134**, 207–221.

Tian, C., Zhao, S. & Lu, Q. 2018 Self-templated synthesis of mesoporous manganese silicates as an electrode material for supercapacitor. *Ceramics International* **44** (14), 17007–17012.

USEPA 1995 Determination of Chlorination Disinfection Byproducts, Chlorinated Solvents, and Halogenated Pesticides/Herbicides in Drinking Water by Liquid-Liquid Extraction and Gas Chromatography with Electron-Capture Detection, Revision 1.0. USEPA, Cincinnati, OH, USA.

USEPA 2004 *Disinfection by-Products in Drinking Water Treatment Systems.* https://cfpub.epa.gov/si/si_public_record_report.cfm?dirEntryId = 18333 (accessed 22 July 2018).

Wang, H., Zhou, A., Peng, F., Yu, H. & Yang, J. 2007 Mechanism study on adsorption of acidified multi-walled carbon nanotubes to Pb(II). *Journal of Colloid Interface Science* **316** (2), 277–283.

Wang, J., Tian, H. & Ji, Y. 2015 Adsorption behavior and mechanism of humic acid on aminated magnetic nanoadsorbent. *Separation Science and Technology* **50** (9), 1285–1293.

Wudthigarn, C. 2015 *Removal of Carbonaceous and Nitrogenous Organic Matters From Surface Water by Activated Carbon Adsorption and Ceramic Membrane Filtration Processes.* Master Thesis, Environmental Engineering, Chiang Mai University, Chiang Mai, Thailand.

Zimmerman, A. R., Goyne, K. W., Chorover, J., Komarneni, S. & Brantley, S. L. 2004 Mineral mesopore effects on nitrogenous organic matter adsorption. *Organic Geochemistry* **35** (3), 355–375.

First received 24 July 2018; accepted in revised form 13 November 2018. Available online 21 November 2018

Adsorption mechanism of dichlorvos onto coconut fibre biochar: the significant dependence of H-bonding and the pore-filling mechanism

Quach An Binh and Puangrat Kajitvichyanukul

ABSTRACT

The adsorption mechanism of dichlorvos onto coconut fibre biochar (CFB) was investigated by the batch adsorption technique. Coconut fibre waste material was synthesised at 600 °C for 4 h under oxygen-limited conditions. The biochar was modified by HCl acid to enhance the specific surface area and porosity. The characteristics of the biochar were analysed by scanning electron microscopy (SEM), Brunauer–Emmett–Teller (BET) specific surface area, and Fourier transform-infrared (FT-IR). The results showed that the BET specific surface area of biochar was 402.4 m^2/g. Experimental data presented a good fit to Langmuir isotherm and the pseudo-second-order model. Langmuir isotherm illustrated that monolayer adsorption of dichlorvos occurred on the surface of CFB, with a maximum adsorption capacity of 90.9 mg/g. The diffusion model confirmed that the liquid film diffusion was the rate-limiting step, and the major diffusion mechanism of dichlorvos onto biochar. The BET result after dichlorvos adsorption demonstrated that pore-filling occurred and occupied 58.27%. The pore-filling and chemical interactions, performed important roles in the adsorption of dichlorvos onto CFB. Chemical adsorption is comprised of two interactions, which are hydrophobic and H-bonding, but the prime is H-bonding. CFB is a very potential material for the removal of dichlorvos and environmental pollutants.

Key words | adsorption, biochar, dichlorvos, diffusion, mechanism, pesticide

Quach An Binh
Puangrat Kajitvichyanukul (corresponding author)
Center of Excellence on Environmental Research and Innovation, Faculty of Engineering,
Naresuan University,
Phitsanulok 65000,
Thailand
E-mail: *puangratk@nu.ac.th*

Quach An Binh
On leave from Dong Nai Technology University,
Dong Nai,
Vietnam

INTRODUCTION

Dichlorvos (2,2-dichlorovinyl dimethyl phosphate) is an organophosphate insecticide that was generally used to control flies, mosquitoes, ticks, cockroaches, clover mites, crickets, cutworms, grasshoppers, and webworms. This insecticide was applied to exterminate insects in domestic, agricultural and industrial areas (Edwards 2006).

Dichlorvos can act as a contact and stomach poison by way of skin penetration, swallowing and respiration (Richter & Corcoran 1997). Dichlorvos was classified by the Environmental Protection Agency (EPA) as highly toxic compound and assigned class I-toxicity. Human contact with dichlorvos over a short period of time can generate symptoms such as megrim, nausea, blurred vision, regurgitation, excessive sweating, and pupil constriction. Chronic exposure to dichlorvos can lead to narcotism or even death (Sahithya *et al.* 2015).

Biochar has been widely applied in the adsorption method. Biochar obtained rich carbon that was synthesised from the biomass pyrolysis process with limited oxygen. Biochar has a great potential to efficiently adsorb contaminants in wastewater. Biomass is available to produce biochar at a lower cost. Additionally, biochar has some special properties that include a large surface area, a porous structure, functional groups, and mineral components (Saleh *et al.* 2016).

After the pyrolysis process, minerals appeared on the surface and blocked the surface pores of the biochar. Hence, acid washing treatment is very necessary to enhance the specific surface area of biochar (Uchimiya *et al.* 2012). If the main contributions on the surface area of the biochar are micropores (<2 nm) and small mesopores (2–20 nm), these pore sizes were suggested to be the primary factor to participate in the uptake of the organic compounds by pore-filling adsorption (Inyang & Dickenson 2015). There were several interactions that occurred between biochar and the organic compounds, including hydrophobic adsorption,

doi: 10.2166/wst.2018.529

H-bonding, electrostatic, π–π electron-donor–acceptor (EDA) interactions, and cation–π bonding (Inyang & Dickenson 2015).

Recently, several adsorbents were used in the adsorption method to resolve the dichlorvos pollutant such as biopolymer modified MMT–CuO composites (Sahithya *et al.* 2015), Super paramagnetic poly (styrene-co-acrylic acid) hydrogel (Alka & Anita 2014), TiO_2 (Lu *et al.* 1996), and activated carbon from groundnut shell (Gimba *et al.* 2010). However, the information of the dichlorvos removal by biochar was rare and the adsorption mechanism of dichlorvos onto the biochar was not reported. This inspirited us to explore the adsorption mechanism between dichlorvos and biochar.

The purpose of this study was to investigate the adsorption mechanism of dichlorvos and modified coconut fibre biochar (CFB). The CFB was synthesised under oxygen-limited conditions and modified by HCl acid. The characteristics of CFB were investigated using different analytical techniques including Brunauer–Emmett–Teller (BET) method, scanning electron microscope (SEM), and Fourier transform-infrared (FT-IR). The batch adsorption method was applied to study the kinetics adsorption model, the adsorption isotherm, the intraparticle diffusion model, and the diffusion mechanism of dichlorvos and CFB. Especially, the adsorption mechanism between the CFB and dichlorvos pollutant were explored and explained in depth.

MATERIALS AND METHODS

Chemicals and materials

In this study, dichlorvos (50% w/v) was provided by Big Giant Company, Thailand. HCl (36.46%, ACI Labscan, Thailand). The dilution of all the solutions was carried out using deionized water that had a resistivity of 18 MΩ.cm. The CFB to produce biochar was collected in Northern Thailand. This material was chosen in this work due to its characteristic as fibrous material with high specific surface area (402.4 m2/g) and porous structure.

Synthesis and washing of CFB

The CFB was chopped into cubiform of approximately 1 cm^2. In order to extrude the low moisture content that was in the CFB, it was dried in an oven at 105 °C for a few hours until a constant weight was achieved. Subsequently, the biomass was transferred to a ceramic crucible and burnt in a furnace (Nabertherm, Germany). The operating temperature in the furnace was slowly increased by 3 °C per minute (Krzesińska & Zachariasz 2007). The pyrolysis process was conducted under oxygen-limited conditions at 600 °C for 4 h.

To perform the characterization and the adsorption experiments of the biochar, a pestle and mortar was used to crush the CFB, then the biochar was passed through a 0.8 mm sieve. The 0.1 M HCl acid was used to remove minerals from the CFB according to the method of Uchimiya *et al.* (2012).

Characterization of biochar

The surface morphology of the biochar was explored by SEM (LEO 1455 VP model, Carl Zeiss Microscopy GmbH, UK). The BET method was used to determine the specific surface area (SSA) and total pore volume (TPV) of the CFB (TriStar II 3020, Micromeritics Instrument Corporation, USA). The functional groups of biochar were explored by the FT-IR method (Frontier, PerkinElmer, Germany). The Boehm titration method was used to determine the surface chemistry of the biochar (Chun *et al.* 2004; Mukherjee *et al.* 2011). Mathematical formulae according to Goertzen *et al.* (2010) were used to calculate its quantity. The batch experiment was used to determine the pHpzc of biochar and followed the method of Tan *et al.* (2016).

Adsorption experiment

The batch adsorption method was applied to the dichlorvos adsorption experiment. The stock of dichlorvos solution was 1,000 mg/L and diluted in deionized water, because the dichlorvos hydrolysis shows a pH higher than 4, HCl acid was used to adjust the pH of the solution ($pH \leq 4$) to control the hydrolysis of the dichlorvos (Oncescu & Oancea 2007). The solutions used in the experiment were in the range of 10–30 mg/L and were diluted from the stock dichlorvos solution. The adsorption experiments were conducted using 250 mL sealed conical glass, and the dosage of CFB was 1.5 g/L.

In this experiment, Erlenmeyer flasks were shaken (MS Orbital Shaker, MS-NOR-30/MS-NOR-3001) for 6 h at 120 rpm in dark conditions. For the duration of the adsorption experiment, the filtrate of dichlorvos solution was passed through Whatman filter paper 42 and the residual concentration was analysed by UV-Vis spectrometer

(Genesys 10S UV-Vis spectrophotometer, Thermo Scientific, USA) at a wavelength of 210 nm (Rahman & Muneer 2005).

Mathematical models of adsorption kinetics and adsorption isotherm

The adsorption kinetics and adsorption isotherm were applied to determine the experimental data:

$$q_t = \frac{V\,(C_0 - C_t)}{W} \tag{1}$$

$$q_e = \frac{V\,(C_0 - C_e)}{W} \tag{2}$$

where q_t (mg/g) and q_e (mg/g) are the amount of dichlorvos adsorption at t and equilibrium time, respectively; C_0 and C_e (mg/L) are the initial and the equilibrium dichlorvos concentration, respectively; W (g) is the mass of biochar; V (L) is the volume of the dichlorvos solution (Liu *et al.* 2015a; Tan *et al.* 2016).

Two kinetic models, the pseudo-first-order (PFO) (Equation (3)) and pseudo-second-order (PSO) (Equation (4)) were applied to analyse the adsorption kinetics of dichlorvos and CFB. The PFO and PSO showed mononuclear and binuclear sorption, respectively (Tan *et al.* 2016), in the equations shown below:

$$\ln\,(q_e - q_t) = \ln q_e - k_1 t \tag{3}$$

$$\frac{t}{q_t} = \frac{1}{k_2 q_e^2} + \frac{1}{q_e}\,t \tag{4}$$

where k_1 (h^{-1}) and k_2 ($g.mg^{-1}.h^{-1}$) are the sorption rate constants of the pseudo-first–order and the pseudo-second-order models, respectively (Tan *et al.* 2016).

To evaluate the adsorption capacities of the dichlorvos on the CFB, Langmuir and Freundlich isotherm models were used to calculate the adsorption experiment data.

Langmuir adsorption isotherms presented monolayer sorption onto a homogeneous surface and are shown in Equation (5)

$$q_e = \frac{Q_0\, K_L\, C_e}{1 + K_L\, C_e} \tag{5}$$

where Q_0 (mg/g) is the maximum monolayer coverage capacity; K_L (L/mg) is the Langmuir isotherm constant (Tan *et al.* 2016).

R_L is the separation factor (Equation (6)) and is calculated from the Langmuir isotherm. This value showed the favourable adsorption between adsorbent and adsorbate ($R_L > 1$: unfavourable, $R_L = 1$: linear, $0 < R_L < 1$: favourable, and $R_L = 0$: irreversible) (Essandoh *et al.* 2017).

$$R_L = \frac{1}{(1 + K_L\, C_0)} \tag{6}$$

Freundlich adsorption isotherm was applied to define the heterogeneous adsorption on the surface, Freundlich adsorption isotherms are presented in Equation (7).

$$q_e = K_f\, C_e^{\frac{1}{n}} \tag{7}$$

where K_f ($mg.g^{-1}$) ($mg.L^{-1}$)$^{-n}$ is the Freundlich isotherm constant and is related in the adsorption capacity; n is the intensity of the adsorption (Tan *et al.* 2016).

Diffusion study

The film diffusion model and the particle diffusion model were applied to identify the rate-limiting step of the adsorption process. The film diffusion model expresses the liquid film that surrounds the biochar surface. The particle diffusion model expresses the diffusion of dichlorvos molecules that enter the biochar pores (Liu *et al.* 2016).

The mathematical model of the particle diffusion model shows as follows:

$$-\ln\,(1 - X^2(t)) = \frac{D_p \pi^2}{r_0^2} t = k_p t \tag{8}$$

where $X(t) = q_t/q_e$; D_p (m^2/s) is the particle diffusion coefficient; r_0 (m) is the biochar radius; k_p (min^{-1}) is the effective intraparticle diffusivity.

$$kp = \frac{D_p \pi^2}{r_0^2} \tag{9}$$

The k_p (min^{-1}) was the slope on the graph, and was made by plotting–ln (1-X^2 (t)) versus t.

The mathematical model of the film diffusion model is as follows:

$$-\ln(1 - X(t)) = k_f.t \tag{10}$$

where k_f (min^{-1}) was the slope on the graph, and was made by plotting the –ln (1-X(t)) versus t; D_f (m^2/s) is the film diffusion coefficient.

$$D_f = \frac{k_f r_0 \delta C_r}{3C_e} \tag{11}$$

C_e (mg/L) is the equilibrium concentration of the dichlorvos in the solution, and C_r (mg/L) is the dichlorvos concentration that is adsorbed by the biochar. δ (10^{-5} m) is the thickness of the liquid film (Liu *et al.* 2016).

RESULTS AND DISCUSSION

Characterization of the CFB

The SEM micrograph of modified CFB synthesised at 600 °C for 4 h is shown in Figure 1(a). The surface morphology of the CFB had tiny heterogeneous and longitudinal pores with diameters in the range of 2.90–10.48 μm. These structures likely manifest a high capacity adsorption of biochar with pollutants, whereas minerals were found on the surface of a non-modified CFB (Figure 1(b)). The presence of minerals on the biochar surface can cause blocking of the pores, and decreases the specific surface area of the biochar (Uchimiya *et al.* 2012). This phenomenon would significantly affect their adsorption ability. Therefore, mineral removal is very important to improve the porosity of biochar (Uchimiya *et al.* 2012).

The adsorption capacity of biochar depended significantly on the specific surface area. The specific surface area (BET) of CFB was 402.4 m^2/g, the large specific surface area of biochar interfaced well with the adsorption of pollutants (Tan *et al.* 2016). The TPV is one of the main factors for the high adsorption of biochar. The TPV and pore volume distributions of the CFB were calculated using the Barrett–Joyner–Halenda (BJH) model. The pore volume distributions in pore size of CFB of micropores (<2 nm) and narrow mesopores (2–20 nm) were 17.40, 69.22%, respectively. The pore size of biochar is very important in achieving the uptake of adsorbate (Wang *et al.* 2018).

The values of the surface functional groups of the CFB were estimated by the Boehm titrations method. The CFB surface functional groups are 0.375 mmol/g carboxyl, 0.250 mmol/g lactonic, 0.625 mmol/g phenolic, 1.250 mmol/g total acidity, and 0.285 mmol/g total basicity. Both acid and base functional groups acted as efficient adsorption sites for the removal of adsorbate. The pHpzc of biochar, determined by the batch experiment, was 0.7.

The FT-IR spectrum of the CFB are indicated in Figure 2 which manifests the presence of eight main functional groups on the surfaces of the CFB. The band at 1,217 cm^{-1} represented the aromatic C-O stretching of phenolic hydroxyl (Chun *et al.* 2004). The band at 1,070 cm^{-1} was assigned to the aromatic C-O stretching of alcohol (Saffari *et al.* 2015). Peaks around 878 and 750 cm^{-1} were ascribed to the aromatic C-H out-of-plane bending mode (Tran *et al.* 2017). The 2,915 and 2,845 cm^{-1} peak area represented C-H stretching of the alkyl group (Kinney *et al.* 2012). The 3,000 to 3,600 cm^{-1} peak areas, and one that peaked at 3,789 were assigned to the hydroxyl groups (O-H) stretching vibrations of carboxylic acid, alcohol, and phenol (Tran *et al.* 2017). The peak at 1,715 cm^{-1} was assigned to the carbonyl group C=O, and could show ketone, carboxylic acid, and ester (Liu *et al.* 2015a). The bands at 2,354 and 2,299 cm^{-1} were ascribed ketone group C=O (Nuithitikul *et al.* 2010). The band at

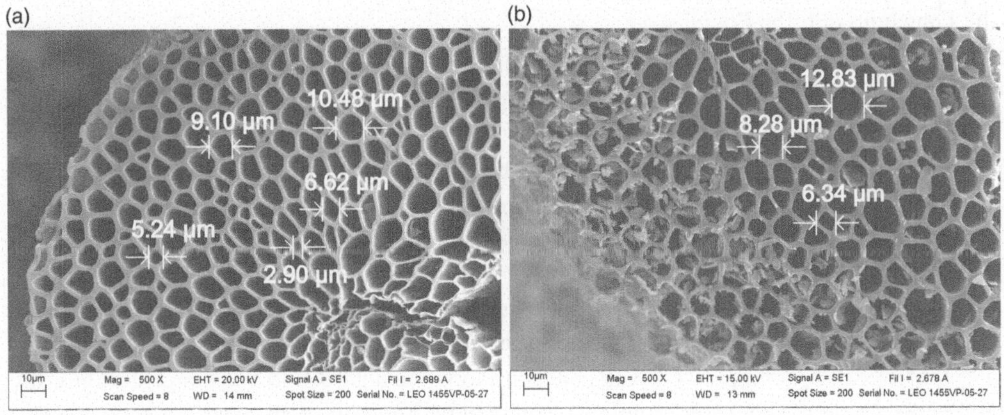

Figure 1 | SEM picture of coconut fibre biochar synthesised at 600 °C for 4 h: (a) non-modified, (b) HCl modified.

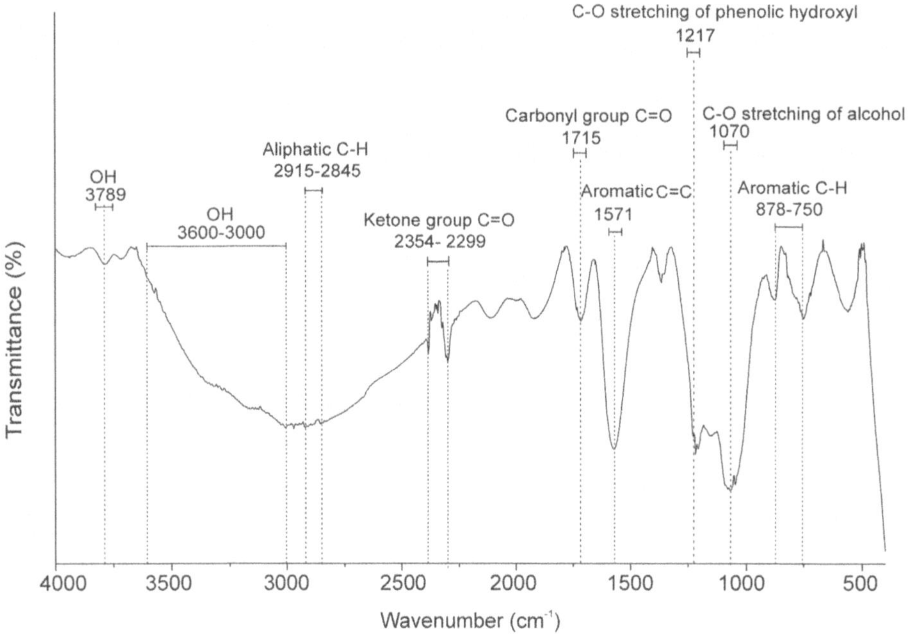

Figure 2 | FT-IR spectrum analysis of coconut fibre biochar.

1,571 cm^{-1} was ascribed to the aromatic C = C (Saffari *et al.* 2015). The functional groups of the CFB are hydroxyl, ketone, carbonyl, C-O stretching of phenolic hydroxyl and alcohol, can form an H-bonding interaction with pollutants (Liu *et al.* 2015a). Additionally, the alkyl groups of biochar with hydrophobic properties (Kinney *et al.* 2012) and can cause the hydrophobic interaction with pollutants.

Adsorption isotherm models of dichlorvos adsorbed on CFB

The adsorption isotherms of dichlorvos and CFB was explored in this study. Table 1 showed that the Langmuir, and the Freundlich models fitted the experimental data. However, the Langmuir model was more pertinent than the Freundlich model. The R^2 of the Langmuir, and the Freundlich models were 0.992 and 0.987, respectively. This demonstrated that monolayer adsorption of dichlorvos occurred on the surface of CFB. From the data indicated in Table 1, the R_L value was 0.0004, and proved to be the favourable adsorption of CFB and dichlorvos (Essandoh *et al.* 2017).

Table 1 | Isotherm parameters for dichlorvos adsorption onto coconut fibre biochar

Parameters of Langmuir isotherm				Parameters of Freundlich isotherm		
Q_0 (mg/g)	K_L (L/mg)	R_L	R^2	K_F (mg g^{-1}) (mg L^{-1})$^{-n}$	1/n	R^2
90.9	0.09	0.0004	0.992	7.99	0.843	0.987

The kinetics adsorption models of dichlorvos adsorption onto CFB

The kinetic parameters of CFB and dichlorvos were assessed including the adsorption capacity, the rate constants, and R^2, and are shown in Table 2. The q_e value calculated from the PFO was significantly dissimilar to the q_e value of the experimental data. The q_e values that were calculated from the PSO were confirmed as fitting the experimental data, and the R^2 values of the PSO were 0.999 to 1.000, whereas the R^2 values of the PFO were lower than the R^2 values of the PSO, which were 0.806 to 0.998. The two values of q_e and R^2 that were analysed above, the adsorption kinetic of dichlorvos was adsorbed onto CFB, totally fitted the PSO.

Effects of the initial dichlorvos concentration on the adsorption process of CFB

The rate of the adsorption process depended on the initial adsorbate concentration. The correlation of the initial dichlorvos concentrations and the adsorption rate constant of the PSO showed an inverse relation. Table 2 demonstrates the adsorption rate constant k_2 (PSO) of dichlorvos concentration at 10 mg/L was faster than the other dichlorvos concentrations. At high dichlorvos concentrations, the intense competition between them fixed the activated surface sites on the biochar, leading to slow diffusion (Weng

Table 2 | Kinetic parameters for dichlorvos adsorption onto coconut fibre biochar

		Pseudo-first-order parameters			Pseudo-second-order parameters		
Initial concentration (mg/L)	**q_e, exp. (mg/g)**	**q_e, calc. (mg/g)**	**k_1 (h^{-1})**	**R^2**	**q_e, calc. (mg/g)**	**k_2 (g mg^{-1} h^{-1})**	**R^2**
10	6.15	0.57	0.85	0.998	6.17	4.40	1.000
15	9.25	0.72	0.72	0.806	9.30	2.95	1.000
20	12.28	1.05	0.60	0.916	12.34	1.87	1.000
25	15.23	1.71	0.58	0.944	15.33	1.12	0.999
30	18.12	2.16	0.48	0.951	18.24	0.81	0.999

et al. 2009). The high initial concentration is the reason that causes a low rate constant k_2 (PSO) and the high rate constant k_2 is the result of a low concentration.

In addition, the experimental data illustrated that the adsorption capacities of dichlorvos were increased, when the initial concentration increased (Table 2 and Figure 3) leading the driving force of the concentration gradient was strengthened, it generated a higher adsorption capacity (Weng *et al.* 2009). Therefore, increasing the initial dichlorvos concentrations brought about the increase in the adsorption capacity.

Intraparticle diffusion (IPD) model of dichlorvos adsorbed onto CFB

To study the diffusion process of dichlorvos adsorption onto CFB, the IPD diffusion model was applied to this experiment. The IPD model signifies the rate of the adsorption process, which depends on the diffusion rate of the adsorbate to move towards the adsorbent (Hafshejani *et al.*

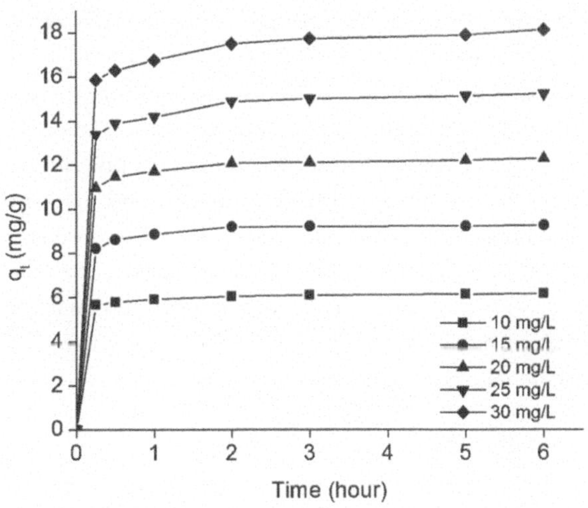

Figure 3 | Adsorption capacities of coconut fibre biochar with dichlorvos (the dosage of biochar = 1.5 g/L, pH = 2).

2016). The IPD model (Weber-Morris model) is used to describe the empirically explored functional relationship of q_t versus $t^{1/2}$. The IPD model is presented below.

$$q_t = k_{id}t^{1/2} + C_i \tag{12}$$

where k_{id} ($mg.g^{-1}.h^{-1/2}$) includes k_{id1}, k_{id2}, k_{id3} and are the diffusion rates. k_{id1}, k_{id2}, and k_{id3} are the rate constant of the external surface, the gradual adsorption, and the equilibrium adsorption stage, respectively. C_i (mg/g) is the intercept, and shows the thickness of the boundary layers effect (Lu *et al.* 2015).

In this study, the parameters of k_{id1}, k_{id2}, k_{id3} for dichlorvos adsorption onto CFB are indicated in Table 3. Figure 4 (q_t versus $t^{1/2}$) of dichlorvos adsorption onto CFB was not a linear line over the whole of the time range, showing the event of more than one kinetic stage in the dichlorvos adsorption process. From the IPD graph of CFB, the dichlorvos adsorption process consists of three stages: the instant adsorption (1), the slow adsorption (2), and the equilibrium adsorption stage (3). In the three stages, the second and third participated in the IPD process (Lu *et al.* 2015). Figure 4 indicated the linear lines of stages 2 and 3 did not pass

Table 3 | Intraparticle diffusion model data for the dichlorvos adsorption onto coconut fibre biochar

The intraparticle diffusion model			
Dichlorvos concentration	**Instant adsorption stage k_{id1} ($mg.g^{-1}.h^{-1/2}$)**	**Slow adsorption stage k_{id2} ($mg.g^{-1}.h^{-1/2}$)**	**Equilibrium adsorption stage k_{id3} ($mg.g^{-1}.h^{-1/2}$)**
10 mg/L	11.30	0.37	0.07
15 mg/L	16.41	0.80	0.05
20 mg/L	21.91	0.87	0.24
25 mg/L	26.79	1.42	0.32
30 mg/L	31.73	1.72	0.52

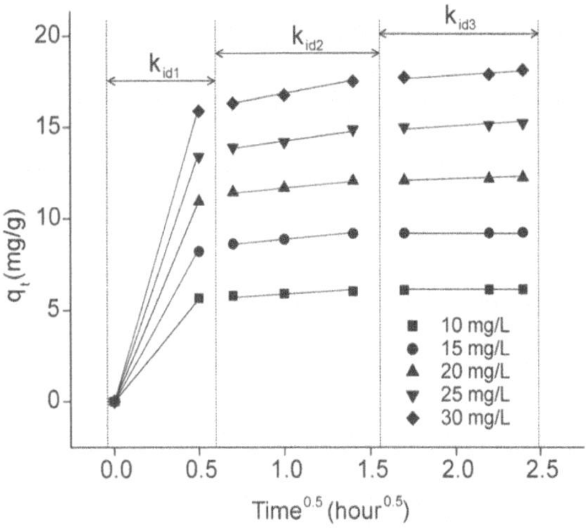

Figure 4 | Intraparticle diffusion graph shows the different dichlorvos concentrations onto coconut fibre biochar (k_{id1}: the instant adsorption stage, k_{id2}: the slow adsorption stage, k_{id3}: the equilibrium adsorption stage).

through the origin, which proved that the rate of the entire dichlorvos adsorption process was not controlled by the intraparticle diffusion (Lu *et al.* 2015).

In the various concentration of the dichlorvos, the k_{id1} were higher than the k_{id2} and k_{id3} (Table 3). This demonstrates that the rate constant of the instant adsorption (k_{id1}) was faster than the rate constant of the gradual adsorption stage (k_{id2}) and the equilibrium adsorption stage (k_{id3}) of the IPD. This determined that the IPD was involved in the dichlorvos adsorption process. However, the IPD was not the only rate-limiting step of the whole adsorption process of dichlorvos. The rate-limiting step of the diffusion mechanism, is a major contribution in the adsorption of dichlorvos and CFB, and was followed by film diffusion, which is clarified in the next section.

Diffusion study

The CFB particle radius (r_0) used in this study was in the range of 100–270 μm. Hence, the average radius of the CFB particles were applied to the particle diffusion and the film diffusion models, was 185 μm.

From the formulae of the film diffusion, and the particle diffusion models, the D_f and D_p values were evaluated to explore the rate-limiting step of dichlorvos adsorption onto CFB. The D_f values of the film diffusion model in this experiment were in the range of 1.01×10^{-12}–8.42×10^{-12} (m^2/s) and D_p values of the particle diffusion model were in the range of 6.01×10^{-13}–1.16×10^{-12} (m^2/s). The particle and film diffusion were the rate-limiting steps, when the D_p values of the particle diffusion model were 10^{-15} to 10^{-18} (m^2/s), and the D_f values of the film diffusion model were 10^{-10} to 10^{-12} (m^2/s) (Michelsen *et al.* 1975). The comparisons show that the values of D_f in this experiment were in the range of the D_f values that Michelsen suggested, but the values of D_p were not. Therefore, the rate-limiting step of the dichlorvos adsorption process onto CFB was the film diffusion. The film diffusion was the main contribution and the major mechanism of the dichlorvos adsorption process onto CFB.

The adsorption mechanism of CFB and dichlorvos

The adsorption mechanism of biochar and organic pollutants comprises of H-bonding, hydrophobic, electrostatic, π-π EDA interactions, and pore-filling (Tan *et al.* 2015). From the evidence in this study, the intensities of FT-IR on the CFB surface and TPV in the CFB before and after the dichlorvos adsorption were different. The proposed adsorption mechanism of the CFB and dichlorvos consists of the pore-filling mechanism and chemical adsorption (Figure 6).

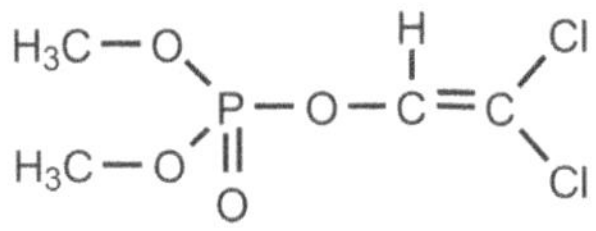

(The molecular structure of dichlorvos; length × width × height: 0.73 × 0.26 × 0.61 nm.)

The CFB had a high distribution of pore volume in micropores (<2 nm) and narrow mesopores (2–20 nm), which were in the range of 17.40 and 69.22%, respectively. The geometry of the molecular structure of dichlorvos was calculated by using the ChemBio3D software (Liu *et al.* 2015b) with the result of 0.73 × 0.26 × 0.61 nm. The molecular structure size of dichlorvos was in the range of the pore size (micropores and narrow mesopores) of the CFB. Therefore, dichlorvos can be adsorbed into the micropores and narrow mesopores in CFB by pore-filling mechanism. The pore-filling mechanism performed when the pore size of biochar is close to the molecular structure size of adsorbate (Lian *et al.* 2016). The micropores and narrow mesopores of biochar were the major factors for diffusion of adsorbate (Wang *et al.* 2018).

In this research, the SSA and the TPV results of the CFB showed before and after dichlorvos adsorption changes. Before and after adsorption, the SSA of the CFB was 402.4 m^2/g and 161.8 m^2/g, and the TPV of the CFB was 0.151 and 0.063 cm^3/g, respectively. The TPV of the

CFB after dichlorvos adsorption was significantly decreased by 58.27%. The reduction of the SSA and the TPV of the CFB after dichlorvos adsorption, the pore size of biochar and the molecular size of dichlorvos are explained above. This indicates that the pore-filling mechanism was performed in the adsorption process between the CFB and dichlorvos.

The formation of the chemical adsorption was formed by the interactions of the functional groups of biochar and the dichlorvos molecules. Based on the results of FT-IR of the CFB before and after dichlorvos adsorption (Figure 5 and Table 4), suggest that the chemical interaction of the CFB and dichlorvos consisted of H-bonding, and hydrophobic interactions.

For the formation of H-bonding, the shift of FT-IR results after dichlorvos adsorption were observed and presented in Figure 5 and Table 4. The FT-IR changes at the peak where C-O stretching of alcohol (1,070 cm^{-1}), C-O stretching of phenolic hydroxyl (1,217 cm^{-1}), O-H of alcohol, phenol, and carboxyl (3,600–3,000 and 3,789 cm^{-1}) suggested that hydrogen in these functional groups of biochar, participated in H-bonding interaction with oxygen from dichlorvos. According to a report by Essandoh *et al.* (2017), carboxylic acid, phenolic group of biochar can act as a hydrogen donor or acceptor, which can perform the H-bonding interaction with the oxygen of pollutants. Additionally, oxygen atoms from the phosphates group of dichlorvos carry a negative charge (Gamoke *et al.* 2009). Hence, H-bonding interaction could be formed between hydrogen from the functional groups of biochar, including carboxylic acid, alcohol, phenolic groups, and oxygen from the phosphate group ($P = O$) of dichlorvos.

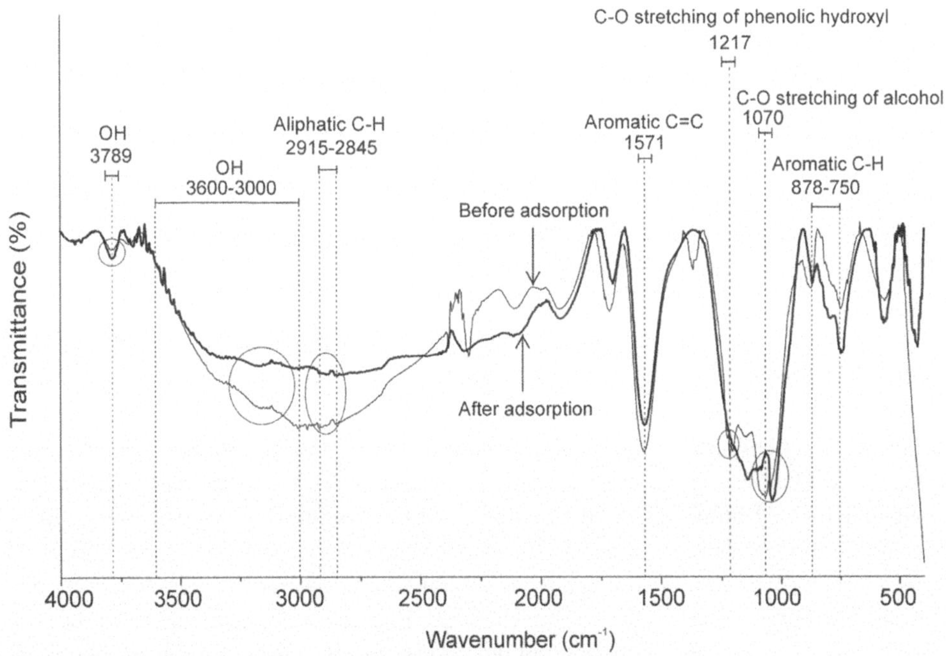

Figure 5 | FT-IR analysis of coconut fibre biochar before and after dichlorvos adsorption.

Table 4 | Intensity change (% transmittance) of FT-IR of coconut fibre biochar before and after dichlorvos adsorption

Wavenumber (cm^{-1})	Ascribed to the type of vibration	The functional group	The intensity changed (% Transmittance)	The proposed interactions
1,070	C-O	C-O stretching of alcohol	0.84	H-bonding
1,217	C-O	C-O stretching of phenolic hydroxyl	0.50	H-bonding
2,915–2,845	C-H	Aliphatic C-H	0.92	Hydrophobic
3,600–3,000 and 3,789	O-H	Alcohol, phenol, carboxylic acid	1.04	H-bonding

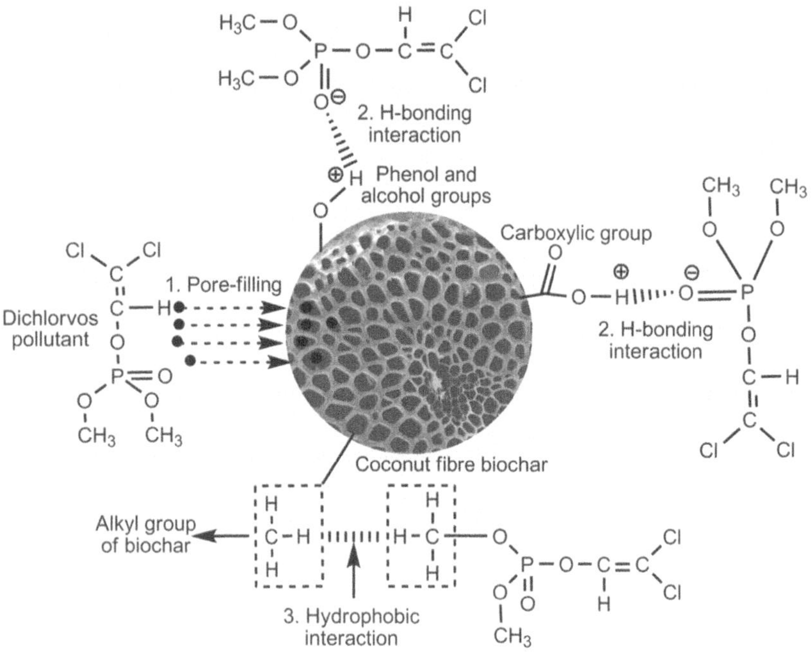

Figure 6 | Suggested adsorption mechanism of the coconut fibre biochar and dichlorvos.

The intensity of the peak areas were 2,915–2,845 cm^{-1} that changed after dichlorvos adsorption, they were ascribed to the alkyl group. The alkyl groups of biochar presented hydrophobic properties (Kinney *et al.* 2012). Hence, the hydrophobic bond can be formed from alkyl side chains (aliphatic) from dichlorvos, and alkyl groups from the CFB.

From the FT-IR results, the functional groups in CFB included alcohol, phenol, and carboxylic acid group which participated in H-bonding interaction. H-bonding interaction occurs between the O atom and H atom. H atom have a positive charge and can attach to O atom which have a higher electronegativity. The O atom bearing negative charge (Gamoke *et al.* 2009) of dichlorvos can have H-bonding interaction with H atom bearing positive charge of phenol, alcohol, and carboxylic groups of biochar. From surface functional groups of the CFB estimated by the Boehm titrations method and the transformation intensity in FT-IR after the dichlorvos adsorption, several functional groups of biochar, including, alcohol, phenol, and carboxylic acid groups, showed the H-bonding interaction, whereas only the alkyl group of biochar presented the hydrophobic interaction (Figure 6). Therefore, the H-bonding interaction was the major chemical interaction. To conclude, the adsorption mechanism between the CFB and dichlorvos consists of the pore-filling sorption mechanism and the chemical interaction. H-bonding was the primary interaction that played an important role in the chemical adsorption.

CONCLUSIONS

Acid washing removed minerals from the biochar surface and achieved the perfect porosity of the CFB. Adsorption isotherm of dichlorvos and the CFB was followed by the Langmuir isotherm model. The adsorption kinetics data were a good fit to the pseudo-second-order model. The liquid film model was the rate-limiting step of the adsorption process. The pore-filling and the chemical interaction were the main adsorption mechanism of dichlorvos onto the CFB. The chemical interaction included H-bonding and the hydrophobic interactions, however, the main contributor was H-bonding. Based on the overall results of this study, CFB had a high efficiency in the removal of dichlorvos. It is a very useful material for the removal of environmental pollutants.

ACKNOWLEDGEMENTS

This work was sponsored by the Thailand Research Fund (Grant No. BRG6180009) and the National Research Council of Thailand (Grant No. R2562B020). The authors express their deepest gratitude to the Faculty of Engineering, and the editing services of Mr Peter Barton (Naresuan University Language Centre, Naresuan University, Thailand).

REFERENCES

Alka, T. & Anita, B. 2014 Effective removal of pesticide (dichlorvos) by adsorption onto super paramagnetic poly (styrene-co-acrylic acid) hydrogel from water. *International Research Journal of Environment Sciences* **3** (11), 41–46.

Chun, Y., Sheng, G., Chiou, C. T. & Xing, B. 2004 Compositions and sorptive properties of crop residue-derived chars. *Environment Science Technology* **38**, 4649–4655.

Edwards, D. 2006 *Interim Reregistration Eligibility Decision (IRED) Document for Dichlorvos (DDVP)*, Report EPA 738-R-06-013, United States Environmental Protection Agency, Washington, DC, USA, 20460.

Essandoh, M., Wolgemuth, D., Pittman Jr, C. U., Mohan, D. & Mlsna, T. 2017 Phenoxy herbicide removal from aqueous solutions using fast pyrolysis switchgrass biochar. *Chemosphere* **174**, 49–57.

Gamoke, B., Neff, D. & Simons, J. 2009 Nature of PO bonds in phosphates. *Journal of Physical Chemistry A* **113** (19), 5677–5684.

Gimba, C. E., Salihu, A. A., Kagbu, J. A., Turoti, M., Itodo, A. U. & Sariyya, A. I. 2010 Study of pesticide (dichlorvos) removal from aqueous medium by *Arachis hypogaea* (groundnut) shell using GC/MS. *World Rural Observations* **2** (1), 1–9.

Goertzen, S. L., Thériault, K. D., Oickle, A. M., Tarasuk, A. C. & Andreas, H. A. 2010 Standardization of the Boehm titration. Part I. CO_2 expulsion and endpoint determination. *Carbon* **48** (4), 1252–1261.

Hafshejani, L. D., Hooshmand, A., Naseri, A. A., Mohammadi, A. S., Abbasi, F. & Bhatnagar, A. 2016 Removal of nitrate from aqueous solution by modified sugarcane bagasse biochar. *Ecological Engineering* **95**, 101–111.

Inyang, M. & Dickenson, E. 2015 The potential role of biochar in the removal of organic and microbial contaminants from potable and reuse water: a review. *Chemosphere* **134**, 232–240.

Kinney, T. J., Masiello, C. A., Dugan, B., Hockaday, W. C., Dean, M. R., Zygourakis, K. & Barnes, R. T. 2012 Hydrologic properties of biochars produced at different temperatures. *Biomass and Bioenergy* **41**, 34–43.

Krzesińska, M. & Zachariasz, J. 2007 The effect of pyrolysis temperature on the physical properties of monolithic carbons derived from solid iron bamboo. *Journal of Analytical and Applied Pyrolysis* **80** (1), 209–215.

Lian, F., Cui, G., Liu, Z., Duo, L., Zhang, G. & Xing, B. 2016 One-step synthesis of a novel N-doped microporous biochar derived from crop straws with high dye adsorption capacity. *Journal of Environmental Management* **176**, 61–68.

Liu, N., Charrua, A. B., Weng, C. H., Yuan, X. & Ding, F. 2015a Characterization of biochars derived from agriculture wastes and their adsorptive removal of atrazine from aqueous solution: a comparative study. *Bioresource Technology* **198**, 55–62.

Liu, H., Zhang, J., Ngo, H. H., Guo, W., Wu, H., Guo, Z., Cheng, C. & Zhang, C. 2015b Effect on physical and chemical characteristics of activated carbon on adsorption of trimethoprim: mechanisms study. *RSC Advances* **5**, 85187–85195.

Liu, N., Zhu, M., Wang, H. & Ma, H. 2016 Adsorption characteristics of Direct Red 23 from aqueous solution by biochar. *Journal of Molecular Liquids* **223**, 335–342.

Lu, M. C., Roam, G. D., Chen, J. N. & Huang, C. P. 1996 Adsorption characteristics of dichlorvos onto hydrous titanium dioxide surface. *Water Research* **30** (7), 1670–1676.

Lu, X., Shao, Y., Gao, N. & Ding, L. 2015 Equilibrium, thermodynamic, and kinetic studies of the adsorption of 2,4-Dichlorophenoxyacetic acid from aqueous solution by MIEX resin. *Journal of Chemical and Engineering Data* **60**, 1259–1269.

Michelsen, D. L., Gideon, J. A., Griffith, G. P., Pace, J. E. & Kutat, H. L. 1975 *Removal of Soluble Mercury From Waste Water by Complexing Techniques*. US Department of Industry, Office of Water Research and Technology, USA, Vol. Bulletin 74.

Mukherjee, A., Zimmerman, A. R. & Harris, W. 2011 Surface chemistry variations among a series of laboratory-produced biochars. *Geoderma* **163** (3), 247–255.

Nuithitikul, K., Srikhun, S. & Hirunpraditkoon, S. 2010 Kinetics and equilibrium adsorption of Basic Green 4 dye on activated carbon derived from durian peel: effects of pyrolysis and post-treatment conditions. *Journal of the Taiwan Institute of Chemical Engineers* **41** (5), 591–598.

Oncescu, T. & Oancea, P. 2007 On the mechanism of dichlorvos hydrolysis. *Revista de Chimie* **58**, 232–235.

Rahman, M. A. & Muneer, M. 2005 Photocatalysed degradation of two selected pesticide derivatives, dichlorvos and phosphamidon, in aqueous suspensions of titanium dioxide. *Desalination* **181**, 161–172.

Richter, P. & Corcoran, J. 1997 *Toxicological Profile for Dichlorvos*. Report Dichlorvos, US Department of Health and Human Services, Public Health Service, Agency for Toxic Substances and Disease Registry, USA.

Saffari, M., Karimian, N., Ronaghi, A., Yasrebi, J. & Ghasemi-Fasaei, R. 2015 Stabilization of nickel in a contaminated calcareous soil amended with low-cost amendments. *Journal of Soil Science and Plant Nutrition* **15** (4), 896–913.

Sahithya, K., Das, D. & Das, N. 2015 Effective removal of dichlorvos from aqueous solution using biopolymer modified MMT–CuO composites: equilibrium, kinetic and thermodynamic studies. *Journal of Molecular Liquids* **211**, 821–830.

Saleh, S., Kamarudin, K. B., Ghani, W. A. W. A. K. & Kheang, L. S. 2016 Removal of organic contaminant from aqueous solution using magnetic biochar. *4th International Conference on Process Engineering and Advanced Materials* **148**, 228–235.

Tan, X., Liu, Y., Zeng, G., Wang, X., Hu, X., Gu, Y. & Yang, Z. 2015 Application of biochar for the removal of pollutants from aqueous solutions. *Chemosphere* **125**, 70–85.

Tan, G., Sun, W., Xu, Y., Wang, H. & Xu, N. 2016 Sorption of mercury (II) and atrazine by biochar, modified biochars and biochar based activated carbon in aqueous solution. *Bioresource Technology* **211**, 727–735.

Tran, H. N., Wang, Y. F., You, S. J. & Chao, H. P. 2017 Insights into the mechanism of cationic dye adsorption on activated charcoal: the importance of π–π interactions. *Process Safety and Environmental Protection* **107**, 168–180.

Uchimiya, M., Wartelle, L. H. & Boddu, V. M. 2012 Sorption of triazine and organophosphorus pesticides on soil and biochar. *Journal of Agricultural and Food Chemistry* **60**, 2989–2997.

Wang, L., Yan, W., He, C., Wen, H., Cai, Z., Wang, Z., Chen, Z. & Liu, W. 2018 Microwave-assisted preparation of nitrogen-doped biochars by ammonium acetate activation for adsorption of acid red 18. *Applied Surface Science* **433**, 222–231.

Weng, C. H., Lin, Y. T. & Tzeng, T. W. 2009 Removal of methylene blue from aqueous solution by adsorption onto pineapple leaf powder. *Journal of Hazardous Materials* **170** (1), 417–424.

First received 31 August 2018; accepted in revised form 18 December 2018. Available online 28 December 2018

Coagulation and adsorption as pretreatments of thin-film composite–forward osmosis (TFC-FO) for ink printing wastewater treatment

W. Maknakorn, P. Jutaporn and W. Khongnakorn

ABSTRACT

This study investigated the performance of coagulation with ferric chloride ($FeCl_3.6H_2O$) and adsorption with activated carbon used as pretreatments prior to thin-film composite–forward osmosis (TFC-FO) filtration of ink printing wastewater. Wastewater samples were collected from a printing factory located in Chao Phraya river basin where zero liquid-discharge standards are regulated. The FO filtration unit was operated in co-current mode with the cross-flow velocity of 600 mL/min using 2 M NaCl as draw solution. The FO membrane achieved 83.9–91.0% chemical oxygen demand and 91.2–99.9% color rejections. FO filtration of raw wastewater, coagulation-treated wastewater, and coagulation and adsorption-treated wastewater resulted in 56.3%, 49.0% and 46.1% of flux decline in 150 min. Scanning electron microscopy and Fourier transform infrared spectrometry results revealed that cake formation caused by colloidal particles and pore blocking from precipitation were the major fouling mechanisms. Cake-enhanced concentration polarization also promoted adsorption of color and pigments at the membrane surface. Both fouling mechanisms were successfully mitigated by adsorption following coagulation pretreatment. Overall, the combined pretreatments and FO have potential for the effective treatment and reuse of ink printing wastewater.

Key words | adsorption, coagulation, forward osmosis, membrane fouling, printing wastewater

W. Maknakorn
W. Khongnakorn (corresponding author)
Center of Excellence in Membrane Science and Technology, Department of Civil Engineering, Faculty of Engineering,
Prince of Songkla University,
Songkhla, 90112,
Thailand
and
Department of Civil Engineering, Faculty of Engineering,
Prince of Songkla University,
Songkhla, 90112,
Thailand
E-mail: *watsa.k@psu.ac.th*

P. Jutaporn
Department of Environmental Engineering, Faculty of Engineering, and Research Center for Environmental and Hazardous Substance Management,
Khon Kaen University,
Khon Kaen,
Thailand
and
Research Program on Development of Appropriate Technologies for Coloring Agent Removal from Textile Dyeing, Pulp & Paper, Sugar Industries for Sustainable Management,
Center of Excellence on Hazardous Substance Management (HSM),
Chulalongkorn University,
Bangkok,
Thailand

INTRODUCTION

The discharge of colored wastewater into natural water bodies can cause serious environmental pollution. Color pollution is one of the environmental challenges as a result of stricter environmental regulations. The ink printing industry is an industry producing a large volume of wastewater from cleaning and washing processes. The printing wastewater is high in color and organic content due to colorant pigments, dyes, machinery fluids, auxiliary solvents (alcohols, glycols and glycol ethers) and additives (driers, waxes, defoamers and dispersing aids) (Ma & Xia 2009). Ink and dye wastewaters are comprised of various organic compounds of complex structure, which are not effectively treated by a conventional biological wastewater treatment process, such as activated sludge. Previous studies have suggested that this wastewater can be treated by a combined flocculation and biological process (Zhang & Liu 2003; Meteš*et al.* 2004). The coagulation and flocculation process has been used to remove a substantial portion of the organic content and color efficiently (Zhang & Liu 2003; Verma *et al.* 2012). Moreover, organic compounds recalcitrant to coagulation can be treated with adsorption (Meteš *et al.* 2004; Noonpui *et al.* 2010; Gisi *et al.* 2016), advanced oxidation (Ma & Xia 2009), and electrocatalytic processes (Tung *et al.* 2013). Even though good performance of chemical oxygen demand (COD) and color removal could be achieved by these processes, color and residual organic compound in effluent is still a major challenge as a result of stringent requirements of wastewater discharge standards in Chao Phraya river basin for zero liquid-discharge in Thailand (Notification of the Ministry of Industry 2008).

doi: 10.2166/wst.2019.100

Therefore, growing concerns about effluent discharge resulted in the development of a more innovative and effective treatment technology for ink printing wastewater.

The invention of synthetic membranes is one of the most significant developments in water and wastewater treatment technologies. Membranes have been previously used to treat wastewater and also to concentrate and recover valuable substances from wastewater (Baker 2012). The ultrafiltration process has been used for treating wastewater from banknote printing works (Zhang & Liu 2003; Zhang *et al.* 2004); however, membrane fouling and high energy consumption remain a significant drawback of this application. In the past few decades, forward osmosis (FO), which uses natural energy in the form of osmotic pressure to transport water through a semi-permeable membrane (Zhao *et al.* 2012), has attracted attention in many potential applications. The driving force for FO separation is osmotic pressure gradient between a 'draw' solution of high concentration and a 'feed' solution of lower concentration. With the ability to draw water with low energy input, the FO process has been extensively studied in various disciplines, such as seawater desalination, wastewater reclamation, industrial wastewater treatment, and power generation (Zhao *et al.* 2012). The advantages of the FO process include low energy consumption, rejection of a wide range of contaminated materials, and high recovery rate and permeate water quality (Huang *et al.* 2017). Han *et al.* (2016a) demonstrated that the use of the FO process to recover water from wastewater resulted in wastewater volume reduction and enhancement of dye concentration.

Many researchers studied the FO process using cellulose triacetate (CTA) membrane, and reported that the CTA membrane had disadvantages including low flux rate and high fouling propensity (Li *et al.* 2013, 2018). Recently, a thin-film composite (TFC) membrane containing carboxyl acid groups, thus exhibiting higher water permeability, was developed for the FO process (Gu *et al.* 2013; Zhang *et al.* 2017; Li *et al.* 2018). However, TFC-FO performance is limited by internal and external concentration polarization (ECP), which is the build-up of concentration gradients around FO membrane resulting in transmembrane flux decline. Concentration polarization falls into two main categories: internal concentration polarization (ICP) and ECP (Gao *et al.* 2014). ICP take places in the porous support layer when solutes inside the support are concentrated as water is drawn through the membrane (Lee *et al.* 2010; Boo *et al.* 2012). ECP takes place at the membrane surface as feed components are accumulated at the membrane active layer (Gao *et al.* 2014). Fouling is often a negative effect of concentration polarization and is always the critical issue that limits membrane application. Therefore, fouling reduction has been the focus of membrane research and application in recent years.

Due to the characteristic of ink printing wastewater, the main foulants are suspended solids and organic substances from ink and dye. Given the demonstrated ability of coagulation to remove suspended solid and the ability of activated carbon adsorption to remove organic substances including color (Meteš *et al.* 2004), it is reasonable to expect that the combined coagulation–adsorption process can potentially minimize fouling in TFC-FO filtration. Accordingly, this study evaluates the role of coagulation and adsorption processes as pretreatments to TFC-FO membrane filtration in the treatment of ink printing wastewater. Fouling mechanisms of raw and pretreated wastewater were investigated. The manuscript suggests a new approach for ink printing wastewater treatment to comply with Thailand's strengthening standard and for long-term stability of the FO filtration.

MATERIAL AND METHODS

Ink printing wastewater

A one-time grab sample of 200 L of wastewater sample was collected from an equalization tank in a wastewater treatment plant at a printing factory (Bangkok, Thailand) in December 2016. The wastewater sample was analyzed for pH, COD, biological oxygen demand (BOD), color, total suspended solid (TSS), total dissolved solid (TDS), total Kjeldahl nitrogen (TKN) and lead (Pb) following the standard methods (APHA/AWWA/WEF 2012). Color was measured at wavelength in the range of 400–700 nm using a UV-spectrophotometer (Proved 300, Merck, Germany) following Method 2120F (APHA/AWWA/WEF 2012).

Pretreatment processes

Coagulation with ferric chloride ($FeCl_3.6H_2O$, Sigma-Aldrich, Munich, Germany) and adsorption with bituminous coal based activated carbon (GB 1000; Carbokarn Co. Ltd, Bangkok, Thailand) were selected for suspended solids and color pigments removal, respectively. The laboratory-scale pretreatment experiments were performed using a jar test apparatus with 1-L jars (Phipps and Bird Inc., Richmond, VA, USA). Optimum coagulant dose was determined at 1.0 g $FeCl_3.6H_2O$/L of wastewater, based on results of a preliminary jar test study that varied coagulant dose from 0.6 to 1.0 g/L. The mixing protocol entailed a rapid mixing at 120 rpm for 2 min, 30 rpm for 30 min, and settling for 45 min. Adsorption kinetic studies were conducted on coagulation-pretreated wastewater with the dosage in the range 0.04–20 g/L and contact time of 15–1,200 min. An optimum activated carbon dosage of 0.5 g/L at 390 min contact time was selected for further FO membrane filtration experiments. The mixing at 100 rpm was provided by an incubator shaker at a constant temperature of 30 °C. Wastewater after coagulation and adsorption pretreatments was analyzed for the parameters mentioned above following the standard method procedure (APHA/AWWA/WEF 2012).

TFC-FO membrane characterization

The commercial TFC-FO Aquaporin Inside™ membrane was supplied by Aquaporin (Kongens Lyngby, Denmark). The effective area (A_m) was 34.8 cm^2. The water permeability and salt rejection of the membranes were measured using a laboratory-scale reverse osmosis filtration unit at a transmembrane pressure of 1 bar. Change in permeate weight was monitored by a digital balance (GF-300, AND, Japan) connected to a computer for automatic data recording. Permeate flux, J_w, (in L/m^2 h, abbreviated as LMH) was calculated from volume change over time per unit membrane area. The water permeability coefficient, A, (in L/(m^2 h bar)), was obtained from the pure water permeation flux using deionized (DI) water as feed (Han et al. 2016b). The salt rejection, R_s, (in %) was determined by measuring conductivity of permeate and the feed using 10 mM NaCl solution as a feed at a pressure of 1 bar. The salt rejection was calculated as:

$$R_s = \left(1 - \frac{C_p}{C_f}\right) \times 100 \tag{1}$$

where C_p and C_f (in mS/cm) is the conductivity of permeate and feed solution, respectively.

The salt permeability coefficient, B, (in LMH) was calculated using the following equation:

$$B = \left(\frac{1 - R_s}{R_s}\right)(A(\Delta P - \Delta\pi)) \tag{2}$$

where ΔP and $\Delta\pi$ are the difference in pressure and osmotic pressure gradient across the membrane (in bar), respectively.

The reverse salt flux, J_s, (in g/m^2 h, abbreviated as gMH) was calculated as (Han et al. 2016b):

$$J_s = \left(\frac{C_{f,t}\,V_{f,t} - C_{f,i}\,V_{f,i}}{A_m \Delta t}\right) \tag{3}$$

where $C_{f,t}$ and $V_{f,t}$ are the conductivity and feed volume at the end of filtration experiment, respectively, while $C_{f,i}$ and $V_{f,i}$ are the initial conductivity and total feed volume at the beginning, respectively. Δt (h) refers to a predetermined time of the experiment.

Bench scale TFC-FO filtration system

Raw wastewater (RW), coagulation-treated wastewater (CW), and coagulation and adsorption-treated wastewater (C + AW) were used as a feed to a bench scale FO filtration system as depicted in Figure 1. The TFM-FO module was operated in the countercurrent mode with the cross-flow rate of 3.5×10^{-3} m/s (600 mL/min) on the feed side. A 2 M NaCl solution with conductivity of 150 mS/cm ± 10% was used as drawing solution on the draw side (Han et al. 2016a), as it provided suitable different osmotic pressure between the draw solution and feed of 1,811 mOsm/kg.

Filtration experiments were conducted in a single-run mode to evaluate membrane permeate flux and in multi-cycle experiments to investigate fouling mechanism. Prior to each single-run filtration experiment, the TFC-FO membrane was stabilized in DI water for 30 min and flux was constant. Concentration of draw solution was maintained constant throughout the tests by conductivity control to minimize the effects of draw solution dilution (Han et al. 2016a). Conductivity of feed and permeate was measure every 5 min. Weight of feed and permeate was monitored every 5 min using a digital balance and data were recorded in real time with a laboratory computer until 150 min of operation time was achieved. The time taken for color to appear in the feed permeate side was considered to be the optimum operation time of the filtration cycle. Permeate flux was calculated as weight of permeate per time per area of membrane. At the end of each filtration run, permeate water quality was analyzed. Differential pressure from a pressure gauge was monitored so that it did not exceed 0.3 bar, to ensure a low driving force of the FO system.

For multi-cycle experiments, prior to each fouling test, the membranes were equilibrated with DI water for several hours. Permeate flux of wastewater filtration continuously was monitored until normalized flux (J/J_o) was equal to 0.5; then physical cleaning of the fouled membrane was

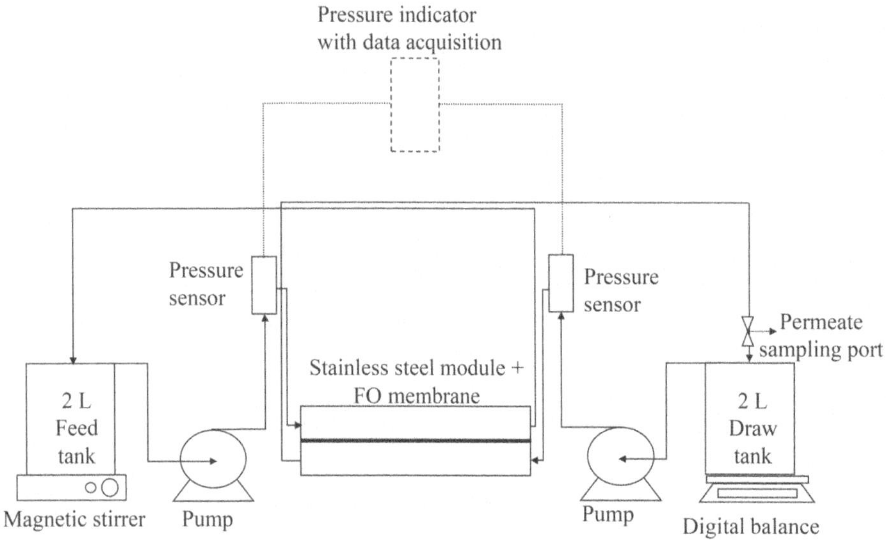

Figure 1 | Systematic diagram of bench scale FO filtration system.

performed using DI water flushing for 30 min without using any chemicals (Han *et al.* 2016a; Khongnakorn & Youravong 2016). Each filtration experiment using RW, CW and C + AW as feed, was operated for three cycles, consecutively. Each cycle lasted 150 min and physical cleaning was performed after each cycle. The effectiveness of physical cleaning was determined by measuring permeate flux of the cleaned membrane. The filtration experiment with each type of feed water was performed in triplicate and the average value was reported.

The virgin and fouled TFC-FO membranes after the third filtration cycle were characterized for the microstructure by scanning electron microscopy (SEM; Quanta400, FEI). The surface chemical structure of membrane samples, RW, and foulant samples scraped from fouled membrane surface were analyzed by a Fourier transform infrared (FTIR) spectrometer (Vertex70, Bruker, Germany) with an OMNI sampler and an attenuated total reflectance accessory (Spectrum One B, Perkin Elmer Inc.). Prior to FTIR analysis, the membrane samples were dried in a vacuum freeze-drying machine for 24 h. Wastewater and foulant samples were analyzed in a solid phase using the freeze-drying protocol following Sun *et al.* (2016).

RESULTS AND DISCUSSION

Wastewater characteristics of FO feed and permeate

Wastewater sampled from the ink printing industry was high in color, COD, and suspended solids and low in biodegradability (BOD/COD = 0.21). Characteristics of feed and permeate from filtration experiments using RW, CW, and C + AW is shown in Table 1. Color constituents in RW mainly came from ink from the machine washing process. Various types of ink were used at the factory, for example, black iron and red iron oxide ink. Coagulation pretreatment achieved 98% removal of COD, 99% removal of TSS, and

Table 1 | Characteristics of raw printing wastewater and pretreated wastewater

Characteristics	pH	COD (mg/L)	BOD (mg/L)	Color (ADMI)	TSS (mg/L)	TDS (mg/L)	TKN (mg/L)	Pb (mg/L)
RW feed	6.6–7.1	32,850	4,350	39,900	21,107	1,500	270	0.913
CW feed	6.5–6.6	491	18	340	32	1,823	123	ND
C + AW feed	6.0–6.4	365	12	250	28	1,671	112	ND
RW permeate	6.9–7.1	5,304	1,114	50.0	47.0	1,450	70	ND
CW permeate	7.0–7.1	44	2.5	22.0	5.0	1,630	35	ND
C + AW permeate	7.0–7.1	37	2.4	22.0	5.0	1,456	35	ND

ND = non-detectable, RW = raw wastewater, CW = coagulation-treated wastewater, C + AW = coagulation and adsorption-treated wastewater.

99% removal of color. After coagulation, adsorption pretreatment could further reduce COD from 491 to 365 mg/L (26%) and color from 340 to 250 ADMI (26%). Pb was removed to a non-detectable level after coagulation pretreatment, suggesting that Pb was successfully trapped in coagulation flocs and thus precipitated.

Due to the high strength of RW from printing industries, even though coagulation and adsorption pretreatments were effective, the color, and BOD of pretreated wastewater (C + AW feed) in Table 1 was still 250 ADMI and 12 mg/L, respectively. These water quality parameters did not meet criteria for wastewater reuse at the factory, which requires non-objectionable color and BOD not more than 3.0 mg/L, to comply with the zero liquid-discharge standard in Thailand. COD and TKN of C + AW feed were also as high as 365 mg/L and 112 mg/L, respectively, indicating that further treatment is required for wastewater reuse.

When the FO system alone was used to treat RW, 83.8% COD rejection, 74.4% BOD rejection and 99.9% color rejection were achieved. However, the RW permeate still contained a considerable amount of COD (5,304 mg/L) and undesirable color (50 ADMI). When C + AW was used as feed water to the FO system, 89.9% COD rejection, 80% BOD rejection and 91.2% color removal were achieved. BOD and color of C + AW permeate did meet the criteria for wastewater reuse at the factory. Overall, the water quality parameters results show that C + AW permeate could be used as make-up water in industrial processes and/or cleaning. Additionally, it is worthwhile to note that there was only marginal difference between the permeate quality of CW and C + AW, and that coagulation alone could be sufficient to pretreat wastewater for FO system. However, these permeate quality results did not provide any detail on permeate flux and membrane fouling. For long-term stability of the FO system, effect of flux decline from different types of feed water should be taken into consideration as it affects operation cost of the system most significantly. More details of membrane flux and fouling will be discussed later on.

TFC-FO membrane characteristic

The transport properties of the obtained TFC-FO membrane compared to other studies are presented in Table 2. The membrane was designed by Aquaporin (Denmark) specifically to have high water permeability for this application. Compared to other previous studies, the TFC-FO membrane in this study had relatively high pure water permeability of 4.87 L/(m^2 h bar) and salt permeability coefficient of 0.64 LMH. At an operating pressure of 1 and 3 bar, the salt rejection was 85% and 93%, respectively.

Table 2 | Transport properties of membranes

Membrane	Water permeability (L/(m^2 h bar))	Salt permeability coefficient (LMH)	Reference
TFC-FO	4.87	0.64	This study
TFC-FO	3.40	0.103	Han *et al.* (2016a)
TFC-FO	2.30	0.206	Han *et al.* (2016b)
TFC-FO	2.67	1.12	Zhang *et al.* (2017)
TFC-FO	1.94	0.52	Li *et al.* (2018)

Water flux and reverse salt flux from single-run experiments

Average permeate water flux (J_w) from single-run experiments using RW (26.65 LMH) was lower than that using CW (31.32 LMH), and C + AW (36.13 LMH) (Figure 2(a)). Given the high strength of RW, it is reasonable to expect that its higher osmotic pressure reduced the net osmotic driving force for permeate water flux. Since higher osmotic pressure near the membrane surface leads to a significant drop in the net driving force, substantial decline in permeate flux was observed (Boo *et al.* 2012). Since osmotic pressure generally increased with TDS increase (Baker 2012), the association between fouling TDS concentration and permeate flux was evaluated. TDS concentration of CW (1,823 mg/L) and C + AW (1,671 mg/L) was higher than that of RW (1,500 mg/L) (Table 1), but water flux of these pretreated waters was greater than that of RW (Figure 2(a)). This result indicates that water flux performance could not be explained by osmotic pressure gradient alone. In the filtration of wastewater with high suspended solids (TSS), permeated flux decline was caused by not only the decrease of net driving force but also particle fouling (Boo *et al.* 2012; Motsa *et al.* 2015). The fact that coagulation pretreatment removed TSS by 99.8% (Table 1) resulted in low TSS in CW and C + AW feed waters; thus the greater permeate water flux was achieved.

The reverse salt diffusion flux (J_s) was measured as an increase in conductivity, due to reverse diffusion of the draw solute (Han *et al.* 2016a). There was no significant difference between J_s of RW (4.48 gMH), CW (4.66 gMH), and C + AW (4.66 gMH) (Figure 2(b)). The water flux/reverse salt flux ratios (calculated as J_w/J_s) of CW (6.73 L/g) and C + AW (7.76 L/g) were slightly higher than that of

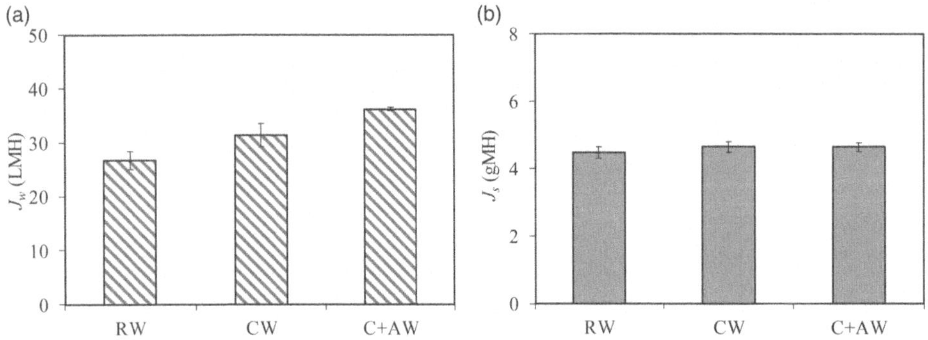

Figure 2 | (a) Membrane permeate flux (J_w) and (b) specific reverse salt flux (J_s) of raw wastewater (RW), coagulation-treated wastewater (CW), and coagulation and adsorption-treated wastewater (C + AW). Error bars correspond to the difference between the averages of results obtained for triplicate filtration experiments.

RW (5.94 L/g). The high J_w/J_s obtained in the CW and C + AW filtration experiments that had higher permeate flux indicated that effect of the decrease in driving pressure by the dilutive draw solution was minimal.

Effect of pretreatments on membrane fouling

Flux decline results from triplicate multi-cycle FO filtration experiments using RA, CW, and C + AW as feed are shown in Figure 3. For each experiment, between 150-min filtration cycles, 30-min physical cleaning cycles by DI water flushing were performed. After the first cycle of filtration, J_w declined drastically from 30.2 LMH to 13.2 LMH (56.3% decrease) for the filtration experiment using RW, while J_w decrease was moderate for the filtration experiment using CW (49.0%) and C + AW (46.1%) (Figure 3). These results indicate that RW caused more severe fouling than did pretreated waters. Thus, coagulation and adsorption pretreatments not only enhanced water flux (Figure 2(a)) but also helped in fouling mitigation.

During these three cycles of filtration, decline in water flux had a similar pattern: it declined appreciably with filtration time up to 100 min at which point it started to plateau. Given that printing wastewater was high in COD and TSS, combined organic and colloidal fouling could govern fouling mechanisms in this study. Previous studies on FO membrane fouling proposed a three-step fouling mechanism to explain combined organic and colloidal fouling behavior. First, the organic and colloidal particles initially attached onto the membrane surface. Second, colloid–colloid interactions occurred and colloidal particles deposited on the membrane surface. Third, the cohesion of the colloidal and adsorption on cake layer occurred at the membrane (Mi & Elimelech 2010; Boo *et al.* 2012; Motsa *et al.* 2015).

After each cycle of filtration, physical cleanings were immediately carried out by flushing with DI water. Average

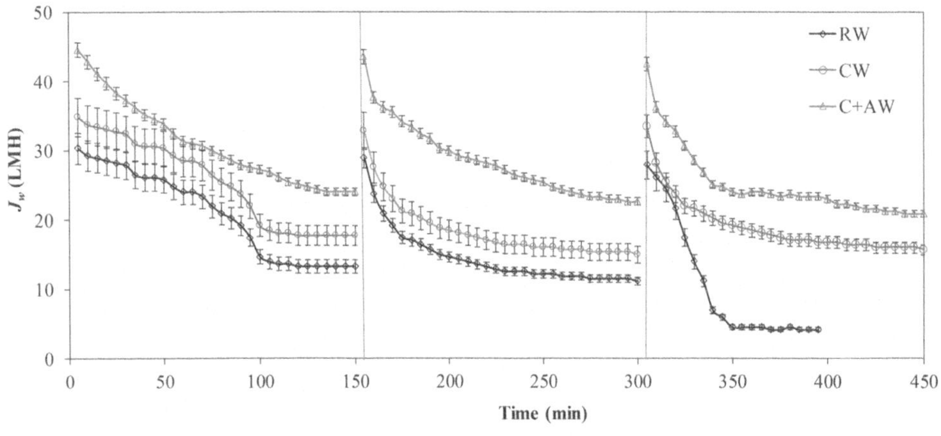

Figure 3 | The reduction of water flux infiltration experiment using raw wastewater (RW), coagulation-treated wastewater (CW), and coagulation and adsorption-treated (C + AW) during multi-cycle filtration experiments.

flux recovery achieved by first and second cleaning cycles was 92%, 95% and 95% for RW, CW and C + AW, respectively. This result suggests that fouling is very reversible and that foulant can be cleaned effectively without any chemical reagents. Similar results were reported by Han *et al.* (2016a) in which fouling from FO filtration experiments using synthetic textile wastewater could be recovered by 98%. CEOP that built up from reverse salt diffusion and occurred near the membrane surface was found to be the major mechanism governing colloidal fouling (Boo *et al.* 2012). CEOP could also be successfully removed by physical cleaning (Boo *et al.* 2012; Motsa *et al.* 2017). Overall, effective flux recovery by physical cleaning in this study confirms that colloidal particles were the major foulants in the FO process of printing wastewater.

It is noteworthy that coagulation and adsorption pretreatments not only were effective in fouling mitigation, but also resulted in overall higher J_w of the FO process. Without any pretreatments, J_w of the RW experiment in the third cycle decreased drastically and color in the wastewater was observed in the draw side at 400 min; thus the experiment was discontinued at that point. Combined coagulation and adsorption pretreatment reduced fouling more effectively than coagulation pretreatment alone (Figure 3). Average ratio of water flux at 150 min to initial water flux at 5 min for all three cycles of filtration experiments was 0.48 and 0.52 for CW and C + AW, respectively. Based on the average flux during three cycles of filtration, CW achieved 38% greater flux compared to RW experiments, while C + AW pretreatments resulted in 87% increase in water flux. Thus, adding adsorption pretreatment to the system produced 49% greater water flux compared to coagulation pretreatment alone. To meet wastewater reuse for zero liquid-discharge requirements, greater wastewater volume reduction is desired. Also, for long-term stability of the FO system, fouling should be mitigated as much as possible. Thus, combined coagulation and adsorption pretreatment should be selected to pretreat printing wastewater before FO application.

Morphological analysis

Since FO experiments with RW resulted in the remarkable flux decline, a membrane sample after the third cycle from RW filtration was selected for SEM analyses to observe difference in morphological structure of the virgin and fouled membrane (Figure 4). Compared to virgin membrane (Figure 4(a) and 4(b)), the pores at the feed side of fouled

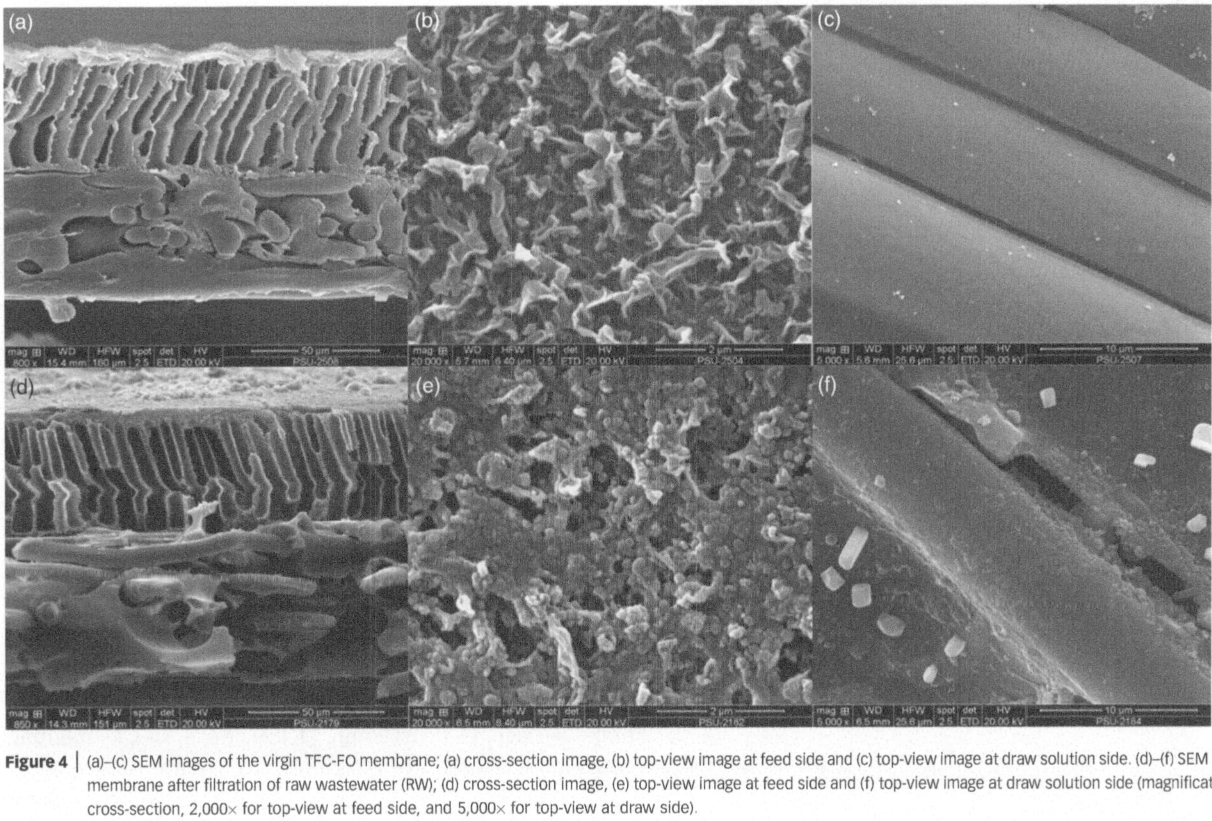

Figure 4 | (a)–(c) SEM images of the virgin TFC-FO membrane; (a) cross-section image, (b) top-view image at feed side and (c) top-view image at draw solution side. (d)–(f) SEM images of FO membrane after filtration of raw wastewater (RW); (d) cross-section image, (e) top-view image at feed side and (f) top-view image at draw solution side (magnification: 800× for cross-section, 2,000× for top-view at feed side, and 5,000× for top-view at draw side).

membranes (Figure 4(d) and 4(e)) were relatively unchanged, but the accumulation of CEOP was observed at the active layer. Thus, cake layer formation from colloidal particles and organic substances was responsible for flux loss, consistent with Motsa *et al.* (2015) who reported formation of a relatively loose cake layer due to the lack of hydraulic pressure in FO. Water flux decreased gradually in the first and second cycles of RW filtration (Figure 3), which also indicates the cake layer fouling mechanism. In the third cycle, drastic decrease of water flux suggested a pore blocking mechanism from reverse salt flux (Zhang *et al.* 2014; Motsa *et al.* 2017), as confirmed by salt crystal formation at the support layer of draw solution side (Figure 4(f)).

Top-view images at the feed side from CW and C + AW experiments are shown in Figure 5(a) and 5(b), respectively.

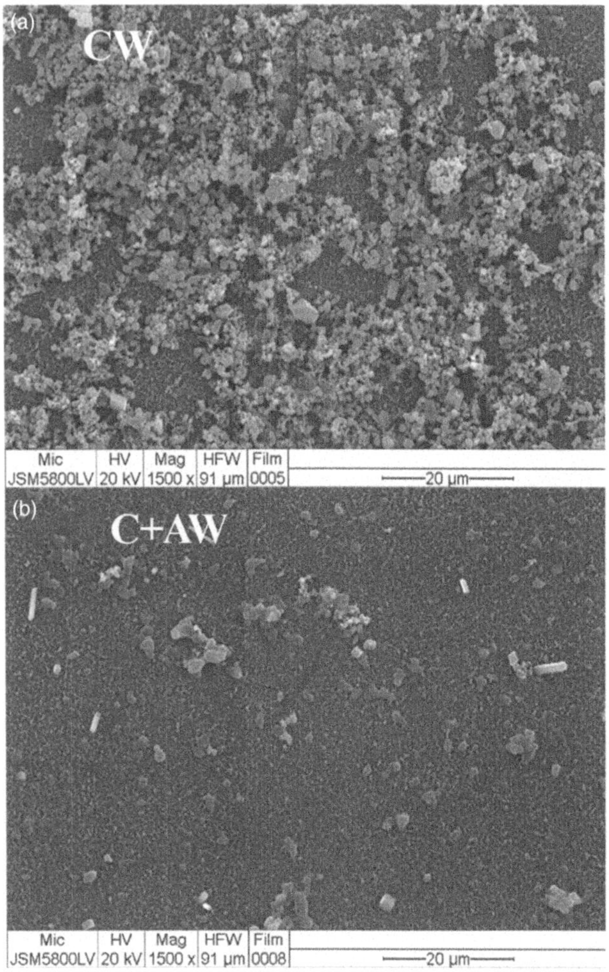

Figure 5 | SEM top-view images at feed side of membrane from (a) coagulation-treated wastewater (CW) and (b) coagulation and adsorption-treated wastewater (C + AW) (magnification: 1,500×).

Compared to RW (Figure 4(e)), the membrane from CW filtration (Figure 5(a)) had distinctively less accumulation of colloidal particles. A combined C + AW pretreatment (Figure 5(b)) reduced the accumulation even further to the point that a minimal number of colloidal particles and organic substances were observed on the surface. This morphological analysis explains why adding adsorption pretreatment to the system was effective at fouling reduction and resulted in higher permeate flux rate. This result is well-supported by the significant amount of suspended solids removal by coagulation and color removal by adsorption.

Thus, the results of the morphological analyses are consistent with the previous section that identified colloidal particles to be the major foulants and that cake layer formation was the main mechanism governing fouling at the active layer side in the initial stage of filtration. A combined coagulation and adsorption pretreatment successfully minimized cake layer formation due to significant removal of colloidal particles, which makes it suitable for this application. Moreover, the morphological results also revealed that the reverse salt flux caused precipitation and pore blocking at the draw solution side, so pore blocking also contributed to fouling especially in the third cycle of filtration. Overall, the cake layer formed by colloidal particles and pore blocking caused by the reverse salt flux were the two mechanisms responsible for flux loss during FO filtration of raw printing wastewater.

FTIR spectroscopy was employed to analyze the chemical nature of fouling on FO membranes after filtration of RW. All samples were analyzed in a solid phase using the freeze-drying protocol. The FTIR spectra of virgin FO membrane compared to fouled membrane, foulant, and RW are shown in Figure 6. The main adsorption peaks of virgin and fouled membranes were detected in the range from 1,000 cm^{-1} to 3,000 cm^{-1}. The main peaks were associated with the vibration of the N–H stretching (3,360 cm^{-1}), C–H stretching (2,965 cm^{-1}), C=O stretching (1,730 cm^{-1}) and C–F stretching (1,159 cm^{-1}) (Silva *et al.* 2013). Most of the peak heights for the fouled membrane were lower compared to the virgin membrane. This signal reduction indicated that the aromatic and carboxylic groups of the membrane were coated with organic compounds, which were the major constituents of printing wastewater.

The vibrational bands associated with foulants could be clearly distinguished from those of the membrane material. The circled areas show foulants accumulated at the membrane surface, C–F stretching (1,159 cm^{-1}), C–Cl stretching (696 cm^{-1}), and C–Br (555 cm^{-1} stretching), which were associated with color and pigments of ink in the printing

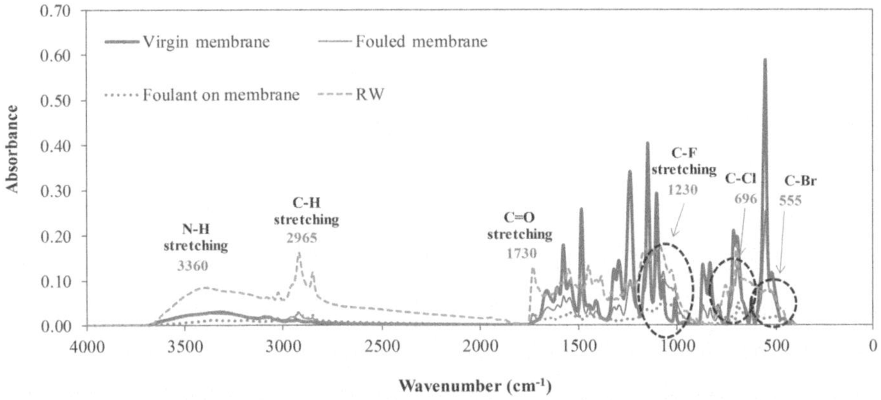

Figure 6 | FTIR spectrum of FO membrane compared to foulant and raw wastewater (RW). Circled regions show foulant accumulated on membrane surface after RW filtration.

process. The FTIR spectrum of RW printing wastewater also contained spectral features produced by additives, such as calcium carbonate, which presented at the vibration around 1,430, 875 and 712 cm^{-1}. The spectrum of RW in this study is consistent with previous studies reporting typical absorption bands of printing ink in 1,800 to 650 cm^{-1} regions (Silva *et al.* 2013; Doncea & Ion 2014). Similar peak regions with RW, but less absorbance, were found in foulant at the membrane surface.

Adsorption of foulants at the membrane surface was the major mechanism in the ECP. The cake layer of colloidal foulants can lead to the CECP mechanism, thus promoting adsorption of color and pigments at the membrane surface. For a dense membrane, CECP is an important fouling mechanism that results in severe flux decline as well as reduced solute rejection (Gao *et al.* 2014). The FTIR spectrum of the membrane fouled by filtration of RW shows distinctive peaks of C–F, C–Cl and C–Br stretching associated with the color of printing ink, thus confirming that organic substances, especially pigments, were also the constituents of foulants. Given that foulants were composed of both colloidal particles and color pigments, both coagulation and activated carbon adsorption should be employed as pretreatments for long-term stability of the FO system in ink printing wastewater treatment.

CONCLUSIONS

In the current work, coagulation and activated carbon adsorption were employed as pretreatment processes for TFC-FO filtration in ink printing wastewater treatment. Fouling mechanisms and fouling reduction performance were evaluated using a three-cycle bench scale FO filtration experiment. Membranes and foulants were examined using SEM and FTIR spectrometry. The following conclusions can be drawn from this study.

(1) Permeate quality from FO filtration using either coagulation pretreatment alone or coagulation followed by adsorption pretreatments could meet requirements for wastewater reuse in the factory to comply with the zero liquid-discharge standard. The FO membrane achieved 83.9–91.0% COD rejection and 91.2–99.9% color rejection.
(2) Cake formation caused by colloidal particles at the active layer and pore blocking from precipitation of reverse salt flux at the support layer were the major fouling mechanisms in the FO process of printing wastewater. Cake layer formation also led to the CECP mechanism, thus promoting adsorption of color and pigments at the membrane surface. FTIR spectroscopy revealed that the composition of foulants was mainly organic compounds combined with color pigments. This fouling is very reversible and can be cleaned effectively without any chemical reagents.
(3) Coagulation pretreatment successfully removed more than 99% of TSS and adsorption further removed 26% of color from wastewater, suggesting their suitability for target foulants removal. Filtration of coagulation-pretreated wastewater had 38% greater flux compared to experiments using RW. Then adding adsorption after coagulation produced 49% greater water flux through three-cycle filtration compared to coagulation pretreatment alone.

Overall, this study demonstrates the potential of FO with coagulation and adsorption pretreatments for wastewater reuse application. The combined pretreatments could enhance permeate water quality, produce greater water flux, and also minimize fouling of FO filtration, thus providing long-term stability of the FO membrane in ink printing wastewater filtration.

ACKNOWLEDGEMENT

This research was supported by a grant (MSD58I0067) from the Thailand Research Fund (TRF) for Research and Researchers for Industries (RRI) program.

REFERENCES

APHA/AWWA/WEF 2012 *Standard Methods for the Examination of Water and Wastewater*, 22nd edn. American Public Health Association/American Water Works Association/Water Environment Federation, Washington, DC, USA.

Baker, R. W. 2012 *Membrane Technology and Applications*, 3rd edn. John Wiley and Sons, Chichester, UK.

Boo, C., Lee, S., Elimelech, M., Meng, Z. & Hong, S. 2012 Colloidal fouling in forward osmosis: role of reverse salt diffusion. *J. Membr. Sci.* **390–391**, 277–284. DOI: 10.1016/j.memsci.2011.12.001.

Doncea, S. M. & Ion, R. M. 2014 FTIR (DRIFT) analysis of some printing inks from the 19th and 20th centuries. *Rev. Roum. Chim.* **59** (3–4), 173–183.

Gao, Y., Wang, Y.-N., Li, W. & Tang, C. Y. 2014 Characterization of internal and external concentration polarizations during forward osmosis processes. *Desalination* **338**, 65–73. DOI: 10.1016/j.desal.2014.01.021.

Gisi, S., Lofrano, G., Grassi, M. & Notarnicola, M. 2016 Characteristics and adsorption capacities of low-cost sorbents for wastewater treatment: a review. *Sustainable Mater. Technol.* **9**, 10–40. DOI: 10.1016/j.susmat.2016.06.002.

Gu, Y., Wang, Y.,-N., Wei, J. & Tang, C. Y. 2013 Organic fouling of thin-film composite polyamide and cellulose triacetate forward osmosis membranes by oppositely charged macromolecules. *Water Res.* **47**, 1867–1874. DOI: 10.1016/j.watres.2013.01.008.

Han, G., Liang, C. Z., Chung, T. S., Weber, M., Staudt, C. & Maletzko, C. 2016a Combination of forward osmosis (FO) process with coagulation/flocculation (CF) for potential treatment of textile wastewater. *Water Res.* **91**, 361–370. DOI: 10.1016/j.watres.2016.01.031.

Han, G., Zhao, B., Fu, F., Chunga, T. S., Weber, M., Staudt, C. & Maletzkod, C. 2016b High performance thin-film composite membranes with mesh-reinforced hydrophilic sulfonated polyphenylenesulfone (sPPSU) substrates for osmotically driven processes. *J. Membr. Sci.* **502**, 84–93. DOI: 10.1016/j.memsci.2015.12.023.

Huang, J., Long, Q., Xiong, S., Shen, L. & Wang, Y. 2017 Application of poly (4-styrenesulfonic acid-co-maleic acid) sodium salt as novel draw solute in forward osmosis for dye-containing wastewater treatment. *Desalination* **421**, 40–46. DOI: 10.1016/j.desal.2017.01.039.

Khongnakorn, W. & Youravong, W. 2016 Concentration and recovery of protein from tuna cooking juice by forward osmosis. *J. Eng. Sci. Technol.* **11**, 962–973.

Lee, S., Boo, C., Elimelech, M. & Hong, S. 2010 Comparison of fouling behavior in forward osmosis (FO) and reverse osmosis (RO). *J. Membr. Sci.* **365**, 34–39. DOI: 10.1016/j.memsci.2010.08.036.

Li, G., Li, X.-M., He, T., Jiang, B. & Gao, C. 2013 Cellulose triacetate forward osmosis membranes: preparation and characterization. *Desalin. Water Treat.* **51** (13–15), 2656–2665. DOI: 10.1080/19443994.2012.749246.

Li, J. Y., Ni, Z. Y., Zhou, Z. Y., Hu, Y. X., Xu, X. H. & Cheng, L. H. 2018 Membrane fouling of forward osmosis in dewatering of soluble algal products: comparison of TFC and CTA membranes. *J. Membr. Sci.* **552**, 213–221. DOI: 10.1016/j.memsci.2018.02.006.

Ma, X. J. & Xia, H. L. 2009 Treatment of water-based printing ink wastewater by Fenton process combined with coagulation. *J. Hazard. Mater.* **162**, 386–390. DOI: 10.1016/j.jhazmat.2008.05.068.

Meteš, A., Kovačević, D., Vujević, D. & Papić, S. 2004 The role of zeolites in wastewater treatment of printing inks. *Water Res.* **38**, 3373–3381. DOI: 10.1016/j.watres.2004.04.012.

Mi, B. & Elimelech, M. 2010 Organic fouling of FO membranes: fouling reversibility and cleaning without chemical reagents. *J. Membr. Sci.* **348**, 337–345. DOI: 10.1016/j.memsci.2009.11.021.

Motsa, M. M., Mamba, B. B. & Verliefde, A. R. D. 2015 Combined colloidal and organic fouling of FO membranes: the influence of foulant–foulant interactions and ionic strength. *J. Membr. Sci.* **493**, 539–548. DOI: 10.1016/j.memsci.2015.06.035.

Motsa, M. M., Mamba, B. B., Thwala, J. M. & Verliefde, A. R. D. 2017 Osmotic backwash of fouled FO membranes: cleaning mechanisms and membrane surface properties after cleaning. *Desalination* **402**, 62–71. DOI: 10.1016/j.desal.2016.09.018.

Noonpui, S., Thiravetyan, P., Nakbanpote, W. & Netpradit, S. 2010 Color removal from water-based ink wastewater by bagasse fly ash, sawdust fly ash and activated carbon. *Chem. Eng. J.* **162**, 503–508. DOI: 10.1016/j.cej.2010.05.051.

Notification of the Ministry of Industry 2008 Measure to control dirtiness of industrial wastewater for restoration of water quality in the Chao-Phraya River, B.E.2551. Thailand Ministry of Industry, Bangkok (in Thai).

Silva, C. S., Borba, F. d. S. L., Pimentel, M. F., Pontes, M. J. C., Honorato, R. S. & Pasquini, C. 2013 Classification of blue pen ink using infrared spectroscopy and linear discriminant analysis. *Microchem. J.* **109**, 122–127. DOI: 10.1016/j.microc.2012.03.025.

Sun, Y., Tian, J., Zhao, Z., Shi, W., Liu, D. & Cui, F. 2016 Membrane fouling of forward osmosis (FO) membrane for municipal wastewater treatment: a comparison between direct FO and OMBR. *Water Res.* **104**, 330–339. DOI: 10.1016/j.watres.2016.08.039.

Tung, C. H., Shen, S. Y., Chang, J. H., Hsu, Y. M. & Lai, Y. C. 2013 Treatment of real printing wastewater with an electrocatalytic process. *Sep. Purif. Technol.* **117**, 131–136. DOI: 10.1016/j.seppur.2013.07.028.

Verma, A. K., Dash, R. R. & Bhunia, P. 2012 A review on chemical coagulation/flocculation technologies for removal of colour from textile wastewaters. *J. Environ. Manage.* **93**, 154–168. DOI: 10.1016/j.jenvman.2011.09.012.

Zhang, G. & Liu, Z. 2003 Membrane fouling and cleaning in ultrafiltration of wastewater from banknote printing works. *J. Membr. Sci.* **211**, 235–249. DOI: 10.1016/S0376-7388(02)00422-2.

Zhang, G., Liu, Z. Z., Song, L. F., Hu, J. Y., Ong, S. L. & Ng, W. J. 2004 Post-treatment of banknote printing works wastewater ultrafiltration concentrate. *Water Res.* **38**, 3587–3595. DOI: 10.1016/j.watres.2004.06.005.

Zhang, X., Ning, Z., Wang, D. K. & Costa, J. C. D. 2014 Processing municipal wastewaters by forward osmosis using CTA membrane. *J. Membr. Sci.* **468**, 269–275. DOI: 10.1016/j.memsci.2014.06.016.

Zhang, X., Tian, J., Gao, S., Shi, W., Zhang, Z., Cui, F., Zhang, S., Guo, S., Yang, X., Xie, H. & Liu, D. 2017 Surface functionalization of TFC FO membranes with zwitterionic polymers: improvement of antifouling and salt-responsive cleaning properties. *J. Membr. Sci.* **544**, 368–377. DOI: 10.1016/j.memsci.2017.09.044.

Zhao, S., Zou, L., Tang, C. Y. & Mulcahy, D. 2012 Recent developments in forward osmosis: opportunities and challenges. *J. Membr. Sci.* **396**, 1–21. DOI: 10.1016/j.memsci.2011.12.023.

First received 13 August 2018; accepted in revised form 8 March 2019. Available online 18 March 2019

Kinetic and equilibrium modeling of biosorption of nickel (II) and cadmium (II) on brewery sludge

Rajeswari M. Kulkarni, K. Vidya Shetty and G. Srinikethan

ABSTRACT

In the current study, utilization of industrial waste brewery sludge for the biosorption of nickel (II) and cadmium (II) has been explored. The suitable conditions for the effective removal of Ni (II) and Cd (II) from aqueous solutions were examined. The kinetic evaluation showed that the biosorption process using the sludge followed pseudo-second order kinetics. In the presence of a metal co-ion, competitive and preferential biosorption was observed. The Langmuir model and Freundlich model were able to describe the sorption equilibrium for biosorption of Ni (II) and Cd (II) ions in single and dual metal systems. The effects of co-ion concentrations onto mono-component isotherm parameters (Langmuir and Freundlich) were studied and the inhibitory effect of co-ion concentration was observed. The effective reusability of biomass was assessed by three cycles of sorption-desorption. The sludge, owing to its high biosorption intensity and large availability from the local supply, is a better biosorbent for the treatment of Ni (II) and Cd (II) contaminated water.

Key words | brewery sludge, cadmium and nickel ions, kinetics, preferential biosorption

Rajeswari M. Kulkarni (corresponding author)
Department of Chemical Engineering,
Ramaiah Institute of Technology,
Bangalore,
India
E-mail: *rmkulkarni@msrit.edu*

Rajeswari M. Kulkarni
K. Vidya Shetty
G. Srinikethan
Department of Chemical Engineering,
National Institute of Technology,
Surathkal, Karnataka,
India

INTRODUCTION

The growing population and industrial development pose a serious threat to water resources and ecosystems. One threat is the accretion of heavy metals, such as cadmium, mercury, nickel, lead, chromium, arsenic and their compounds. These are very toxic and some are subjected to bio-magnification (Orhan & Büyükgüngör 1993; Aksu & Dönmez 2006; Ahmaruzzaman 2011). Nickel and cadmium are of special interest since they are widely used together in many industries such as electroplating, smelting, alloy preparation, electronics, and battery manufacturing (Abu Al-Rub *et al.* 2003; Febrianto *et al.* 2009). The concentrations of these metals in the groundwater of certain areas of the state of Karnataka in India are high (KSPCB; Singh *et al.* 2010). According to the Central Pollution Control Board (India), the permissible discharge level of Ni (II) and Cd (II) with many of the industrial effluents into inland water is 3 mg/L and 2 mg/L respectively (CPCB).

The technology used in the removal of metal ions from wastewater in the process industries has evolved from conventional methods like chemical precipitation to adsorption processes using biosorbents. Recently, there has been a considerable amount of research regarding the use of low-cost industrial waste as biosorbents for detoxification of water contaminated with metal ions (Table 1). Microorganisms such as bacteria and yeast are used in breweries and food industries to aid fermentation. These industries are facing problems related to safe sludge disposal (Kuyucak 1990). In the current study, assessing the possible utilization of industrial waste brewery sludge, for the removal of Ni (II) and Cd (II) ions from aqueous solutions, has been explored. This study also addresses the effect of a co-ion presence during simultaneous biosorption of metal ions on the sludge.

METHODS

Preparation of biosorbent

Industrial waste sludge containing a yeast biomass of *Saccharomyces carlsbergensis* was obtained from United Breweries Limited, Nelamangala, India. The wet sludge was sun-dried for 72 h and later oven dried at 60 °C for 24 h. The dry sludge was powdered in a mortar and the yeast powder was sieved through 80–100 mesh to obtain an estimated average particle size of 150 μm.

doi: 10.2166/wst.2019.090

Table 1 | Biosorption capacity of different low cost and industrial waste biosorbents for Ni (II) and Cd (II) removal

Metal	Biomass type	Metal uptake (mg/g)	Reference
Ni (II)	Fly ash	0.03	Rao *et al.* (2002)
	Baker's yeast	11.40	Padmavathy *et al.* (2003)
	Bagasse fly ash	1.12	Gupta *et al.* (2003)
	Tea factory waste	15.26	Malkoc & Nuhoglu (2005)
	Activated sludge	4.06	Ajaykumar *et al.* (2009)
	Brewer's yeast	5.34	Cui *et al.* (2010)
	Sugarcane bagasse	2.00	Alomá *et al.* (2012)
	Saccharomyces cerevisiae	1.57	Galedar & Younesi (2013)
	Nanocellulose fibers of rice straw	8.55	Kardam *et al.* (2014)
	Elaeagnus angustifolia	2.42	Amiri *et al.* (2014)
	Brewery sludge	**7.87**	**Present study**
Cd (II)	*Aspergillus niger*	4.0	Kapoor & Viraraghavan (1998)
	Bagasse fly ash	1.24	Gupta *et al.* (2003)
	Activated sludge	18.08	Ajaykumar *et al.* (2009)
	Brewer's yeast	10.17	Cui *et al.* (2010)
	Aspergillus aculeatus	14.24	Panday & Banerjee (2012)
	Baker's yeast	2.36	Wang (2012)
	Saccharomyces cerevisiae	3.74	Galedar & Younesi (2013)
	Nanocellulose fibers of rice straw	9.7	Kardam *et al.* (2014)
	Elaeagnus angustifolia	1.97	Amiri *et al.* (2014)
	Brewery sludge	**13.96**	**Present study**

Preparation of metal ion solution and metal ion analysis

Metal ion solutions of Ni (II) and Cd (II) of desired concentrations were prepared by suitably diluting the stock solutions. Stock solutions of 100 mg/L of Ni (II) and 100 mg/L of Cd (II) were prepared in 1,000 mL standard volumetric flasks by dissolving 0.4479 g of nickel sulphate [$NiSO_4.6H_2O$] and 0.2744 g of cadmium nitrate [$Cd(NO_3)_2.4H_2O$] in distilled water. For mixed metal studies, desired combinations of Ni (II) and Cd (II) were prepared by mixing suitable volumes of stock solutions of both metals. The initial pH of the metal solution was adjusted by using dilute sodium hydroxide or sulphuric acid solution.

The initial metal ion concentration and the residual concentration of Ni (II) and Cd (II) in the biosorption medium after biosorption was determined by atomic absorption spectrometer (GBC 932 plus) at a wavelength of 351.5 nm for Ni (II) and 326.1 nm for Cd (II). Before the analysis of residual metal ion concentrations, the samples were centrifuged, filtered using Whatman Grade 42 filter paper and the filtrate was analyzed using an atomic absorption spectrometer.

Batch biosorption experiments

The biosorption process was studied through batch experiments, to examine the influence of different factors such as initial pH (2.0–9.0), contact time (10–180 min), sludge dosage (2–40 g/L), initial metal ion concentration (10–50 mg/L), temperature (30–50 °C) and co-ion concentrations on biosorption of Ni (II) and Cd (II) on the sludge. The batch experiments were performed in 250 mL Erlenmeyer flasks with a working volume of 50 mL in an incubator orbital shaker at 150 rpm. All the experiments were conducted in duplicate and the mean values were shown in the results. Error bars are provided in all the related figures.

RESULTS AND DISCUSSION

Effect of initial pH

As shown in Figure 1, barely any biosorption was seen at a pH of 2.0 as there was a high concentration of protons, which compete with metal ions for the binding sites of the sludge. The percentage of biosorption for Ni (II) increased from 1% to 35% when the pH of the solution was increased from 2.0 to 5.0. Similarly, in the case of the Cd (II) ion, the percentage of biosorption was found to have increased from 49% to 73%, at pH of 2.0 to 6.0. At higher pH, the metal ions became more competitive with hydrogen ions, resulting in an increased biosorption (Bermúdez *et al.* 2011; Amiri *et al.*

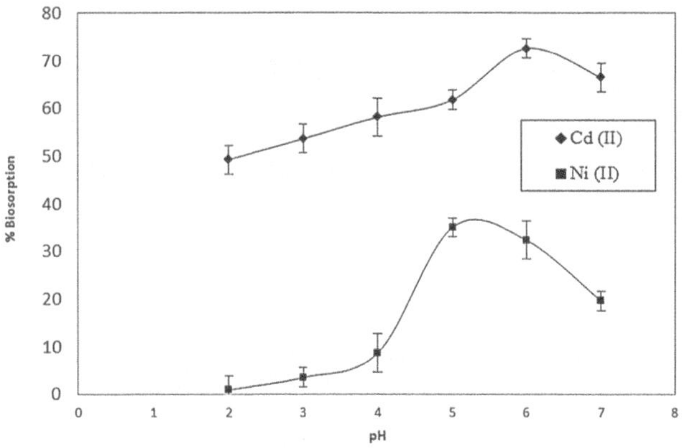

Figure 1 | Percentage metal biosorption vs pH during biosorption of Ni (II) and Cd (II) on the sludge in a single metal ion system at a temperature of 30 ± 2 °C, 24 h contact time, 50 mg/L initial metal ion concentration, 4 g/L biosorbent dosage, average particle size 150 μm.

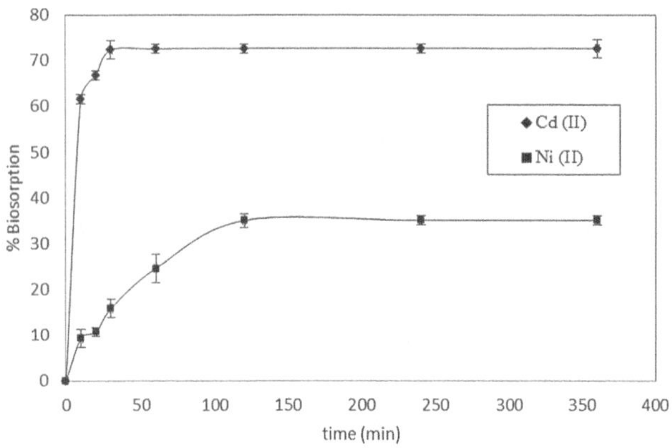

Figure 2 | Effect of contact time on % biosorption of Ni (II) and Cd (II) on the sludge in a single metal ion system at a temperature of 30 ± 2 °C, 50 mg/L initial metal ion concentration, 4 g/L biosorbent dosage, average particle size 150 μm, pH 5.0 for Ni (II) and pH 6.0 for Cd (II).

2014). The negative charge of the functional groups on the sludge promoted biosorption with metal cations. In an alkaline condition, formation of metal hydroxide complex inhibits the contact of metal ions to the biomass. The maximum biosorption of Ni (II) and Cd (II) on the yeast biomass was observed at a pH of 5.0 for Ni (II) and at pH 6.0 for Cd (II).

Effect of contact time and biosorption kinetics

As shown in Figure 2, the maximum removal rate for Ni (II) was 35% and 73% for Cd (II) at 120 min when equilibrium was achieved. The applicability of the pseudo-first order (Equation (1)) and pseudo-second order kinetic models (Equation (2)) were tested using the data shown in Figure 2.

$$\log(q_e - q) = \log q_e - \left(\frac{k_1}{2.303}\right)t \quad (1)$$

$$\frac{t}{q} = \frac{1}{k_2 q_e^2} + \frac{t}{q_e} \quad (2)$$

where, q and qe are metal uptake (mg/g) at time t and at equilibrium time, k_1 is the pseudo-first order rate constant (min^{-1}), t is the contact time (min), and k_2 is the pseudo-second order rate constant ($g\,mg^{-1}\,min^{-1}$).

The calculated model parameters are presented in Table 2. Ni (II) and Cd (II) biosorption onto the sludge follows a pseudo-second order kinetic model ($R^2 \approx 0.99$) but not a pseudo-first order model. This indicates that the surface reaction through sharing or exchange of ions is the rate controlling step for biosorption of Ni (II) and Cd (II) ions by the sludge (Ho & McKay 1999; Alomá *et al.* 2012).

Effect of sludge dosage

The increase in the sludge dosage increases the number of binding sites for biosorption. It is seen from Figure 3 that as the sludge dosage increased from 2 g/L to 40 g/L, the percentage biosorption was found to increase from 26% to 39% for Ni (II) and 66% to 79% for Cd (II). The dosage effect (2–40 g/L) is similar for both Ni (II) and Cd (II) biosorption, with a total increase in the removal rate of around 13%. The increase in the sludge dosage can be attributed to an increase in the active biosorbent sites (Akar *et al.* 2009; Kulkarni *et al.* 2018).

Table 2 | Pseudo-first order and pseudo-second order kinetic parameters for Ni (II) and Cd (II) biosorption on the sludge with goodness of fit of model

		First order kinetics			Second order kinetics		
Metal ion	**Experimental q_e (mg/g)**	**Predicted q_e (mg/g)**	**k_1 (min^{-1})**	**R^2**	**Predicted q_e (mg/g)**	**k_2 (g/mg.min)**	**R^2**
Ni (II)	4.37	2.28	0.026	0.8493	4.95	0.005	0.991
Cd (II)	9.07	0.31	0.029	0.7155	9.10	0.111	0.999

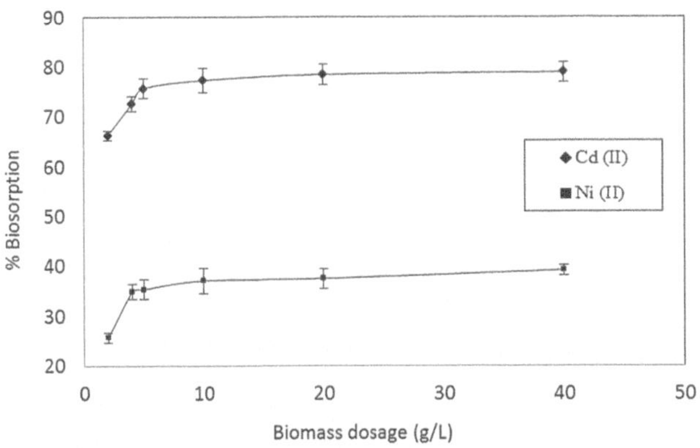

Figure 3 | Effect of biosorbent dosage on % biosorption of Ni (II) and Cd (II) on the sludge in a single metal ion system at a temperature of 30 ± 2 °C, 120 min contact time, 50 mg/L initial metal ion concentration, average particle size 150 μm, pH 5.0 for Ni (II) and pH 6.0 for Cd (II).

Effect of initial Ni (II) and Cd (II) metal ion concentration and temperature

The percentage biosorption decreased from 49% to 35% at 30 °C, 41% to 31% at 40 °C and 33% to 23% at 50 °C, with an increase in Ni (II) ion concentration from 10 mg/L to 50 mg/L (Figure 4). Similarly, the percentage biosorption decreased from 86% to 73% at 30 °C, 81% to 65% at 40 °C and 74% to 54% at 50 °C with an increase in Cd (II) ion concentration from 10 mg/L to 50 mg/L (Figure 5). The decrease in biosorption is due to the increased number of metal ions competing for the available active sites on the sludge (Guyo *et al.* 2016). It is evident from Figures 4 and 5 that the biosorption of Ni (II) and Cd (II) ions onto the sludge is controlled by an exothermic process. The maximum uptake occurred at 30 °C in the selected temperature range of 30 °C to 50 °C.

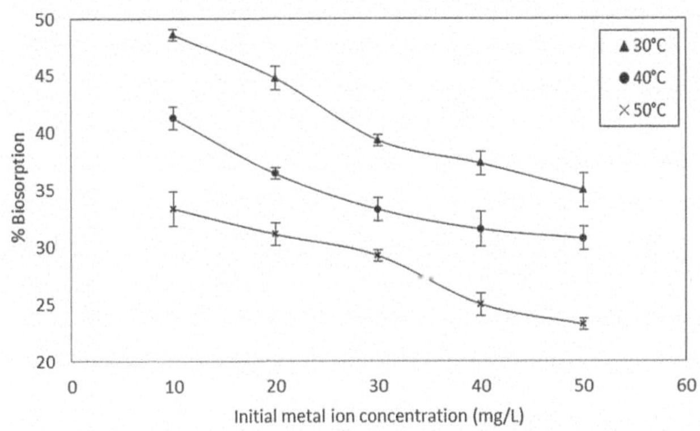

Figure 4 | Effect of initial Ni (II)ion concentration and temperature on % biosorption of Ni (II) on the sludge in a single metal ion system at pH 5.0, 120 min contact time, 4 g/L biosorbent dosage, average particle size 150 μm.

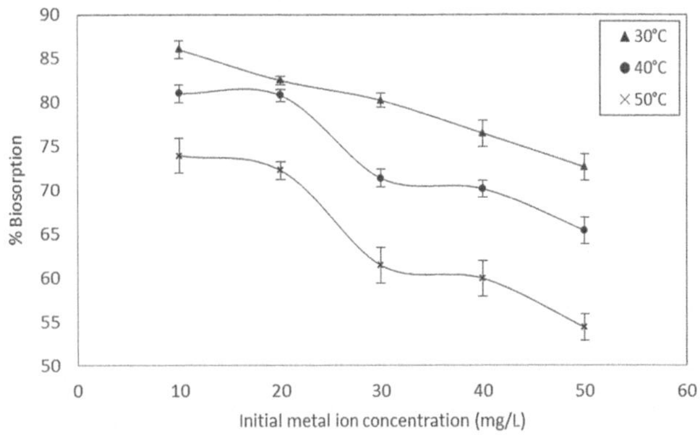

Figure 5 | Effect of initial Cd (II)ion concentration and temperature on % biosorption of Cd (II) on the sludge in a single metal ion system at pH 6.0, 120 min contact time, 4 g/L biosorbent dosage, average particle size 150 μm.

Isotherms for single metal biosorption

The Langmuir and Freundlich models were used to describe the equilibrium biosorption data (Table 3). The linearized form of the Langmuir isotherm (Equations (3) and (4)) and the Freundlich isotherm (Equation (5)) are

$$\frac{1}{q_e} = \frac{1}{q_{max} b} \frac{1}{C_e} + \frac{1}{q_{max}} \quad (3)$$

$$R_L = \frac{1}{1 + bC_0} \quad (4)$$

$$\ln q_e = \frac{1}{n} \ln C_e + \ln K \quad (5)$$

where q_{max} is the maximum monolayer biosorption capacity of the biosorbent (mg/g), b is biosorption affinity in L/mg, K and n are equilibrium constants indicative of biosorption capacity (mg g^{-1} (L mg – $1)^{1/n}$) and biosorption affinity respectively, R_L is the Hall separation factor, C_0 is the initial metal ion concentration in mg/L.

The values of the coefficient of determination being near to one (>0.97) suggest that both models provide a good fit to the equilibrium biosorption data at all the studied temperatures. Table 1 provides a summary of the sludge biosorption capacity compared with the different low cost and industrial waste biosorbents for Ni (II) and Cd (II) removal. It is clearly evident from the comparison that the maximum biosorption capacity of the sludge is fairly high compared to some of the reported biosorbents.

Effect of co-ions in biosorption of nickel (II) and cadmium (II)

Biosorption of Ni (II) and Cd (II) ions from the binary solutions of metal ions on the sludge was examined and

Table 3 | Langmuir and Freundlich isotherm parameters for Ni (II) and Cd (II) biosorption on brewery sludge

		Langmuir isotherm				Freundlich isotherm		
Metal ion	Temperature (°C)	q_{max} (mg/g)	b (L/mg)	R_L	R^2	K	n	R^2
Ni (II)	30	7.874	35.45×10^{-3}	0.36	0.9986	0.40	1.45	0.9956
	40	7.092	28.67×10^{-3}	0.41	0.9961	0.28	1.36	0.9995
	50	6.514	22.17×10^{-3}	0.47	0.9977	0.22	1.40	0.9827
Cd (II)	30	13.966	12.53×10^{-2}	0.13	0.9989	1.80	1.56	0.9924
	40	12.886	10.04×10^{-2}	0.16	0.9839	1.51	1.64	0.9659
	50	10.193	8.64×10^{-2}	0.18	0.9885	1.15	1.71	0.9686

contrasted with that of single metal ion biosorption (Tables S1 and S2, available with the online supplement, as are other tables and figures numbered with the prefix S). The metal uptake was found to decrease with an increase in co-ion concentration, indicating competitive biosorption and interference of co-ions during biosorption. Competition between metal ions for vacant sites resulted in the decrease in metal uptake by the sludge.

Biosorption isotherms – effect of co-ion on mono component isotherm parameters

Biosorption isotherms for the biosorption of Ni (II) and Cd (II) on the sludge were analyzed in both single and binary metal solutions to study the effect of co-ion concentrations on the mono-component isotherm parameters (Table 4). This helps in studying the alteration of equilibrium in the presence of co-ions. The values of the Langmuir and Freundlich constants, q_{max} and K were found to decrease with an increase in co-ion concentrations. The b and n values obtained were found to be higher for Cd (II) than for Ni (II), indicating a higher affinity of the sludge toward Cd (II). The sequence of selectivity followed the order of decreasing electronegativity, increasing ionic radius and atomic weight and decreasing hardness (Aksu & Dönmez 2006; Kulkarni *et al.* 2014). Cd (II) will be preferentially biosorbed over Ni (II) due to greater affinity for the carboxylic groups existing on the sludge (Abu Al-Rub *et al.* 2003; Kleinübing *et al.* 2011). It was observed that the biosorption capacity of the sludge in the dual metal system was lower than the single metal ion system.

Table 4 | Biosorption isotherm constants for Ni (II) and Cd (II) biosorption on the sludge in the absence and in the presence of other metal co-ions

		Langmuir constants			Freundlich constants		
Metal ion	Co-ion mg/L	q_{max} (mg/g)	b (L/mg)	R^2	K	n	R^2
Ni (II)	$C_{i\ Cd\ (II)} = 0$	7.87	0.035	0.998	0.405	1.453	0.996
	10	7.25	0.028	0.994	0.281	1.363	0.992
	30	6.66	0.026	0.993	0.244	1.347	0.997
	50	6.48	0.020	0.999	0.194	1.341	0.996
Cd (II)	$C_{i\ Ni\ (II)} = 0$	13.96	0.125	0.999	1.774	1.564	0.992
	10	13.16	0.092	0.994	1.360	1.533	0.985
	30	12.54	0.058	0.988	0.799	1.334	0.989
	50	11.34	0.056	0.978	0.696	1.328	0.988

Characterization of biomass

The surface images of the sludge before and after Ni (II) and Cd (II) biosorption (Figure S1) were obtained using a scanning electron microscope (JEOL, Japan) with energy dispersive X-ray spectroscopy (EDX). The biosorbent surface was observed to be rough before biosorption. Biosorption of metal ions has led to the formation of aggregates on the sludge surface, which may be due to complexation of metal ions with the cell surface. EDX analysis showed the presence of Ni (II) and the presence of Cd (II) on the biomass surface along with other components of the biomass sample (C,N,O,P,S) after Ni (II) and Cd (II) biosorption.

Infrared spectra of biosorbents loaded with and without metal ions were recorded using a Fourier transform-infrared (FTIR) spectrometer (Bruker alpha) in the range of 4,000–500 cm^{-1}, with a resolution of 4 cm^{-1} (Figure S2). After nickel biosorption, the band at 2,919.7 cm^{-1} shifts to 2,925.6 cm^{-1}, 1,641.5 cm^{-1} to 1,638.5 cm^{-1}, 1,532.6 cm^{-1} to 1,535.5 cm^{-1} and 1,037.3 cm^{-1} to 1,039.7 cm^{-1}. Similar observations were made with the FTIR spectra of the sludge after Cd (II) biosorption with a band shift from 2,919.7 cm^{-1} to 2,923.6 cm^{-1}, 1,641.5 cm^{-1} to 1,637 cm^{-1}, 1,532.6 cm^{-1} to 1,531.5 cm^{-1} and 1,037.3 cm^{-1} to 1,034.6 cm^{-1}, which indicates that the $-CH_2$-, -OH, -C = O-, $-NH_2$ and C-O anionic functional groups present on the sludge are involved in the biosorption.

Desorption, regeneration and reuse studies

The application of biosorbent towards detoxification of metal bearing effluents depends not only on its biosorption capacity, but also on how well the metal may be recovered from the biosorbent surface; the possibility of biomass regeneration and reuse. The metal loaded biomass, after biosorption was contacted with 50 mL of 0.1N HCl elutant to study desorption of metal ions (Table S3).

The reusability of the sludge was evaluated in three consecutive cycles of biosorption-desorption using HCl elutant and regeneration using distilled water in a batch system. Acid treatment with 0.1 M HCl was found to be the most effective, where more than 90% of metal was desorbed from the spent sludge throughout the three cycles. Only 12.7% decrease in sorption of Ni (II) and 5.62% decrease in sorption of Cd (II) by the biomass were observed in three cycles, indicating effectiveness of the desorption – regeneration process (Table S4).

CONCLUSIONS

This study identified brewery sludge as a suitable biosorbent to be utilized for treating Ni (II) and Cd (II) ions from metal contaminated water. The optimal biosorption conditions for the selected metal removal were determined using the batch mode of studies. The biosorption mechanism followed the pseudo-second order kinetic model. Preferential biosorption of Cd (II) over Ni (II) was observed in both cases of single and dual metal ion biosorption on to the sludge. The metal uptake was found to decrease with increase in co-ion concentration, indicating competitive biosorption and interference of co-ions during biosorption. Brewery sludge possesses advantages such as abundant availability, good biosorption capacity and intensity and ease of regeneration with possibilities of metal recovery. This makes biosorption by the sludge an attractive treatment option for Ni (II) and Cd (II) bearing water.

ACKNOWLEDGEMENT

We are grateful to Ramaiah Institute of Technology, Bangalore and NITK, Surathkal, for their continuous support.

CONFLICT OF INTEREST

The authors have declared no conflict of interest.

REFERENCES

Abu Al-Rub, F. A., Kandah, M. & Al-Dabaybeh, N. 2003 Competitive adsorption of nickel and cadmium on sheep manure wastes: experimental and prediction studies. *Separation Science and Technology* **38** (2), 483–497.

Ahmaruzzaman, M. 2011 Industrial wastes as low-cost potential adsorbents for the treatment of wastewater laden with heavy metals. *Advances in Colloid and Interface Science* **166** (1), 36–59.

AjayKumar, A. V., Darwish, N. A. & Hilal, N. 2009 Study of various parameters in the biosorption of heavy metals on activated sludge. *World Applied Sciences Journal (Special Issue for Environment)* **5** (5), 32–40.

Akar, T., Kaynak, Z., Ulusoy, S., Yuvaci, D., Ozsari, G. & Akar, S. T. 2009 Enhanced biosorption of nickel (II) ions by silica-gel-immobilized waste biomass: biosorption characteristics in batch and dynamic flow mode. *Journal of Hazardous Materials* **163** (2), 1134–1141.

Aksu, Z. & Dönmez, G. 2006 Binary biosorption of cadmium (II) and nickel (II) onto dried *Chlorella vulgaris*: co-ion effect on mono-component isotherm parameters. *Process Biochemistry* **41** (4), 860–868.

Alomá, I., Martín-Lara, M. A., Rodríguez, I. L., Blázquez, G. & Calero, M. 2012 Removal of nickel (II) ions from aqueous solutions by biosorption on sugarcane bagasse. *Journal of the Taiwan Institute of Chemical Engineers* **43** (2), 275–281.

Amiri, M. J., Fadaei, E., Baghvand, A. & Ezadkhasty, Z. 2014 Removal of heavy metals Cr (VI), Cd (II) and Ni (II) from aqueous solution by bioabsorbtion of *Elaeagnus angustifolia*. *International Journal of Environmental Research* **8** (2), 411–420.

Bermúdez, Y. G., Rico, I. L. R., Bermúdez, O. G. & Guibal, E. 2011 Nickel biosorption using *Gracilaria caudata* and *Sargassum muticum*. *Chemical Engineering Journal* **166** (1), 122–131.

CPCB. Central Pollution Control Board. http://www.cpcb.nic.in/Industry_Specific_Standards.php.

Cui, L., Wu, G. & Jeong, T. S. 2010 Adsorption performance of nickel and cadmium ions onto brewer's yeast. *The Canadian Journal of Chemical Engineering* **88** (1), 109–115.

Febrianto, J., Kosasih, A. N., Sunarso, J., Ju, Y. H., Indraswati, N. & Ismadji, S. 2009 Equilibrium and kinetic studies in adsorption of heavy metals using biosorbent: a summary of recent studies. *Journal of Hazardous Materials* **162** (2), 616–645.

Galedar, M. & Younesi, H. 2013 Biosorption of ternary cadmium, nickel and cobalt ions from aqueous solution onto *Saccharomyces cerevisiae* cells: batch and column studies. *American Journal of Biochemistry and Biotechnology* **9** (1), 47–60.

Gupta, V. K., Jain, C. K., Ali, I., Sharma, M. & Saini, V. K. 2003 Removal of cadmium and nickel from wastewater using bagasse fly ash – a sugar industry waste. *Water Research* **37** (16), 4038–4044.

Guyo, U., Sibanda, K., Sebata, E., Chigondo, F. & Moyo, M. 2016 Removal of nickel (II) from aqueous solution by *Vigna unguiculata* (Cowpea) pods biomass. *Water Science and Technology* **73** (10), 2301–2310.

Ho, Y. S. & McKay, G. 1999 Pseudo second order model for sorption processes. *Process Biochemistry* **34**, 451–465.

Kapoor, A. & Viraraghavan, T. 1995 Fungal biosorption – an alternative treatment option for heavy metal bearing wastewaters: a review. *Bioresource Technology* **53** (3), 195–206.

Kapoor, A. & Viraraghavan, T. 1998 Biosorption of heavy metals on *Aspergillus niger*: effect of pretreatment. *Bioresource Technology* **63** (2), 109–113.

Kardam, A., Raj, K. R., Srivastava, S. & Srivastava, M. M. 2014 Nanocellulose fibers for biosorption of cadmium, nickel, and lead ions from aqueous solution. *Clean Technologies and Environmental Policy* **16** (2), 385–393.

Kleinübing, S. J., Da Silva, E. A., Da Silva, M. G. & Guibal, E. 2011 Equilibrium of Cu (II) and Ni (II) biosorption by marine alga Sargassum filipendula in a dynamic system: competitiveness and selectivity. *Bioresource Technology* **102** (7), 4610–4617.

KSPCB. Karnataka State Pollution Control Board. http://kspcb.kar.nic.in/annual_reports.html.

Kulkarni, R. M., Shetty, K. V. & Srinikethan, G. 2014 Cadmium (II) and nickel (II) biosorption by *Bacillus laterosporus* (MTCC 1628). *Journal of the Taiwan Institute of Chemical Engineers* **45** (4), 1628–1635.

Kulkarni, M. R., Bhagyalakshmi, C., Anand, D. & Herle, R. N. 2018 Removal of remazol brilliant blue dye from aqueous solutions using water hyacinth root powder. *Desalination and Water Treatment* **122**, 331–338.

Kuyucak, N. 1990 Feasibility of biosorbents application. In: *Biosorption of Heavy Metals* (B. Volesky, ed.). CRC Press, Boca Raton, FL, pp. 371–378.

Malkoc, E. & Nuhoglu, Y. 2005 Investigations of nickel (II) removal from aqueous solutions using tea factory waste. *Journal of Hazardous Materials* **127** (1), 120–128.

Orhan, Y. & Büyükgüngör, H. 1993 The removal of heavy metals by using agricultural wastes. *Water Science and Technology* **28** (2), 247–255.

Padmavathy, V., Vasudevan, P. & Dhingra, S. C. 2003 Biosorption of nickel (II) ions on Baker's yeast. *Process Biochemistry* **38** (10), 1389–1395.

Panday, A. & Banerjee, D. 2012 Biosorption of cadmium (II) using discarded biomass of *Aspergillus aculeatus* DBF9. *Terrestrial and Aquatic Environmental Toxicology* **6** (1), 8–13.

Rao, M., Parwate, A. V. & Bhole, A. G. 2002 Removal of Cr^{6+} and Ni^{2+} from aqueous solution using bagasse and fly ash. *Waste Management* **22** (7), 821–830.

Singh, M. J., Somashekar, R. K., Prakash, K. L. & Shivanna, K. 2010 Investigation of heavy metals in crystalline aquifer groundwater from different valleys of Bangalore, Karnataka. *Journal of Geography and Regional Planning* **3** (10), 262–270.

Wang, Y. 2012 Optimization of cadmium, zinc and copper biosorption in an aqueous solution by *Saccharomyces cerevisiae*. *International Journal of Chemistry* **01**, 1–13.

First received 13 July 2018; accepted in revised form 23 February 2019. Available online 7 March 2019

Assessing local materials for the treatment of wastewater in open drains

Priyanka Jamwal, Daniel Phillips and Kim Karlsrud

ABSTRACT

In the present study, three low-cost filter aggregate materials were tested and compared for organic matter and fecal coliform (FC) removal at the laboratory scale. Setups were subjected to synthetic wastewater at two hydraulic loading rates (HLR), i.e. 4 cm/day and 40 cm/day. The hydraulic retention time (HRT) at the two HLRs varied from 4 days to 12 h, respectively. The result obtained shows that the biochemical oxygen demand (BOD_5) removal efficiency of aggregate materials decreased with the increase in HLR. Both at high and low HLR, the terracotta aggregate material exhibited maximum BOD_5 loading removal and without significant difference for the case of FC removal efficiency for all the three aggregate materials. At higher HLR, cell debris and biofilm loss from the aggregate material contributed to the chemical oxygen demand (COD) levels in the treated water. The terracotta aggregate material provided best organic matter removal at both HLRs. The study demonstrates the potential of incorporating inexpensive and readily available local materials into decentralized, frugal green infrastructure interventions capable of lowering the quantum of harmful biological contaminants in open storm water channels in rapidly urbanizing cities of developing countries, and that the terracotta aggregate material provided best organic removal at both HLRs.

Key words | filtration, green infrastructure, megacities, storm drains, terracotta, wastewater

Priyanka Jamwal (corresponding author)
Centre for Environment and Development,
Ashoka Trust for Research in Ecology and the Environment (ATREE),
Jakkur, 560064,
India
E-mail: *priyanka.jamwal@atree.org*

Daniel Phillips
COMMONstudio (http://www.thecommonstudio.com/), With support from the U.S. Fulbright Program (Fulbright-Nehru Fellowship), which is co-sponsored by USIEF (United States-India Educational Foundation)

Daniel Phillips
Kim Karlsrud
School for Environment and Sustainability (SEAS),
University of Michigan,
State Street,
Ann Arbor, MI 48109,
USA

INTRODUCTION

Explosive urban growth throughout India and elsewhere in the global south continues to exert extreme pressure on urban wastewater systems, threatening human health, natural ecosystems, and livelihoods. Despite numerous laws, regulations and apocalyptic headlines, the widespread unchecked dumping of contaminated wastewater from domestic and industrial sources into urban lakes and streams will likely continue into the foreseeable future, as will it's associated impacts on public health (Jumbe & Nandini 2009; Ramachandra *et al.* 2015; Jamwal 2017). The interconnected nature of urban hydrological infrastructure means that many common domestic and industrial contaminants such as excess organic matter, biological oxygen demand (BOD_5), nutrients (nitrogen and phosphorus), trace metals, fecal coliforms (FC) and disease causing microorganisms tend to hyper-accumulate as they make their way through the system, causing a range of human and environmental health impacts downstream (Suthar *et al.* 2010; Jamwal *et al.* 2015). This includes lake eutrophication, fish kills, toxic foam, fire events, and downstream agricultural contamination. The implications of degraded urban surface water impacts stakeholders from across the socio-economic spectrum, and may impact the very viability of many burgeoning megacities in the near future (Bhasthi 2017). Larger infrastructural responses are required, and in some cases being implemented, but are often not scaling fast enough to keep pace with this growing crisis (Jamwal 2018). Inadequate investment, training and maintenance often limits the performance of existing large scale, technologically intensive infrastructures such as sewage treatment plants (Morel 2006; Jamwal *et al.* 2015). These realities underscore the urgent need for more frugal, flexible, decentralized responses aimed at supplementing existing efforts from the bottom up.

Elsewhere in the world, the notion of green infrastructure (GI) has gained much as a response to the issues of urban watershed contamination (Mell 2010; Kato 2011). GI is often modelled after naturally occurring wetland ecologies

doi: 10.2166/wst.2019.105

and deployed to reduce peak flows and reduce harmful pollutants in urban wastewater (Pisoeiro *et al.* 2016). Much of the existing GI research of water quality tends to focus on the performance of introduced vascular plants and macrophytes as tertiary treatment systems (Horppila & Nurminen 2001; Demırezen & Aksoy 2004; Malarvizhi *et al.* 2010; Pisoeiro *et al.* 2016). However, in developing countries such as India, urban storm water and domestic wastewater systems are often interwoven into an informal combined sewer overflow (CSO) system in which primary and secondary treatment is either non-existent or inconsistently provided (Morel 2006; Tortajada 2016). Moreover, urban wastewater systems in many Indian megacities is subject to high levels of 'infrastructural disarray', marked by many hidden failures and inefficiencies at various scales (Alley *et al.* 2018). In such a system, conventional approaches to GI are inadequate and inappropriate. New approaches are necessary that account for the myriad constraints presented by South Asian megacities.

What alternative approaches to GI might be possible for the developing world? How might we design more frugal, flexible and inclusive modes of GI which could be easily installed by local communities to compliment top-down infrastructures such as conventional wastewater treatment plants? Existing literature in the field of water science demonstrates that significant reductions in BOD_5 and total suspended solids (TSS) could be achieved even in unplanted (i.e. 'control') constructed wetland systems, which were likely achieved through the ability of basic material substrates to accommodate the growth of biofilms (Weerakoon *et al.* 2013; Pisoeiro *et al.* 2016). These insights suggest that the intentional introduction of 'disruptive' material aggregates into uncovered urban wastewater channels (referred to locally as 'Nallahs') may have the potential to provide immediate benefits to water quality (Prominski *et al.* 2012; Bhatnagar 2017). The objective of this study is therefore to add robustness to this emerging field of GI innovation in the Indian context by evaluating the comparative decontamination performance of a variety of basic, readily available aggregate materials which could be cheaply and easily introduced into urban wastewater channels. The novel approach explored in this study is to better understand how higher and more variable hydraulic loading rate (HLR) and hydraulic retention time (HRT) impact the removal efficiency of these filtration processes, and thus their likely performance in the context of real-world 'instream' interventions. Insights from this laboratory-based research are being leveraged to inform the design and implementation of small scale watershed interventions within the southern Indian city of Bangalore.

MATERIAL AND METHODS

Experimental setup

Subsurface filtration systems were set up to evaluate the contaminant removal efficiencies of three aggregate materials: terracotta, gravel, and cinder. These materials are inexpensive and readily available in the market and have distinct characteristics which make them suitable for treating wastewater. Terracotta is a lightweight porous material, cinder is a low strength material with substantial porosity, and gravel is a high-density material with low porosity (Yang *et al.* 2017). These materials have been used in bio-filters for treating domestic wastewater in the past (Lekang & Kleppe 2000; Tekerlekopoulou & Vayenas 2008).

The study was conducted by running synthetic wastewater through these material substrates at a variety of HLRs to compare their relative decontamination performance and potential. The experimental setup is illustrated in Figure 1, consisting of two overhead tanks, one for mixing and one for distribution (150 L and 200 L respectively). These tanks were gravity fed into three 'filter units', comprised of heavy duty rectangular plastic tubs measuring 940 mm (L) × 630 mm (W) × 535 mm (H) filled uniformly with material aggregates. Aggregate pieces ranged from 40–90 mm in approximate size and were all obtained from local construction sites at a nominal cost.

Synthetic wastewater

Synthetic wastewater was prepared using the mixture of urea, sugar, ammonium chloride, potassium phosphate, and FC culture prepared at ATREE's Water and Soil Laboratory. The quantities of constituents added per liter of water are presented in Table 1. For the preparation of synthetic wastewater, these constituents were thoroughly mixed and added to overhead storage tanks used in the experimental setup.

Table 2 presents the characteristics of the synthetic wastewater. The BOD_5, chemical oxygen demand (COD), ammonia-N, nitrates, and FC levels in the synthetic wastewater were similar to the theoretical levels estimated using the stoichiometric equations.

System operation

The BOD_5, COD, ammonia-N and FC levels in the synthetic wastewater represented the wastewater quality encountered in the open storm drains in the city of Bangalore (Jamwal

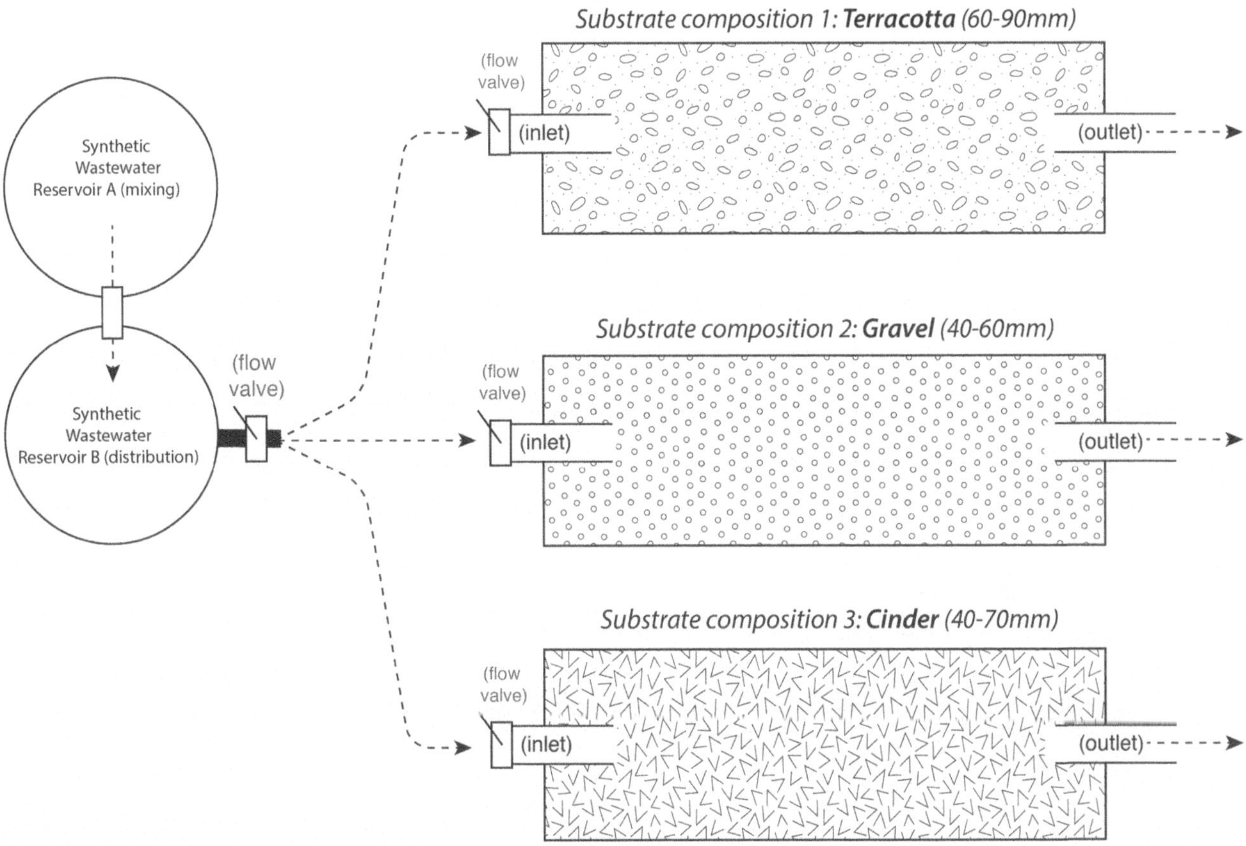

Figure 1 | The experimental setup.

et al. 2015). The effectiveness of filtration setups was evaluated at two HLRs: 4 cm/day and 40 cm/day. The HRT was estimated considering the porosity of the filter media (0.60). Table 3 presents the HRT at different HLRs.

The experimental setup consists of two tanks: the storage tank and the distribution tank (Figure 1). The storage tank was connected to the distribution tank with a check valve. The check valve was mounted on the distribution tank to maintain the constant HLR in the filtration setups. Throughout the experiment, the water depth in the distribution tank remained constant, and the storage tank was filled manually after a fixed interval of time.

Table 1 | Composition of the synthetic wastewater

Compounds	Quantity added/L
Sugar	0.53 g
Urea	0.005 g
NH_4Cl	0.117 g
KH_2PO_4	0.010 g
FC (suspended in NaCl)	0.1 ml

For acclimatization, the filtrations setups were fed with synthetic wastewater initially for three weeks. After the third week, the filtration setups were operated at two HLRs, i.e. 4 cm/day and 40 cm/day for six weeks each. The evaluation period (six weeks) selected was determined according to those observed in studies used by Chen *et al.* (2006), Chyan *et al.* (2013) and Ghosh & Gopal (2010).

Table 2 | Characteristics of the synthetic wastewater at two hydraulic loading rates

	Synthetic wastewater	
Water quality	Mean ($n = 6$) ± Std. Dev (HLR-4 cm/day)	Mean ($n = 4$) ± Std. Dev (HLR-40 cm/day)
pH	6.67 ± 0.56	7.67 ± .08
Conductivity (μS/cm)	1,733 ± 40	1,645 ± 29
BOD_5 (mg/L)	273 ± 46	250 ± 25
COD (mg/L)	396.0 ± 46.7	397.5 ± 35.8
Ammonia – N (mg/L)	28.8 ± 6.4	25.5 ± 4.5
Nitrate – N (mg/L)	6.90 ± 7.8	9.30 ± 10.5
$Log_{10}FC$ (MPN/100 mL)	5.67 ± 0.82	5.60 ± 0.68

Table 3 | Experimental design for efficiency evaluation of filtration setup

HLR (cm/day)	Influent flow rate (l/day)	HRT (day)	Maximum loading rate (MLR) Ammonia – N (g/m²/day)	BOD$_5$ (g/m²/day)	COD (g/m²/day)	FC (MPN m²/day)	Overhead tank refilling frequency
4	22.6	4	1.1	8	13.2	$8*10^5$	Once in two days
40	226	0.5	10.3	112	168	$3*10^9$	Two times in a day

Assessment of organic matter and FC removal efficiencies

Pollutant removal efficiencies were evaluated by collecting water samples weekly twice at the inlet and the outlet of the three filtration setups. Samples were analyzed for BOD$_5$, COD, ammonia-N, nitrates and FC concentration. Standard methods for the examination of water and wastewater were followed to test the water quality parameters (APHA 2005). Removal efficiencies were calculated as a percentage reduction in the concentration from influent to the effluent samples using Equation (1). The corresponding mass loading rates (MLR) and mass removal rates (MRR) were also estimated. The MLR and MRR for the BOD$_5$, COD and FC were calculated using Equations (2) and (3).

$$\text{Removal efficiency (RE)} = \frac{(C_i - C_o)}{C_i} \times 100 \qquad (1)$$

$$\text{Mass loading rate} = (C_i - C_o) \times (\text{HLR}) \qquad (2)$$

$$\text{Mass removal rate (MRR)} = (C_i - C_o) \times (\text{HLR}) \times 100 \qquad (3)$$

where C_i = concentration of contaminant in the influent and C_o = concentration of the contaminant in the effluent.

Statistical tests

Significant differences in the contaminated removal efficiencies of three aggregate materials subjected to different HLRs were evaluated using one-way analysis of variance (ANOVA) test for the normally distributed data at 0.05 significant levels. The correlation between the MLR and the MRR were identified using linear regression.

RESULTS AND DISCUSSION

Wastewater characteristics and effluent quality

The influent characteristics of synthetic wastewater are presented in Table 2. Significant variation in the influent quality was observed at lower HLR. This could be attributed to the longer refilling time of synthetic wastewater into the overhead tank. The biological transformation in the overhead and the distribution tanks may have caused variations in the quality of synthetic wastewater. Similar variations were reported in the studies where in synthetic sewage is used for carrying out experiments under controlled conditions (Weerakoon *et al.* 2013; Xu *et al.* 2018).

The effluent characteristics from three filter media setup are shown in Table 4. A statistical test (Kolmogorov-Smirnov) showed that the influent and effluent characteristics at different HLRs were normally distributed. The average, maximum and minimum contaminant levels in influent and the effluent of the setups were estimated. For FC, the data were converted into the log normal form for the analysis.

Figure 2 shows the levels of BOD$_5$, COD, ammonia-N, and FC in the influent and effluent during the course of the experiment. The BOD$_5$ represents the biodegradable component of organic matter and COD represents both the biodegradable and non-biodegradable component along with chemically oxidisable compounds present in the wastewater (Kim 1989). Ammonia-N represents the excess nutrients levels commonly present in domestic grey and blackwater. High ammonia-N levels in the effluent indicate poor nutrient removal capabilities of the treatment system. Presence of FC indicates the possible presence of pathogens in the source water.

The synthetic sewage represents characteristics of wastewater flowing in the open storm water channels. The average BOD$_5$, COD and FC levels are similar to the levels reported in Vrishabhavathy stream in the city of Bangalore (Jamwal *et al.* 2015). Considering the use of aggregate material for the removal of contaminants *in situ*, the pilot setup was exposed to very high to normal flow rates, i.e. ranging from 226 L/day to 22.6 L/day. These flow rates replicate the wastewater flows in the open storm water drains and corresponds to the 40 cm/day and 4 cm/day HLRs, respectively. Generally, the performance of subsurface treatment systems is tested at low contaminant loading rates (Mashauri *et al.* 2000; Chen *et al.* 2006; Ghosh & Gopal 2010). However, considering the application of this technology in treating highly

Table 4 | Water quality characteristics of effluent from three filtration setups

Water quality	Outlet 1 – Terracotta		Outlet 2 – Gravel		Outlet 3 – Cinder	
Parameter	HLR – 4 cm/m² Mean ± Std. Dev	HLR – 40 cm/m² Mean ± Std. Dev	HLR – 4 cm/m² Mean ± Std. Dev	HLR – 40 cm/m² Mean ± Std. Dev	HLR – 4 cm/m² Mean ± Std. Dev	HLR – 40 cm/m² Mean ± Std. Dev
pH	7.34 ± 0.33	7.3 ± 0.88	7.9 ± 0.38	7.5 ± 0.65	7.4 ± 0.44	7.5 ± 0.90
Conductivity (μS/cm)	1,779 ± 102.5	1,644 ± 121	1,669 ± 77	1,649 ± 94	1,704 ± 43	1,700 ± 140
BOD_5 (mg/L)	26 ± 15	69 ± 22	34 ± 24	81 ± 18	127 ± 43	102 ± 50
COD (mg/L)	104.2 ± 24.2	232.5 ± 84.2	115.8 ± 45.8	292.0 ± 100.1	143.3 ± 75.8	260 ± 117.4
Ammonia – N (mg/L)	17.0 ± 6.0	19.5 ± 2.0	14.5 ± 6.3	16.5 ± 1.2	22 ± 10	19.5 ± 1
Nitrate – N (mg/L)	1.7 ± 0.88	2.7 ± 2.0	1.2 ± 0.45	3.6 ± 4.2	1.0 ± 0.50	0.70 ± 1.17
$Log_{10}FC$ (MPN/100 mL)	3.3 ± 0.8	4.5 ± 1.0	3.7 ± 0.8	4.5 ± 0.6	3.7 ± 0.8	4.3 ± 1.0

contaminated flows in the channeled sections of the open storm water drains, the setups were exposed to medium to very high contaminants loading rates (Table 3).

Significant reductions in, BOD_5, COD and FC levels were observed in all three setups. Increase in the contaminants removal efficiency was observed at the higher HRT.

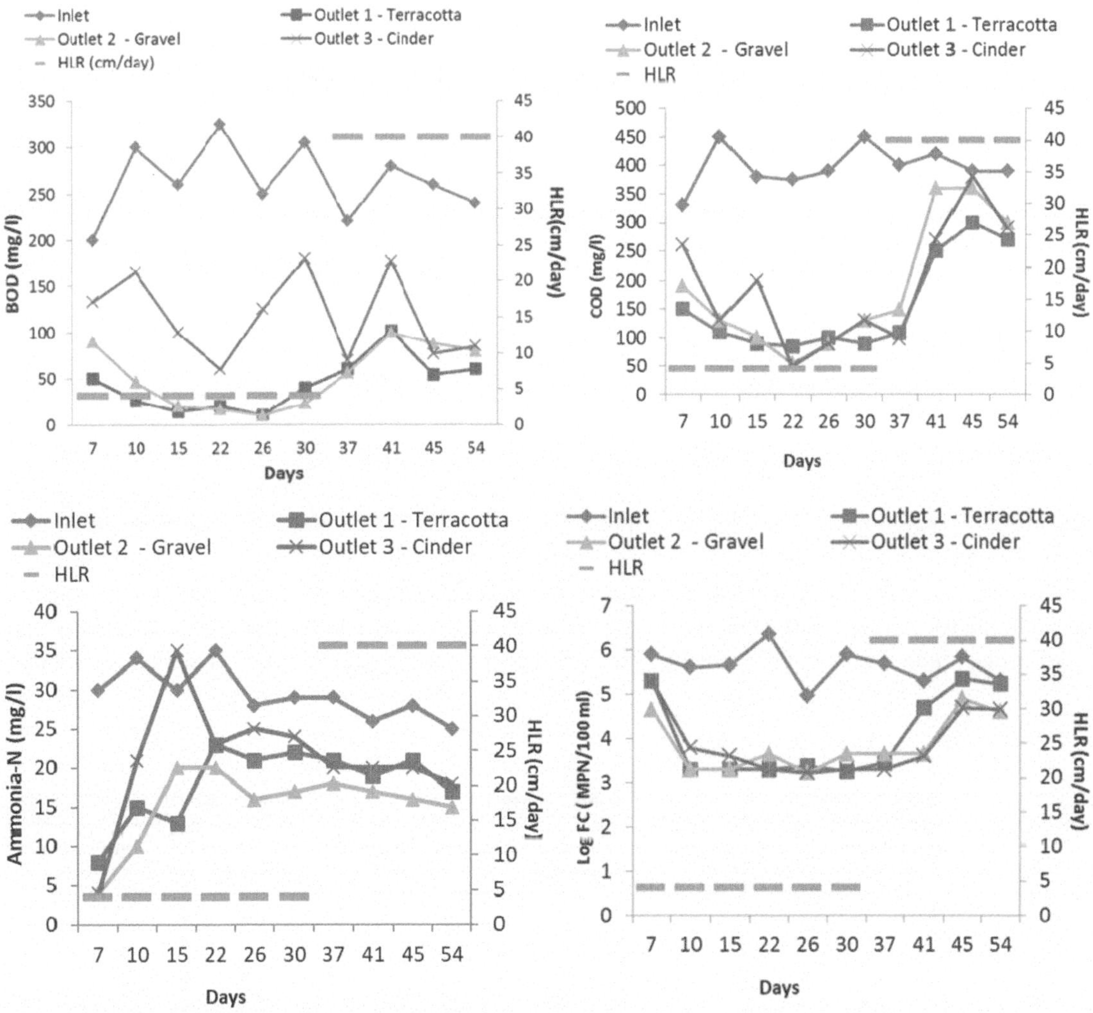

Figure 2 | Effluent quality from three filtration setup during the experiment at different HLR.

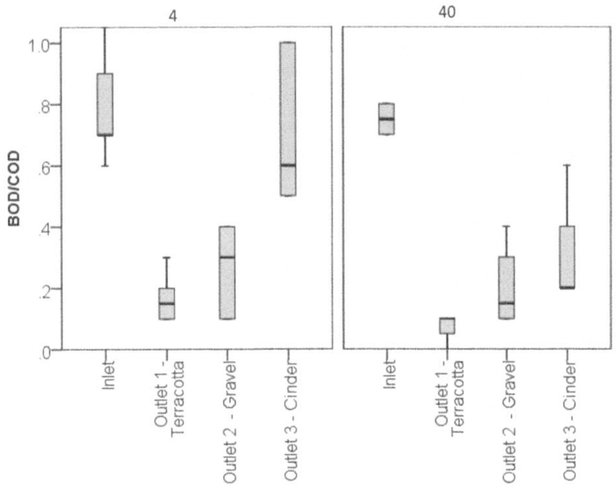

Figure 3 | BOD_5/COD ratio observed at two HLRs.

The COD levels observed in the influent and effluents were much higher than the BOD_5 levels. Given that glucose contributed primarily to the BOD_5 whereas both glucose and other chemically oxidisable compounds contributed to COD, the levels of COD observed were greater than the BOD_5. Various physical, chemical, and biological processes help in the removal of contaminants from the wastewater. Sedimentation and biological decomposition is mainly responsible for the removal of organic matter, i.e. COD and BOD_5.

The contaminant levels observed in the effluent from three setups followed a similar trend. As compared to lower HLR, poor effluent quality was observed at higher HLR, i.e. 40 cm/day. This could be attributed to a) increased flow velocities at high HLR and b) reduced contact time of contaminants with the filter media. At higher flow velocities, biofilm abrasion and removal of cell debris contributes to the COD. The effluent debris and biofilms are composed of complex organic matter that is hard to decompose and hence contributes to the COD (Behnke *et al.* 2011). Also, reduced contact time lowers the microbial degradation of organic matter present within the filtration system. BOD_5/COD ratio indicates the proportion of readily oxidisable organic matter present in the effluent. Figure 3 presents the BOD_5/COD ratio in the influent and effluent at different HLR. A drop in the average BOD_5/COD ratio was observed for gravel and cinder based filtration setups, whereas for terracotta, the BOD_5/COD ratio remain same at both high and low HLRs. This indicates the stable and abrasion resistant nature of biofilms formed on the terracotta material. Few studies have reported the impact of terracotta balls in removal of bacterial species and organic matter. One of the studies suggested that the biofilm formation on the terracotta balls in the slow sand filtration process is stable and effectively removes biodegradable organic compounds from the source water (Le Dantec *et al.* 2002).

Except for the cinder based filtration setup, higher reduction in the ammonia-N levels were observed at the low HLR (Table 5). The reduction of ammonia-N in the setups could be attributed to the biofilm formation within the system. The low levels of nitrate-N observed in the effluents suggests limited conversion of ammonia-N into nitrate-N. This suggests that in the absence of nitrification and plants, uptake of ammonia-N by heterotrophic bacteria for biofilm formation is major reason for its reduction (Ribot *et al.* 2013). Both terracotta and gravel setup exhibited similar removal efficiencies at lower HLR. However, at higher HLR, gravel setup exhibited the highest ammonia-N removal efficiency.

Out of the three materials tested, the terracotta based filtration setup produced the best overall quality effluent both at low and high HLR. Efficacy of terracotta material in treating surface and groundwater for drinking water purposes is well established. Various studies reported the effectiveness of terracotta material in reducing the organic carbon and coliphage in the drinking water (Low 2002; Franz 2005). At lower HLR, the BOD_5 and COD levels in the effluent from terracotta filtration system met effluent

Table 5 | Contaminant removal efficiency of three filtration setup at different HLRs

Removal efficiency	Outlet 1 - Terracotta		Outlet 2 - Gravel		Outlet 3 - Cinder	
Parameter	HLR – 4 cm/m^2 % Removal	HLR – 40 cm/m^2 % Removal	HLR – 4 cm/m^2 % Removal	HLR – 40 cm/m^2 % Removal	HLR – 40 cm/m^2 % Removal	HLR – 40 cm/m^2 % Removal
BOD_5 (mg/L)	90	76	86	68	54	59
COD (mg/L)	73	41	70	27	62	35
Ammonia – N (mg/L)	45	28	53	39	28	27
Log_{10}FC (MPN/100 mL) removal*	2	1.2	1.6	1.5	1.8	1.5

discharge standards for secondary treated water. No significant difference was observed in the FC levels of the effluent from all the three filtration setups. The ability of solar disinfection SODIS in reducing the *Escherichia coli* (*E. coli*) levels by 5 log orders in 2.5 h is well documented in literature (Lonnen *et al.* 2005). Given that the three setups were located outdoors and were equally exposed to the sunlight, the effect of ultraviolet radiation in removal of FC is likely more significant than the other mechanisms such as sedimentation, adsorption and natural die-off (Mashauri *et al.* 2000).

Pollutant removal efficiencies

Table 5 illustrates the Ammonia-N, BOD_5, COD and FC removal efficiencies of three filtration setups over the entire study period. The HRT at respective HLRs is reported in Table 3. With the exception of the cinder material filtration unit (where no significant difference was observed in the ammonia-N removal efficiencies at both low and high HLR), ammonia-N, BOD_5, COD and FC removal efficiency followed a similar trend, i.e. decrease in pollutant removal efficiency at higher HLR.

The organic matter removal mechanism in the filter media includes adsorption, sedimentation and microbial metabolism. Filter media with high porosity provides additional capacity as well as space and a favorable environment for microbial growth for better organic matter removal (Chyan *et al.* 2013). Further, porosity increases the adsorption of organic matter on the surface thereby improves the organic matter removal efficiency. The terracotta setup provided the highest BOD_5 and COD removal efficiency at both HLRs. In the case of gravel based filtration system, a significant drop was observed in BOD_5 and COD removal efficiency at high HLR. Of all three experimental setups, the cinder setup was found to be the least efficient. The poor efficiency of cinder aggregate material could be attributed to a significant drop in the BOD_5/COD ratio at high HLR. Also, the frequent peaks in the effluent BOD_5 and COD indicates the periodic abrasion of biofilm from the surface of cinder material (Figure 3). Ammonia-N removal is the function of biofilm response to N uptake on exposure to ammonia-N (Ribot *et al.* 2013). Given that both terracotta and gravel (to a lesser extent) exhibits stable and abrasion resistant biofilm, at lower HLR the ammonia-N removal efficiency in both is significantly greater than that of the cinder aggregate material.

Coliform removal mechanism in the filtration systems includes physical processes such as filtration, sedimentation,

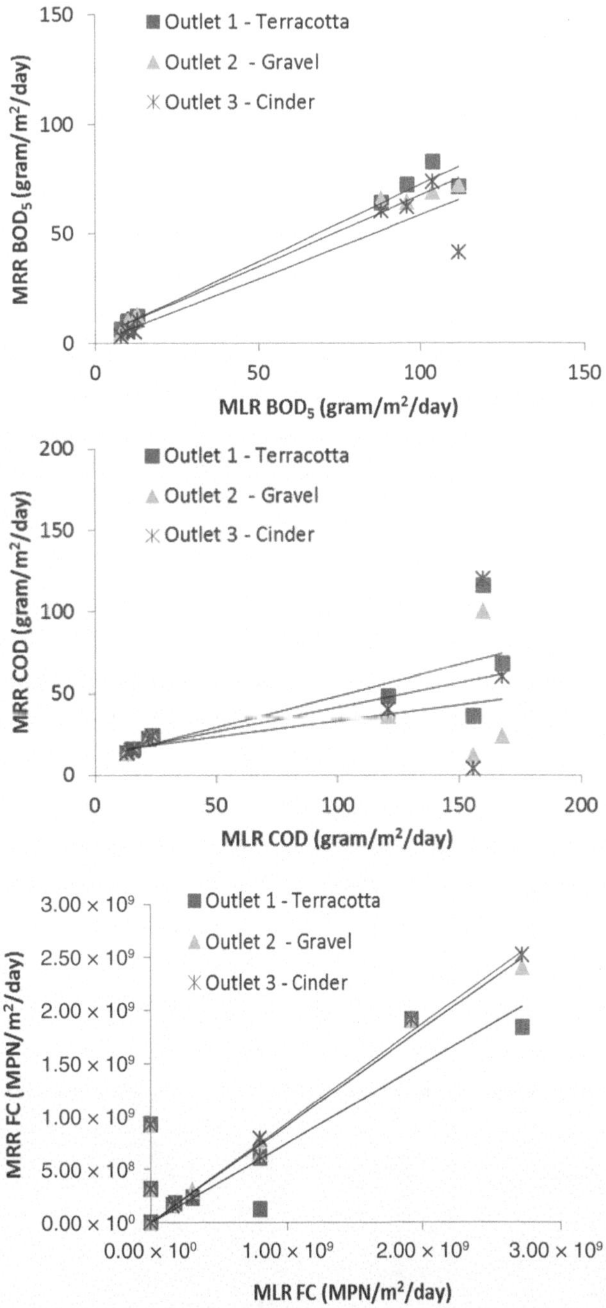

Figure 4 | Relationship between contaminant mass loading rate and mass removal rate.

ultraviolet radiation; chemical processes such as absorption, oxidation, natural die-off and die-off due to toxins and biological activities, as well as ingestion by nematodes and protozoa (Karim *et al.* 2004; Wu *et al.* 2016). It is believed that porosity and HRT has a significant impact on the settling of microorganisms within the filtration system. All the factors are interrelated, and apportioning contribution of each factor to the removal of FC is complex. At low

Table 6 | Mass loading rate and mass removal rate and corresponding R^2 values

Parameter	Units	MLR Influent	MRR Outlet 1 – Terracotta (R^2)	Outlet 2 – Gravel (R^2)	Outlet 3 – Cinder (R^2)
BOD_5	$g/m^2/day$	8–112	6–82 (0.99)	4.4–72 (0.98)	2.7–73 (0.88)
COD	$g/m^2/day$	13.2–168	7.2–116 (0.66)	5.6–100 (0.26)	2.8–120 (0.37)
FC	(MPN m^2/day)	8–27,200*10^5	7.2–19,088 *10^5 (0.84)	9.92000–24,040 *10^5 (0.82)	9.9–25,240 *10^5 (0.68)

HLR, terracotta based filtration system provided maximum FC removal, whereas at high HLR all three filtration setups provided similar FC removal levels. An ANOVA test was conducted to test the impact of filter material on FC removal efficiency. The results showed that filtration material does not have any effect on the FC removal efficiency. Factors such as exposure to sunlight and natural die-off facilitated FC removal in all three setups.

Figure 4 shows the relationship between incoming contaminant MLR and MRR for BOD_5, COD and FC. The relationship between MLR and MRR was determined to evaluate the effectiveness of filter media under varying contaminant loadings. Compared to previous studies, in this study, filtration setups were subjected to moderate to very high BOD_5 and COD loading rates (Table 6) (Ghosh & Gopal 2010; Weerakoon *et al.* 2013). The incoming contaminant load was increased significantly during the study period with HLR increment from 4 to 40 cm/day. Except for COD, a good linear correlation was observed between the incoming BOD_5 MLR (8–112 $g/m^2/day$) and the MRR with an R^2 value greater than 0.90. A moderate correlation was observed between the incoming COD MLR and MRR for the terracotta filtration setup ($R^2 = 0.67$), which could be attributed to the presence of stable and abrasion resistant biofilm. The FC removal rates (MPN/m^2/day) in all the three filtration setups showed a very strong linear correlation to the incoming loads with corresponding R^2 values of more than 0.98. The relationship suggests that for all aggregate material the BOD_5 and FC removal per unit of the BOD_5 and FC load applied is significantly higher than that of the COD.

Subsequent site based testing

The study described here is one part of an ongoing interdisciplinary effort currently underway in the city of Bangalore. Strategic Instream Systems (STRAINS) proposes decentralized, cost-effective solutions to the ongoing challenge of contamination within rapidly growing urban contexts (Figure 5). The results of these laboratory based tests suggest that the use of low-cost terracotta fragments may be an appropriate aggregate material to incorporate into larger scale interventions within the urban storm water channels ('Nallahs'). Subsequent intermediate scale demonstration sites will incorporate terracotta material as an integral part of their function. Subsequent field based tests and regular

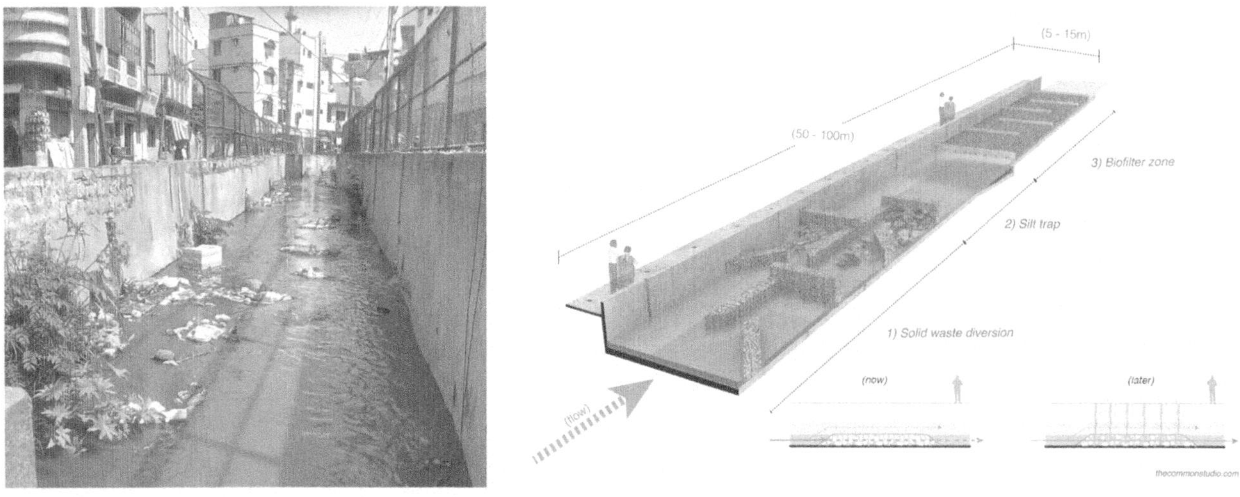

Figure 5 | Typical urban Nallah condition (left) and conceptual rendering of a three stage STRAINS intervention including terracotta filtration zone in the final stage.

water-quality monitoring will follow, which will provide deeper insights into its potentials and limitations of this approach.

CONCLUSIONS

The objective of the study was to assess the organic matter and FC removal efficiencies of horizontal subsurface filtration systems composed of three different aggregate materials: terracotta, gravel, and cinder. The setups were tested using synthetic wastewater prepared in the laboratory and contaminant removal efficiency was evaluated at different HLRs. From the results, the following was concluded:

(a) All three filtration materials followed similar trends in effluent quality. Significant reduction in the effluent quality was observed at high HLR, which could be attributed to reduced contact time of contaminants within the system due to increased flow velocities.
(b) At both high (40 cm/day) and low (4 cm/day) HLRs, the terracotta based filtration setup provided maximum reduction in COD and BOD_5 levels. Although at low HLR gravel based filtration system provided BOD_5 reduction similar to that of the terracotta based system, the terracotta based filtration system was found to be more efficient at higher HLRs.
(c) Both terracotta and gravel setup exhibited similar ammonia-N removal efficiencies at lower HLR. However, at higher HLR, gravel setup exhibited the ammonia-N removal efficiency. Out of three, the cinder based filter setup exhibited lowest removal efficiency at both low and high HLR.
(d) No significant difference was observed in the FC removal efficiencies of three filtration systems, indicating that filter material does not affect the FC removal from the wastewater.
(e) Except for COD, a strong correlation was observed in input BOD_5 and FC mass loading rate with the mass removal rate. The release of biofilm and cell debris at higher HLR would have contributed to considerable variation in the effluent COD from the three filtration systems.
(f) Significant reductions in organic pollutants may be possible using affordable aggregate biofilters in open storm water channels which are vectors for greywater, and that these reductions may be significant even in unplanted configurations. This suggests a low-maintenance approach as compared to traditional modes of GI or constructed wetlands, which typically promote the presence and maintenance of plants.

ACKNOWLEDGEMENTS

This research was supported by a research grant from the Royal Norwegian Embassy (RNE) and the Ashoka Trust for Research in Ecology and the Environment (ATREE). The authors would like to thank the Praveen Kumar (laboratory analyst), and Sowmya Murugan (intern from IISER) at ATREE for helping us with the analysis of water samples and maintenance and upkeep of the experimental setup.

REFERENCES

Alley, K. D., Barr, J. & Mehta, T. 2018 Infrastructure disarray in the clean Ganga and clean India campaigns. *Wiley Interdisciplinary Reviews: Water* **5**, e1310.

APHA 2005 *Standard Methods for the Examination of Water and Wastewater*, 21st edn. APHA, Washington, DC, USA

Behnke, S., Parker, A. E., Woodall, D. & Camper, A. K. 2011 Comparing the chlorine disinfection of detached biofilm clusters with those of sessile biofilms and planktonic cells in single- and dual-species cultures. *Applied and Environmental Microbiology* **77**, 7176–7184. https://doi.org/10.1128/AEM.05514-11.

Bhasthi, D. 2017 City of burning lakes: experts fear Bangalore will be uninhabitable by 2025. Cities, The Guardian, 1 March.

Bhatnagar 2017 *INTACH Pilot Project Reduces Assi River Pollution in Varanasi by 70% in Two Months, at Least Cost, without any Structures*. SANDRP.

Chen, T. Y., Kao, C. M., Yeh, T. Y., Chien, H. Y. & Chao, A. C. 2006 Application of a constructed wetland for industrial wastewater treatment: a pilot-scale study. *Chemosphere* **64**, 497–502.

Chyan, J.-M., Senoro, D.-B., Lin, C.-J., Chen, P.-J. & Chen, I.-M. 2013 A novel biofilm carrier for pollutant removal in a constructed wetland based on waste rubber tire chips. *International Biodeterioration & Biodegradation* **85**, 638–645.

Demırezen, D. & Aksoy, A. 2004 Accumulation of heavy metals in *Typha angustifolia* (L.) and *Potamogeton pectinatus* (L.) living in Sultan Marsh (Kayseri, Turkey). *Chemosphere* **56**, 685–696.

Franz, A. 2005 *A Performance Study of Ceramic Candle Filters in Kenya Including Tests for Coliphage Removal*. PhD Thesis, Massachusetts Institute of Technology, Cambridge, MA, USA.

Ghosh, D. & Gopal, B. 2010 Effect of hydraulic retention time on the treatment of secondary effluent in a subsurface flow constructed wetland. *Ecological Engineering* **36**, 1044–1051. https://doi.org/10.1016/j.ecoleng.2010.04.017.

Horppila, J. & Nurminen, L. 2001 The effect of an emergent macrophyte (*Typha angustifolia* (L.)) on sediment resuspension in a shallow north temperate lake. *Freshwater Biology* **46**, 1447–1455.

Jamwal, P. 2017 Fate of nutrients in human dominated ecosystems. *Resonance* **22**, 279–290.

Jamwal, P. 2018 Remediation of Contaminated Urban Streams: A Decentralized Ecological Wastewater Treatment Approach. In: *Water Remediation*. Springer, Singapore, pp. 29–41.

Jamwal, P., Md Zuhail, T., Raje Urs, P., Srinivasan, V. & Lele, S. 2015 Contribution of sewage treatment to pollution abatement of urban streams. *Current Science* **108**, 677–685.

Jumbe, A. S. & Nandini, N. 2009 Heavy metals analysis and sediment quality values in urban lakes. *American Journal of Environmental Sciences* **5**, 678.

Karim, M. R., Manshadi, F. D., Karpiscak, M. M. & Gerba, C. P. 2004 The persistence and removal of enteric pathogens in constructed wetlands. *Water Research* **38**, 1831–1837. https://doi.org/10.1016/j.watres.2003.12.029.

Kato, S. 2011 Green infrastructure for Asian cities: the spatial concepts and planning strategies. In: *Journal of the 2011 International Symposium on City Planning*. Korea Planners Association, pp. 161–170.

Kim, B. R. 1989 Effect of ammonia on COD analysis. *Journal (Water Pollution Control Federation)*, 614–617.

Le Dantec, C., Duguet, J.-P., Montiel, A., Dumoutier, N., Dubrou, S. & Vincent, V. 2002 Occurrence of mycobacteria in water treatment lines and in water distribution systems. *Applied and Environmental Microbiology* **68**, 5318–5325. https://doi.org/10.1128/AEM.68.11.5318-5325.2002.

Lekang, O.-I. & Kleppe, H. 2000 Efficiency of nitrification in trickling filters using different filter media. *Aquacultural Engineering* **21**, 181–199.

Lonnen, J., Kilvington, S., Kehoe, S. C., Al-Touati, F. & McGuigan, K. G. 2005 Solar and photocatalytic disinfection of protozoan, fungal and bacterial microbes in drinking water. *Water Research* **39**, 877–883. https://doi.org/10.1016/j.watres.2004.11.023.

Low, C. S. 2002 *Appropriate Microbial Indicator Tests for Drinking Water in Developing Countries and Assessment of Ceramic Water Filters*. PhD Thesis, Massachusetts Institute of Technology, Cambridge, MA, USA.

Malarvizhi, R., Venkateswarlu, Y. & Begum, S. S. 2010 Studies on removal of chromium (VI) from water using chitosan coated Cyperus pangorei. *Water Science and Technology* **62**, 2435–2441.

Mashauri, D. A., Mulungu, D. M. M. & Abdulhussein, B. S. 2000 Constructed wetland at the university of Dar es Salaam. *Water Research* **34**, 1135–1144. https://doi.org/10.1016/S0043-1354(99)00238-9.

Mell, I. C. 2010 *Green Infrastructure: Concepts, Perceptions and its use in Spatial Planning*. PhD Thesis, Newcastle University, Newcastle, UK. https://theses.ncl.ac.uk/dspace/handle/10443/914.

Morel, A. 2006 *Greywater Management in low and Middle-Income Countries*. Swiss Federal Institute of Aquatic Science and Technology, Dubenforf, Switzerland.

Pisoeiro, J., Galvão, A., Ferreira, F. & Matos, J. 2016 Potential for CSO treatment with horizontal flow constructed wetlands: influence of hydraulic load, plant presence and loading frequency. *Environmental Science and Pollution Research* **23**, 20591–20599.

Prominski, M., Stokman, A., Stimberg, D., Voermanek, H. & Zeller, S. 2012 *River. Space. Design: Planning Strategies, Methods and Projects for Urban Rivers*. Birkhäuser, Basel, Switzerland.

Ramachandra, T. V., Bhat, S., Cincy, V., Asulabha, K. S., Lakkangoudar, K. & Rahaman, M. F. 2015 *Fish Mortality In Jakkur Lake: Causes and Remedial Measures (ENVIS Technical Report No. 86)*. Centre for Ecological Sciences, Indian Institute of Science, Bangalore, India.

Ribot, M., von Schiller, D., Peipoch, M., Sabater, F., Grimm, N. B. & Martí, E. 2013 Influence of nitrate and ammonium availability on uptake kinetics of stream biofilms. *Freshwater Science* **32**, 1155–1167. https://doi.org/10.1899/12-209.1.

Suthar, S., Sharma, J., Chabukdhara, M. & Nema, A. K. 2010 Water quality assessment of river Hindon at Ghaziabad, India: impact of industrial and urban wastewater. *Environmental Monitoring and Assessment* **165**, 103–112.

Tekerlekopoulou, A. G. & Vayenas, D. V. 2008 Simultaneous biological removal of ammonia, iron and manganese from potable water using a trickling filter. *Biochemical Engineering Journal* **39**, 215–220.

Tortajada, C. 2016 Policy dimensions of development and financing of water infrastructure: the cases of China and India. *Environmental Science & Policy* **64**, 177–187.

Weerakoon, G., Jinadasa, K., Herath, G. B. B., Mowjood, M. I. M. & Van Bruggen, J. J. A. 2013 Impact of the hydraulic loading rate on pollutants removal in tropical horizontal subsurface flow constructed wetlands. *Ecological Engineering* **61**, 154–160.

Wu, S., Carvalho, P. N., Müller, J. A., Manoj, V. R. & Dong, R. 2016 Sanitation in constructed wetlands: a review on the removal of human pathogens and fecal indicators. *Science of the Total Environment* **541**, 8–22. https://doi.org/10.1016/j.scitotenv.2015.09.047.

Xu, W., Chen, J., Jian, Y., Pan, Z. & Mou, Z. 2018 Treatment of sewage using a constructed soil rapid infiltration system combined with pre-denitrification. *International Journal of Environmental Research and Public Health* **15**, 2005.

Yang, X. J., Tong, Y. M., Song, Y., Qi, W. K., Li, Y. Y. & Guo, Y. L. 2017 Domestic sewage and secondary effluent treatment using vertical submerged biological filter. In: *IOP Conference Series: Earth and Environmental Science*. IOP Publishing, Bristol, UK, p. 012067.

First received 11 July 2018; accepted in revised form 11 March 2019. Available online 21 March 2019

Degradation of paraquat from contaminated water using green TiO_2 nanoparticles synthesized from *Coffea arabica* L. in photocatalytic process

Patcharaporn Phuinthiang and Puangrat Kajitvichyanukul

ABSTRACT

The TiO_2 nanoparticles synthesized from the extract of *Coffea arabica* L. (or TiO_2/C) were used to remove paraquat from contaminated water in heterogeneous photocatalysis process. In this work, the sol-gel process using *Coffea arabica* L. as the solvent chemical were performed to obtain the TiO_2 nano-catalyst. The value of pH_{pzc} of TiO_2/C was 2.9 which caused a highly acidic surface of catalyst. The paraquat is effectively removed in alkaline medium due to the adsorption ability of paraquat on the surface of TiO_2/C. The paraquat degradation followed the pseudo-first-order model with the apparent rate constants of 5.84×10^{-2}, 4.08×10^{-2}, and 2.28×10^{-2} min^{-1} for TiO_2/C, TiO_2, and without TiO_2, respectively, under the presence of ultraviolet (UV) and H_2O_2. The combined TiO_2/C with UV and H_2O_2 was the most efficient process, exhibiting a maximum 66.3% degradation of 50 mg/L over 90 min at pH 10.

Key words | chemical oxidation, green chemistry, nanoparticles, paraquat, TiO_2, water treatment

Patcharaporn Phuinthiang
Puangrat Kajitvichyanukul (corresponding author)
Center of Excellence on Environmental Research and Innovation, Faculty of Engineering, Naresuan University, Phitsanulok 65000, Thailand
E-mail: *puangratk@nu.ac.th*

INTRODUCTION

Titanium dioxide (TiO_2) has proved to be a highly useful and active photocatalytic material to remove or degrade recalcitrant organic contaminants through the chemical oxidation reaction for water and air purification. Due to its physical and chemical properties, the application of TiO_2 with ultraviolet (UV) has been widely used, and it has been considered as one type of advanced oxidation processes (Chen *et al.* 2016). The organic pollutants can be degraded by photocatalytic oxidation in which photo-induced holes in TiO_2 oxidize OH^- or water molecules adsorbed on the surface of the particles to produce $HO^{\bullet}$ and $O_2^{-\bullet}$ which subsequently attack adsorbed organic molecules (Hoffman *et al.* 1995). Photocatalytic reactions have been applied to degrade several pesticides including atrazine, bentazon, monuron, dichlorvos, propyzamide, dicloran and triadimefon (Pelizzetti *et al.* 1989; Pelizzetti *et al.* 1993; Pramuro *et al.* 1993; Minero *et al.* 1997).

The emerging of 'green chemistry' leads to the new synthesizing method of nanoparticles that can improve the physico-chemical properties of the materials. These green nanoparticles are expected to provide higher efficiency in contaminant removal (Humayun *et al.* 2017). However, few works reported in detail about the kinetics and degradation efficiency of the pollutant using those green nanomaterials, especially the TiO_2. In this area, there has also been increasing interest in identifying environmentally friendly materials that are multifunctional and provide good properties of materials. In this work, coffee extract was chosen to be a solvent for TiO_2 formation following the green chemistry synthesis method. Caffeine/polyphenols from coffee extracts can form complexes with metal ions in solution and reduce them to the corresponding metals (Nadagouda *et al.* 2010). The caffeine/polyphenols contain molecules bearing alcoholic functional groups which can be exploited for reduction, as well as stabilization of the nanoparticles. Thus, these chemicals act as reducing agent and stabilizing agent in TiO_2 synthesis. The chemicals from coffee extracts are also the capping agent or dispersing agent of TiO_2 that can prevent the agglomeration of the nanoparticles and enhance the photocatalytic efficiency of the nanomaterials (Nadagouda *et al.* 2010; Varma 2012).

This work aims to synthesize the TiO_2 using green methods by applying the green solvent obtained from the extract of *Coffea arabica* L. (or TiO_2/C) and applying it in removal of paraquat from aqueous solution. Paraquat (1,1′-dimethyl-4,4′-bipyridinium dichloride) is a widely used commercial herbicide applied in many agricultural activities. This chemical has a high toxicity to mammals and is

doi: 10.2166/wst.2018.493

recognized as a potent human poison. The U.S. EPA has classified paraquat as a possible carcinogen and has recommended the maximum concentration of paraquat in drinking water at 200 μg/L (Rossi 1997). The removal of paraquat using TiO_2 has been published previously (Zahedi *et al.* 2015). However, the application of green photocatalytic to remove paraquat is not reported elsewhere.

In this work, the details in kinetics and paraquat degradation efficiency were calculated and reported. The paraquat degradation performance using this new type of TiO_2/C in comparison with that using the conventional TiO_2 was also discussed. This approach addresses several factors that play an important role in paraquat removal enhancement by this new type of green TiO_2 nanomaterial.

MATERIALS AND METHODS

Materials

Titanium tetraisopropoxide (TTiP), paraquat, and glacial acetic acid (CH_3COOH) were purchased from Aldrich Chemicals, Thailand. Nitric acid (HNO_3), sulphuric acid (H_2SO_4) and sodium hydroxide (NaOH) were obtained from Merck Chemicals, Thailand. All chemicals were used as received for preparing the nanocatalysts and conducting photocatalytic experiments. The 18 MΩ deionised water (H_2O) was used for the preparation of all the solutions. All reagents used were of analytical grade and employed as received.

Tio_2 catalyst synthesis

All experiments were conducted at the Center of Excellence on Environmental Research and Innovation, Naresuan University. The typical sol–gel synthesis of TiO_2-loaded nanoparticles was used in this work. The method is as follows: TTIP (4.83 mL), glacial acetic acid (15.35 mL) and water were used in the molar ratio of 1:10:300.16 TTIP was mixed with glacial acetic acid in an ice bath. When this mixture turned to a sol, it was stirred at room temperature to form a gel before undergoing the drying process. After drying at 100 °C for 90 min, the powder was collected and calcined at 600 °C. The nanoparticles obtained from this process were shortly denoted as 'TiO_2'.

For the green TiO_2 nanoparticles synthesized from *Coffea arabica* L., 1 g of coffee was boiled in 50 mL of water and filtered through a filter paper (Whatman, No. 41). Then 5 mL of coffee extract was added to the mixture solution of TTiP and glacial acetic acid. The drying process was the same procedure as above. The obtained TiO_2 nanoparticles were denoted as 'TiO_2/C'.

Adsorption and photocatalytic of paraquat

Photocatalysis experiments were performed in a batch reactor. The reactor was cylindrical with a volume of 1.1 L made from quartz glass (ACE Glass Co. 7841-06; Vineland, NJ, USA). The UV light source was a 10 W (Philips) with 254 nm wavelength. The initial concentration of paraquat was in the range of 5–50 mg/L, and the catalyst was 0.6 g/L. In this photocatalytic experiment, the paraquat solution was injected into the reactor and treated in batch operation mode with a steady temperature maintained at 25 ± 2 °C throughout the test in an oxygen atmosphere. The liquid was allowed to equilibrate in the dark for 30 min. The paraquat adsorption data were collected. After reaching equilibrium, the reaction was started by switching on the light at t = 0, and the initial concentration of paraquat was designated as C_0. The initial pH of solution was 7. It was adjusted to pH 3 by adding 36.8 N H_2SO_4 and to pH 10 by adding NaOH before the photoreaction experiment unless otherwise specified. At a chosen interval of irradiation time, aliquots of the reaction mixture were withdrawn and filtered through a membrane filter (0.1 μm). Paraquat concentration was detected by a colorimetric method using UV-Vis spectrophotometer (Model Lambda 650, Perkin Elmer, USA) by reducing paraquat to its blue radical. A solution sample in the test tube was added with 0.1% sodium dithionite in 0.1 M sodium hydroxide. The mixture was gently mixed and measured for light absorbance within 1 min. The detection was done at 600 nm wavelength using a spectrophotometer. The paraquat concentrations in the final solution for each experiment were also confirmed by liquid chromatography–mass spectrometry (Model Single Quadrupole LC/MS, Agilent, USA). Total organic carbon (TOC) analysis was done by TOC carbon analyzers (Model TOC-L, Shimadzu, Japan). The Brunauer–Emmett–Teller (BET) surface area for all nanoparticles were done by BET surface area analyzer (Model SA-9600, Horiba, Japan). Analysis of pH at point of zero charge (pH_{pzc}) for all nanoparticles was measured by Zetasizer (Model ZS90, Malvern, UK).

RESULTS AND DISCUSSION

Adsorption of paraquat onto TiO_2 surface

Preliminary adsorption experiments revealed that in the absence of TiO_2, no noticeable change in paraquat concentration occurred during a 60-min experimental period. Adsorption of paraquat using TiO_2/C at different pH was evaluated as shown in Figure 1. The adsorption was rapid in the first few minutes and reached equilibrium within 20 min for all pH conditions. The paraquat adsorption was highest at pH 10 with the maximum adsorption capacity of 20 mg/g. At pH 3 and pH 7, the maximum adsorption capacities were 16.8 and 12.2 mg/g, respectively. Our results are in good agreement with previous work. Florencio *et al.* (2004) reported that the degradation of diquat and paraquat did not take place in acid medium but was pronounced in an alkaline solution. Less adsorption of paraquat on the surface of TiO_2 resulted in low degradation of the chemicals in acidic and moderate degradation in neutral pH medium (Florencio *et al.* 2004).

The paraquat adsorption of TiO_2/C was compared with that of TiO_2 from chemical-based process as shown in Figure 2. The BET surface areas of TiO_2/C (75.8 m^2/g) was slightly changed when compared with TiO_2 (73.2 m^2/g). While the values of pHpzc were 2.9 and 4.6 for TiO_2/C, and TiO_2, respectively. The lower the pHpzc causes the higher affinity for cations to adsorb on the surface of nanoparticles. Thus, using the *Coffea Arabic* L. for TiO_2 synthesis mainly affected on the surface charge rather than the surface area and, consequently, resulting in higher adsorption ability of TiO_2/C over TiO_2. At pH 10, the surface charges of the catalyst were negative, favouring the adsorption of cations like paraquat. This finding also occurred for other types of catalyst such as TiO_2/SBA-15 and Cu-TiO_2/SBA-15 (Sorolla II *et al.* 2012).

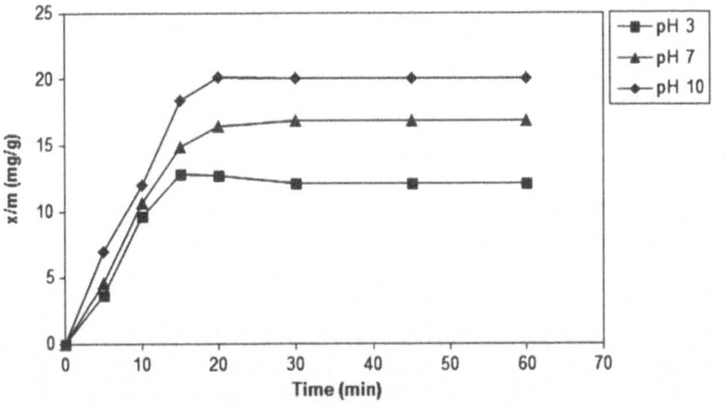

Figure 1 | Adsorption of paraquat using TiO_2/C at different pHs. Experimental condition: $[TiO_2] = 0.6$ g/L, and [PQ] = 50 ppm.

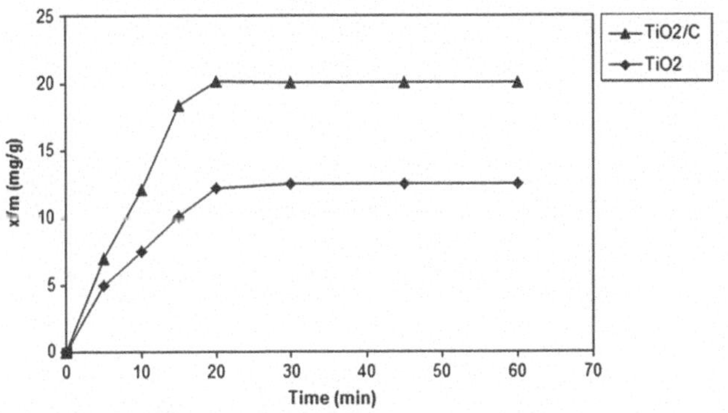

Figure 2 | Comparison of paraquat adsorption using TiO_2/C, and TiO_2 at pH 10. Experimental condition: $[TiO_2] = 0.6$ g/L, [PQ] = 50 ppm and pH 10.

Paraquat removal by TiO_2/C photocatalytic reaction

Paraquat degradation using TiO_2/C with UV in different concentrations of paraquat is shown in Figure 3. The removal percentages of paraquat with initial concentrations of 5, 10, 20, 35, and 50 mg/L were 76.3, 52.1, 31.7, 27.9, and 4.3%, respectively. These results clearly indicate that low concentration paraquat solutions were easily degraded by photocatalytic reactions. However, as the initial concentration of paraquat increased (higher than 35 ppm), the TiO_2/C with UV hardly degraded paraquat within 1 h. It was previously reported that the paraquat removal using TiO_2 was feasible for the concentration lower than 30 mg/L since it is chemically stable in nature (Moctezuma *et al.* 1999). The paraquat solutions remained stable in the temperature range of 20–40 °C and after standing for 23 days (Florencio *et al.* 2004).

To enhance the paraquat degradation, H_2O_2 was added to the experiment. Solutions of 50 ppm paraquat were irradiated under four different conditions: H_2O_2/UV, TiO_2/C with UV (without H_2O_2), TiO_2/C with UV + H_2O_2, TiO_2 with UV (without H_2O_2), and TiO_2 with UV + H_2O_2 systems. Figure 4 shows the changes in the degree of degradation of aqueous paraquat versus irradiation time. It is noted that there was no paraquat degradation by the direct photolysis with a 10 W UV lamp during 90 min. As shown previously in Figure 3, the 50 mg/L of paraquat was hardly decreased using TiO_2/C with UV irradiation and the removal efficiency was only 4.3%. The paraquat removal increased up to 36.2%

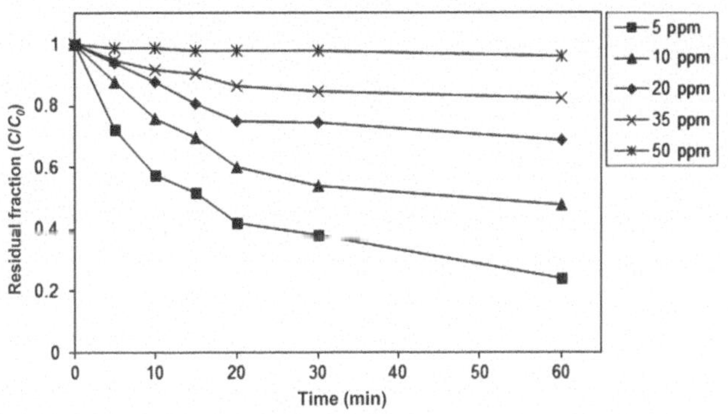

Figure 3 | Paraquat degradation by photocatalytic process using different concentration of TiO_2/C. Experimental condition: $[TiO_2] = 0.6$ g/L, UV 10 watt and pH 10.

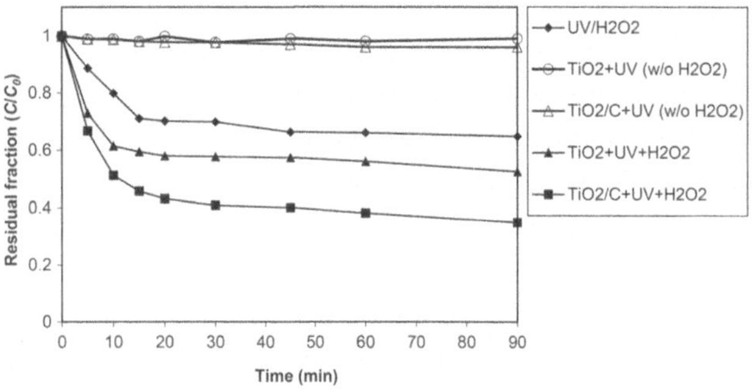

Figure 4 | Paraquat degradation using different processes, (a) UV/H_2O_2, (b) TiO_2+ UV (without H_2O_2), (c) TiO_2/C+ UV (without H_2O_2), (d) TiO_2/C + UV + H_2O_2, and (e) TiO_2 + UV + H_2O_2, Experimental condition: [TiO_2] = 0.6 g/L, [PQ] = 50 ppm, [H_2O_2] = 10 mM, UV = 10 W, and pH 10.

in the presence of 10 mM of H_2O_2 under UV irradiation (without TiO_2/C). Approximately 66.3% of paraquat was degraded within 90 min when 0.6 g/L TiO_2/C was added with UV and 10 mM H_2O_2.

In comparison, the paraquat degradation using TiO_2/C was compared with TiO_2 in the presence of H_2O_2 under UV irradiation as shown in the same figure. The TiO_2/C provided the higher photocatalytic efficiency for paraquat degradation than the TiO_2. The high efficiency in paraquat removal using TiO_2/C partially due to the high adsorption of paraquat onto its surface as explained earlier, enhancing more paraquat to react with OH radicals. The least amount of degradation was observed for chemical based TiO_2, which has slightly lower surface area, higher value of pH_{pzc} and the lower adsorption of paraquat on its surface as seen in Figure 2. The overall percentages of degradation for paraquat after 90 min of irradiation were 66.3, 48.7, 36.2, and 4% for TiO_2/C with UV + H_2O_2, TiO_2 with UV + H_2O_2, UV/H_2O_2, and TiO_2/C +UV (without H_2O_2), respectively.

The reaction mechanisms during the TiO_2 in the presence of H_2O_2 under UV irradiation are summarized as follows (Barakat *et al.* 2005; Gao *et al.* 2015):

$$TiO_2 + hv \rightarrow e^- + h^+ \tag{1}$$

$$H_2O_2 + hv \rightarrow 2OH^{\cdot} \tag{2}$$

$$OH^- + h^+ \rightarrow OH^{\cdot} \tag{3}$$

$$O_2 + e^- \rightarrow O_2^{-\cdot} \tag{4}$$

$$H_2O_2 + e^- \rightarrow OH^{\cdot} + OH^- \tag{5}$$

$$H_2O_2 + O_2^{-\cdot} \rightarrow OH^{\cdot} + OH^- + O_2 \tag{6}$$

$$RH + OH^{\cdot} \rightarrow R^{\cdot} + HO_2 \rightarrow \text{further oxidation} \tag{7}$$

where RH refers to the paraquat compound.

In the system that contained only TiO_2, the electron (e^-) and hole (h^+) on the TiO_2 surface were generated upon UV irradiation (Equation (1)). Equations (3), (4), and (7) express that the hydroxyl radicals were formed under light excitation when the positive holes reacted with OH^- on the TiO_2 surface. When the H_2O_2 was presented in the system, the UV/H_2O_2 reaction can generate $OH^{\cdot}$ (Equation (2)), enhancing the rate of photocatalytic reaction. H_2O_2 was a better electron acceptor than oxygen in other mechanisms (Equation (5)). Thus, the amount of electron recombination decreased, and the paraquat degradation using this photocatalytic process was enhanced as shown in Figure 4.

To describe the kinetics of this paraquat degradation, the Langmuir–Hinshelwood model (or L–H model) was selected as it demonstrated a good fit in photocatalysis process in many previous works (Rajeshwar *et al.* 2008). The development of this model to obtain a kinetic equation depends on the limiting step (adsorption, surface reaction, or desorption). The L–H model relates the degradation rate (r) to the concentration of the organic reactant (C), as shown in Equation (8).

$$r = -\frac{dC}{dt} = \frac{k_r K_{ad} C}{1 + K_{ad} C} \tag{8}$$

where k_r is the rate constant, and K_{ad} is the adsorption equilibrium constant. When the adsorption is relatively weak, or the concentration of the reactant is low, Equation (8) can be simplified to be the first-order kinetics with an apparent rate constant k_{app} (Equation (9)):

$$\ln\left(\frac{C}{C_o}\right) = -k_r K_{ad} t = -k_{app} t \tag{9}$$

Plotting $-ln(C/C_0)$ vs. reaction time (t) generates a straight line with a slope equal to k_{app} (Chiou *et al.* 2008). The apparent rate constants (k_{app}), initial degradation rates (r), and correlation coefficients (R^2) for paraquat degradation using different types of TiO_2 are shown in Table 1.

Table 1 | Apparent rate constants (k_{app}), initial degradation rates (r), and correlation coefficients (R^2) for paraquat degradation using different types of TiO_2

TiO_2	Photocatalytic condition	k_{app} (min^{-1})	r (mM/min)	R^2
–	UV + 10 mM H_2O_2	0.0228	0.0044	0.9990
TiO_2	UV + 10 mM H_2O_2	0.0408	0.0106	0.8391
TiO_2/C	UV only	0.0014	0.0004	0.9080
TiO_2/C	UV + 10 mM H_2O_2	0.0584	0.0130	0.9203

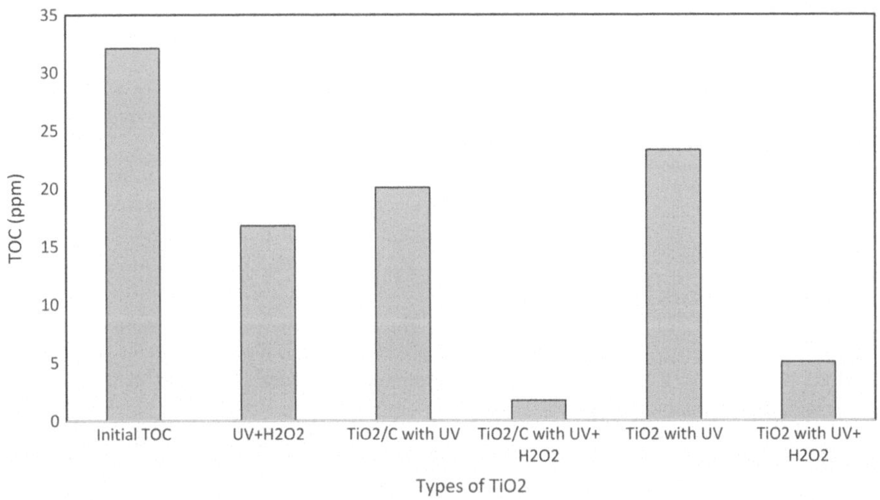

Figure 5 | TOC concentration of paraquat in photocatalytic process using different processes. Experimental condition: $[TiO_2] = 0.6$ g/L, [PQ] = 50 ppm, $[H_2O_2] = 10$ mM, UV = 10 W, and pH 10.

The initial reaction rate of the TiO_2/C with $UV + H_2O_2$ (5.84×10^{-2} mM/min) in paraquat removal was nearly forty-two fold that of the TiO_2/C with UV but without the H_2O_2 system (1.4×10^{-3} mM/min). This difference is attributed to the contributions of the hydroxyl radicals produced by H_2O_2 during the TiO_2/C with the $UV + H_2O_2$ process.

The degradation percentage of paraquat regarding mineralization of TOC in the presence of $UV + H_2O_2$, TiO_2 with UV and TiO_2 with $UV + H_2O_2$ was also investigated. The initial concentration of paraquat was 50 ppm with an initial TOC value of 32.08 ppm. The degradation percentage and mineralization to the TOC of paraquat using different processes are shown in Figure 5. All cases were expressed as mean ± standard deviation. An analysis of variance (ANOVA) was used to test the significance of the results with $p < 0.05$ and $F = 4.34$ was statistically significant. From the results, mineralization of paraquat for TiO_2/C in the presence of UV with H_2O_2 showed the highest yield. By contrast, the TiO_2 from chemical-based synthesis provided lowest TOC percentage removal.

CONCLUSIONS

In this work, a green, low-cost, and reproducible of TiO_2 nanoparticles were used to remove paraquat from contaminated water in heterogeneous photocatalysis process. The novel 'green' nanoparticles were obtained by the sol-gel process using coffee as the solvent chemical. The paraquat degradation using the TiO_2/C was conducted in comparison with the TiO_2 from chemical-based synthesis. Both TiO_2 with UV and TiO_2 with UV/H_2O_2 have been investigated. From overall experiments, the combined TiO_2/C with UV and H_2O_2 was the most effective photocatalyst, exhibiting a maximum 66.3% degradation of 50 mg/L over 90 min at pH 10. The paraquat removal depended on the pH of the medium. The TiO_2/C provided the highest adsorption percentage of paraquat due to its dominated negative charge on the surface. The BET surface areas of TiO_2/C (75.8 m^2/g) was slightly changed when compared with TiO_2 (73.2 m^2/g); however, the values of pH_{pzc} of both materials were pronouncedly different. The pH_{pzc} of TiO_2 were 2.9, and 4.6 for TiO_2/C, and TiO_2, respectively. Thus, the surface charge of particles played a major role for the adsorption of positive ions of paraquat in this work. The high adsorption of paraquat on TiO_2/C led to high efficiency in paraquat removal. The Langmuir–Hinshelwood model could well describe the photodegradation of paraquat during all three catalytic oxidation processes. The apparent first-order rate constants were 5.84×10^{-2}, 4.08×10^{-2}, and 2.28×10^{-2} min^{-1} for TiO_2/C, TiO_2, and UV, respectively, under the presence of UV and H_2O_2. Results from this work offer the benefit of using the green nanoparticles (such as TiO_2/C) in pollutant removal for water and wastewater treatment.

ACKNOWLEDGEMENT

This research received funding support from National Research Council, Thailand (NRCT) through Naresuan University under grant No. R2562B021 and Thailand Research Fund under grant No. BRG6180009.

REFERENCES

Barakat, M. A., Tseng, J. M. & Huang, C. P. 2005 Hydrogen peroxide-assisted photocatalytic oxidation of phenolic compounds. *Applied Catalysis B: Environmental* **59**, 99–104.

Chen, J., Cen, J., Xu, X. & Li, X. 2016 The application of heterogeneous visible light photocatalysts in organic synthesis. *Catalysis Science and Technology* **6**, 349–362.

Chiou, C. H., Wu, C. Y. & Juang, R. S. 2008 Photocatalytic degradation of phenol and *m*-nitrophenol using irradiated TiO_2 in aqueous solution. *Separation and Purification Technology* **62** (3), 559–564.

Florencio, M. H., Pires, E., Castro, A. L., Nunes, M. R., Borges, C. & Costa, F. M. 2004 Photodegradation of diquat and paraquat in aqueous solutions by titanium dioxide: evolution of degradation reactions and characterization of intermediates. *Chemosphere* **55**, 345–355.

Gao, M., Zhu, L., Ong, W. L., Wang, J. & Ho, G. W. 2015 Structural design of TiO_2-based photocatalyst for H2 production and degradation applications. *Catalytic Science Technology* **5**, 4703–4726.

Hoffman, M. R., Martin, S. T., Choi, W. & Bahnemann, D. W. 1995 Environmental applications of semiconductor photocatalysis. *Chemical Reviews* **95** (1), 69–96.

Humayun, M., Raziq, F., Khan, A. & Luo, W. 2017 Modification strategies of TiO_2 for potential applications in photocatalysis: a critical review. *Green Chemistry Letters and Reviews* **11**, 86–102.

Minero, C., Maurino, V. & Pelizzetti, E. 1997 Heterogeneous photocatalytic transformations of s-triazine derivatives. *Research on Chemical Intermediates* **23** (4), 291–310.

Moctezuma, E., Leyva, E., Monreal, E., Villegas, N. & Infante, D. 1999 Photocatalytic degradation of the herbicide paraquat. *Chemosphere* **39** (3), 511–517.

Nadagouda, M. N., Castle, A. B., Murdock, R. C., Hussain, S. M. & Varma, R. S. 2010 *In vitro* biocompatibility of nanoscale zerovalent iron particles (NZVI) synthesized using tea polyphenols. *Green Chemistry* **12**, 114–122.

Pelizzetti, E., Maurino, V., Minero, C., Zerbinati, O. & Borgarello, E. 1989 Photocatalytic degradation of bentazon by TiO_2 particles. *Chemosphere* **18** (8), 1437–1445.

Pelizzetti, E., Minero, C., Borgarello, E., Tinucci, L. & Serpone, N. 1993 Photocatalytic activity and selectivity of titania colloids and particles prepared by the sol-gel technique: photooxidation of phenol and atrazine. *Langmuir* **9** (11), 2995–3001.

Pramuro, E., Vincenti, M., Augugliaro, V. & Palmisano, L. 1993 Photocatalytic degradation of monuron in aqueous TiO_2 dispersions. *Environmental Science & Technology* **27** (9), 1790–1795.

Rajeshwar, K., Osugi, M. E., Chanmanee, W., Chenthamarakshan, C. R., Zanoni, M. V. B., Kajitvichyanukul, P. & Krishnan-Ayer, R. 2008 Heterogeneous photocatalytic treatment of organic dyes in air and aqueous media. *Journal of Photochemistry and Photobiology C: Photochemistry Reviews* **9** (4), 171–192.

Rossi, L. 1997 *Reregistration Eligibility Decision (RED): Paraquat Dichloride.* Report EPA 738-F-96-018, United States Environmental Protection Agency, Washington, DC, USA.

Sorolla II, M. G., Dalida, M. L., Khemthong, P. & Grisdanurak, N. 2012 Photocatalytic degradation of paraquat using nano-sized Cu-TiO_2/SBA-15 under UV and visible light. *Journal of Environmental Sciences* **24** (6), 1125–1132.

Varma, R. S. 2012 Greener approach to nanomaterials and their sustainable applications. *Current Opinion in Chemical Engineering* **1**, 123–128.

Zahedi, F., Behpour, M., Ghoreishi, S. M. & Khalilian, H. 2015 Photocatalytic degradation of paraquat herbicide in the presence TiO_2 nanostructure thin films under visible and sun light irradiation using continuous flow photoreactor. *Solar Energy* **120**, 287–295.

First received 26 August 2018; accepted in revised form 26 November 2018. Available online 6 December 2018

Efficient degradation of tetracycline by ultraviolet-based activation of peroxymonosulfate and persulfate

Jiamin Hu, Jing Zhang, Qingguo Wang, Qian Ye, Hao Xu, Guanyu Zhou and Jinfeng Lu

ABSTRACT

In this study, the difference in oxidative capacity for removing antibiotics and the mechanism between the Cu(II)/peroxymonosulfate (PMS)/UV and Cu(II)/persulfate (PDS)/UV systems were compared under various conditions. The optimal Cu(II) concentration in the Cu(II)/PMS/UV system was 30 μM, and in the Cu(II)/PDS/UV system was 50 μM. With the PMS or PDS concentration increasing, higher tetracycline (TC) degradation in these two systems occurred. Investigation on the mechanism revealed that •OH was the primary radical in the Cu(II)/PMS/UV system, while SO_4^-• was the primary radical in the Cu(II)/PDS/UV system where •OH also played an important role. In these two systems, it was observed that Cu(I) was generated by PMS or PDS activated via UV illumination; however, oxygen alone could not promote TC removal. The degradation of TC was increased with the increasing pH level. In addition, TC degradation in the Cu(II)/PMS/UV system followed the pseudo-first-order kinetics model during the entire reaction period. It was found that the TC degradation kinetics in the Cu(II)/PDS/UV system can be divided into two parts (0 to 7 min and 10 to 50 min) and these two parts had good agreement with the pseudo-first-order kinetics model, respectively.

Key words | copper, hydroxyl radicals, sulfate radical, tetracycline, UV

Jiamin Hu
Jing Zhang (corresponding author)
Qingguo Wang
Qian Ye
Hao Xu
Guanyu Zhou
College of Architecture & Environment,
Sichuan University,
Chengdu 610065,
China
E-mail: *zjing428@163.com*

Jinfeng Lu
College of Environmental Science and Engineering,
Nankai University,
Tianjin 300071,
China

INTRODUCTION

There is increasing attention to the presence of pharmaceuticals in the aquatic environment, due to the great negative impact on human health and the ecosystem (Kuster & Adler 2014; Natija *et al.* 2017). Among various pharmaceutics, tetracycline (TC), a common type of antibiotics, is widely used to treat human and animal diseases and in the animal husbandry and fisheries fields (Shi *et al.* 2017; Zhou *et al.* 2017). However, TC is difficult to be metabolized completely by humans and animals and finds its way into the water system (Natija *et al.* 2017; Shi *et al.* 2017). Traditional methods such as electrochemistry and adsorption can remove TC (Oturan *et al.* 2013; Zhang *et al.* 2011; Wu *et al.* 2016). Additionally, ultraviolet (UV)-based advanced oxidation processes (AOPs) have been shown to effectively degrade pharmaceuticals. TC photolytic degradation has also been reported in recent years, attracting more and more attention because of the high efficiency in removing refractory compounds (Zhang & Li 2014; Ai *et al.* 2015; Sun *et al.* 2016; Xiao *et al.* 2017).

Sulfate radical (SO_4^-•), generated from the activation of peroxymonosulfate (PMS) by UV, transition metals and heat (Anipsitakis & Dionysiou 2004; Cui *et al.* 2016; Jorge *et al.* 2017; Yang *et al.* 2017), have been shown to be effective in treating pharmaceuticals due to the high redox potential (2.5–3.1 V), compared with hydroxyl radical (•OH, 1.9–2.7 V) (Buxton *et al.* 1988; Yang *et al.* 2014; Zhang *et al.* 2015; Ye *et al.* 2017). Many studies found that the combination of ultraviolet and transition metals like Fe(II) and Co(II) can strongly enhance the oxidation capacity of a single system (Khan *et al.* 2016; Jorge *et al.* 2017). For instance, Fe(II) activates PMS to produce Fe(III) at first, and then Fe(II) regenerated via the reaction between Fe(III) and UV, indicating that UV promotes the transformation of the iron-redox cycle. Meanwhile, SO_4^-• is produced and aims to remove the organic compounds. The reaction

doi: 10.2166/wst.2019.034

is grown in Equations (1) and (2) as follows (Khan *et al.* 2016):

$$Fe^{2+} + HSO_5^- \rightarrow Fe^{3+} + SO_4^- \cdot + OH^- \ (k = 3.0 \times 10^4 \ M^{-1} \ s^{-1}) \quad (1)$$

$$Fe(OH)^{2+} + h\nu \rightarrow \cdot OH + Fe^{2+} \quad (2)$$

Although iron is the most common transition metal in wastewater treatment, there are some drawbacks in iron-based AOPs. The most obvious one is the limitation of the pH level due to the precipitation of Fe^{2+}/Fe^{3+} (Khan *et al.* 2016). Other metals like copper have drawn wide concern recently in overcoming this shortcoming, because the solubility product (k_{sp}) of $Cu(OH)_2$ (1.6×10^{-19}) is much higher than $Fe(OH)_3$ (2.5×10^{-35}) (Meighan *et al.* 2008; Kang *et al.* 2014). It shows that copper is not easy to precipitate at neutral or alkaline conditions compared with iron. Moreover, Verma *et al.* (2016) reported that the addition of Fe^{3+} in a PMS/UV system was found to be more efficient in the degradation of anatoxin-a. Furthermore, cuprous copper (Cu(I)) is considered as a promising catalyst and has a high potential to activate the other oxidation agents like oxygen (O_2), hydrogen peroxide (H_2O_2) and PMS, but it is easily oxidized to cupric copper (Cu(II)) (Yuan *et al.* 2012, 2013). Thus, copper is seldom used in wastewater treatment.

Similar to PMS, peroxydisulfate (PDS) can also be activated by UV to generate $SO_4^-\bullet$. Zhang's study showed that the AOP's performance in removing pharmaceuticals was higher with PDS/UV than H_2O_2/UV (Zhang *et al.* 2015). In addition, the sulfate radical was produced through combining Fe(II) with PDS according to Equation (3).

$$S_2O_8^{2-} + Fe^{2+} \rightarrow SO_4^- \cdot + Fe^{3+} + SO_4^{2-} \ (k = 2 \times 10 \ M^{-1} \ s^{-1}) \quad (3)$$

Most previous studies used reducing agents like hydroxyl amine (HA) to reduce the transition metals (Lee *et al.* 2016). Based on Equation (2), it has been verified that UV is also capable of reducing high valence transition metal ions to low valence transition metal ions (Bedoui *et al.* 2011; Liu *et al.* 2016a, 2016b). As oxidation agents and transition metals simultaneously exist, on the one hand, the addition of UV directly activates oxidation agents like PMS and PDS to produce free radicals (Lou *et al.* 2016; Liu *et al.* 2016a, 2016b). On the other hand, UV continuously reduces transition metals that are oxidized by oxidation agents, forming a metal cycle with high and low valence ions (Wu *et al.* 1999; Bedoui *et al.* 2011). Thus, it is reasonable to consider that Cu(II) may be reduced to Cu(I) via UV first, and then Cu(I) activates PMS or PDS to promote the production of radicals like $SO_4^-\bullet$ and •OH, which further enhance the oxidative ability of the Cu(II)/PMS or Cu(II)/PDS systems.

This study compares and evaluates the efficiency of removing the pharmaceuticals TC in Cu(II)/PMS/UV and Cu(II)/PDS/UV systems. The objects of this study are: (1) to compare the oxidative capacity for removing TC in different systems; (2) to examine the effect of Cu(II) concentration, PMS and PDS concentration in the two systems; (3) to determine the mechanism of radicals generation and investigate the species of radicals in different systems.

MATERIALS AND METHODS

Materials

Potassium peroxymonosulfate (PMS) and copper sulfate pentahydrate ($CuSO_4 \cdot 5H_2O$, ≥99.0%) were supplied by Sigma-Aldrich. Tetracycline (TC) was bought from Aladdin. Sodium persulfate (PDS, ≥99.0%), neocuproine hemihydrate (NCP, ≥98%), sulfuric acid (H_2SO_4, 75%), sodium hydroxide (NaOH, ≥96%) and coumarin (≥98%) were supplied by Sinopharm Chemical Reagent Co., Ltd. Tert-butyl alcohol (TBA, ≥99.5%) and ethyl alcohol (EA, 75%) were bought from Kelong chemical reagent factory. All of the reagents above were of analytic purity. Pure oxygen (O_2, ≥99.2%) and pure nitrogen (N_2, ≥99.9%) were stored in special high-pressure gas cylinders. Methanol and ammonium acetate were chromatographically pure and were bought from Sinopharm Chemical Reagent Co., Ltd (China).

Experimental procedure

All experiments were performed in a 500 mL tailor-made columniform glass container with a quartz tube of 35 mm diameter. A 25 W low pressure (LP) ultraviolet lamp peaking at 254 nm was placed in the quartz tube. During the experiments, samples in the container were surrounded and cooled by condensate water in order to keep the temperature of the samples constant at the room temperature. Cu(II) ($CuSO_4 \cdot 5H_2O$) and TC at the desired concentrations were spiked into 500 mL pure water. The initial pH was adjusted by 0.5 M H_2SO_4 or 1.0 M NaOH. Oxygen or nitrogen was bubbled into the samples at 1.0 L•min^{-1} for 20 min before the experiments, and was then set at 0.5 L•min^{-1} during the experiments. Each run was started by PMS or PDS addition and UV light emition. TBA and EA, the scavengers of •OH and SO_4^{-}•, were added to the samples immediately after UV emition in the quenching experiments, respectively. Samples with coumarin were quenched by $NaNO_2$ after being withdrawn and then were detected by high performance liquid chromatography (HPLC).

Analysis methods

The concentration of TC was detected at 357 nm by a UV-vis spectrometer (MAPADA, UV-1800) (Shi *et al.* 2017). The concentration of Cu(I) was spectrophotometrically measured by the neocuproine method (American Public Health Association 2005). In addition, the concentration of dissolved oxygen (DO) was determined by a dissolved oxygen meter (JPB-607A). A pH meter (pHs-25) was used to detect the pH level and the dates changed less than ±0.1. Each experiment was replaced three times at the same conditions and the standard deviation was less than 2%.

The product of coumarin and •OH, 7-hydroxycoumarin (7OHC), was determined by HPLC (Waters e2695, USA) equipped with a reverse-phase C18 column (4.6 × 150 mm). The mobile phases were methanol and 0.1% ammonium acetate (50:50, v/v). The analysis conditions were as follows: the excitation wavelength: 346 nm; the determination wavelength: 456 nm; the column temperature: 30 °C; the injection speed: 1 mL/min (Louit *et al.* 2005; Takuya *et al.* 2011; Zhou *et al.* 2019).

RESULTS AND DISCUSSION

Oxidative degradation of TC

To investigate the efficiency of different systems, experiments for removing TC were examined at pH 3.5 in the PMS, PDS, UV, Cu(II), Cu(II)/UV, PMS/UV, PDS/UV, Cu(II)/PMS/UV and Cu(II)/PDS/UV systems, respectively. As shown in Figure 1, TC removal was 49.3%, 71.8% and 93.1% in the PMS, PMS/UV and Cu(II)/PMS/UV systems (top). Through these results, it was found that the addition of UV (PMS/UV) increased TC degradation of the PMS system more than 20%. Besides, TC degradation of the Cu(II)/PMS/UV system was increased by over 20% than that in the PMS/UV system, indicating that the addition of Cu(II) and UV could strongly enhance the oxidative capacity of the PMS system. The degradation of TC in the PDS, PDS/UV and Cu(II)/PDS/UV systems (bottom) were 17.1%, 40.7% and 67.1%, which was similar with the former results of PMS. Moreover, single Cu(II), single UV and Cu(II)/UV systems could not remove TC during the 50 min reaction period.

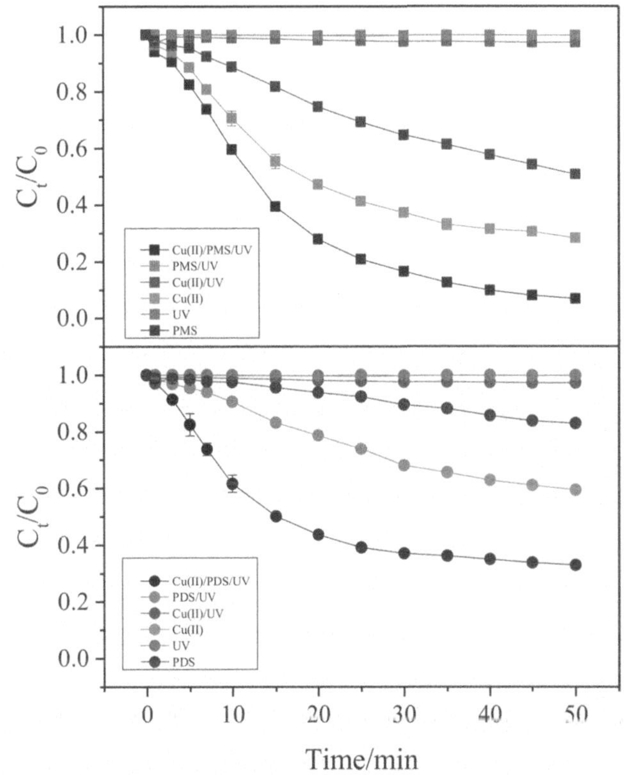

Figure 1 | Oxidative degradation of TC by PMS (top) and PDS (bottom) activated. Experimental conditions: $[Cu(II)]_0 = 20\ \mu M$; $[PMS]_0 = [PDS]_0 = 0.50$ mM; $[TC]_0 =$ 9 mg/L; pH = 3.5; $T_0 = 25\ °C$; power of UV light = 25 W; reaction time = 50 min.

Effect of Cu(II) concentration

As shown in Figure 1, TC degradation in the PMS/UV and PDS/UV systems with Cu(II) were both higher than that without Cu(II), indicating that the Cu(II) concentration may be one of key factors to influence the oxidative capacity of such systems. As shown in Figure 2, the degradation of TC was 84.0%, 86.9%, 95.4%, 94.7% and 92.8% at 50 min in the Cu(II)/PMS/UV system (top), when the Cu(II) concentration was 5, 15, 30, 50 and 100 μM, respectively. It was found that TC degradation increased with the increased Cu(II) concentration, but TC degradation was decreased slightly when the Cu(II) concentration exceeded 30 μM. Thus, the optimal Cu(II) concentration was 30 μM. Likewise, as shown in Figure 2, the TC degradation followed

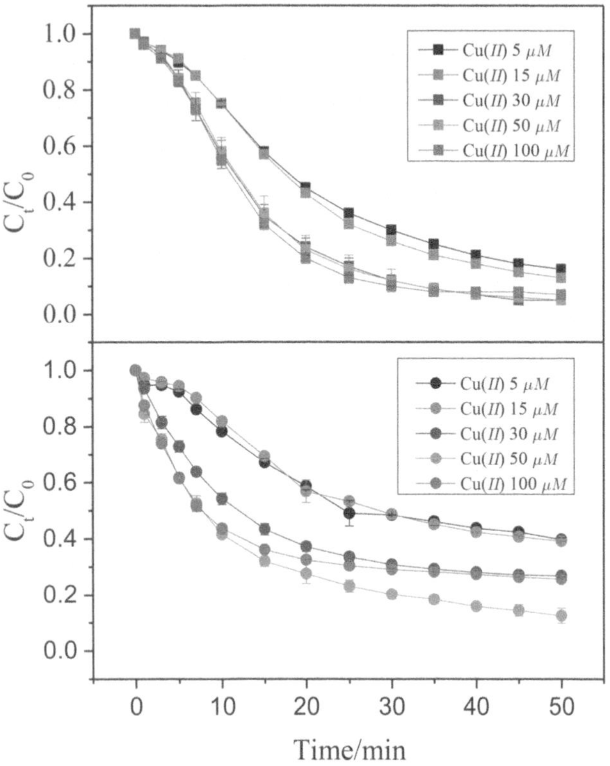

Figure 2 | Effect of Cu(II) concentration in the Cu(II)/PMS/UV (top) and Cu(II)/PDS/UV (bottom) systems. Experimental conditions: $[Cu(II)]_0 = 5$, 15, 30, 50, 100 μM; $[PMS]_0 = [PDS]_0 = 0.50$ mM; $[TC]_0 = 9$ mg/L; pH = 3.5; $T_0 = 25$ °C; power of UV light = 25 W; reaction time = 50 min.

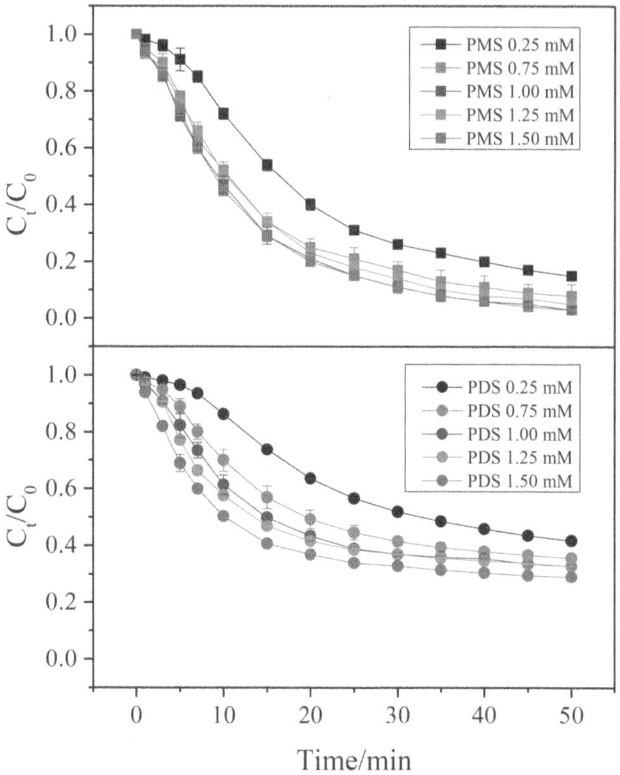

Figure 3 | Effect of oxidative agents concentration in the Cu(II)/PMS/UV (top) and Cu(II)/PDS/UV (bottom) systems. Experimental conditions: $[Cu(II)]_0 = 20$ μM; $[PMS]_0 = [PDS]_0 = 0.25$, 0.75, 1.00, 1.25 and 1.50 mM; $[TC]_0 = 9$ mg/L; pH = 3.5; $T_0 = 25$ °C; power of UV light = 25 W; reaction time = 50 min.

the same rule in the Cu(II)/PDS/UV system (bottom), where the optimal Cu(II) concentration was 50 μM. The enhanced degradation rate with higher Cu(II) concentration for these two systems is likely attributed to the higher amount of Cu(I) and radicals. The phenomenon of much higher Cu(II) concentration may be because excessive Cu(I) reacted with radicals, following Equations (4) and (5) (Buxton *et al.* 1988; Liang & Su 2009).

$$Cu(I) + SO_4^{-}\cdot \rightarrow Cu(II) + SO_4^{2-} \quad (4)$$

$$Cu(I) + \cdot OH \rightarrow Cu(II) + OH^{-} \quad (5)$$

Effect of PMS and PDS concentration

To find the effect of PMS and PDS concentration on TC degradation, different PMS and PDS concentrations were examined in their own systems. As shown in Figure 3, TC removal was 85.0%, 92.1%, 96.6%, 95.2% and 96.8% during 50 min in the Cu(II)/PMS/UV system (top). Clearly, TC degradation was increased with the increasing PMS concentration, which may be because higher PMS concentration could be activated to generate more radicals by UV and Cu(I) at the same time. Based on the above, more free radicals such as $SO_4^{-}\cdot$ and $\cdot OH$ could remove more TC due to its strong oxidative capacity. Similar to PMS, when the PDS concentration was increased from 0.25 mM to 1.50 mM, TC degradation in the Cu(II)/PDS/UV system (bottom) was increased as well. In general, TC was removed more completely in the Cu(II)/PMS/UV system compared with the Cu(II)/PDS/UV system at the same PMS or PDS concentration.

The reaction mechanism of the Cu(II)/UV system with PMS and PDS

Quenching studies for radical identification

Based on the results above, it is established that the oxidative capacity of the Cu(II)/PMS/UV system is higher than that of Cu(II)/PDS/UV system at the same conditions. One of reasons may be related to the different species of the free radicals that exist in these two systems (Liang & Su 2009; Verma *et al.* 2016). In order to identify the species of radicals in the Cu(II)/PMS/UV and Cu(II)/PDS/UV systems, experiments involving the addition of TBA and EA were undertaken. TBA, an effective scavenger for •OH, has a higher reaction rate with •OH ($k=6.0\times10^{8}\ M^{-1}\,s^{-1}$) (Buxton *et al.* 1988) and slightly lower reaction rate with SO_4^{-}• ($k=8.0\times10^{5}\ M^{-1}\,s^{-1}$) (Neta *et al.* 1988). In contrast, EA is effective for both •OH and SO_4^{-}•, whose reaction rates are 1.2×10^{9}–$2.8\times10^{9}\ M^{-1}\,s^{-1}$ and 1.6×10^{7}–$7.7\times10^{7}\ M^{-1}\,s^{-1}$, respectively (Buxton *et al.* 1988; Neta *et al.* 1988). As shown in Figure 4(a), TC degradation was greatly decreased from 93.1% to 28.8% and 23.8% in the Cu(II)/PMS/UV system with TBA and EA (top), respectively, which indicated that •OH was the primary free radical in the Cu(II)/PMS/UV system. TC removal was 54.0% and 27.8% in the Cu(II)/PDS/UV system with TBA and EA compared with 67.1% in the Cu(II)/PDS/UV system. This result showed that a great amount of SO_4^{-}• and a small amount of •OH exist in the Cu(II)/PDS/UV system, following Equation (6) (He *et al.* 2013).

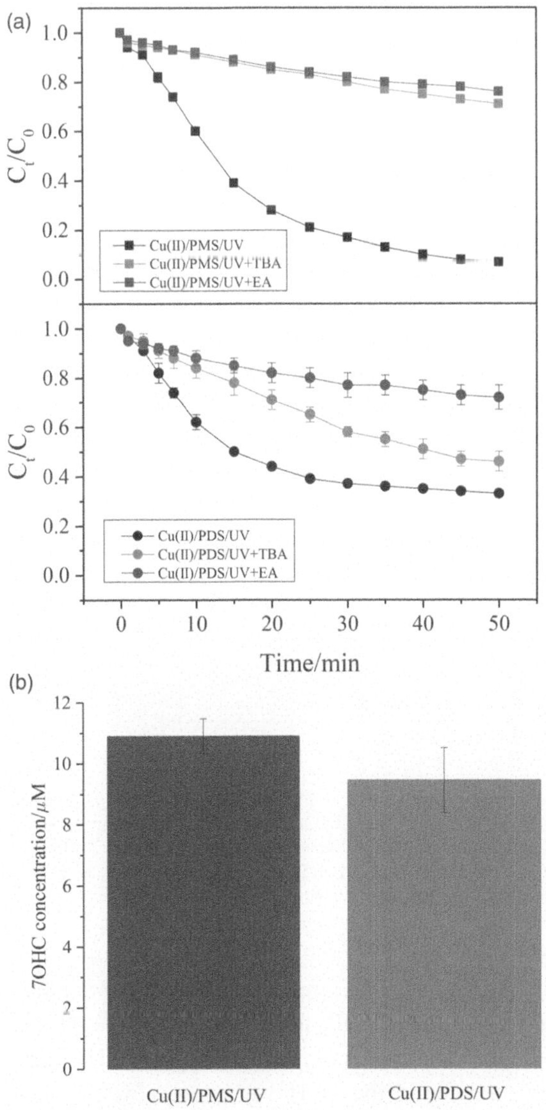

Figure 4 | (a) Effect of the addition of different radical scavengers in the Cu(II)/PMS/UV (top) and Cu(II)/PDS/UV (bottom) systems. (b) The concentration of 7OHC. Experimental conditions: $[Cu(II)]_0=20\ \mu M$; $[PMS]_0=[PDS]_0=0.50$ mM; $[TBA]_0=50$ mM; $[EA]_0=50$ mM; $[coumarin]_0=2$ mM; $[TC]_0=9$ mg/L; pH = 3.5; $T_0=25\ ^{\circ}C$; power of UV light = 25 W; reaction time = 50 min.

Besides, from the results of Figure 4(a), •OH may play an important role in these two systems. In order to further determine •OH, coumarin was chosen to capture •OH to generate the fluorescent 7-hydroxycoumarin (7OHC), following Equation (7) (Takuya *et al.* 2011). However, coumarin cannot capture •OH absolutely, so it is a semi-quantitative method to estimate the generation of •OH through the concentration of 7OHC (Zhou *et al.* 2019). As shown in Figure 4(b), the concentration of 7OHC in the Cu(II)/PMS/UV and Cu(II)/PDS/UV systems at 50 min were 10.9 and 9.45 μM, respectively, indicating that •OH was favorable for TC degradation.

$$S_2O_8^{2-}+h\nu\rightarrow SO_4^{-}\cdot+SO_4^{-}\cdot \tag{6}$$

$$\text{Coumarin } C_9H_6O_2+2\cdot OH \rightarrow 7-\text{hydroxycoumarin } C_9H_6O_3+H_2O \tag{7}$$

The generation of Cu(I)

It has been proposed that the production of free radicals catalyzed by the copper-redox cycle could strongly enhance the oxidative capacity of the PMS/UV or PDS/UV systems, as shown in Figure 1. Therefore, the generation of Cu(I) may be a key factor to influence the oxidative capacity. The concentrations of Cu(I) in the Cu(II)/PMS/UV, Cu(II)/PDS/UV and Cu(II)/UV systems were detected in order to examine the generation of Cu(I) without the addition of the target compound. As shown in Figure 5, the concentration of Cu(I) in the Cu(II)/PMS/UV and Cu(II)/PDS/UV systems were similar and higher than that in the Cu(II)/UV system, indicating that UV can reduce Cu(II) to Cu(I) at first, and then PMS and PDS can subsequently react with Cu(I) to generate radicals, while Cu(I) is oxidized to Cu(II). Because of this, the copper-redox cycle is formed. According to Equations (1) and (3), the reactions of these two systems can be described by Equations (8)–(10):

$$Cu(II)\overset{h\upsilon}{\rightarrow}Cu(I) \tag{8}$$

$$Cu(I)+HSO_5^{-}\rightarrow Cu(II)+SO_4^{2-}+\cdot OH \tag{9}$$

$$Cu(I)+S_2O_8^{2-}\rightarrow Cu(II)+SO_4^{-}\cdot+SO_4^{2-}(10) \tag{10}$$

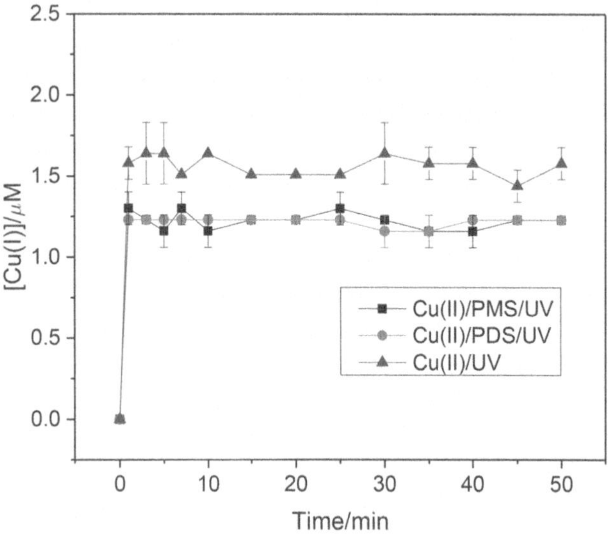

Figure 5 | The concentrations of Cu(I) in the Cu(II)/PMS/UV, Cu(II)/PDS/UV and Cu(II)/UV systems. Experimental conditions: $[Cu(II)]_0 = 20\ \mu M$; $[PMS]_0 = [PDS]_0 = 0.50$ mM; pH = 3.5; $T_0 = 25\ ^\circ C$; power of UV light = 25 W; reaction time = 50 min.

The role of oxygen

Many studies have shown that oxygen can be activated by UV and transition metals to generate H_2O_2, which is further decomposed to produce •OH (Wen *et al.* 2014; Lee *et al.* 2016). In order to examine the role of oxygen in the Cu(II)/PMS/UV (top) and Cu(II)/PDS/UV (bottom) systems, O_2 or N_2 was continuously bubbled into these two systems. As shown in Figure 6(a), compared with the Cu(II)/PMS/UV (or Cu(II)/PDS/UV) and Cu(II)/PMS/UV/N_2 (or Cu(II)/PDS/UV/N_2) systems, it was observed that the addition of O_2 could enhance the oxidative capacity and increase TC degradation, which was consistent with the former studies. Nevertheless, the degradation of TC was almost the same in the Cu(II)/PMS/UV, Cu(II)/PMS/UV/O_2 and Cu(II)/PMS/UV/N_2 systems, indicating that PMS rather than O_2 is the primary oxidizing agent in the Cu(II)/PMS/UV system. The same results were found in the Cu(II)/PDS/UV system as well. Based on the description above, the reaction mechanism of the Cu(II)/UV system with PMS and PDS can be shown in Figure 6(b).

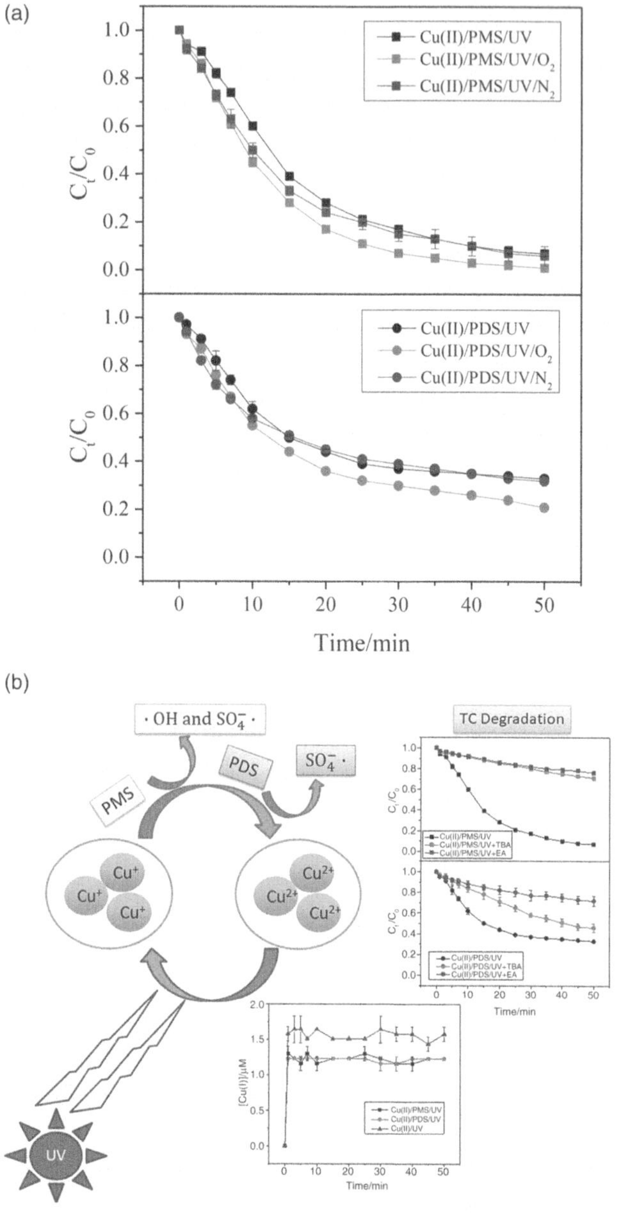

Figure 6 | (a) The role of oxygen in the Cu(II)/PMS/UV (top) and Cu(II)/PDS/UV (bottom) systems. (b) The reaction mechanism of the Cu(II)/UV system with PMS and PDS. Experimental conditions: $[Cu(II)]_0 = 20\ \mu M$; $[PMS]_0 = [PDS]_0 = 0.50$ mM; $[TC]_0 = 9$ mg/L; pH = 3.5; $T_0 = 25\ ^\circ C$; power of UV light = 25 W; O_2 flow rate = 0.5 L•min^{-1}; N_2 flow rate = 0.5 L•min^{-1}; reaction time = 50 min.

Effect of pH

In order to further investigate the reaction mechanism of the Cu(II)/PMS/UV and Cu(II)/PDS/UV systems, the degradation of TC was examined at different initial pH values ranging from 2.0 to 6.0. As shown in Figure 7, TC degradation was increased with the increasing pH level in the Cu(II)/PMS/UV and Cu(II)/PDS/UV systems, which may be related to the species of the free radicals. According to the results in the section, Quenching studies for radical identification, compared with the Cu(II)/PMS/UV system, TC degradation was higher than that in the Cu(II)/PDS/UV system, indicating that •OH was more effective than SO_4^{-}• in removing TC. Many previous studies showed that SO_4^{-}• could be further converted to •OH at high pH, following Equation (11) (Buxton *et al.* 1988; Guan *et al.* 2011). Thus, more TC was removed at pH 5 and 6 due to the great

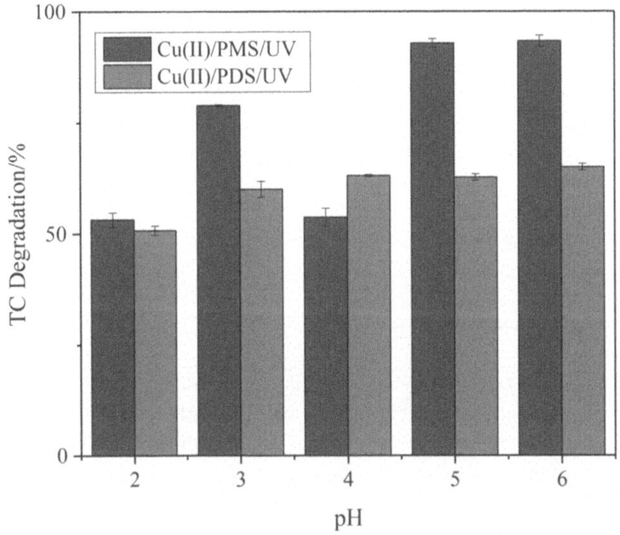

Figure 7 | Effect of different pH in the Cu(II)/PMS/UV and Cu(II)/PDS/UV systems at 50 min. Experimental conditions: $[Cu(II)]_0 = 20\ \mu M$; $[PMS]_0 = [PDS]_0 = 0.50$ mM; $[TC]_0 = 9$ mg/L; pH = 2, 3, 4, 5 and 6; $T_0 = 25\ °C$; power of UV light = 25 W; reaction time = 50 min.

amount of •OH formed in these two systems. Ji *et al.* (2016) found that deprotonated, non-dissociated TC was more reactive for the generation of SO_4^-• because TC has three p*K*a (p$K_{a1} = 3.3$, p$K_{a2} = 7.7$ and p$K_{a3} = 9.7$) and PDS was activated by TC itself. It can be noted that TC degradation at pH 3.0 in the Cu(II)/PMS/UV was higher than that at pH 4.0, which may be likely that pH 3.0 was close to pK_{a1} and the form of TC at pH 3.0 in the system would generate considerable SO_4^-•.

$$SO_4^{\cdot -} + OH^- \rightarrow SO_4^{2-} + \cdot OH \tag{11}$$

Degradation kinetics

To investigate the degradation kinetics of TC in the Cu(II)/PMS/UV and Cu(II)/PDS/UV systems, different concentrations of TC ranging from 3 to 15 mg/L, were introduced. As shown in Figure 8(a) and 8(b), the TC degradation in these two systems was decreased with the increasing TC concentration, in accordance with the general rule. As shown in Figure 8(a), the degradation of TC during a 50 min period in the Cu(II)/PMS/UV system can be described by the pseudo first-order kinetics model, and the kinetics equation is grown in Equation (12). The rate constant (*k*) and correlation coefficient (R^2) in the Cu(II)/PMS/UV system are summarized in Table 1. Moreover, as shown in Figure 8(b), according to reaction time, the TC degradation kinetics in the Cu(II)/PDS/UV system can be divided into two parts: one from 0 to 7 min and the other from 10 to 50 min. These two parts were in good agreement with the pseudo-first-order kinetics model, respectively. R^2 and *k* values in the Cu(II)/PDS/UV system are summarized in Table 2, where the rate constants during

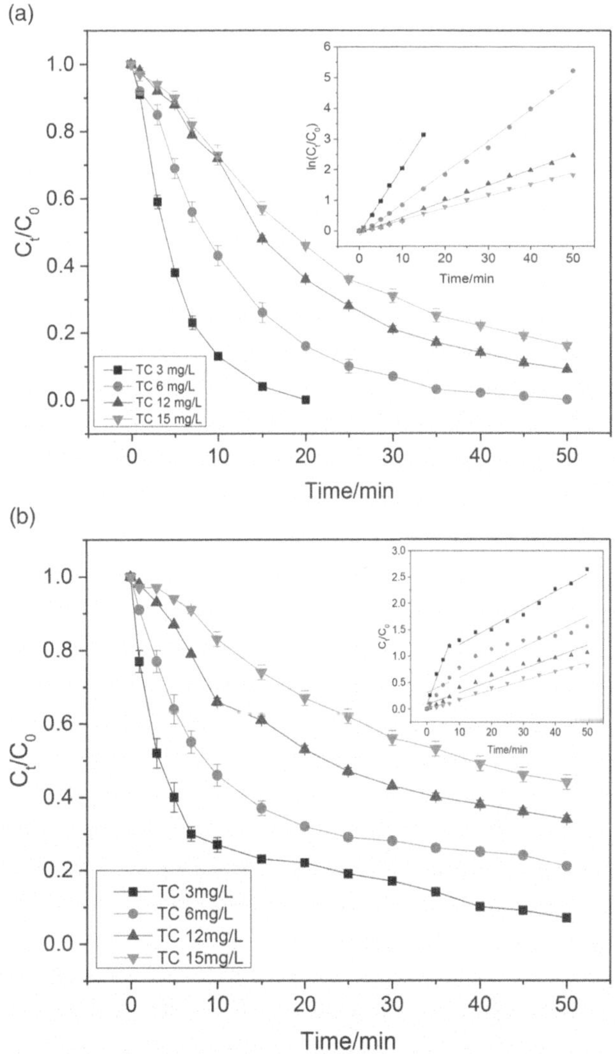

Figure 8 | (a) Effect of TC concentration and linear fit of rate constants from kinetic models for TC degradation in the Cu(II)/PMS/UV system. (b) Effect of TC concentration and linear fit of rate constants from kinetic models for TC degradation in the Cu(II)/PDS/UV system. Experimental conditions: $[Cu(II)]_0 = 20\ \mu M$; $[PMS]_0 = [PDS]_0 = 0.50$ mM; $[TC]_0 = 3$, 6, 12 and 15 mg/L; pH = 3.5; $T_0 = 25\ °C$; power of UV light = 25 W; reaction time = 50 min.

Table 1 | Pseudo-first-order rate constants for TC degradation in the Cu(II)/PMS/UV system

[TC], mg/L	3	6	12	15
k, min^{-1}	0.206	0.098	0.050	0.038
R^2	0.996	0.993	0.993	0.994

Experimental conditions: $[Cu(II)]_0 = 20\ \mu M$; $[PMS]_0 = 0.50$ mM; $[TC]_0 = 3$, 6, 12 and 15 mg/L; pH = 3.5; $T_0 = 25\ °C$; power of UV light = 25 W; reaction time = 50 min.

Table 2 | Pseudo-first-order rate constants for TC degradation in the Cu(II)/PDS/UV system

	0–7 min				10–50 min			
[TC], mg/L	3	6	12	15	3	6	12	15
k, min^{-1}	0.181	0.086	0.031	0013	0.034	0.017	0.017	0.016
R^2	0.971	0.998	0.975	0.948	0.974	0.934	0.968	0.981

Experimental conditions: $[Cu(II)]_0 = 20\ \mu M$; $[PDS]_0 = 0.50$ mM; $[TC]_0 = 3$, 6, 12 and 15 mg/L; pH = 3.5; $T_0 = 25\ ^\circ C$; power of UV light = 25 W; reaction time = 50 min.

0 to 7 min were higher than that during 10 to 50 min. In addition, compared with the Cu(II)/PDS/UV system, k in the Cu(II)/PMS/UV system was much higher, indicating that TC was removed more quickly in the Cu(II)/PMS/UV system. These results are consistent with the phenomenon in Figures 1 and 4.

$$-\ln(C_t/C_0) = kt \tag{12}$$

CONCLUSIONS

This study investigated TC degradation with Cu(II) and UV in order to compare with the activation of PMS and PDS. The results showed that TC degradation was much higher in the Cu(II)/PMS/UV and Cu(II)/PDS/UV systems via comparison with other different systems. The TC degradation was increased with the increasing Cu(II) concentration, but a much higher Cu(II) concentration may result in more Cu(I) being generated, which could quench the radicals and inhibit TC removal. It was observed that the optimal Cu(II) concentration of the Cu(II)/PMS/UV and Cu(II)/PDS/UV systems were 30 μM and 50 μM, respectively. With the PMS or PDS concentration increasing in their own systems, the TC degradation was increased due to more radicals being generated during the same time period. Quenching studies indicated that •OH was primary free radicals in the Cu(II)/PMS/UV system. However, SO_4^-• was the main free radical in the Cu(II)/PDS/UV system where •OH also played an important role. The degradation of TC was increased with increasing pH level. Besides, the degradation of TC in these two systems can be described by the pseudo-first-order kinetics model.

ACKNOWLEDGEMENTS

This work was supported by the National Natural Science Foundation of China (No. 51508353, 51741807, 51878357).

REFERENCES

Ai, C. L., Zhou, D. D., Wang, Q., Shao, X. W. & Lei, Y. J. 2015 Optimization of operating parameters for photocatalytic degradation of tetracycline using In_2S_3 under natural solar radiation. *Solar Energy* **113**, 34–42.

American Public Health Association/American Water Works Association/Water Environment Federation 2005 *Standard Methods for the Examination of Water and Wastewater*, Washington, DC, USA.

Anipsitakis, G. P. & Dionysiou, D. D. 2004 Radical generation by the interaction of transition metals with common oxidants. *Environmental Science & Technology* **38**, 3705–3712.

Bedoui, A., Limem, E., Ahmed, A. & Bensalah, N. 2011 Photo-fenton treatment of actual agro-industrial wastewaters. *Industrial & Engineering Chemistry Research* **50**, 6673–6680.

Buxton, G. V., Greenstock, C. L., Helman, W. P. & Ross, A. B. 1988 Critical review of rate constants for reactions of hydrated electrons, hydrogen atoms and hydroxyl radicals (·OH/·O^-) in Aqueous Solution. *Journal of Physical and Chemical Reference Data* **17**, 513–886.

Cui, C. Z., Jin, L., Jiang, L., Han, Q., Lin, K. F., Lu, S. G., Zhang, D. & Cao, G. M. 2016 Removal of trace level amounts of twelve sulfonamides from drinking water by UV-activated peroxymonosulfate. *Science of the Total Environment* **572**, 244–251.

Guan, Y. H., Ma, J., Li, X. C., Fang, J. Y. & Chen, L. W. 2011 Influence of pH on the formation of sulfate and hydroxyl radicals in the UV/peroxymonosulfate system. *Environmental Science & Technology* **45**, 9308–9314.

He, X. X., Cruz, A. A. & Dionysiou, D. D. 2013 Destruction of cyanobacterial toxin cylindrospermopsin by hydroxyl radicals and sulfate radicals using UV-254 nm activation of hydrogen peroxide, persulfate and peroxymonosulfate. *Journal of Photochemistry and Photobiology A: Chemistry* **251**, 160–166.

Ji, Y. F., Shi, Y. Y., Dong, W., Wen, X., Jiang, M. D. & Lu, J. H. 2016 Thermo-activated persulfate oxidation system for tetracycline antibiotics degradation in aqueous solution. *Chemical Engineering Journal* **298**, 225–233.

Jorge, R. C., Carlos, A., Tania, S., Dionysios, D. D., Gianluca, L. P., Marco, S. L. & Jose, A. P. 2017 Treatment of winery wastewater by sulphate radicals: HSO_5^-/transition metal/UV-A LEDs. *Chemical Engineering Journal* **310**, 473–483.

Kang, D. E., Lim, C. S., Kim, J. Y., Kim, E. S., Chun, H. J. & Cho, B. R. 2014 Two-Photon probe for Cu^{2+} with an internal reference: quantitative estimation of Cu^{2+} in human tissues by two-photon microscopy. *Analytical Chemistry* **86**, 5353–5359.

Khan, S., He, X. X., Khan, H. M., Boccelli, D. & Dionysiou, D. D. 2016 Efficient degradation of lindane in aqueous solution by iron (II) and/or UV activated peroxymonosulfate. *Journal of Photochemistry and Photobiology A: Chemistry* **316**, 37–43.

Kuster, A. & Adler, N. 2014 Pharmaceuticals in the environment: scientific evidence of risks and its regulation. *Philosophical Transactions of the Royal Society B* **369**, 1–8.

Lee, H., Lee, H. J., Seo, J., Kim, H. E., Shin, Y. K., Kim, J. H. & Lee, C. 2016 Activation of oxygen and hydrogen peroxide by copper(II) coupled with hydroxylamine for oxidation of organic contaminants. *Environmental Science & Technology* **50**, 8231–8238.

Liang, C. J. & Su, H. W. 2009 Identification of sulfate and hydroxyl radicals in thermally activated persulfate. *Industrial & Engineering Chemistry Research* **48** (11), 5558–5562.

Liu, W. Q., Fang, C. L., Huang, Y., Ai, L. Y., Yang, F., Wang, Z. H. & Liu, J. S. 2016a Is UV/Ce(IV) process a chloride-resistant AOPs for organic pollutants decontamination? *The Royal Society of Chemistry* **6**, 93558–93563.

Liu, Y. Q., He, X. X., Duan, X. D., Fu, Y. S., Despo, F. & Dionysiou, D. D. 2016b Significant role of UV and carbonate radical on the degradation of oxytetracycline in UV-AOPs: kinetics and mechanism. *Water Research* **95**, 195–204.

Lou, X. Y., Xiao, D. X., Fang, C. L., Wang, Z. H., Liu, J. S., Guo, Y. G. & Lu, S. Y. 2016 Comparison of UV/hydrogen peroxide and UV/peroxydisulfate processes for the degradation of humic acid in the presence of halide ions. *Environmental Science and Pollution Research* **23**, 4778–4785.

Louit, G., Foley, S. & Cabillic, J. 2005 The reaction of coumarin with the OHradical revisited: hydroxylation product analysis determined by fluorescence and chromatography. *Radiation Physics and Chemistry* **72**, 119–124.

Meighan, M., MacNeil, J. & Falconer, R. 2008 Determining the solubility product of $Fe(OH)_3$: an equilibrium study with environmental significance. *Journal of Chemical Education* **85** (2), 254–255.

Natija, B., Hugo, O., Nihal, O., David, H., Abdellatif, G., Salah, A., Enric, B. & Mehmet, A. O. 2017 Kinetics of oxidative degradation/mineralization pathways of the antibiotic tetracycline by the novel heterogeneous electro-Fenton process with solid catalyst chalcopyrite. *Applied Catalysis B: Environmental* **209**, 637–647.

Neta, P., Huie, R. E. & Ross, A. B. 1988 Rate constants for reactions of inorganic radicals in aqueous solution. *Journal of Physical and Chemical Reference Data* **17** (3), 1027–1284.

Oturan, N., Wu, J., Zhang, H., Sharma, V. K. & Oturan, M. A. 2013 Electrocatalytic destruction of the antibiotic tetracycline in aqueous medium by electrochemical advanced oxidation processes: effect of electrode materials. *Applied Catalysis B: Environmental* **140–141**, 92–97.

Shi, W. L., Guo, F. & Yuan, S. L. 2017 *In situ* synthesis of Z-scheme $Ag_3PO_4/CuBi_2O_4$ photocatalysts and enhanced photocatalytic performance for the degradation of tetracycline under visible light irradiation. *Applied Catalysis B: Environmental* **209**, 720–728.

Sun, P. Z., Tyree, C. & Huang, C. H. 2016 Inactivation of *Escherichia coli*, bacteriophage MS2, and *Bacillus* spores under UV/H_2O_2 and UV/peroxydisulfate advanced disinfection conditions. *Environmental Science & Technology* **50**, 4448–4458.

Takuya, M., Masahiro, T. & Makoto, S. 2011 Hydroxyl radical concentration profile in photo-Fenton oxidation process: generation and consumption of hydroxyl radicals during the discoloration of azo-dye Orange II. *Chemosphere* **82**, 1422–1430.

Verma, S., Nakamura, S. & Sillanpaa, M. 2016 Application of UV-C LED activated PMS for the degradation of anatoxin-a. *Chemical Engineering Journal* **284**, 122–129.

Wen, G., Wang, S. J., Ma, J., Huang, T. L., Liu, Z. Q., Zhao, L. & Xu, J. L. 2014 Oxidative degradation of organic pollutants in aqueous solution using zero valent copper under aerobic atmosphere condition. *Journal of Hazardous Materials* **275**, 193–199.

Wu, K. Q., Xie, Y. D., Zhao, J. C. & Hidaka, H. 1999 Photo-Fenton degradation of a dye under visible light irradiation. *Journal of Molecular Catalysis A* **144**, 77–84.

Wu, S. K., Shen, X. P., Zhou, H., Zhu, G. X., Wang, R. Y., Ji, Z. Y., Chen, K. M. & Chen, C. J. 2016 Morphological synthesis of Prussian blue analogue $Zn_3[Fe(CN)_6]_2\ xH_2O$ micro-/nanocrystals and their excellent adsorption performance toward methylene blue. *Journal of Colloid and Interface Science* **464**, 191–197.

Xiao, R. Y., Gao, L. W., Wei, Z. S., Spinney, R., Luo, S., Wang, D. H., Dionysiou, D. D., Tang, C. J. & Yang, W. C. 2017 Mechanistic insight into degradation of endocrine disrupting chemical by hydroxyl radical: an experimental and theoretical approach. *Environmental Pollution* **231**, 1446–1452.

Yang, Y., Pignatello, J. J., Ma, J. & Mitch, W. A. 2014 Comparison of halide impacts on the efficiency of contaminant degradation by sulfate and hydroxyl radical-based advanced oxidation processes (AOPs). *Environmental Science & Technology* **48**, 2344–2351.

Yang, Z. H., Su, R. K., Spinney, R L. S., Cai, M. Q., Xiao, R. X. & Wei, Z. S. 2017 Comparison of the reactivity of ibuprofen with sulfate and hydroxyl radicals: an experimental and theoretical study. *Science of the Total Environment* **590–591**, 751–760.

Ye, T. T., Wei, Z. S., Spinney, R., Tang, C. J., Luo, S., Xiao, R. Y. & Dionysiou, D. D. 2017 Chemical structure-based predictive model for the oxidation of trace organic contaminants by sulfate radical. *Water Research* **116** (1), 106–115.

Yuan, X., Pham, A. N., Xing, G. W., Rose, A. L. & Waite, T. D. 2012 Effects of pH, chloride, and bicarbonate on Cu(I) oxidation kinetics at circumneutral pH. *Environmental Science & Technology* **46**, 1527–1535.

Yuan, X., Pham, A. N., Miller, C. J. & Waite, T. D. 2013 Copper-catalyzed hydroquinone oxidation and associated redox cycling of copper under conditions typical of natural saline waters. *Environmental Science & Technology* **47**, 8355–8364.

Zhang, A. & Li, Y. M. 2014 Removal of phenolic endocrine disrupting compounds from waste activated sludge using UV, H_2O_2, and UV/H_2O_2 oxidation processes: effects of reaction conditions and sludge matrix. *Science of the Total Environment* **493**, 307–323.

Zhang, L., Song, X. Y., Liu, X. Y., Yang, L. Y., Pan, F. & Lv, J. N. 2011 Studies on the removal of tetracycline by multi-walled carbon nanotubes. *Chemical Engineering Journal* **178**, 26–33.

Zhang, R. C., Sun, P. Z., Boyer, T. H., Zhao, L. & Huang, C. H. 2015 Degradation of pharmaceuticals and metabolite in synthetic human urine by UV, UV/H_2O_2, and UV/PDS. *Environmental Science & Technology* **49**, 3056–3066.

Zhou, K. F., Xie, X. D. & Chang, C. T. 2017 Photocatalytic degradation of tetracycline by Ti-MCM-41 prepared at room temperature and biotoxicity of degradation products. *Applied Surface Science* **416**, 248–258.

Zhou, P., Zhang, J. & Zhang, G. C. 2019 Degradation of dimethyl phthalate by activating peroxymonosulfate using nanoscale zero valent tungsten: mechanism and degradation pathway. *Chemical Engineering Journal* **359**, 138–148.

First received 24 July 2018; accepted in revised form 11 January 2019. Available online 23 January 2019

Electro-oxidation of landfill leachate using boron-doped diamond: role of current density, pH and ions

F. Agustina, A. Y. Bagastyo and E. Nurhayati

ABSTRACT

Electro-oxidation using a boron-doped diamond (BDD) anode can be used as an alternative to leachate treatment. Aside from the hydroxyl radical, BDDs are capable of generating chloride and sulfate radical species that play significant roles in the oxidation of pollutants. This research investigated the role of $Cl^-:SO_4^{2-}$ ions at molar ratios of 237:1, 4:1 and 18:1, and the influence of applied current density (i.e. 50, 75 and 100 mA cm^{-2}) on the removal of organic and ammonium contaminants. The results show that current density had considerable effects on chemical oxygen demand (COD) and colour removal, while ion composition of $Cl^-:SO_4^{2-}$ at pH 3, 5 and 8.5 (original pH) gave different effects on COD and ammonium removal. The pH had a significant effect on the COD removal at the ratio of 237:1, but showed no dramatic effect at the ratio of 18:1, giving ~40% of COD removal at all pHs tested. This indicates that electro-oxidation at the ratio of 18:1 could be effectively conducted at a wide range of pH. Furthermore, the optimum ammonium removal was obtained at pH 8.5 with the ratio of 237:1. This process was found to be ineffective in increasing the biodegradability index of the leachate; instead, it exhibited mineralization of organic content.

Key words | boron-doped diamond anode, electrochemical oxidation, leachate

F. Agustina
A. Y. Bagastyo (corresponding author)
E. Nurhayati
Department of Environmental Engineering, Faculty of Civil, Environmental, and Geo Engineering,
Institut Teknologi Sepuluh Nopember,
Surabaya 60111,
Indonesia
E-mail: *bagastyo@enviro.its.ac.id*

INTRODUCTION

Leachate generated from municipal solid waste landfill is a wastewater containing organic and inorganic contaminants, including some refractory and toxic compounds. Due to its complex composition, landfill leachate is one of the major environmental problems concerning water pollution (Fernandes *et al.* 2015) in the context of solid waste management. Landfill leachates have been identified as potential sources of ground and surface water contamination, if they are not properly collected, treated and safely disposed (Tatsi *et al.* 2003).

The composition of landfill leachates varies depending on landfill characteristics, such as waste type and its composition, water content and its degree of compaction, landfilling technique (waterproof covers, liner requirements such as clay and geotextiles), climate and seasonal weather variation, and age of landfill leachate (Renou *et al.* 2008). As the landfill age increases, most of the biodegradable organic matter contributed to the leachate matrix; that is, determined as biochemical oxygen demand (BOD) concentration, is decomposed in the stabilization process. As a result, the remaining organic matter contained in the leachate will be dominantly non-biodegradable. Therefore, the organic matter determined as chemical oxygen demand (COD) will be the dominant non-biodegradable fractions, and will largely remain unchanged (Fernandes *et al.* 2015), whereas BOD will be decreased. This leads to a lower ratio of BOD/COD in the leachate generated from a 'mature' landfill site. This type of leachate has recently become of concern to governments, particularly in developing countries, as an advanced treatment process is required to support the existing biological processes.

Electrochemical oxidation is one of the effective physico-chemical methods applied to remove organic pollutant and inorganic nutrients, such as ammonia, particularly contained in landfill leachate wastewater. The efficiency of electro-oxidation depends on various parameters such as current density, type of anode, pH, as well as the type and concentration of electrolyte (i.e. ionic compound) used during the process (Anglada *et al.* 2011). In terms of applied current density, Zhou *et al.* (2016) reported that COD removal increases with increasing current density. Increasing current density during electro-oxidation yielded

doi: 10.2166/wst.2019.040

enhancement of pollutant removal, which may be attributed to a higher amount of the formed hydroxyl radicals at the anode surface or by the accelerated generation of chlorine/hypochlorite compounds (Mandal *et al.* 2017).

Based on the different heterogeneous species formed from water discharge, anodes were classified as 'active' and 'non-active' (Comninellis 1994). Active anodes interact strongly with $OH^{\cdot}$, forming chemisorbed 'active oxygen' species that lead to electrochemical conversion of organic pollutants. On the other hand, non-active anodes, such as boron-doped diamond (BDD), interact weakly with $OH^{\cdot}$ to form physisorbed $OH^{\cdot}$; that is, one of the strongest oxidants. This leads to mineralization of organic compounds. Therefore, non-active anodes, such as BDD, are attractive to be implemented for the treatment of wastewater containing recalcitrant organic contaminants like leachate wastewater.

In addition, BDD is one of the non-active anodes having the capability of generating reactive oxygen species (ROS), such as heterogeneous $OH^{\cdot}$, H_2O_2, and O_3, from the electrochemical process. Besides ROS, BDD anodes are also capable of generating inorganic radical species such as chloride ($Cl^{\cdot}$, $Cl_2^{\cdot-}$) and sulfate ($SO_4^{\cdot-}$) in the presence of their corresponding ion (Davis *et al.* 2014; Farhat *et al.* 2017). During the electrochemical process, the sulfate ion can be oxidized to form $SO_4^{\cdot-}$. This species may demonstrate 10–15 fold higher removal rates of several persistent organic contaminants compared to $OH^{\cdot}$ based anodic oxidation (Farhat *et al.* 2015). The addition of the chloride ion gradually decreases the rate constant of organic compound removal measured as total organic carbon (TOC) in Na_2SO_4 anolyte, as reported by Farhat *et al.* (2017). This inhibiting effect of chloride is a consequence of more pronounced scavenging of $SO_4^{\cdot-}$ by Cl^- (Equation (1)). In the presence of chloride, the electro-generated $OH^{\cdot}$ can be expected to be scavenged by the Cl^- and converted to chloride radical species.

$$SO_4^{\bullet-} + Cl^- \rightarrow Cl^{\bullet} + SO_4^{2-} \quad (1)$$

It is interesting to note that the presence of ionic compounds, for example chloride and sulfate ions, have significant roles in the removal of organic pollutants, which can be affected by their concentration and pH of the wastewater. Therefore, this study aims to investigate the role of sulfate and chloride ions that are available in different ratios on the oxidation of landfill leachate wastewater. The treatment efficiency was evaluated based on the removal of COD and ammonia. Furthermore, BOD and TOC concentrations were also determined to evaluate the oxidation mechanism of organic pollutants in the leachate.

METHODS

Samples and reactor materials

The leachate used in this study was sampled from the influent of the leachate treatment facility of a municipal solid waste landfill located in Gresik, Indonesia. The characteristics of the leachate are summarized in Table 1. The experiments were conducted in a single cell acrylic reactor (internal dimensions: $4 \times 2 \times 10$ cm, with an actual volume of 80 mL). The reactor was run in a batch recirculation mode. Boron-doped diamond was used as anode and silver was used as cathode with each surface area 8 cm^2 and an electrode gap of 20 mm.

Experimental set-up for electrochemical oxidation

This study was comprised of two experimental steps; that is, (i) an experiment to investigate the role of current density and (ii) the main experiment to investigate the role of the supporting ion and pH. The first step of the experiment was conducted by evaluating three different constant current densities; that is, 50, 75, and 100 $mAcm^{-2}$ without pH adjustment (pH 8.5) and electrolyte addition. In this experiment, 1 L landfill leachate was pumped into the reactor and recirculated for 6 hours at a flow rate of 2 $mL\ s^{-1}$ (Figure 1). Samples taken after 0,2,4 and 6 h of electro-oxidation were filtered using a PTFE 0.45 μm syringe filter (CNW Technologies) for further analyses. The concentration of COD, true colour intensity, pH and conductivity were measured.

The main experiment of landfill leachate electro-oxidation was conducted at a constant current density of 75 $mA\ cm^{-2}$. In this experiment, 1 L landfill leachate was

Table 1 | Characteristics of landfill leachate used in this study

Parameter	Unit	Value
pH	–	8.5
Conductivity	$mS\ cm^{-1}$	10.74
COD_{total}	$mg\ O_2\ L^{-1}$	2,040
Cl^-	$mg\ L^{-1}$	1,400
NH_4^+	$mg\ L^{-1}$	535
SO_4^{2-}	$mg\ L^{-1}$	40.58
BOD_5	$mg\ O_2\ L^{-1}$	350
BOD_5/COD	–	0.17

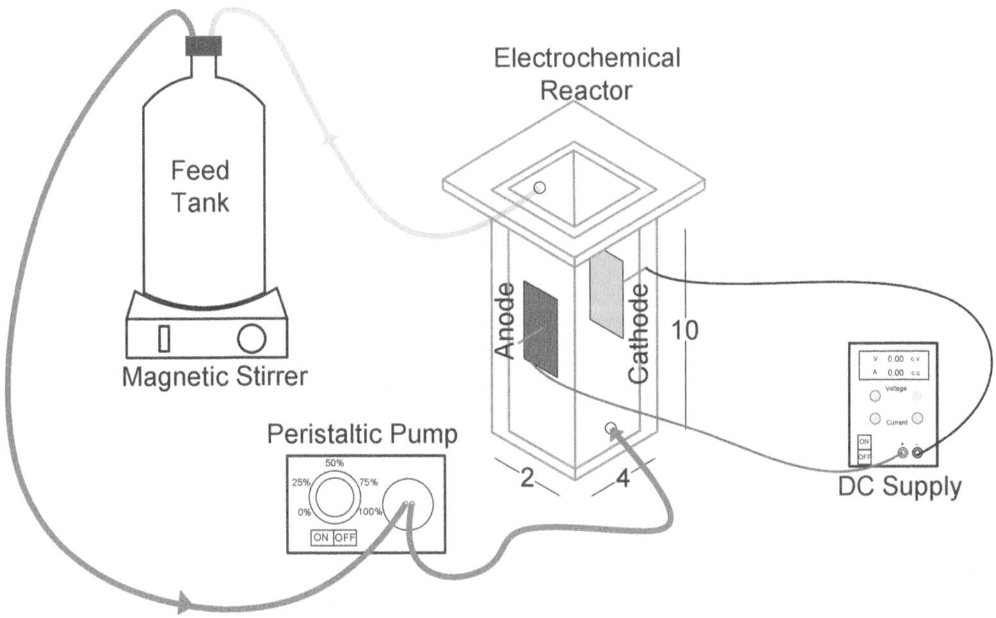

Figure 1 | Schematic diagram of experimental set-up for electro-oxidation of landfill leachate.

continuously recirculated for 7 hours at a flow rate of 2 mL s^{-1}. Similarly, samples taken after 0, 1, 2, 3, 5 and 7 h of electro-oxidation were filtered by using a PTFE 0.45 μm syringe filter (CNW Technologies). The experiments were conducted at pH 3, pH 5 and the original pH (pH 8.5). Nitric acid was added to adjust the leachate pH to 3 and 5. In order to assess the role of generated active species such as radical chloride and radical sulfate, the supporting ions (i.e. NaCl and Na_2SO_4) were added to the leachate. The supporting ions added were 0.05 M NaCl, 0.05 M Na_2SO_4, and 0.025 NaCl + 0.025 M Na_2SO_4. This addition was intended to reach molar ratios of Cl^-: SO_4^{2-} in the leachate 237:1, 4:1 and 18:1, respectively. The concentration of COD, BOD_5, ammonium, sulfate, chloride, TOC, and total chlorine were measured.

Analytical methods

COD was measured using closed reflux and colourimetric method (Merck and Spectroquant NOVA). True colour intensity was measured by means of absorbance at 420 nm wavelength using the Genesys20 spectrophotometer. Ammonium was measured using the Genesys20 spectrophotometer (Nessler method). BOD_5 was measured using the Winkler method. Total chlorine was measured using DPD (N,N-diethyl-p-phenylenediamine) method (Hach). pH and conductivity were monitored by using a pH meter and electrical conductivity meter (Lutron).

RESULTS AND DISCUSSION

Effect of current density on the removal of COD and the decrease of colour intensity

To determine the efficiency of electro-oxidation in degrading organic compounds, the removal of COD and the decrease of colour intensity were observed. Figure 2 shows the normalized COD mass [COD/COD_0] and colour absorbance decrease during electro-oxidation of leachate at the unadjusted pH in three different constant current densities. At higher applied current density; that is, 100 mA cm^{-2}, removal of COD was increased up to 49% at the end of the experiment, although there was a slight increase in COD observed after 2 h electro-oxidation. Similarly, up to 36% decrease of colour intensity was observed.

Zhou *et al.* (2016) reported that electro-oxidation of landfill leachate at a current density of 50 and 100 mA cm^{-2} under pH 5.2 removed 0.66 g and 0.68 g COD, respectively, after 6 hours' operation. A similar result was obtained in this study, where 0.43, 0.53 and 0.72 g COD were removed when current densities of 50, 75 and 100 mA cm^{-2} were applied, respectively, after 6 h of electro-oxidation. Both results were comparable since the electro-oxidation of leachate was conducted at the same current density, although the pH difference might cause the difference in COD degradation performance.

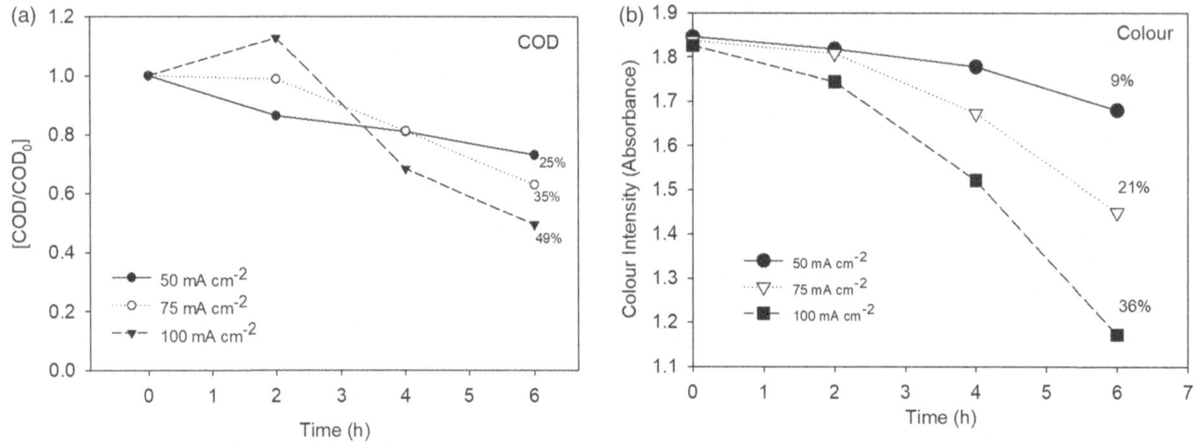

Figure 2 | Removal of (a) COD and (b) true colour intensity, by applying current density of 50, 75, and 100 mA cm^{-2}, without pH adjustment.

Higher current density applied to the system indeed increases specific energy consumption. After 6 h electro-oxidation at 50, 75 and 100 mA cm^{-2} the specific energy consumptions were 22.1, 42.2 and 62.4 Wh L^{-1}, respectively, which is equal to an energy consumption of 51.5, 80.1, and 86.4 Wh g^{-1} COD removed, respectively. It can be noted that increasing the current density from 75 mA cm^{-2} to 100 mA cm^{-2} did not considerably affect the energy consumption to remove COD mass. A similar result was reported by Anglada *et al.* (2011). To eliminate 30% of COD, energy consumption of 102 kWh kg^{-1} COD and 134 kWh kg^{-1} COD were needed at 8.4 A and 18 A, respectively. This also indicates that at higher current density of 100 mA cm^{-2}, energy was more consumed not only for oxidation of organic compounds, but also for oxidation of other ions in the leachate (i.e. dominantly chloride ions), leading to the indirect oxidation of organics, measured as COD.

Effect of Cl^- and SO_4^{2-} on the removal of COD and ammonium without pH adjustment

In order to assess the role of formed active species such as chloride and sulfate radicals on electro-oxidation, the initial leachate was adjusted to simulate three different molar ratios of Cl^- and SO_4^{2-} ions without pH adjustment as shown in Table 2. Evaluation of the electro-oxidation treatment was based on the removal of COD and ammonium. Figure 3 shows the normalized mass of COD [COD/COD_0] and ammonium [$NH_4^+/NH_4^+{}_0$] measured up to 7 h of electro-oxidation process at a constant current density of 75 mA cm^{-2}.

Table 2 | Initial ratio of Cl^- : SO_4^{2-} in the leachate

Cl^- (mg L^{-1})	SO_4^{2-} (mg L^{-1})	Molar ratio Cl^- : SO_4^{2-}
3,550	41	237:1
1,500	900	4:1
2,500	367	18:1

Overall, at the final electro-oxidation of 4.56 Ah L^{-1}, COD removal at a molar ratio of 237:1 was the lowest (i.e. 0.48 g COD) amongst the other two molar ratios. Electro-oxidation of leachate at molar ratios of 4:1 and 18:1 showed a similar mass of COD removed; that is, 0.74 g and 0.73 g, respectively. This implies that the presence of sulfate ions may increase organic removal in the leachate due to the formation of sulfate reactive species during electrochemical process. Farhat *et al.* (2015, 2017) reported that the removal of organic pollutants was 10–15 fold higher when sulfate ions were present in the bulk system compared to $OH^\bullet$-based anodic oxidation via indirect oxidation, due to the formation of sulfate reactive species such as $SO_4^{\bullet-}$.

As mentioned earlier, electro-oxidation of organic matter can occur not only on the anode but also through indirect oxidation by oxidant/reactive species formed during the process. In this study, the concentration of chloride ions in the leachate is relatively high; that is, 1.50–3.55 g L^{-1} (Table 2). Thus, the formation of chlorine reactive species was expected as the dominant oxidative species in the anodic oxidation of leachate. Table 3 shows the total chlorine (the sum of free and combined chlorine compounds) formed after 7 h of electro-oxidation (4.56 Ah L^{-1}) for all the experiments.

It can be confirmed that the formation of total chlorine increases with the increasing of chloride concentration at all pH, except at pH 3. In acidic conditions, the formed chlorine species can possibly be stripped out, decreasing its presence in the bulk system (Deborde & von Gunten 2008).

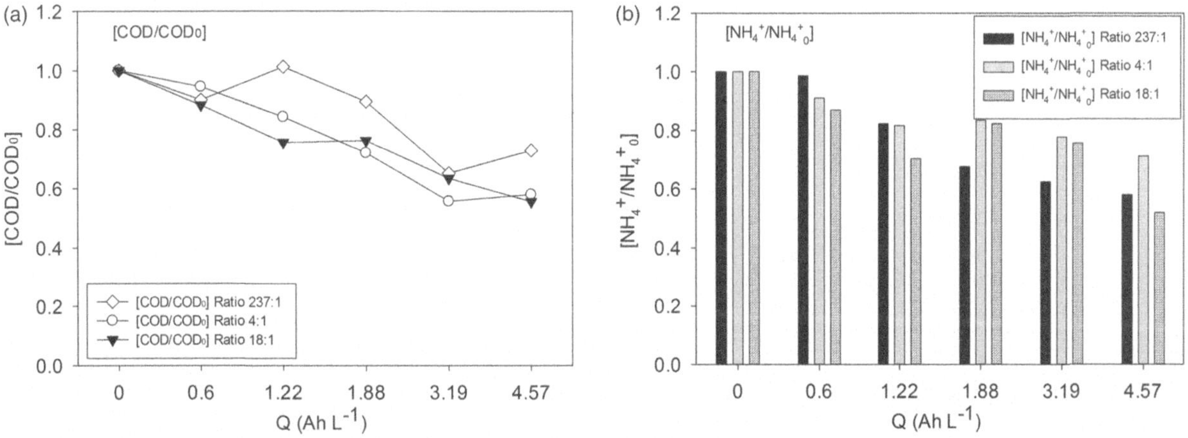

Figure 3 | (a) [COD/COD_0] and (b) [$NH_4^+/NH_4^+{}_0$] during 7 h of electro-oxidation without pH adjustment.

Thus, the total chlorine measured at pH 3 was relatively lower than at other pH conditions. It is expected that by increasing the chloride concentration, the indirect oxidation can be improved (Cabeza *et al.* 2007). However, it is also expected that higher chloride concentration may lead to competition between oxidation of the intended organic pollutant and oxidation of chloride at the anode surface.

Furthermore, besides contributing to the COD removal, chlorine species also play an important role in the removal of ammonium. In this study, the removal of ammonium at molar ratio of 237:1, 4:1, and 18:1 after 4.56 Ah L^{-1} of electro-oxidation was 0.245 g, 0.122 g and 0.215 g, respectively. It can be noted that higher removal of ammonium was obtained at higher chloride concentration; that is, approximately 50–60% removal at a molar ratio of 237:1 and 18:1. Ammonium is mainly removed through the reaction with chlorine/hypochlorite generated during electro-oxidation (Chu *et al.* 2008; Mandal *et al.* 2017). In addition, ammonium is oxidized to form chloramines and other species; that is, NH_2Cl, $NHCl_2$, NOH, N_2 and NO_3^- (Mandal *et al.* 2017). Cabeza *et al.* (2007) suggest that the kinetic of ammonium removal was improved with increasing chloride concentration in the electro-oxidation process. As reported in the study, when the chloride concentration was increased from 1,420 mg L^{-1} to 8,570 mg L^{-1}, an increase of ammonium removal was achieved from around 35% to a nearly complete removal.

Table 3 | Total chlorine formed after 7 h of electro-oxidation

Experiment	Cl_2 (mg L^{-1})
Molar ratio 237:1 at original pH (pH 8.5)	202
Molar ratio 4:1 at original (pH 8.5)	54
Molar ratio 18:1 at original (pH 8.5)	179
Molar ratio 237:1 at pH 5	133
Molar ratio 4:1 at pH 5	27
Molar ratio 18:1 at pH 5	87
Molar ratio 237:1 at pH 3	56
Molar ratio 4:1 at pH 3	55
Molar ratio 18:1 at pH 3	112

Effect of pH on the removal of COD and ammonium at pH 3 and pH 5

The removal of COD at a molar ratio of 237:1 was enhanced in acidic conditions. The electro-oxidation process at a ratio of 237:1 and pH 5 obtained higher COD removal than in the case of the original pH 8.5, whereas at pH 3 COD removal was improved to twofold higher than electro-oxidation at pH 5 (Figures 3 and 4(a)). The COD mass removals at ratio 237:1 under original pH, pH 5 and pH 3 were 0.48 g, 0.58 g and 1.19 g, respectively. The difference in performance of COD removal when the presence of chloride was dominant in the electro-oxidation process is due to the presence of different chlorine species. Chlorine species would be varied as a function of pH value. Chlorine species that is dominant at pH 1–3 is Cl_2, at pH 5–7 is HClO, and at pH above 7 is OCl^- (Deborde & von Gunten 2008). Zhou *et al.* (2016) also observed that removal rates of COD in acidic conditions were higher than in alkali conditions; that is, 87.5% and 65.02% at pH 5.16 and 10, respectively. In the experiment with the ratio of 4:1, there was a slight increase in COD removal when the pH was adjusted to acidic. The COD removal at original pH, pH 5 and pH 3 was 0.74 g, 0.9 g and 1 g, respectively. Bagastyo *et al.* (2012)

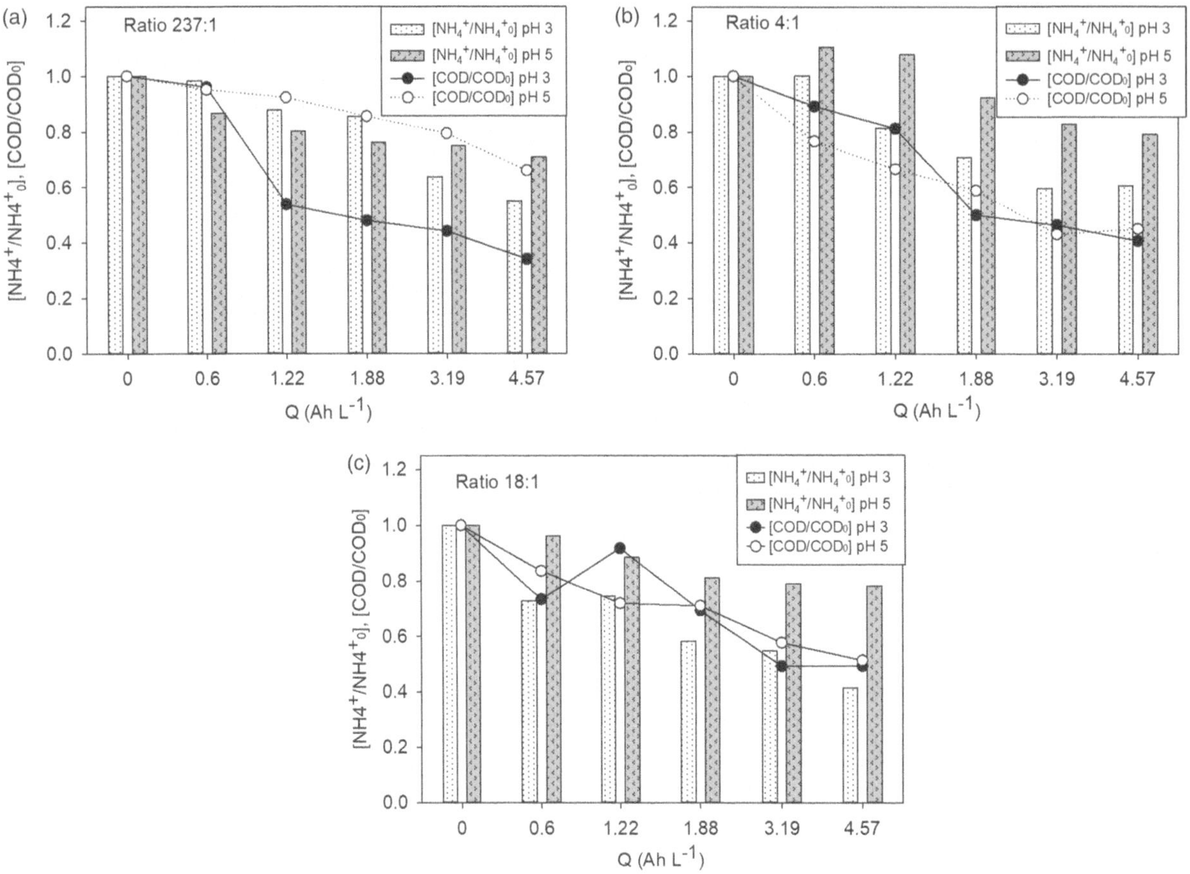

Figure 4 | Normalized mass of COD and ammonium at pH 3 and 5 with the molar ratio of: (a) 237:1, (b) 4:1 and (c) 18:1.

also previously reported that COD removal at acidic pH was faster than at circumneutral. With an initial COD of 136 ± 8 mg L^{-1}, a complete COD removal was obtained after 5.2 and 6.6 Ah L^{-1} at acidic and circumneutral pH, respectively. However, in the present study, pH has no considerable effect at a ratio of 18:1. The COD removals at original pH, pH 3 and pH 5 were 0.73 g, 0.72 g, and 0.73 g, respectively. Therefore, the electro-oxidation process of leachate that has a ratio of 18:1 could possibly be conducted in a wide range of pH values.

In contrast, ammonium removal seems to be affected by pH value (Figure 4). In the experiment at ratios of 4:1 and 18:1, removal of ammonium at pH 3 was higher than at pH 5. Ammonium removal at pH 3 and 5 at a ratio of 4:1 was 0.195 g and 0.09 g and similarly at a ratio of 18:1 was 0.2 g and 0.088 g, respectively. However, in the electro-oxidation process at a ratio of 237:1, the ammonium removal at pH 3 was lower than at pH 5. This result indicates that there is competition for COD and ammonium removal during the electro-oxidation process, signified by the high removal of COD obtained at pH 3 with a ratio of 237:1. In all experiments, the electro-oxidation process at the original pH of the leachate (Figure 3) obtained higher ammonium removal than at pH 5 (Figure 4). Chu *et al.* (2008) also mentioned that the best ammonium removal was obtained at pH 8. Higher ammonium removal was obtained at alkali pH due to the volatilization of ammonium to form NH_3 (Deborde & von Gunten 2008).

Changes of biodegradability index (BOD_5:COD ratio) of landfill leachate

There are two oxidation mechanisms of pollutants in an electrochemical system; that is, (i) electrochemical conversion in which organics are transformed to be more biodegradable and (ii) electrochemical combustion, in which organics are completely oxidized to CO_2 and H_2O (Comninellis 1994). BOD_5 was measured to evaluate the electrochemical conversion of this process. Table 4 shows the changes of biodegradability index after 7 h electro-oxidation for all the experiments.

Table 4 | Changes of biodegradability index

	BOD$_5$:COD	
Experiment	**0 Ah L^{-1}**	**4.57 Ah L^{-1}**
Molar ratio 237:1 at original pH (pH 8.5)	0.41	0.16
Molar ratio 4:1 at original (pH 8.5)	0.37	0.16
Molar ratio 18:1 at original (pH 8.5)	0.38	0.14
Molar ratio 237:1 at pH 5	0.39	0.22
Molar ratio 4:1 at pH 5	0.29	0.16
Molar ratio 18:1 at pH 5	0.31	0.14
Molar ratio 237:1 at pH 3	0.37	0.48
Molar ratio 4:1 at pH 3	0.29	0.21
Molar ratio 18:1 at pH 3	0.43	0.29

BOD$_5$ value was decreased in all experiments, as well as the BOD$_5$:COD ratio, except at the ratio of 237:1 at pH 3, when there was a slight increase of BOD$_5$:COD ratio. In the experiments at original pH and pH 5, all ratios of Cl^-:SO_4^{2-} resulted in similar BOD$_5$:COD ratios. Labiadh *et al.* (2016) reported that in anodic oxidation of reverse osmosis concentrate of landfill leachate using a BDD anode, there was a slight increase in BOD$_5$:COD from 0.43 to 0.51 after 24 h. In other words, the experimental condition of the present electro-oxidation study was not suitable for enhancing the biodegradability index of landfill leachate. However, the decrease of the BOD$_5$:COD ratio indicates that electrochemical combustion may take place.

To evaluate the electrochemical combustion process, TOC was measured to indicate the mineralization of organic pollutants (Table 5). TOC removals of 16–31% were achieved after 4.57 Ah L^{-1}. The maximum TOC removal was obtained at the ratio of 4:1 under original pH, whereas the minimum TOC removal was obtained at a ratio of 237:1, also under original pH; that is, 208.7 mg and 128 mg TOC, respectively. For all of the experiments, the COD removal was higher than TOC removal. A previous study suggested that faster COD removal than DOC removal indicates the formation of organic by-product compounds, such as chlorinated organic compounds (Bagastyo *et al.* 2012).

Energy consumption

This electro-oxidation process requires electrical energy. To analyse the energy consumption under different experimental conditions, the specific energy consumption (g Wh^{-1}) was calculated using Equation (2) below:

$$E_{sp} = \frac{\Delta X}{V.I.t} \tag{2}$$

where ΔX is mass removed (g), V is the cell voltage (volts), I was the applied current intensity (A) and t was the duration of electro-oxidation process (hours). This specific energy consumption was evaluated based on the mass of COD and ammonium removed (Table 6).

Based on Table 6, the optimum energy consumption based on COD removal was obtained at the ratio of 237:1 and pH 3, whereas the optimum energy consumption regarding ammonium removal was attained at the ratio of 237:1 at the original pH of the leachate. Overall, the optimum energy consumption for COD was achieved at pH 3 at all ratios; meanwhile, the optimum energy consumption of ammonium was obtained at the ratio of 237:1.

CONCLUSIONS

In this study, the role of current density and variation of ratio Cl^-:SO_4^{2-} at different pH were investigated. Current

Table 5 | Removal of TOC after 4.57 Ah L^{-1}

Experiment	TOC removal (g)	%R TOC
Molar ratio 237:1 at original pH (pH 8.5)	128	16%
Molar ratio 4:1 at original (pH 8.5)	208.7	29%
Molar ratio 18:1 at original (pH 8.5)	183.3	27%
Molar ratio 237:1 at pH 5	172	22%
Molar ratio 4:1 at pH 5	160.9	26%
Molar ratio 18:1 at pH 5	163.2	26%
Molar ratio 237:1 at pH 3	133.4	19%
Molar ratio 4:1 at pH 3	175	31%
Molar ratio 18:1 at pH 3	158.5	29%

Table 6 | Estimation of specific energy consumption

Experiment	g COD Wh^{-1}	g NH_4^+ Wh^{-1}
Molar ratio 237:1 at original pH (pH 8.5)	0.0140	0.0071
Molar ratio 4:1 at original (pH 8.5)	0.0214	0.0035
Molar ratio 18:1 at original (pH 8.5)	0.0201	0.0059
Molar ratio 237:1 at pH 5	0.0159	0.0060
Molar ratio 4:1 at pH 5	0.0261	0.0026
Molar ratio 18:1 at pH 5	0.0205	0.0025
Molar ratio 237:1 at pH 3	0.0302	0.0048
Molar ratio 4:1 at pH 3	0.0277	0.0053
Molar ratio 18:1 at pH 3	0.0208	0.0057

density plays an important role, as proven in that the higher current density could increase the removal of COD and true colour intensity of the landfill leachate up to 49% and 36%, respectively. However, higher current density leads to an increase of specific energy consumption. In the electro-oxidation process for landfill leachate under original pH, the presence of sulfate ions may increase the removal of COD, while higher ammonium removal was achieved at higher chloride concentrations. On the other hand, the role of $Cl^-:SO_4^{2-}$ at different pH of leachate gives a different effect on COD and ammonium removal. In the electro-oxidation process with a ratio of 237:1, the COD removal increase at acidic conditions, whereas at the ratio of 4:1, there was a slight increase of COD removal in acidic conditions. However, at the ratio of 18:1, no substantial effect of pH on the COD removal was observed, indicating that the electro-oxidation process at this ratio could be conducted within a wider range of pH conditions. In contrast, ammonium removal seems to be more affected by pH value. The optimum ammonium removal was obtained at the ratio of 237:1 at original pH 8.5, due to the volatilization of ammonium to form ammonia. Furthermore, this electro-oxidation system was found to be effective in mineralization of organics but ineffective in increasing the index of biodegradability of landfill leachate.

ACKNOWLEDGEMENTS

This research was supported by Ministry of Research, Technology and Higher Education of the Republic of Indonesia through PDUPT Research Grant Scheme 2018.

REFERENCES

Anglada, Á., Urtiaga, A., Ortiz, I., Mantzavinos, D. & Diamadopoulos, E. 2011 Boron-doped diamond anodic treatment of landfill leachate: evaluation of operating variables and formation of oxidation by-products. *Water Research* **45**, 828–838.

Bagastyo, A. Y., Batstone, D. J., Kristiana, I., Gernjak, W., Joll, C. & Radjenovic, J. 2012 Electrochemical oxidation of reverse osmosis concentrate on boron-doped diamond anodes at circumneutral and acidic pH. *Water Research* **46** (18), 6104–6112.

Cabeza, A., Urtiaga, A., Rivero, M.-J. & Ortiz, I. 2007 Ammonium removal from landfill leachate by anodic oxidation. *Journal of Hazardous Materials* **144**, 715–719.

Chu, Y., Zhang, Q. & Xu, D. 2008 Advanced treatment of landfill leachate from a sequencing batch reactor (SBR) by electrochemical oxidation process. *Journal of Environmental Engineering Science.* **7**, 627–633.

Comninellis, C. 1994 Electrocatalysis in the electrochemical conversion/combustion of organic pollutants for waste water treatment. *Electrochimica Acta* **39**, 1857–1862.

Davis, J., Baygents, J. C. & Farrell, J. 2014 Understanding persulfate production at boron doped diamond film anodes. *Electrochimica Acta* **150**, 68–74.

Deborde, M. & von Gunten, U. 2008 Reactions of chlorine with inorganic and organic compounds during water treatment – kinetics and mechanisms: a critical review. *Water Research* **42**, 13–51.

Farhat, A., Keller, J., Tait, S. & Radjenovic, J. 2015 Removal of persistent organic contaminants by electrochemically activated sulfate. *Environmental Science and Technology* **49**, 14326–14333.

Farhat, A., Keller, J., Tait, S. & Radjenovic, J. 2017 Assessment of the impact of chloride on the formation of chlorinated by-products in the presence and absence of electrochemically activated sulfate. *Chemical Engineering Journal* **330**, 1265–1271.

Fernandes, A., Pacheco, M. J., Ciríaco, L. & Lopes, A. 2015 Review on the electrochemical processes for the treatment of sanitary landfill leachates: present and future. *Applied Catalysis B: Environmental* **176–177**, 183–200.

Labiadh, L., Fernandes, A., Ciríaco, L., Pacheco, M. J., Gadri, A., Ammar, S. & Lopes, A. 2016 Electrochemical treatment of concentrate from reverse osmosis of sanitary landfill leachate. *Journal of Environmental Management* **181**, 515–521.

Mandal, P., Dubey, B. K. & Gupta, A. K. 2017 Review on landfill leachate treatment by electrochemical oxidation: drawbacks, challenges and future scope. *Waste Management* **69**, 250–273.

Renou, S., Givaudan, J. G., Poulain, S., Dirassouyan, F. & Moulin, P. 2008 Landfill leachate treatment: review and opportunity. *Journal of Hazardous Materials* **150**, 468–493.

Tatsi, A. A., Zouboulis, A. I., Matis, K. A. & Samaras, P. 2003 Coagulation–flocculation pretreatment of sanitary landfill leachates. *Chemosphere* **53**, 737–744.

Zhou, B., Yu, Z., Wei, Q., Long, H., Xie, Y. & Wang, Y. 2016 Electrochemical oxidation of biological pretreated and membrane separated landfill leachate concentrates on boron doped diamond anode. *Applied Surface Science* **377**, 406–415.

First received 10 August 2018; accepted in revised form 17 January 2019. Available online 30 January 2019

Degradation and transformation of natural organic matter accountable for disinfection byproduct formations by UV photolysis and UV/chlor(am)ine

Chanathip Hirun-Utok and Songkeart Phattarapattamawong

ABSTRACT

This research aimed to investigate the degradation of natural organic matter responsible for the formation of trihalomethane (THM), haloacetic acid (HAA) and haloacetonitrile (HAN) during ultraviolet (UV) photolysis and a co-exposure of UV with chlorine (UV/chlorine) and chloramine (UV/chloramine). Low pressure UV (LPUV) and vacuum UV (VUV) lamps were used for photolysis. VUV and LPUV irradiation changed aromatic/unsaturated structures to aliphatic ones, resulting in decreased THM and HAN formation. Following irradiation for 60 min, LPUV decreased THM and HAN by 16% ± 2% and 20% ± 6%, respectively. VUV decreased THM and HAN formation by 23% ± 3% and 20% ± 8%, respectively. HAA formation increased following photolysis. UV/chlorine treatment decreased THM, HAA and HAN. Higher chlorine doses had an inversely proportional relationship with THM and HAN formation. A chlorine dose of 4 $mg{\cdot}L^{-1}$ led to the greatest reductions, corresponding to 42% ± 2%, 10% ± 10% and 18% ± 6% for THM, HAA and HAN, respectively. UV/chloramine decreased the formation of THM more than UV/chlorine. With a chloramine dose of 4 $mg{\cdot}L^{-1}$, THM, HAA and HAN formation decreased by 74% ± 10%, 10% ± 10% and 11% ± 10%, respectively. This study showed the potential use of UV/chlor(am)ine for controlling the formation of THM, HAA and HAN.

Key words | advanced oxidation process, disinfection byproducts, UV/chlor(am)ine, UV photolysis

Chanathip Hirun-Utok
Songkeart Phattarapattamawong (corresponding author)
Department of Environmental Engineering, Faculty of Engineering,
King Mongkut's University of Technology Thonburi,
Bangkok,
Thailand
E-mail: *songkeart.pha@kmutt.ac.th*

Songkeart Phattarapattamawong
Research Program in Water Scarcity Resilience,
Center of Excellence on Hazardous Substance Management,
Bangkok,
Thailand

INTRODUCTION

The presence of trihalomethane (THM), haloacetic acid (HAA) and haloacetonitrile (HAN) in water have been strictly regulated by drinking water quality guidelines since they are categorized as potentially carcinogenic and mutagenic disinfection byproducts (DBPs) (Richardson *et al.* 2007; Yan *et al.* 2018). These substances can form through reactions between disinfectants (e.g. chlorine and chloramine) and natural organic matter (NOM). Although NOM is typically considered as a precursor for THM, HAA and HAN, the specific organic materials that contribute to the formation of each group is different. Most organic substances can influence the formation of THM and HAA (Li *et al.* 2017), while nitrogenous organic compounds, a minor component of NOM, are key for HAN formation (Reckhow *et al.* 2001; Westerhoff & Mash 2002; Yang *et al.* 2010; Huang *et al.* 2016). Because the precursor for HAN is present in smaller amounts compared with precursors for THM and HAA, control of THM and HAA formation is expected to be easier than that for HAN formation.

In Thailand, several researchers have studied ways to reduce the formation potential of THM, HAA and HAN. Conventional treatment processes, such as coagulation, flocculation, sedimentation and sand filtration were shown to be ineffective at decreasing HAN formation (Srithep & Phattarapattamawong 2017). Enhanced coagulation with polyaluminium chloride at 40 $mg{\cdot}L^{-1}$ reduced THM formation by 48%, while the addition of polymers and powder activated carbon further reduced THM formation by 83% (Musikavong *et al.* 2013). Ozonation with an ozone dose of 1–5 $mg{\cdot}L^{-1}$ was ineffective for reducing the formation of THM and HAA. Adding hydrogen peroxide during ozonation (i.e. the peroxone process) only decreased the formation of THM (Deeudomwongsa *et al.* 2017). However, a comparative study on control of THM, HAA and HAN has not been explored.

Ultraviolet (UV) photolysis has been used to disinfect water for more than two decades (Kruithof *et al.* 1992; Linden *et al.* 2002; Choi & Choi 2010). A low-pressure UV

doi: 10.2166/wst.2018.496

lamp (LPUV) emits UV radiation at a wavelength of 254 nm. This can degrade and transform large organic compounds into smaller ones (Parkinson *et al.* 2001), causing decrease in DBP formation. In addition, exposure to a UV intensity of 1,140 $mJ \cdot cm^{-1}$ can transform the structure of NOM, lowering its reactivity with chlorine. This can decrease THM formation by 15% (Lamsal *et al.* 2011). Also, UV irradiation at 220 $mJ \cdot cm^{-2}$ was shown to decrease HAN formation by 44% (Sirinthip & Phattarapattamawong 2017). A vacuum UV lamp (VUV) is another type of UV-generating source that simultaneously emits wavelengths of 254 nm and 185 nm. VUV can effectively degrade recalcitrant organic compounds (Imoberdorf & Mohseni 2011). The amount of recalcitrant organic compounds removed by VUV irradiation is typically greater than the amount removed by LPUV irradiation. To the best of our knowledge, a comparison between the kinetic reactions of LPUV and VUV for the control of DBP formation has not been reported.

Recently, an advanced oxidation process (AOP) involving co-exposure between a chlorinating agent (i.e. chlorine or chloramine) and UV light has been proposed (Wang *et al.* 2015a; Chuang *et al.* 2017). Such UV and chlor(am)ine co-exposure (UV/chlorine or UV/chloramine) generates and uses reactive radicals (e.g. •OH, •Cl) for the removal of recalcitrant compounds (Chuang *et al.* 2017). The •OH radical is a strong and unselective oxidant, while •Cl is a selective oxidant that preferentially reacts with substituted aromatics such as phenols, benzoic acid, toluene and aniline (Wu *et al.* 2017). UV/chlor(am)ine co-exposure can be simply incorporated into a water treatment system following pre-chlorination. Chloramine with UV light at a wavelength of 254 nm has the greatest extinction coefficient ($\varepsilon = 354\ M^{-1} \cdot cm^{-1}$), followed by chlorine ($\varepsilon = 60\ M^{-1} \cdot cm^{-1}$) and hydrogen peroxide ($\varepsilon = 18.6\ M^{-1} \cdot cm^{-1}$) (Zhang *et al.* 2015; Chuang *et al.* 2017). This implies that co-exposure with chlor(am)ine and UV potentially produces more reactive radicals (i.e. •OH, •Cl) than typical AOPs (UV/H_2O_2), which mainly rely on •OH alone. A previous study found that the removal of carbamazepine, benzoate and 1,4-dioxane when using UV/chlorine was higher than when using UV/H_2O_2 (Chuang *et al.* 2017). However, information on the control of DBPs by UV/chlorine and UV/chloramine is limited.

This research aimed to investigate the degradation of NOM responsible for disinfectant byproduct (DBP) formation during UV photolysis and the UV/chlor(am)ine process. UV photolysis was tested over different time periods and with different UV lamps. UV/chlorine and UV/chloramine processes were carried out with different chlorinating agent doses (4–20 $mg \cdot L^{-1}$) and contact times (0–60 min). The study investigated which DBPs were of most concern in terms of their removal difficulty. In addition, achieving control of DBPs by UV photolysis or UV/chlor(am)ination is useful information for the future development of conventional water treatment processes.

METHODOLOGY

Chemical and water samples

Standard solutions of THM, HAA and HAN were prepared using analytical grade chemicals purchased from Sigma-Aldrich. Four types of THM (chloroform, dichloromethane, chlorodibromomethane and bromoform), four types of HAA (chloroacetic acid, dichloroacetic acid, trichloroacetic acid and dibromoacetic acid) and one type of HAN (dichloroacetonitrile) were determined. Sodium hypochlorite (NaOCl, 7%) was obtained from Carlo Erba, France. Chloramine solution was prepared by mixing NaOCl and ammonium chloride at a Cl_2:N ratio of 5:1 in a phosphate buffer solution at pH 8. The Cl_2 concentration of the chloramine stock solution was 2,300–2,400 $mg \cdot L^{-1}$.

To avoid variations in water characteristics, a 200-L water sample was collected and used for each of the different treatment conditions. Because turbidity and suspended solids can reduce the efficiency of UV irradiation, the water sample (referred to as RSF) water) was collected following a series of treatments, including coagulation – flocculation, sedimentation and rapid sand filtration. The sample characteristics are shown in S1 (supplementary information, available with the online version of this paper). A 20-L polypropylene tank was used for sample collection and storage at 4 °C. The samples were stored at room temperature (30–32 °C) for more than 5 h before conducting the experiments.

Experimental design

UV photolysis was conducted in a 5-L glass reactor covered with aluminum foil. A 15 W LPUV lamp (Aqua Rich Technology, Thailand) and a 30 W VUV lamp (Universal Lights Source, Inc., USA) were used. The illuminating lamps with quartz sleeves were installed 5 cm above the water surface. The lamp was warmed for 20 min before starting the reaction. The intensities of 254 nm wavelengths, determined by the iodine/iodate method (Rahn *et al.* 2006), were 0.10 and 0.54 $mW \cdot cm^{-2}$ for LPUV and VUV, respectively. The sample was adjusted to pH 7. Samples were collected following UV contact times of 0, 2, 5, 10, 30 and 60 min. For chlorination, 4 $mg{\cdot}L^{-1}$ as Cl_2 was used for determination of THM, HAA, and HAN. The residual chlorine after 24 h was approximately 1 $mg{\cdot}L^{-1}$. The final pH of the chlorinated water was 7 ± 0.2.

The reactor was also used for the UV/chlorine and UV/chloramine experiments. To avoid degradation of chlorinating agents (i.e. chlorine and chloramine) by UV wavelengths of 185 nm, the LPUV was used to provide UV light. The concentration of chlorinating agents was controlled at 4, 10 and 20 $mg{\cdot}L^{-1}$ as Cl_2 (Figure 1). 4 $mg{\cdot}L^{-1}$ as Cl_2 is a typical chlorinating dose for chlorination, whereas higher concentrations (10 and 20 $mg{\cdot}L^{-1}$ as Cl_2) were used to study the effect of chlorinating agents. The reaction was started once chlorinating agents were added. The contact times were 0, 2, 5, 10, 30 and 60 min. Since DBPs formed during the reaction, chlor(am)ine-added water without UV irradiation was taken as the control sample. To stop chlor(am)ination, sodium sulfite (Na_2SO_3) was used as a dechlorinating agent (Wang *et al.* 2015b).

Analysis

Dissolved organic carbon (DOC) and total nitrogen were measured using a DOC analyzer (TOC-V, Shimadzu) connected with a total nitrogen analyzer (Shimadzu). The absorbance at wavelength 254 nm (UV254), an indicator of unsaturated/aromatic compounds, was determined using a UV-Vis spectrophotometer (DR 4000U, Hach). The specific UV absorbance (SUVA), calculated from the UV254 absorbance divided by the DOC concentration, indicates the amount of organic transformation between the aromatic and aliphatic fractions. The concentration of dissolved organic nitrogen (DON) was obtained by subtracting the inorganic nitrogen (i.e. nitrite, nitrate and ammonium) from the total nitrogen. Inorganic nitrogenous substances were separated and determined using ion chromatography (Shimadzu) with an IC-A3 column.

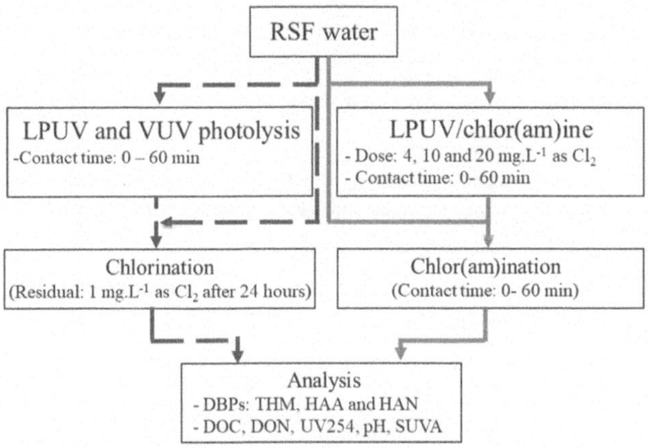

Figure 1 | A schematic diagram of the experimental design.

THM and HAN were concentrated in methyl-tert butyl ether (MTBE) solution. 2.5 mL of MTBE was added to 25 mL of the sample. The mixture was shaken for 5 min, then 2 mL of supernatant was collected. Sodium bicarbonate was used to remove any remaining water. For HAA preparation, 40 mL of sample was extracted with 4 mL of MTBE. To derivatize the HAA structure, the MTBE solution was acidified with methanol solution. Gas chromatography with electron capture detector (GC-ECD) was used for determination of THM, HAA and HAN. The limits of detection and qualification were reported in previous studies (Deeudomwongsa *et al.* 2017; Srithep & Phattarapattamawong 2017).

RESULTS AND DISCUSSION

Effect of UV photolysis in controlling THM, HAA and HAN formation

Following chlorination, the quantities of THM, HAA and HAN formed were 40 ± 1.5, 8 ± 0.7, and 3 ± 0.3 $\mu g{\cdot}L^{-1}$, respectively. VUV irradiation was more effective at controlling THM formation compared with LPUV irradiation (Figure 2(a)). After exposure for 0 to 10 min, VUV decreased THM formation by 22% ± 5%, while LPUV achieved just a 15% ± 4% decrease. Extending the contact time to 60 min slightly enhanced THM reduction by 16% ± 2% and 23% ± 3% for LPUV and VUV irradiation, respectively. Ozonation with 1–5 $mg{\cdot}L^{-1}$ showed a similar THM reduction (10–23%) (Wang & Chen 2014), while the peroxone process (H_2O_2/O_3) enhanced THM reduction by 40% (Deeudomwongsa *et al.* 2017). The reduced formation of THM may result from a reduction in aromatic/unsaturated compounds, since LPUV and VUV irradiation decreased the UV254 absorbance by 13% and 23%, respectively (S1). Aromatic compounds have been reported to be compounds that yield high amounts of THM formation (Buchanan *et al.* 2006; Wang *et al.* 2015b). Although a reduction in UV254 absorbance was seen, the photolysis did not result in complete mineralization of NOMs, as indicated by the unchanged DOC concentrations. This led to a reduction in SUVA from 1.72 to 1.49 and 1.31 $L{\cdot}mg^{-1}{\cdot}m^{-1}$ for LPUV and VUV irradiation, respectively. This SUVA reduction

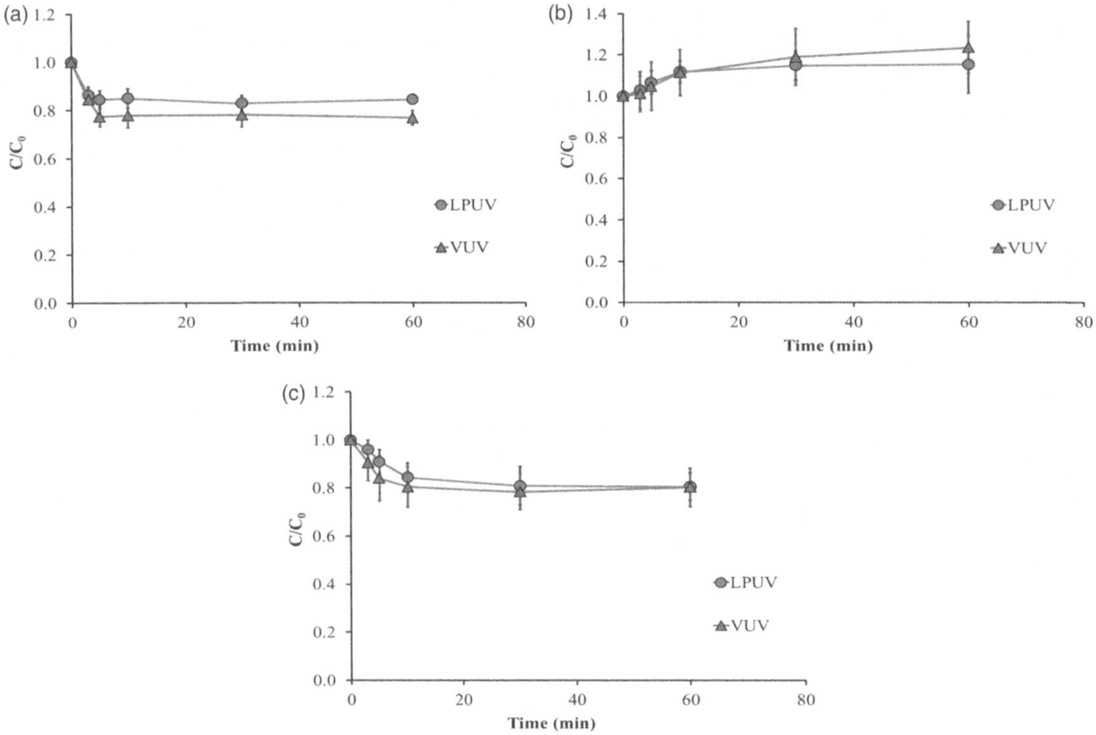

Figure 2 | Formation of THM (a), HAA (b) and HAN (c) following LPUV and VUV photolysis.

indicated a degradation of NOM structures from unsaturated/aromatics to aliphatic ones.

Surprisingly, an increase in HAA formation was observed for both VUV and LPUV. After irradiation for 60 minutes, HAA levels were elevated by 15% ± 14% and 23% ± 13% for LPUV and VUV irradiation, respectively (Figure 2(b)). An increased aliphatic fraction in NOMs tends to increase the formation of HAA (Liang & Singer 2003; Bond *et al.* 2009; Wang *et al.* 2015b). The precursor for HAA was present at favorable levels in the hydrophilic fraction, in which the UV254 absorbance was lower than in the hydrophobic fraction. Previous studies have reported that aliphatic and hydrophilic substances preferentially promoted the formation of HAA (Bond *et al.* 2009; Zhang *et al.* 2009; Wang *et al.* 2015b). Wang *et al.* (2015b) also found that LPUV photolysis of natural water increased HAA formation.

The concentration of HAN declined at the initial contact time (Figure 2(c)). LPUV and VUV irradiation decreased HAN formation by 16% ± 6% and 20% ± 8%, respectively. A longer contact time led to marginally lower HAN formation (20% ± 6% and 20% ± 8% for LPUV and VUV, respectively). The reduction in HAN was plausibly explained by a decrease in DON following LPUV and VUV irradiation (S1). UV photolysis degraded and transformed DON to nitrate. While DON was degraded by 8% and 12% by LPUV and VUV, respectively, the nitrate concentration increased by 7% and 12%. The similar efficiencies seen for both LPUV and VUV may be attributed to a HAN formation pathway. Apart from the reaction with organic nitrogen (called a decarboxylation pathway), HAN formation can result from the aldehyde pathway where a nitrogen atom is obtained from chloramine (Yang *et al.* 2010; Bond *et al.* 2011). This pathway becomes the predominant reaction if the C-to-N ratio is higher (Yang *et al.* 2012). The DOC/DON ratios were increased by 8% and 14% for LPUV and VUV, respectively (S1). Greater reductions of THM and HAN indicated that their precursors were susceptible to UV photolysis. Under the study conditions, the control of HAA formation by UV photolysis was more difficult to achieve, since the transformation product seemed to preferentially yield HAA. A further study on using longer UV exposures may show enhanced reduction of HAA.

The kinetic reaction was fitted to a pseudo-first-order kinetic equation (Equation (1)). To determine the reaction rate constant, an initial rate method was used.

$$Ln\left(\frac{C}{C_0}\right) = -k't \tag{1}$$

Table 1 | The pseudo-first-order reaction constant for LPUV and VUV photolysis ($n = 4$)

	k' (min^{-1})		R^2	
DBPs	**LPUV**	**VUV**	**LPUV**	**VUV**
THM	0.0351	0.0518	0.9144	0.9944
HAA	−0.0114	−0.0113	0.9855	0.9599
HAN	0.0175	0.0215	0.9867	0.8896

where C_0 and C are the substance concentrations at the initial time (t = 0) and time t. The k' value is the pseudo-first-order reaction constant.

The pseudo-first-order reaction constant (k'), together with the determination coefficient (R^2) are shown in Table 1. For LPUV irradiation, the k' values of THM and HAN were 0.0351 ($R^2 = 0.9144$) and 0.0175 min^{-1} ($R^2 = 0.9867$), respectively. The k' value of HAN was similar to that seen in our previous study ($k' = 0.0180$ min^{-1}, $R^2 = 0.980$) (Srithep & Phattarapattamawong 2017). The k' values of THM and HAN for VUV irradiation were approximately 1.5- and 1.2-fold higher than those for LPUV photolysis. The k' values of THM and HAN for VUV irradiation were 0.0518 ($R^2 = 0.9944$) and 0.0215 min^{-1} ($R^2 = 0.8896$), respectively. Instead of presenting the kinetic constant for the reduction of DBP precursors, the negative value indicated the kinetic formation rate of HAA, which resulted from the reaction between chlorine with the degraded products from photolysis. The k' values of HAA for LPUV and VUV irradiation were 0.0114 ($R^2 = 0.9855$) and 0.0113 min^{-1} ($R^2 = 0.9599$), respectively. These similar values implied that the structure of degraded products resulting from LPUV and VUV photolysis may be similar since their formation rates were comparable.

A comparison between UV/chlorine and UV/chloramine for controlling the formation of THM, HAA and HAN

Due to the chlorine involved in the reactions of UV/chlorine and UV/chloramine, the formation of THM, HAA and HAN increased, depending on the contact time. Thus, the removal efficiency was calculated from different concentrations between the treated water and the control sample (chlorination alone). The main reduction occurred within a 10-min contact time, while longer contact times only slightly enhanced the performance. After 60-min reaction time, UV/chlorine with a dose of 4 mg·L^{-1} suppressed THM formation by 42% ± 2%, compared with the control sample (chlorination at 4 mg·L^{-1}) (Figure 3(a)). A higher chlorine dose decreased the efficiency of controlling THM formation. Chlorine doses at 10 and 20 mg·L^{-1}, with LPUV irradiation, reduced THM formation by 26% ± 4% and 14% ± 2%, respectively. The chlorine residuals following the UV/chlorine reaction are shown in S2 (supplementary information, available with the online version of this paper). The excessive residuals of chlorine may have reacted with either NOM to promote the THM formation or reactive radicals (e.g. •OH, •Cl), acting as a scavenger (Fang *et al.* 2014). Thus, optimization between chlorine dose and UV exposure is a key factor for the control of THM. While both LPUV and VUV photolysis tended to increase HAA formation, UV/chlorine decreased HAA formation. UV/chlorine doses of 4–20 mg·L^{-1} decreased HAA formation in the range of 8% to 11% (Figure 3(b)). The reduction in HAN formation was likely inversely proportional to the applied chlorine dose. UV/chlorine at doses of 4, 10 and 20 mg·L^{-1} reduced the formation of HAN by 18% ± 6%, 17% ± 3% and 10% ± 7%, respectively (Figure 3(c)). Among the three species, UV/chlorine preferentially decreased THM formation compared with HAN and HAA formation. Because THM, HAA and HAN formed during the reaction with chlorine, the reactive radicals were presumably reactive to THM, HAA, HAN and their precursors. Since the reduction in THM and HAN formation was inversely proportional to chlorine dose, a chlorine dose of 4 mg·L^{-1}, which is anticipated to produce the most reactive radicals, was recommended for the control of THM, HAA and HAN. Compared with UV photolysis, the UV/chlorine treatment was more effective for controlling THM, HAA and HAN formation. It is worth nothing that there was a high yield of THM, HAA and HAN during the 60-min contact time for chlor(am)-ination. Prolonging their contact time to 24 h slightly increased their formation. A previous study found that the use of an AOP (O_3/UV) decreased chloroform and HAA formation more than UV photolysis (Chin & Bérubé 2005).

For UV/chloramine, a reduction in THM formation was highly dependent on contact time, rather than chloramine dose (Figure 3(d)). At the initial contact time (10 min), a UV/chloramine dose of 4 mg·L^{-1} decreased THM formation by just 21% ± 12%, compared with the control sample (i.e. chloramination at the same dose). Extending the contact time to 60 min greatly decreased THM formation, by 74% ± 10%. Increasing the chloramine dose to 10 and 20 mg·L^{-1} slightly lowered the efficiency, corresponding to 73% ± 4% and 72% ± 4% reductions, respectively. UV/chloramine was therefore much more effective than UV/chlorine at controlling THM. The extinction coefficient for chloramine with UV at a wavelength of

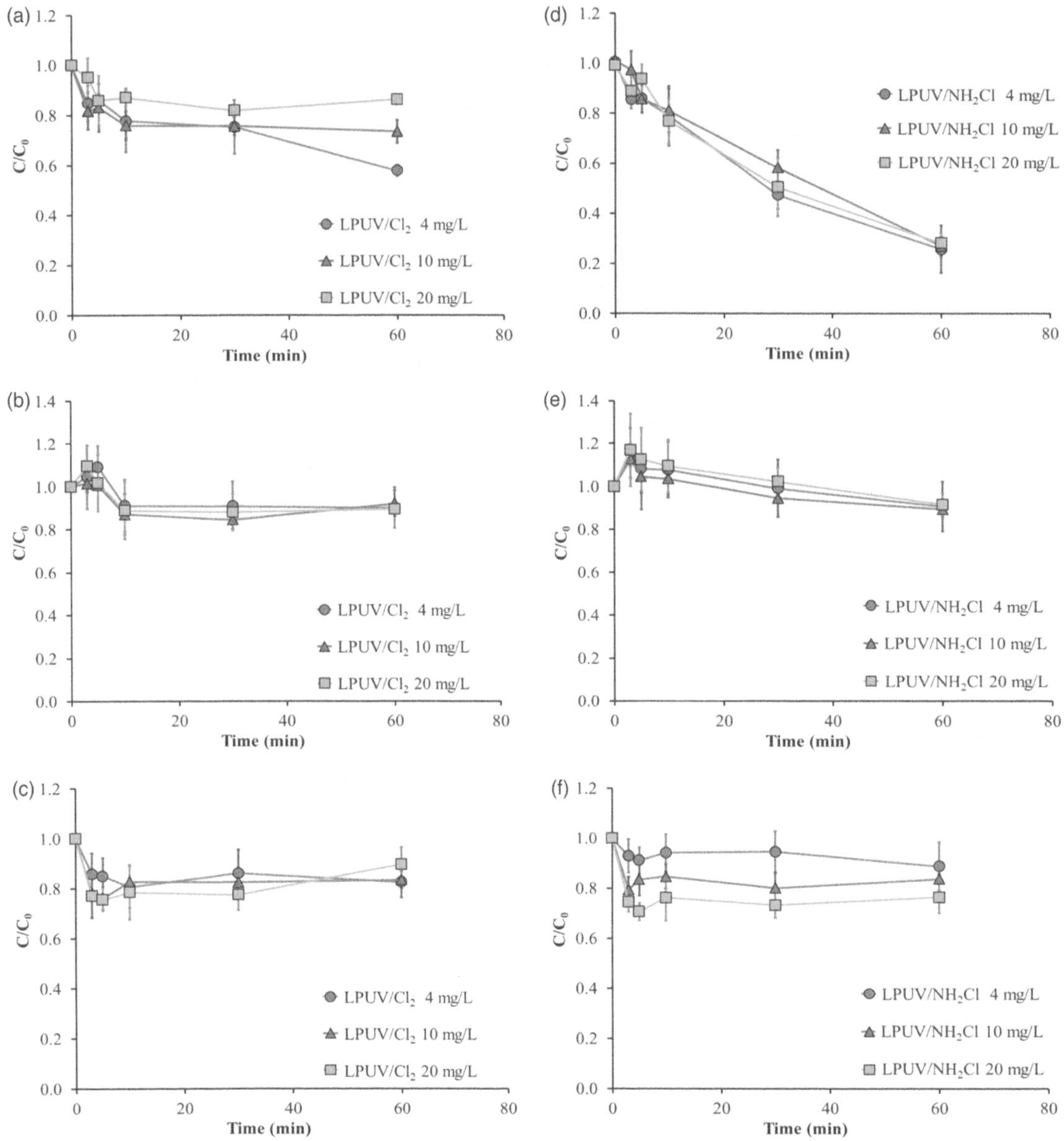

Figure 3 | Formation of THM (a), (d), HAA (b), (e) and HAN (c), (f) following UV/chlorine or UV/chloramine treatment.

254 nm has been reported to be approximately six-fold higher than that of chlorine (Zhang *et al.* 2015). This may result in greater production of reactive radicals. The removal of THM and its precursors by UV/chloramine was greater than for the others (i.e. HAA, HAN and their precursors). The second major reduction was found for HAN (Figure 3(f)). Although chloramine was anticipated to act as a precursor for HAN formation (Wang *et al.* 2007; Yang *et al.* 2010; Tian *et al.* 2013), UV/chloramine decreased HAN formation, depending on the chloramine dose. HAN formation declined by 11% ± 10%, 16% ± 2% and 24% ± 6% for chloramine doses of 4, 10 and 20 $mg \cdot L^{-1}$, respectively. The presence of reactive radicals caused decreased formation not only of THM and HAN but also HAA. Decreases in HAA formation were, however, the least, corresponding to 8–11% (Figure 3(e)). The effect of chloramine dose was the opposite to chlorine dose, indicating that chloramine residuals were less sensitive to reactive radicals. In addition, THM, HAA and HAN formed during UV/chloramine treatment were much lower than during UV/chlorine treatment. The reactivity of chloramine with NOM was lower than that of chlorine, resulting in the formation of fewer DBPs (Yang *et al.* 2007; Doederer *et al.* 2014; Hua *et al.* 2015). Compared with UV/chlorine

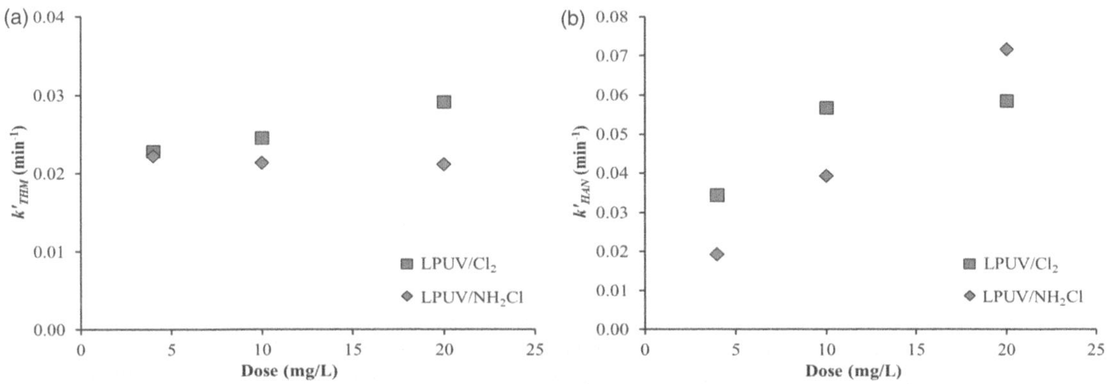

Figure 4 | The pseudo-first-order reaction constant of (a) THM and (b) HAN reduction with UV/chlorine and UV/chloramine ($n = 3$).

treatment, UV/chloramine resulted in 2.6–3.9 times lower THM concentration, 1.3–1.5 times lower HAA concentration, and 1.1–1.2 times lower HAN concentration, respectively. To minimize DBPs, a chloramine dose of 4 mg·L^{-1} was recommended for UV/chloramine application to control THM, HAA and HAN.

The k' value of THM reduction after UV/chlor(am)ine was likely to have been independent of chlorine and chloramine doses (Figure 4(a)). The k' values of THM reduction were in the range of 0.0228–0.0291 min^{-1} for UV/chlorine and 0.0211–0.0222 min^{-1} for UV/chloramine. Although the k' values for UV/chlorine were higher than those for UV/chloramine, the reduction in THM formation was lower. Since the observed concentration in treated water was normalized by the control sample, the k' values indicate competitiveness between the degradation rate and the formation rate. The higher reactivity of chlorine (than chloramine) implied that the reaction between chlorine with intermediates that occurred later was higher than with chloramination. In addition, several previous studies found that chlorination induced THM and HAN formation more than chloramination (Lu *et al.* 2009; Tian *et al.* 2013; Hu *et al.* 2018). The k' value of HAN reduction seemed to depend on the chlorine and chloramine doses (Figure 4(b)). The k' values of HAN reduction for UV/chlorine at the doses of 4, 10 and 20 mg·L^{-1} were 0.0343 ($R^2 = 0.8787$), 0.0567 ($R^2 = 0.8982$) and 0.0584 ($R^2 = 0.8876$) min^{-1}, respectively. The k' values of HAN reduction for UV/chloramine at the doses of 4 and 10 mg·L^{-1} were lower than those for UV/chlorine, while a chloramine dose of 20 mg·L^{-1} caused a higher k' value. The k' values of HAN reduction with UV/chloramine at doses of 4, 10 and 20 mg·L^{-1} were 0.0192 ($R^2 = 0.9544$), 0.0391 ($R^2 = 0.6611$) and 0.0715 ($R^2 = 0.9317$) min^{-1}, respectively. The data for HAA reduction did not fit with the pseudo-first-order reaction.

CONCLUSIONS

LPUV and VUV photolysis preferentially suppressed THM formation, followed by HAN and HAA. The use of VUV irradiation was more effective than LPUV irradiation to control the formation of THM and HAN. Decreases in THM and HAN formation were 23% ± 3% and 20% ± 8% for VUV illumination, while LPUV irradiation achieved 16% ± 2% and 20% ± 6% for THM and HAN reductions, respectively. The precursors for THM and HAN were quickly degraded by UV photolysis during the initial treatment time (10 min). Neither LPUV and VUV photolysis achieved much HAA reduction, because the degradation products (aliphatic-like compounds) yielded high levels of HAA.

The co-exposure of UV and chlorine (or chloramine) improved the performance for controlling THM, HAA and HAN, compared with standalone UV photolysis. Apart from THM and HAN, the use of UV/chlorine was able to suppress HAA formation. The performance of UV/chlorine on the control of THM, HAA and HAN was independent of contact time. The efficiency for controlling THM and HAN was inversely proportional to the chlorine dose. The maximum reductions in THM, HAA and HAN were achieved by the use of UV/chlorine at a chlorine dose of 4 mg·L^{-1}, responsible for 42% ± 2%, 10% ± 10% and 18% ± 6% reductions, respectively. UV/chloramine suppressed the formation of THM more than UV/chlorine. The control of THM formation depended on the contact time, rather than the chloramine dose. To control THM, HAA and HAN through the use of 4 mg·L^{-1} chloramine with UV irradiation, the greatest efficiency found in the study was 74% ± 10%, 10% ± 10% and 11% ± 10%, respectively. This study showed the potential use of UV/chlor(am)ine for controlling the formation of THM, HAA and HAN.

REFERENCES

Bond, T., Henriet, O., Goslan, E., Parsons, S. A. & Jefferson, B. 2009 Disinfection by-product formation and fractionation behaviour of natural organic matter surrogates. *Environmental Science and Technology* **43**, 5982–5989.

Bond, T., Huang, J., Templeton, M. R. & Graham, N. 2011 Occurrence and control of nitrogenous disinfection by-products in drinking water – a review. *Water Research* **45**, 4341–4354.

Buchanan, W., Roddick, F. & Porter, N. 2006 Formation of Hazardous by-products resulting from the irradiation of natural organic matter: comparison between UV and VUV irradiation. *Chemosphere* **63**, 1130–1141.

Chin, A. & Bérubé, P. R. 2005 Removal of disinfection by-product precursors with ozone-UV advanced oxidation process. *Water Research* **39**, 2136–2144.

Choi, Y. & Choi, Y.-J. 2010 The effects of UV disinfection on drinking water quality in distribution systems. *Water Research* **44**, 115–122.

Chuang, Y.-H., Chen, S., Chinn, C. J. & Mitch, W. A. 2017 Comparing the UV/monochloramine and UV/free chlorine Advanced Oxidation Processes (AOPs) to the UV/hydrogen peroxide AOP under scenarios relevant to potable reuse. *Environmental Science and Technology* **51**, 13859–13868.

Deeudomwongsa, P., Phattarapattamawong, S. & Lin, K.-Y. A. 2017 Control of disinfection byproducts (DBPs) by ozonation and peroxone process: role of chloride on removal of DBP precursors. *Chemosphere* **184**, 1215–1222.

Doederer, K., Gernjak, W., Weinberg, H. S. & Farré, M. J. 2014 Factors affecting the formation of disinfection by-products during chlorination and chloramination of secondary effluent for the production of high quality recycled water. *Water Research* **48**, 218–228.

Fang, J., Fu, Y. & Shang, C. 2014 The roles of reactive species in micropollutant degradation in the UV/free chlorine system. *Environmental Science and Technology* **48**, 1859–1868.

Hu, J., Chu, W., Sui, M., Xu, B., Gao, N. & Ding, S. 2018 Comparison of drinking water treatment processes combinations for the minimization of subsequent disinfection by-products formation during chlorination and chloramination. *Chemical Engineering Journal* **335**, 352–361.

Hua, G., Reckhow, D. A. & Abusallout, I. 2015 Correlation between SUVA and DBP formation during chlorination and chloramination of NOM fractions from different sources. *Chemosphere* **130**, 82–89.

Huang, H., Wu, Q.-Y., Tang, X., Jiang, R. & Hu, H.-Y. 2016 Formation of haloacetonitriles and haloacetamides and their precursors during chlorination of secondary effluents. *Chemosphere* **144**, 297–303.

Imoberdorf, G. & Mohseni, M. 2011 Degradation of natural organic matter in surface water using vacuum-UV irradiation. *Journal of Hazardous Materials* **186**, 240–246.

Kruithof, J. C., van der Leer, R. C. & Hijnen, W. A. M. 1992 Practical experiences with UV disinfection in the Netherlands. *Journal of Water Supply: Research and Technology-Aqua* **41** (2), 88–94.

Lamsal, R., Walsh, M. E. & Gagnon, G. A. 2011 Comparison of advanced oxidation processes for the removal of natural organic matter. *Water Research* **45**, 3263–3269.

Li, C., Wang, D., Xu, X. & Wang, Z. 2017 Formation of known and unknown disinfection by-products from natural organic matter fractions during chlorination, chloramination, and ozonation. *Science of the Total Environment* **587–588**, 177–184.

Liang, L. & Singer, P. C. 2003 Factors influencing the formation and relative distribution of haloacetic acids and trihalomethanes in drinking water. *Environmental Science and Technology* **37** (13), 2920–2928.

Linden, K. G., Shin, G.-A., Faubert, G., Cairns, W. & Sobsey, M. D. 2002 UV disinfection of *Giardia lamblia* cysts in water. *Environmental Science and Technology* **36** (11), 2519–2522.

Lu, J., Zhang, T., Ma, J. & Chen, Z. 2009 Evaluation of disinfection by-products formation during chlorination and chloramination of dissolved natural organic matter fractions isolated from a filtered river water. *Journal of Hazardous Materials* **162** (1), 140–145.

Musikavong, C., Inthanuchit, K., Srimuang, K., Suksaroj, T. T. & Suksaroj, C. 2013 Reduction of fractionated dissolved organic matter and their trihalomethane formation potential with enhanced coagulation. *ScienceAsia* **39**, 56–66.

Parkinson, A., Barry, M. J., Roddick, F. A. & Hobday, M. D. 2001 Preliminary toxicity of water treatment with UV irradiation and UVC/H_2O_2. *Water Resources* **36** (15), 3656–3664.

Rahn, R. O., Bolton, J. R. & Stefan, M. I. 2006 The Iodide/Iodate actinometer in UV disinfection: determination of the fluence rate distribution in UV reactors. *Photochemistry and Photobiology* **82** (2), 611–615.

Reckhow, D. A., Platt, T. L., MacNeill, A. L. & McClellan, J. N. 2001 Formation and degradation of dichloroacetonitrile in drinking waters. *Journal of Water Supply: Research and Technology-Aqua* **50**, 1–13.

Richardson, S. D., Plewa, M. J., Wagner, E. D., Schoeny, R. & DeMarini, D. M. 2007 Occurrence, genotoxicity, and carcinogenicity of regulated and emerging disinfection by-products in drinking water: a review and roadmap for research. *Mutation Research* **636** (1–3), 178–242.

Srithep, S. & Phattarapattamawong, S. 2017 Kinetic Removal of haloacetonitrile precursors by photo-based advanced oxidation processes (UV/H_2O_2, UV/O_3, and UV/O_3/H_2O_2). *Chemosphere* **176**, 25–31.

Tian, C., Liu, R., Liu, H. & Qu, J. 2013 Disinfection by-products formation and precursors transformation during chlorination and chloramination of highly-polluted source water: significance of ammonia. *Water Research* **47**, 5901–5910.

Wang, Y.-H. & Chen, K.-C. 2014 Removal of disinfection by-products from contaminated water using a synthetic goethite catalyst via catalytic ozonation and a biofiltration system. *International Journal of Environmental Research and Public Health* **11**, 9325–9344.

Wang, L. S., Hu, H. Y. & Wang, C. 2007 Effect of ammonia nitrogen and dissolved organic matter fractions on the genotoxicity of wastewater effluent during chlorine disinfection. *Environmental Science and Technology* **4**, 160–165.

Wang, D., Bolton, J. R., Andrews, S. A. & Hofmann, R. 2015a Formation of disinfection by-products in the ultraviolet/chlorine advanced oxidation process. *Science of the Total Environment* **518–519**, 49–57.

Wang, H., Zhu, Y., Hu, C. & Hu, X. 2015b Treatment of NOM fractions of reservoir sediments: effect of UV and chlorination on formation of DBPs. *Separation and Purification Technology* **154**, 228–235.

Westerhoff, P. & Mash, H. 2002 Dissolved organic nitrogen in drinking water supplies: a review. *Journal of Water Supply: Research and Technology-Aqua* **51**, 415–448.

Wu, Z., Guo, K., Fang, J., Yang, X., Xiao, H., Hou, S., Kong, X., Shang, C., Yang, X., Meng, F. & Chen, L. 2017 Factors affecting the roles of reactive species in the degradation of micropollutants by the UV/chlorine process. *Water Research* **126**, 351–360.

Yan, M., Roccaro, P., Fabbricino, M. & Korshin, G. V. 2018 Comparison of the effects of chloramine and chlorine on the aromaticity of dissolved organic matter and yields of disinfection byproducts. *Chemosphere* **191**, 477–484.

Yang, X., Shang, C. & Westerhoff, P. 2007 Factors affecting formation of haloacetonitriles, haloketones, chloropicrin and cyanogen halides during chloramination. *Water Research* **41** (6), 1193–1200.

Yang, X., Fan, C. H., Zhao, Q. & Shang, C. 2010 Nitrogenous disinfection byproducts formation and nitrogen origin exploration during chloramination of nitrogenous organic compounds. *Water Research* **44**, 2691–2702.

Yang, X., Shang, C., Shen, Q., Chen, B., Westerhoff, P., Peng, J. & Guo, W. 2012 Nitrogen origins and the role of ozonation in the formation of haloacetonitriles and halonitromethanes in chlorine water treatment. *Environmental Science and Technology* **46** (23), 12832–12838.

Zhang, H., Qu, J. H., Liu, H. J. & Zhao, X. 2009 Characterization of isolated fractions of dissolved organic matter from sewage treatment plant and the related disinfection by-products formation potential. *Journal of Hazardous Materials* **164**, 1433–1438.

Zhang, X., Li, W., Blatchley III, E. R., Wang, X. & Ren, P. 2015 UV/chlorine process for ammonia removal and disinfection by-product reduction: comparison with chlorination. *Water Research* **68**, 804–811.

First received 31 July 2018; accepted in revised form 24 November 2018. Available online 6 December 2018

The efficiency of UV/$S_2O_8^{2-}$ photo-oxidation process in the presence of Al_2O_3 for the removal of dexamethasone from aqueous solution: kinetic studies

Reza Shookohi, Hossein Faraji, Abolfazl Arabkohsar, Mehdi Salari and Mohammad Molla Mahmoudi

ABSTRACT

This study aimed to investigate the efficiency of the UV/$S_2O_8^{2-}$ photocatalytic process in the presence of Al_2O_3 nanoparticles for the removal of dexamethasone from aqueous solution. In this experimental study, the variables pH, persulfate concentration, initial concentration of dexamethasone, the catalyst dose were studied in order to investigate the process efficiency. Furthermore, the efficiency of UV/$S_2O_8^{2-}$ in the presence and absence of catalyst was investigated. The Al_2O_3 nanoparticle catalyst was characterized using X-ray diffraction (XRD) and Fourier transform infrared spectroscopy (FTIR) analyses and scanning electron microscopy (SEM) image. The results showed that a decrease in pH and the initial concentration of dexamethasone increased the process efficiency. Given the increased concentrations of the persulfate and Al_2O_3, the removal efficiency was partially increased. In UV/$S_2O_8^{2-}$/Al_2O_3 under optimum conditions (pH = 3, t = 30 minutes, dexamethasone concentration = 20 mg/L, 0.5 mM of persulfate, and UV radiation = 55 watts), 94% of the dexamethasone was removed. The kinetic response showed that the reaction data corresponded to the pseudo-first-order kinetic model. The results showed that the UV/$S_2O_8^{2-}$ photochemical process can efficiently remove dexamethasone from aqueous solution in the presence of Al_2O_3 catalyst and the mineralization efficiency reached about 98%. Therefore, this process is recommended due to its high efficiency and availability for the removal of pharmaceutical compounds.

Key words | Al_2O_3, dexamethasone, persulfate, photo-oxidation process

Reza Shookohi
Hossein Faraji
Abolfazl Arabkohsar
Mehdi Salari
Mohammad Molla Mahmoudi (corresponding author)
Department of Environmental Health Engineering, School of Public Health,
Hamadan University of Medical Sciences,
Hamadan,
Iran
E-mail: *ms.mahmoudi65@gmail.com*

INTRODUCTION

In the last two decades, there has been a great deal of evidence about the discharge of wastewater containing emerging pollutants into water resources and the environment. Pharmaceutical pollutants are known to be a group of emerging pollutants that may enter the environment after consumption or even without consumption (Gros *et al.* 2010; Santos *et al.* 2010). These compounds can cause genetic mutations in living organisms and sexual disorders in aquatic animals (Arsand *et al.* 2013). Corticosteroid is one of the biggest classes of these drugs, and is widely used in the treatment of human and animal diseases. Corticosteroids are very strong drugs used to relieve inflammation in the human body. Dexamethasone is one of the most commonly used corticosteroids (Table 1). This is mostly found in the wastewater of medical centers and hospitals (Mohseni *et al.* 2016).

doi: 10.2166/wst.2019.109

Different methods have been used to remove pharmaceutical substances from wastewater, including membrane bioreactors (Yang *et al.* 2006; Deegan *et al.* 2011), carbon nanotubes (Wang & Wang 2016) and anaerobic reactors (Chelliapan *et al.* 2011). In order to remove antibiotics, especially dexamethasone as a subclass of pharmaceutical substances, the electrical coagulation process (Arsand *et al.* 2013) and the adsorption process (Yu *et al.* 2016; Ahmed 2017) are further used. In the biological process, the treatment time is long and the amount of sludge is high in the electrical coagulation method (Arsand *et al.* 2013). Finally, in the adsorption process, waste disposal is not cost-effective.

On the other hand, the advanced oxidation process (AOPs) is paid much attention for wastewater treatment due to its ability in the degradation and mineralization of

Table 1 | Dexamethasone properties

Chemical formula	$C_{22}H_{29}FO_5$
λmax	254 nm
Molar mass	392.461 g mol^{-1}
Chemical structure	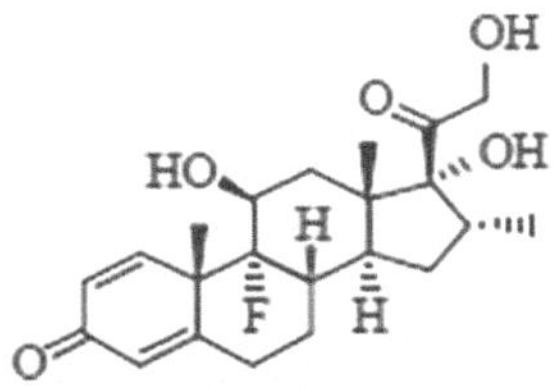

many organic compounds resistant to conventional biological and chemical treatment methods. Generally, the AOPs are based on strong oxidants such as ozone, hydrogen peroxide (H_2O_2), ultrasonic, electron beam and some other methods (Moghaddam *et al.* 2014). The peroxidesulfate anion ($S_2O_8^{2-}$), known as the last and most recent oxidant, has been shown to be able to degrade organic pollutants. $S_2O_8^{2-}$ is a potent oxidizing agent with two electrons and a 2.01-volt potential. Sulfate radicals can be produced from peroxydisulfate PDS or peroxymonosulfate activated through heat, photochemical processes, radiolysis, or redox decomposition (Equations (1)–(4)) (Vicente *et al.* 2011).

$$S_2O_8^{2-} \xrightarrow{hv/heat} 2SO_4^{-} \quad (1)$$

$$S_2O_8^{2-} + e_{CB}^{-} \rightarrow SO_4^{-\cdot} + SO_4^{2-} \quad (2)$$

$$S_2O_8^{2-} + M^{n+} \rightarrow SO_4^{-\cdot} + SO_4^{2-} + M^{(n+1)+} \quad (3)$$

$$HSO_5^{-} \xrightarrow{hv/heat} SO_4^{-\cdot} + OH^{\cdot} \quad (4)$$

$$S_2O_8^{2-} + H^{+} \rightarrow HS_2O_8^{-} \quad \text{Under acidic conditions} \quad (5)$$

$$HS_2O_8^{-} \rightarrow SO_4^{-} + SO_4^{2-} + H^{+} \quad (6)$$

$$SO_4^{-} + OH^{-} \rightarrow SO_4^{2-} + {}^{\cdot}OH \quad \text{Under basic conditions} \quad (7)$$

The sulfate radical is one of the strongest oxidizing agents in aquatic environments with a potential of 2.6 V, which is similar to the hydroxyl radical (2.7 V). Among the strongest oxidizers, the sulfate radical has unique properties, including high kinetic constants, higher stability compared to the hydroxyl radical, and a lower dependence on the type of organic substance followed with a greater effect on organic substances (Waldemer *et al.* 2007). Therefore, the sulfate radical-based advanced oxidation process has attracted much attention (Khan *et al.* 2013). It is desirable to activate the $S_2O_8^{2-}$ and produce $SO_4^{-\cdot}$ causing a high oxidation rate of pollutants. The most commonly used activation methods include ultraviolet (UV) light, heat, pH regulation, or the use of intermediate metals to produce $SO_4^{-\cdot}$ (Matzek & Carter 2016). In the activating process through intermediate metals, the proxy sulfate is converted into sulfate radical by using a sufficient amount of metal cation as an electron donor (Khan *et al.* 2013; Mazloomi *et al.* 2016). Several metals, including copper, cobalt, nickel and manganese, have been used to catalyze the oxidation process of pollutants (Surpăţeanu & Zaharia 2004; Xu *et al.* 2004; Hussain *et al.* 2012). In the present study, the intention was to explore the catalytic effect of Al_2O_3, with advantages such as low cost and easy production, combined with UV radiation in the activation of persulfate. Alumina was found to be an effective catalyst for the ozonation of 2-chlorophenol and for refractory organic compounds such as oxalic, acetic, salicylic and succinic acids (Ni *et al.* 2001). In recent years, researchers have realized that the efficiency of ozonation catalyzed by aluminum oxides is mainly a result of their ability to transform ozone into $^{\cdot}OH$ radicals, which are the strongest oxidants in water (Qi *et al.* 2008). Therefore, this study aimed to investigate the effect of photocatalytic reaction of UV/S_2O_8 in the presence of Al_2O_3 in the removal of the dexamethasone from aqueous solution.

MATERIALS AND METHODS

Chemicals and photoreactor

The chemicals used in the experiments include persulfate ($S_2O_8^{2-}$), sodium hydroxide (NaOH) and sulfuric acid (H_2SO_4) provided by Sigma Aldrich and Merck Co., Germany. The Al_2O_3 nanoparticle powder was purchased from ASPI Co, and the sodium phosphate dexamethasone was procured from Darou Pakhsh Pharmaceutical Co. A stainless steel photoreactor with a 2-liter volume that equipped with a quartz glass and a 55-watt low-pressure lamp (Philips Co.), was used for experiments. A pump was also applied to mix the sample continuously (all details of the reactor are given in Figure 1).

Experimental procedures

This experimental study was carried out at a laboratory scale in a batch flow system. In this study, the effects

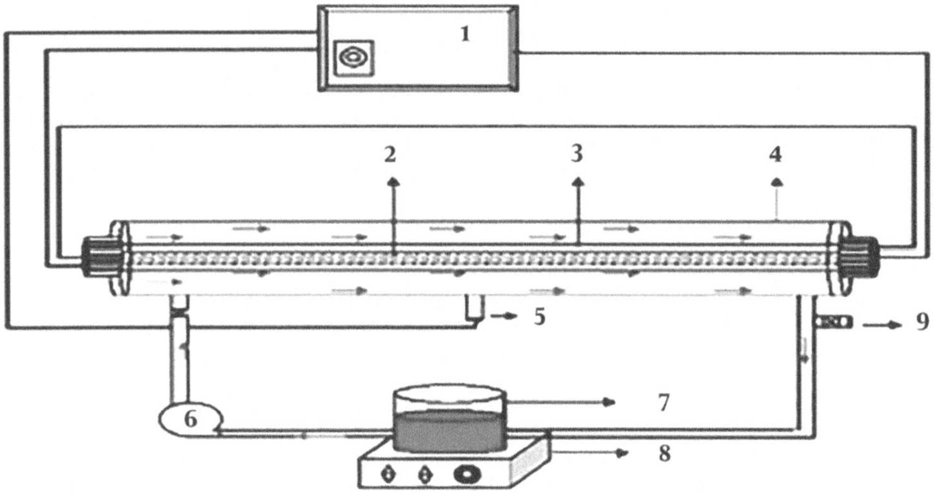

Figure 1 | Schematic view of the photoreactor parts: (1) transformer, (2) low-pressure mercury-vapor UV lamp, (3) quartz cover (length: 85 cm; diameter: 5 cm), (4) stainless steel box (L:85 cm D: 8 cm), (5) photocell, (6) pump (50–400 ml/min), (7) beaker (2 liters), (8) shaker, and (9) sampling tube.

of pH (3–11), persulfate concentration (0.1–7.5 mM), Al_2O_3 dose (0.01–0.2 g/L), time (0–30 minutes) and the initial concentration of dexamethasone (5, 10, 15, 20 mg/L), on the process efficiency were investigated at a constant intensity of ultraviolet light. This concentration range was selected based on previous studies (Forteza *et al.* 1989; Mohseni *et al.* 2016). All experiments were repeated three times at constant temperature (25 °C), and finally the mean values were reported. The first- and second-order linear kinetics were used to find the best-fitted model for the degradation reaction. In the experiments, the main solution was prepared at the 100 mg/l concentration. This stock solution was diluted in the different ratios to prepare appropriate concentrations. In order to measure the concentration of dexamethasone in solution, a UV/Vis spectrophotometer (DR 5000, Hach Co., Germany) was used at a wavelength of 241 nm. Total organic carbon (TOC) analysis by means of a Viro TOC element Co. instrument was applied to measure the mineralization potential of dexamethasone over the treatment process. Also, COD (chemical oxygen demand) analysis and residual sulfate amount were measured according to the standard method (APHA/AWWA/WEF 2005). The digital pH meter device (Hach) was used to measure pH. The X-ray diffraction (XRD) analysis in the range $2\theta = 5$–100 was used to characterize the Al_2O_3 nanoparticles using a Rigaka X-ray diffractometer. The field emission scanning electron microscopy and Fourier transform infrared spectroscopy (FTIR) analysis were also used to scan the morphology of the surface of the nanoparticles and to investigate the functional groups, respectively. Finally, Excel 2013 was used to plot all the charts and analyze the data.

RESULTS AND DISCUSSION

Catalyst characterization

In order to characterize the morphology of nanoparticles, scanning electron microscopy (SEM) analysis was used before the removal process (Figure 2). In the SEM image, it is evident that the aggregates of aluminum nanoparticles yield a porous spongy structure. In the XRD analysis (Figure 3), two peaks were observed in the $2\theta = 33.8$ and 68.6 in the amorphous shape. The accessible surface and the diameter of the catalyst pores were 155 mg/m^2 and 58 Å, respectively. The FTIR spectrum was used to identify and investigate the functional groups on the catalyst

Figure 2 | SEM image of Al_2O_3 nanoparticles.

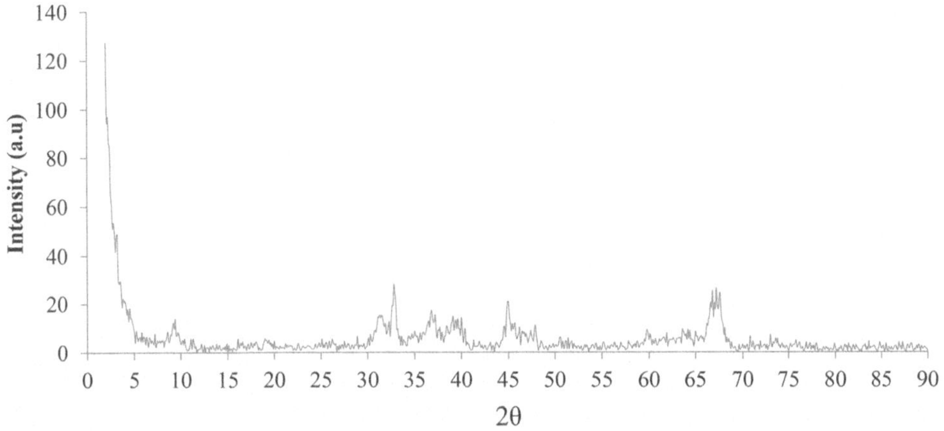

Figure 3 | XRD analysis of Al_2O_3 nanoparticle.

surface (Figure 4). The major peaks appearing in the FTIR spectra may be assigned to the following vibrational modes: (i) –OH stretching vibrations of the surface bonded (or) adsorbed water, (ii) –OH stretching vibrations of structural water corresponding to M-OH stretching, (iii) –OH bending vibrations of structural water, corresponding to M-OH bending and (iv) Al–O vibrations. The peaks 3,468 cm^{-1} and 1,636 cm^{-1} observed in Figure 3 belong to -OH stretching bands and -OH bending vibration (these functional groups play a crucial role in the adsorption process) (Sarkar *et al.* 2007; Muthukumari *et al.* 2009).

The effect of initial pH

Previous studies have shown that pH plays an important role in the advanced oxidation process (Hussain *et al.* 2012). Therefore, in order to investigate the effect of the initial pH of the solution, the pH was arranged in the range of 3–11 and persulfate concentration was fixed at 0.1 mM. The results are shown in Figure 5. The highest efficiency of $UV/S_2O_8^{2-}/Al_2O_3$ process was obtained to be 88% at pH = 3. The experiment showed that once the pH is increased, the removal efficiency decreased dramatically, where the efficiency diminished 52% after 20 minutes of reaction time at pH 11. Because the efficacy of dexamethasone removal becomes almost constant after 20 min operating time, and after 30 min, no degradation happened, the contact time of 0 to 30 min was selected for the rest of the analyses. Hussain *et al.* regarded the acidic conditions as optimum conditions in the removal of nonylphenol using the $S_2O_8^{2-}$/nZVI/BC process. This can be attributed to the degradation of $S_2O_8^{2-}$ in acidic conditions and to

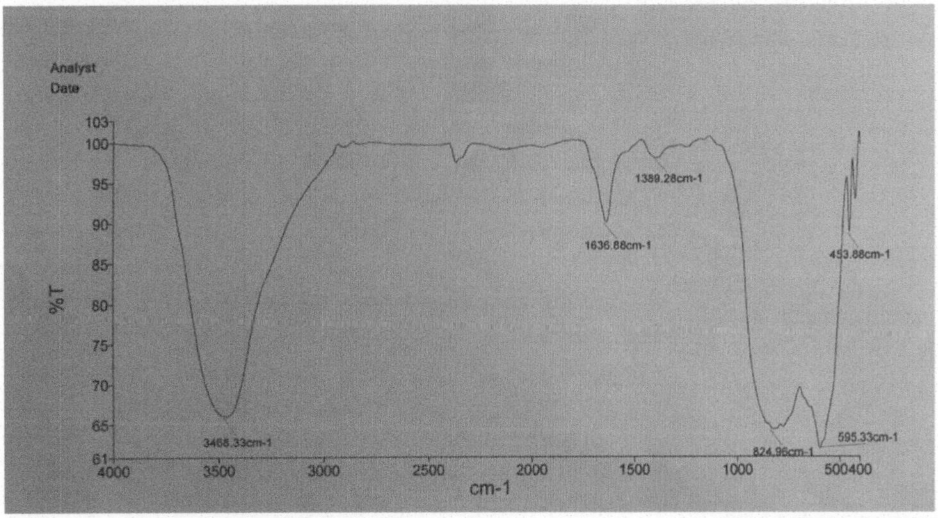

Figure 4 | FTIR spectra of Al_2O_3 nanoparticle.

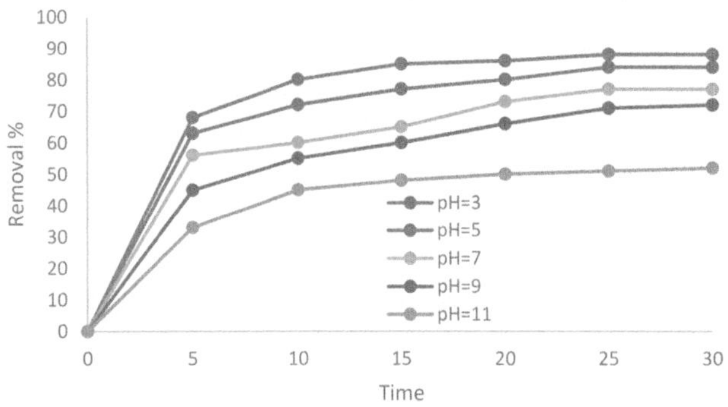

Figure 5 | The effect of pH in dexamethasone removal efficiency (persulfate dose = 0.1 mM, dexamethasone concentration = 20 mg/L and Al_2O_3 dose = 0.05 g/L).

the fact that it is converted to active radicals $SO_4^{-\bullet}$ (Equations (5) and (6)) (Hussain *et al.* 2017). In addition, it has been reported that the accumulation of nanoparticles in acidic conditions decreases compared to neutral conditions, resulting in an increase in the effective surface area of the catalyst. This can lead to increased photocatalytic degradation in acidic conditions (French *et al.* 2009).

The effect of $S_2O_8^{2-}$ dosage

The dexamethasone removal efficiency in different initial concentrations of persulfate at pH 3, dexamethasone concentration of 20 mg/L and a catalyst dose of 0.05 g/l is shown in Figure 6. The results of the experiments show that an increase in the $S_2O_8^{2-}$ concentration from 0.1 to 0.5 mM causes an increase in the removal efficiency from 88% to 94%, respectively. Then, the removal efficiency decreases when the concentration of $S_2O_8^{2-}$ increases to a concentration higher than 5 mM. The increase of removal efficiency can be explained by the generation of sulfate and hydroxyl radicals improving the process performance at concentrations of less than 0.5 mM of $S_2O_8^{2-}$ (Lau *et al.* 2007). However, when the concentration of $S_2O_8^{2-}$ increases, the efficiency decreases slowly. According to studies by Qiao *et al.*, at higher doses of $S_2O_8^{2-}$, the excessive generation of hydroxyl (Equations (8) and (9)) results in the generation of H_2O_2, which has less oxidation potential (Equation (10)) (Qiao *et al.* 2005). Therefore, the oxidation of dexamethasone decreases at high concentrations of $S_2O_8^{2-}$ (Salari *et al.* 2008). Alternatively, studies suggest there is an optimum amount of persulfate above which degradation is inhibited by the reaction of excess persulfate with sulfate radicals and itself (Equations (11) and (12)), reducing the effectiveness organic oxidation (Matzek & Carter 2016).

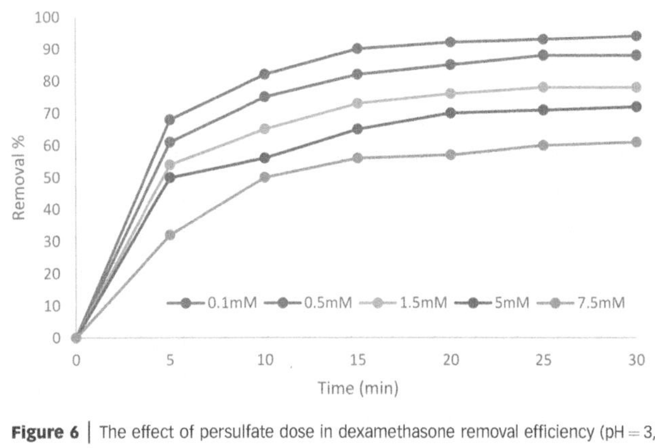

Figure 6 | The effect of persulfate dose in dexamethasone removal efficiency (pH = 3, dexamethasone concentration = 20 mg/L and Al_2O_3 dose = 0.05 g/L.

$$S_2O_8^{2-} + h\upsilon \rightarrow 2SO_4^{-\bullet} \quad (8)$$

$$SO_4^{-\bullet} + H_2O \rightarrow HSO_4^{-} + OH^{\bullet} \quad (9)$$

$$2OH^{\bullet} \rightarrow H_2O_2 \text{ (only in acidic to neutral pH)} \quad (10)$$

$$SO_4^{-\bullet} + S_2O_8^{2-} \rightarrow SO_4^{2-} + S_2O_8^{-\bullet} \quad (11)$$

$$SO_4^{-\bullet} + SO_4^{-\bullet} \rightarrow S_2O_8^{2-} \quad (12)$$

The effect of initial concentration of dexamethasone

From a practical point of view, it is important to investigate the effect of the initial concentration of the pollutant on its removal. The effect of the initial concentration of dexamethasone on its reaction rates in the $UV/S_2O_8^{2-}/Al_2O_3$ process is shown in Figure 7. The results indicated that the increased concentration of dexamethasone (from 5 to 30 mg/L) decreased the process reaction rates. Accordingly, the process rate constant reached 0.056 min^{-1} at a concentration of 30 mg/L, while k for the initial concentration of 5 mg/L was 0.29 min^{-1}. The decrease in the removal rate by increasing the concentration of dexamethasone can be attributed to the fact that at all concentrations, the amount of nanoparticles, contact time and pH are the same. As a result, the amount of radicals generated in each of the four concentrations is the same. Dexamethasone removal is therefore expected to be higher in samples with lower concentrations. On the other hand, the decrease in the reaction rates by increasing the initial concentration of dexamethasone can be attributed to the fact that there is a great deal of surface on the catalyst at lower concentrations, which removes dexamethasone. Additionally, by increasing the concentration of dexamethasone, the ultraviolet light

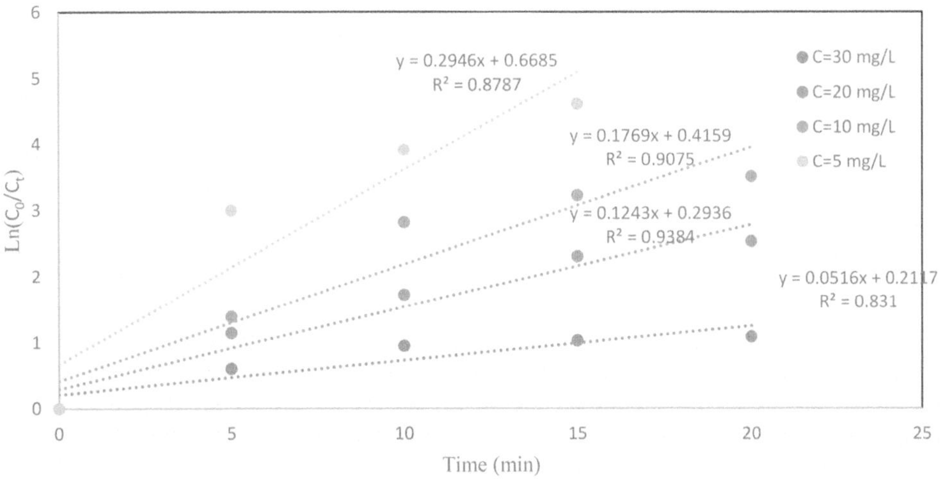

Figure 7 | Kinetic data of the degradation of dexamethasone in different initial concentrations (pH = 3, persulfate dose = 0.5 mM and Al_2O_3 dose = 0.05 g/L).

penetration is decreased. As a result, the process reaction rates are decreased (Modirshahla & Behnajady 2006). Similar results can be found in other studies (Salari *et al.* 2008; Yoon *et al.* 2012).

The effect of catalyst dose Al_2O_3

Figure 8 shows the effect of aluminum oxide on the efficiency of dexamethasone removal in the photo-oxidation process. Obviously, by increasing the dose of Al_2O_3 (0.01 to 0.05 g/L), the removal efficiency rises from 87% to 94%. In addition, by increasing the catalyst dose to more than 0.05 g/l, the efficiency decreases. This can be attributed to reduced light penetration, light diffusion, Al_2O_3 accumulation and sedimentation in high concentrations of catalyst. The results of this study are in good agreement with the reports presented by other researchers. Chen *et al.* showed that by increasing the Fe^{2+} concentration over optimum conditions, the MTBE degradation decreases by the persulfate oxidation process (Chen *et al.* 2009; Rastogi *et al.* 2009).

TOC analysis and residuals of the process

Since dexamethasone is initially converted to other degradation byproducts, which may all be organic compounds. Measurement of TOC was therefore considered, which might further suggest the removal of toxicity. Also the goal of the degradation of any pollutant is not only the pollutant degradation but also to obtain mineralization. In this perspective, DEX mineralization was analyzed by measuring the TOC concentration from samples taken at the beginning and end of the process. Figure 9 displays the TOC and COD content and residual sulfate of the sample at the selected condition. It is observed that the initial TOC of the sample has been measured to be 53.8 mg/L,

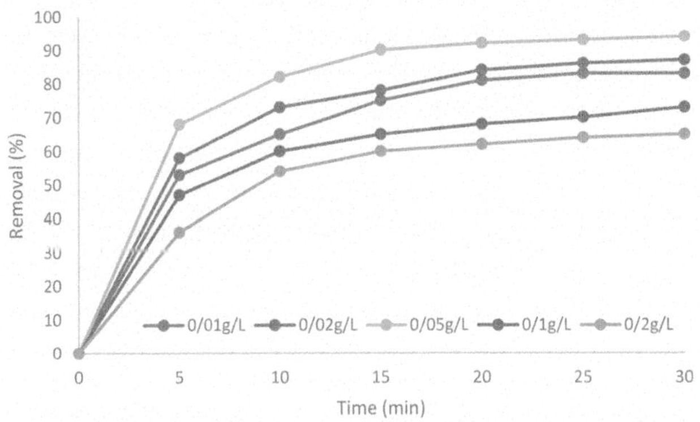

Figure 8 | The effect of Al_2O_3 dose in the dexamethasone removal efficiency (initial dexamethasone concentration = 20 mg/L pH = 3 and persulfate dose = 0.5 mM).

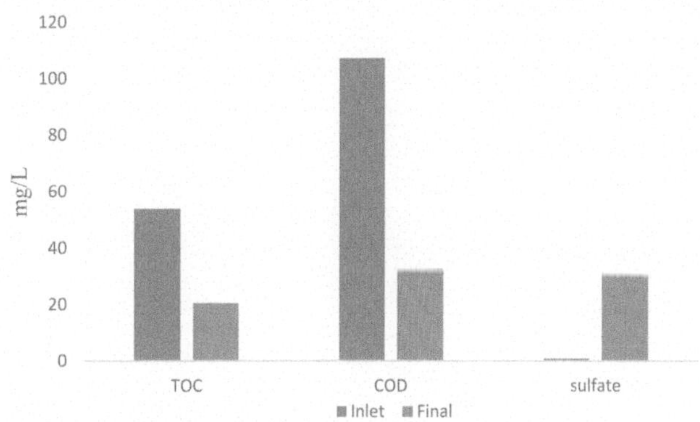

Figure 9 | TOC and COD removal and residuals of sulfate (initial dexamethasone concentration = 20 mg/L pH = 3, persulfate dose = 0.5 mM and Al_2O_3 dose = 0.05 g/L).

while this reached 20.5 mg/L after conducting the UV/ $S_2O_8^{2-}/Al_2O_3$ process for a 30 min operating time. Thus, the mineralization efficiency was calculated to be equal to about 62%. Also, the COD reduction efficiency was obtained as 69%. It is worth noting that the removal efficiency of dexamethasone for the same contact time was found to be 94%. As result, to attain a higher mineralization efficiency, a larger operating time is needed. For example, the TOC reduction efficiency reached 98% at 120 min operating time (not shown in Figure 9). In other words, ecological toxicity of intermediate products could be avoided when the contact time was sufficient to mineralize the target pollutants into non-toxic CO_2 and H_2O by $UV/S_2O_8^{2-}/Al_2O_3$ process. However, from an environmental point of view, intermediate products must be further identified or eco-toxicological tests conducted within the time-scale. Finally, the sulfate amount of the solution after conducting the process at the selected condition (Figure 9) was determined to be 30 mg/L, which was placed within the permissible limit for effluent discharge and even drinking waters.

Comparing the processes

The $UV/S_2O_8^{2-}$ process was performed in the presence and absence of the Al_2O_3 catalyst. The UV and UV/Al_2O_3 processes were also investigated for comparison with each other. As shown in Figure 10, the UV has a low efficiency about 8% after 30 minutes of radiation. The UV/Al_2O_3 process efficiency is about 17%, which may be attributed to the low adsorption in aluminum oxide. However, by comparing the $UV/S_2O_8^{2-}$ photo-oxidation process and the $UV/S_2O_8^{2-}/Al_2O_3$ process, there is clearly a difference in the efficiency of dexamethasone removal. The efficiency of the $UV/S_2O_8^{2-}$ process alone is 66%, while it increases to 94% in the presence of aluminum oxide. The activity of aluminum oxide in catalyzing oxidation decay was relative to the surface acid–base properties of the oxide. Because of the unsaturated state of surface electrons on the aluminum oxide, the surface would adsorb water molecules. This behavior may result in the formation of surface hydroxyl groups. Surface hydroxyl groups play an important role in the acid–base properties of aluminum oxide. It was thought that surface hydroxyl groups made a significant contribution to the ability of aluminum oxide to catalyze decomposition (Qi *et al.* 2008). Thus, the efficiency of dexamethasone removal has increased in the presence of aluminum oxide.

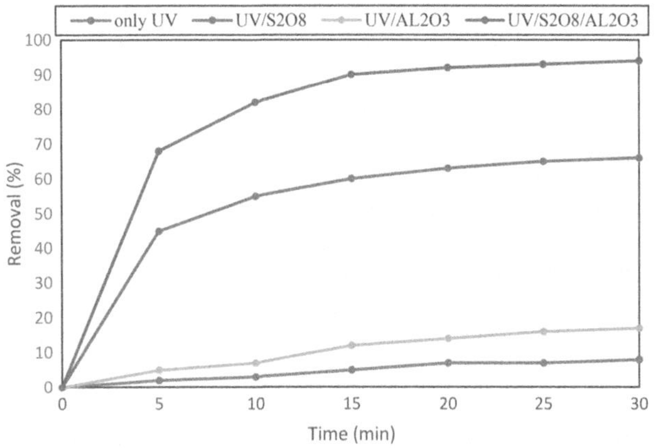

Figure 10 | Comparing the mechanisms of UV, $UV/S_2O_8^{2-}$, UV/Al_2O_3 and $UV/S_2O_8^{2-}/Al_2O_3$ at the optimum condition in the removal efficiency of dexamethasone.

Investigating the process kinetics

In this study, since the catalyst concentration is high, the Langmuir-Hinshelwood equation (Equation (13)) was used to study dexamethasone removal over the $UV/S_2O_8^{2-}/Al_2O_3$ process.

$$\frac{dC}{dt} = -k[C_{Pollutant}][C_{Catalyst}] \tag{13}$$

Assuming the catalyst concentration is constant during the experiment, Equation (13) was reduced to the pseudo-first-order (PFO) kinetic equation given in Equation (14).

$$\frac{d[pollutant]}{dt} = -k_{app}\, C_{pollutant} \tag{14}$$

K_{aap} represents the PFO rate constant (min^{-1}) and is determined by plotting (Ln_{Ct}/Ln_{C0}) against time values based upon the following Equation (15):

$$\text{In}\ \frac{C_{HA\,t}}{C_{HA\,t=0}} = -k_{app}t \tag{15}$$

Kinetic study information on the process is given in Table 2 and Figure 11. The results clearly illustrate that the reaction constant for $UV/S_2O_8^{2-}/Al_2O_3$ is much larger than the other process in the removal of dexamethasone. This occurrence confirmed the excellent synergistic effect of the combined $UV/S_2O_8^{2-}/Al_2O_3$ process compared to UV, Al_2O_3 and UV/Al_2O_3 alone.

CONCLUSION

This study was carried out in order to remove dexamethasone, one of the most widely used anti-inflammatory drugs,

Table 2 | The obtained coefficients of the pseudo-first-order kinetic model for the removal of dexamethasone by UV, Al_2O_3, $UV/S_2O_8^{2-}$ and $UV/S_2O_8^{2-}/Al_2O_3$ processes

	UV		Al_2O_3		$UV/S_2O_8^{2-}$		$UV/S_2O_8^{2-}/Al_2O_3$	
Process	R^2	K	R^2	K	R^2	K	R^2	K
Pseudo-first-order model	0.9621	0.0028	0.965	0.0063	0.733	0.031	0.8724	0.0878

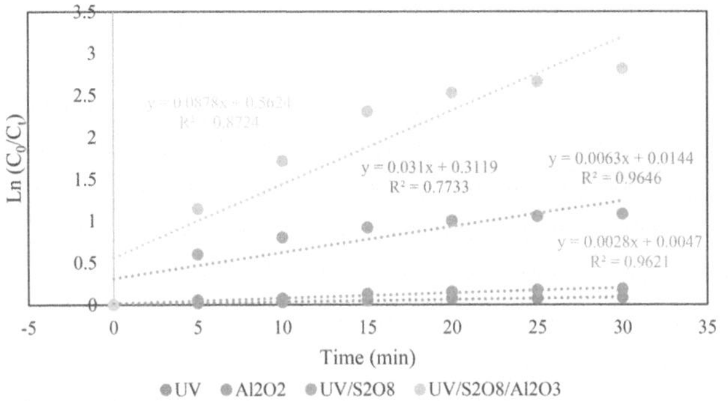

Figure 11 | Pseudo-first-order kinetic model for the removal of dexamethasone by UV, Al_2O_3, $UV/S_2O_8^{2-}$ and $UV/S_2O_8^{2-}/Al_2O_3$ processes.

using the AOT process, which is based on the generation of sulfate and hydroxyl radicals. The results showed that dexamethasone degradation in the $UV/S_2O_8^{2-}/Al_2O_3$ process increased by decreasing the pH and initial concentration of dexamethasone. By increasing the concentration of sulfate and aluminum oxide nanoparticles up to 0.5 mmol and 0.05 gr/L, respectively, the process efficiency also increased. The maximum efficiency for dexamethasone removal was obtained under the following condition: pH = 3, concentration of persulfate = 0.5 mM, dose of Al_2O_3 = 0.05 g/L, initial concentration = 20 mg/l and contact time = 30 minutes. The kinetic response showed that the reaction data corresponded to the pseudo-first-order kinetic model. The results showed that the $UV/S_2O_8^{2-}$ photochemical process can efficiently remove dexamethasone from aqueous solution in the presence of the Al_2O_3 catalyst and that the mineralization efficiency reached about 98%. Additionally, the results showed that the $UV/S_2O_8^{2-}$ process has a higher efficiency in the presence of the Al_2O_3 catalyst than the $UV/S_2O_8^{2-}$ process and that the activity of aluminum oxide in catalyzing oxidation decay was relative to the surface acid–base properties of the oxide. Due to laboratory limitations, it is recommended to use LC mass to propose the pathways and byproducts of the process. Considering the appropriate removal efficiency of the investigated process at optimum conditions, this process is recommended to be used for the removal of pharmaceutical compounds.

ACKNOWLEDGEMENT

Sincere thanks and appreciation go to Hamedan University of Medical Sciences and the experts of the Faculty of Water and Wastewater Laboratory for their financial support and help to conduct this research.

REFERENCES

Ahmed, M. J. 2017 Adsorption of non-steroidal anti-inflammatory drugs from aqueous solution using activated carbons. *Journal of Environmental Management* **190**, 274–282.

APHA/AWWA/WEF 2005 *Standard Methods for the Examination of Water and Wastewater*. American Public Health Association (APHA), Washington, DC, USA.

Arsand, D. R., Kümmerer, K. & Martins, A. F. 2013 Removal of dexamethasone from aqueous solution and hospital wastewater by electrocoagulation. *Science of the Total Environment* **443**, 351–357.

Chelliapan, S., Wilby, T., Yuzir, A. & Sallis, P. J. 2011 Influence of organic loading on the performance and microbial community structure of an anaerobic stage reactor treating pharmaceutical wastewater. *Desalination* **271**, 257–264.

Chen, K., Kao, C., Wu, L., Surampalli, R. & Liang, S. 2009 Methyl tert-butyl ether (MTBE) degradation by ferrous ion-activated persulfate oxidation: feasibility and kinetics studies. *Water Environment Research* **81**, 687–694.

Deegan, A., Shaik, B., Nolan, K., Urell, K., Oelgemöller, M., Tobin, J. & Morrissey, A. 2011 Treatment options for wastewater effluents from pharmaceutical companies. *International Journal of Environmental Science & Technology* **8**, 649–666.

Forteza, M., Galán, E. & Cornejo, J. J. A. C. S. 1989 Interaction of dexamethasone and montmorillonite – adsorption-degradation process. *Applied Clay Science* **4**, 437–448.

French, R. A., Jacobson, A. R., Kim, B., Isley, S. L., Penn, R. L. & Baveye, P. C. 2009 Influence of ionic strength, pH, and cation valence on aggregation kinetics of titanium dioxide nanoparticles. *Environmental Science & Technology* **43**, 1354–1359.

Gros, M., Petrović, M., Ginebreda, A. & Barceló, D. 2010 Removal of pharmaceuticals during wastewater treatment and environmental risk Assessment using hazard indexes. *Environment International* **36**, 15–26.

Hussain, I., Zhang, Y., Huang, S. & Du, X. 2012 Degradation of p-chloroaniline by persulfate activated with zero-valent iron. *Chemical Engineering Journal* **203**, 269–276.

Hussain, I., Li, M., Zhang, Y., Li, Y., Huang, S., Du, X., Liu, G., Hayat, W. & Anwar, N. J. C. E. J. 2017 Insights into the mechanism of persulfate activation with nZVI/BC nanocomposite for the degradation of nonylphenol. *Chemical Engineering Journal* **311**, 163–172.

Khan, J. A., He, X., Khan, H. M., Shah, N. S. & Dionysiou, D. D. 2013 Oxidative degradation of atrazine in aqueous solution by $UV/H_2O_2/Fe^{2+}$, $UV/S_2O_82-/Fe^{2+}$ and $UV/HSO5-/Fe^{2+}$ processes: a comparative study. *Chemical Engineering Journal* **218**, 376–383.

Lau, T. K., Chu, W. & Graham, N. J. 2007 The aqueous degradation of butylated hydroxyanisole by UV/S_2O_82-: study of reaction mechanisms via dimerization and mineralization. *Environmental Science & Technology* **41**, 613–619.

Matzek, L. W. & Carter, K. E. J. C. 2016 Activated persulfate for organic chemical degradation: a review. *Chemosphere* **151**, 178–188.

Mazloomi, S., Nasseri, S., Nabizadeh, R., Yaghmaeian, K., Alimohammadi, M., Nazmara, S. & Mahvi, A. H. 2016 Remediation of fuel oil contaminated soils by activated persulfate in the presence of MnO_2. *Soil & Water Research* **11**, 131–138.

Modirshahla, N. & Behnajady, M. A. 2006 Photooxidative degradation of Malachite Green (MG) by UV/H_2O_2: influence of operational parameters and kinetic modeling. *Dyes and Pigments* **70**, 54–59.

Moghaddam, S. K., Rasoulifard, M., Vahedpour, M. & Eskandarian, M. 2014 Removal of tylosin from aqueous solution by UV/nano $Ag/S_2O_8^{2-}$ process: influence of operational parameters and kinetic study. *Korean Journal of Chemical Engineering* **31**, 1577–1581.

Mohseni, S., Amooey, A., Tashakkorian, H. & Amouei, A. 2016 Removal of dexamethasone from aqueous solutions using modified clinoptilolite zeolite (equilibrium and kinetic). *International Journal of Environmental Science and Technology* **13**, 2261–2268.

Muthukumari, B., Selvam, K., Muthuvel, I. & Swaminathan, M. 2009 Photoassisted hetero-Fenton mineralisation of azo dyes by Fe (II)-Al_2O_3 catalyst. *Chemical Engineering Journal* **153**, 9–15.

Ni, C. & Chen, J. J. W. S.Technology 2001 Heterogeneous catalytic ozonation of 2-chlorophenol aqueous solution with alumina as a catalyst. *Water Science and Technology* **43**, 213–220.

Qi, F., Chen, Z., Xu, B., Shen, J., Ma, J., Joll, C. & Heitz, A. 2008 Influence of surface texture and acid–base properties on ozone decomposition catalyzed by aluminum (hydroxyl) oxides. *Applied Catalysis B: Environmental* **84**, 684–690.

Qiao, R.-P., Li, N., Qi, X.-H., Wang, Q.-S. & Zhuang, Y.-Y. 2005 Degradation of microcystin-RR by UV radiation in the presence of hydrogen peroxide. *Toxicon* **45**, 745–752.

Rastogi, A., Al-Abed, S. R. & Dionysiou, D. D. 2009 Sulfate radical-based ferrous–peroxymonosulfate oxidative system for PCBs degradation in aqueous and sediment systems. *Applied Catalysis B: Environmental* **85**, 171–179.

Salari, D., Daneshvar, N., Niaei, A., Aber, S. & Rasoulifard, M. 2008 The photo-oxidative destruction of CI Basic Yellow 2 using $UV/S_2O_8^{2-}$ process in an annular photoreactor. *Journal of Environmental Science and Health, Part A* **43**, 657–663.

Santos, L. H., Araújo, A. N., Fachini, A., Pena, A., Delerue-Matos, C. & Montenegro, M. 2010 Ecotoxicological aspects related to the presence of pharmaceuticals in the aquatic environment. *Journal of Hazardous Materials* **175**, 45–95.

Sarkar, D., Mohapatra, D., Ray, S., Bhattacharyya, S., Adak, S. & Mitra, N. 2007 Synthesis and characterization of sol–gel derived ZrO_2 doped Al_2O_3 nanopowder. *Ceramics International* **33**, 1275–1282.

Surpăţeanu, M. & Zaharia, C. 2004 Advanced oxidation processes for decolorization of aqueous solution containing acid Red G azo dye. *Open Chemistry* **2**, 573–588.

Vicente, F., Santos, A., Romero, A. & Rodriguez, S. 2011 Kinetic study of diuron oxidation and mineralization by persulphate: effects of temperature, oxidant concentration and iron dosage method. *Chemical Engineering Journal* **170**, 127–135.

Waldemer, R. H., Tratnyek, P. G., Johnson, R. L. & Nurmi, J. T. 2007 Oxidation of chlorinated ethenes by heat-activated persulfate: kinetics and products. *Environmental Science & Technology* **41**, 1010–1015.

Wang, J. & Wang, S. 2016 Removal of pharmaceuticals and personal care products (PPCPs) from wastewater: a review. *Journal of Environmental Management* **182**, 620–640.

Xu, X.-R., Li, H.-B., Wang, W.-H. & Gu, J.-D. 2004 Degradation of dyes in aqueous solutions by the Fenton process. *Chemosphere* **57**, 595–600.

Yang, W., Cicek, N. & Ilg, J. 2006 State-of-the-art of membrane bioreactors: worldwide research and commercial applications in North America. *Journal of Membrane Science* **270**, 201–211.

Yoon, S.-H., Jeong, S. & Lee, S. 2012 Oxidation of bisphenol A by $UV/S_2O_8^{2-}$: comparison with UV/H_2O_2. *Environmental Technology* **33**, 123–128.

Yu, F., Li, Y., Han, S. & Ma, J. 2016 Adsorptive removal of antibiotics from aqueous solution using carbon materials. *Chemosphere* **153**, 365–385.

First received 26 August 2018; accepted in revised form 8 March 2019. Available online 22 March 2019

Co_3O_4/TiO_2 hetero-structure for methyl orange dye degradation

Mahabubur Chowdhury, Sarah Kapinga, Franscious Cummings and Veruscha Fester

ABSTRACT

Advanced oxidation processes based on sulphate radical generated by peroxymonosulphate (PMS) activation is a promising area for environmental remediation. One of the biggest drawbacks of heterogeneous PMS activation is catalyst instability and metal ion leaching. In this study, a simple organic binder mediated route was explored to substitute Ti^{4+} ions into the Co_3O_4 host lattice structure to create a Co-O-Ti bond to minimise cobalt leaching during methyl orange degradation. The catalyst was characterised by X-ray diffraction, and scanning and transmission electron microscopy. The as-prepared catalysts with Co_3O_4:TiO_2 ratio of 70:30 exhibited minimal leaching (0.9 mg/L) compared to other ratios studied. However, the pristine Co_3O_4 exhibited highest catalytic activity (rate constant = 0.41 min^{-1}) and leaching (26.7 mg/L) compared to composite material (70:30 Co_3O_4:TiO_2). Interestingly, the morphology of the composite and leaching of Co^{2+} ions were found to be temperature dependent, as an optimum temperature ensured strong Co-O-Ti bond for prevention of Co^{2+} leaching. The classical quenching test was utilised to determine the presence and role of radical species on methyl orange degradation. The fabricated catalyst also exhibited good catalytic activity in degrading mixed dyes and good recyclability, making it a potential candidate for commercial application.

Key words | advanced oxidation processes (AOPs), Co_3O_4, heterogeneous reaction, leaching, PMS, TIO_2

Mahabubur Chowdhury (corresponding author)
Sarah Kapinga
Veruscha Fester
Flow Process and Rheology Centre,
Cape Peninsula University of Technology,
Cape Town 8000,
South Africa
E-mail: *chowdhurym@cput.ac.za*

Mahabubur Chowdhury
Sarah Kapinga
Department of Chemical Engineering,
Cape Peninsula University of Technology,
Cape Town 8000,
South Africa

Franscious Cummings
Electron Microscope Unit,
University of the Western Cape,
Bellville 7535,
South Africa

INTRODUCTION

The expansion of industries such as textiles, polycarbonates, nylon, cleansers, herbicides, resin paint, wood, paper and pulp, and pharmaceuticals has brought about significant water pollution issues due to the release of organic pollutants into the water body (Fleeger *et al.* 2003). A large portion of these organic toxins are carcinogenic; hence water treatment is of significant importance. Hydroxyl radical (˙OH) based advanced oxidation processes have received much consideration due to their high efficiency compared to other chemical, physical and biological processes for the removal of recalcitrant and harmful organic contaminants (Wei *et al.* 2015). Recently sulphate based radicals ($SO4^{\bullet-}$) have been shown to be a promising alternative over hydroxyl radicals in light of the fact that $SO4^{\bullet-}$ has a higher reduction potential (2.5–3.1 V) compared to ˙OH (1.8–2.7 V) (Yang *et al.* 2009). Sulphate radicals are generally generated by both homogeneous and heterogeneous activation of peroxymonosulphate (PMS). PMS activation via homogeneous routes is usually more efficient compared to heterogeneous catalysis (Miyamura *et al.* 2007). Various transition metals such as Co(II), Ru(III), Fe(II), Ce(III), Mn(II), Fe(III), and Ni(II) have been used to activate PMS for $SO4^{\bullet-}$ generation in homogeneous systems, with Co(II) the best activator for PMS (Anipsitakis & Dionysiou 2004). A Co^{2+} based PMS activation reaction follows the mechanism given in Equation (1) below (Chan & Chu 2009):

$$Co^{2+} + HSO_5^- \rightarrow Co^{3+} + SO_4^{\bullet-} + HO^- \quad (1)$$

It must be noted that Co^{2+} ions are toxic and hence Co^{2+} based heterogeneous catalysts, i.e. Co_3O_4, are of interest. Pristine Co_3O_4 and Co_3O_4 supported on various metal oxide supports have been studied as a heterogeneous catalyst for PMS activation for degradation of organic

doi: 10.2166/wst.2018.383

pollutants (Chen *et al.* 2008; Shukla *et al.* 2010a, 2010b; Zhang *et al.* 2010; Liang *et al.* 2012; Muhammad *et al.* 2012; Saputra *et al.* 2012; Saputra *et al.* 2013a; Chowdhury *et al.* 2015a). Incorporation of Co ions into the support structure to form a bond is the most important factor in Co^{2+} leaching during the PMS activation reaction (Yang *et al.* 2007; Zhu *et al.* 2013). Among various supports, titanium dioxide (TiO_2) exhibited stronger cobalt-support interaction (Yang *et al.* 2008). One-dimensional Co_3O_4 nanomaterials have attracted special interest due to their promising application in rechargeable lithium ion battery materials as anode material (Hui *et al.* 2008; Tan *et al.* 2016). With the surge in usage of mobile and handheld devices, large amounts of Co_3O_4 waste materials are to be produced in future. However, a wet impregnation method is mostly used in literature to synthesise TiO_2 supported Co_3O_4 material (Yang *et al.* 2007, 2008; Zhu *et al.* 2013). This technique might not be useful when utilising the waste Co_3O_4 from battery materials, to create TiO_2 supported Co_3O_4 material, for obvious reason.

In this study, we have used an organic binder mediated route to synthesise TiO_2/Co_3O_4 hetero-structure catalyst from separate Co3O4 and TiO_2 material instead of the wet impregnation method. This will allow an application to be found for waste Co_3O_4 produced from battery materials in future.

EXPERIMENTAL METHOD

Chemicals and materials

PMS – commercially known as oxone – methyl orange (MO), absolute ethanol (C_2H_6O; EtOH), tert-butyl alcohol ($C_4H_{10}O$, TBA), urea (CH_4N_2O), maleic acid ($C_4H_4O_4$), cobalt chloride ($CoCl_2$) and TiO_2 (anatase) were purchased from Sigma Aldrich South Africa. All reagents were of analytical grade and used without further purification.

Synthesis of $Co(OH)_2$

Cobalt hydroxide $Co(OH)_2$ was synthesised using a previously reported method (Wang *et al.* 2009). In a typical synthesis procedure, 18 g of $CoCl_2.6H_2O$ was dissolved in 300 mL of deionised water by stirring at room temperature. In another beaker 0.09 g of urea was dissolved in 300 mL water. The urea solution was then added dropwise to the $CoCl_2$ solution while stirring. The mixed solution was moved into a Teflon-lined autoclave. The autoclave was kept at 105 °C for 6 h and afterwards was permitted to cool at room temperature. The resulting pink-coloured precipitates were separated by centrifuging, followed by washing with distilled water and absolute ethanol for several times, followed by drying in an oven at 60 °C for 24 h. The role of urea was to release OH^- ions at elevated temperature and pressure in the reactor, which reacts with Co^{2+} to precipitate $Co(OH)_2$. Decomposition of urea produces NH_3 and CO_2 followed by the release of OH^- (Ibupoto *et al.* 2014).

Synthesis of TiO_2/Co_3O_4 hetero-structure material

An organic linker mediated route was used to create a chemically bonded interface between TiO_2 and Co_3O_4. The method of synthesis was adopted from our previous studies (Oghenochuko *et al.* 2015; Chowdhury *et al.* 2015b, 2016). Seven different ratios (wt%) of Co_3O_4/TiO_2 composite, namely 98:2, 95:5, 90:10, 85:15, 80:20, 70:30 and 60:40, were prepared. In a typical synthesis $Co(OH)_2$ and TiO_2 (anatase) were independently suspended in 30 mL of ethanol (labelled as suspension A and B respectively); then 10 mL of 0.1 mol/L maleic acid was added to suspension A. The suspensions were stirred at room temperature for around 5 h. Suspension A was then added to suspension B and the mixture was stirred for 24 h at room temperature. The role of maleic acid was to anchor the TiO_2 on the $Co(OH)_2$ surface through its dicarboxylic functional group (Kim *et al.* 2009). The resulting catalyst was collected by centrifuging and washing with distilled water and absolute ethanol for a few times and dried in a vacuum oven at 60 °C overnight. After the anchoring process the dried powder was annealed at elevated temperature of 600 °C in air for 4 h. Catalytic activity data reported in this study were obtained from the catalyst synthesised at 600 °C unless otherwise stated.

Evaluation of catalytic activity

The system was made out of a 250 mL jacketed glass reactor containing 200 mL of MO (40 mg/L). A specific amount of Co_3O_4/TiO_2 catalyst was dispersed via sonication in the MO solution. The solution was left under stirring for 30 min to reach adsorption equilibrium. A certain amount of oxone was added to the solution. A 2 mL sample of the solution was regularly withdrawn from the reactor. The withdrawn sample was immediately diluted with equivalent volume of EtOH to quench ongoing catalytic reaction. Afterwards the collected sample was centrifuged to remove any solid catalyst. The concentration of MO without catalyst was analyzed by UV–Vis spectrophotometry (Shimadzu UV-2550) at the maximum absorption band (486 nm).

Characterisation

The crystal structures of the synthesised products were determined using X-ray diffraction (XRD); a Phillips PW 3830/40 Generator with Cu-Kα radiation was used. The structure of the catalysts was studied using a Tecnai $G^2$20 field emission gun transmission electron microscope (TEM) operated at 200 kV in bright-field mode. A Fischione high angular annular dark-field detector (HAADF) was used to collect dark-field scanning transmission electron microscopy (STEM) micrographs of the catalysts. A Gatan GIF-2001 electron energy spectrometer was used to collect electron energy loss spectroscopy (EELS) data, whereas all selected area electron diffraction (SAED) patterns were obtained by using a 200 nm diameter aperture. Energy dispersive X-ray spectroscopy data were collected in STEM mode, using an EDAX liquid nitrogen cooled Si(Li) detector.

RESULTS AND DISCUSSION

Structural evaluation of the catalyst

Temperatures of 350 and 600 °C were used to evaluate the role of temperature in the formation of Co_3O_4/TiO_2 composite material. Figure 1(a) shows the bright-field TEM image of the TiO_2/Co_3O_4 sample synthesised at 600 °C. Unlike the case when synthesised at 350 °C (Figure S1, supplementary information, available with the online version of this paper), the TiO_2 and Co_3O_4 do not exist as separate materials; instead, as shown, the nanostructures form rods of individual nanoparticles. It can be postulated that at higher temperature individual particles diffuse together to form rod-like structures. A graphical illustration of the rod-like structure formation is presented in the supplementary information (Figure S2, available online). The high-resolution micrograph of Figure 1(b) shows a uniform lattice fringe size (d-spacing) for the individual nanoparticles constituting the nanowires. The energy dispersive X-ray spectroscopy (EDS) spectrum of Figure 1(c), however, shows the elemental presence of Ti, Co and O in the nanowires, suggesting that at 600 °C the TiO_2 exists as a possible dopant in the Co_3O_4 matrix or forms a TiO_2/Co_3O_4 'alloy' or hybrid material. The SAED pattern of Figure 1(d) suggests the latter, as the diffraction pattern once more exhibits both the cubic crystal structure of Co_3O_4 and the tetragonal crystal phase of anatase TiO_2. From the analysis of this pattern, the lattice constant of Co_3O_4 matrix is determined as $a =$ 0.852 nm, with the lattice constants of the tetragonal TiO_2 determined as $a = 0.292$ nm and $c = 0.923$ nm. It was observed that there is an increase in the lattice constant of the cubic Co_3O_4 matrix synthesised at 600 °C compared to the case at 350 °C ($a = 0.820$ nm), accompanied by a decrease in the a and c axes (0.346 and 0.943 nm, respectively) of the tetragonal TiO_2 lattice. This implies that during the synthesis process at 600 °C, the TiO_2 lattice experiences a compressive stress in all three lattice axes, thereby inducing a smaller unit cell volume, which ultimately promotes the diffusion process.

HAADF-STEM coupled with EDS analysis was used to further confirm the inter-diffusion of TiO_2 and Co_3O_4 (Figure 2). The STEM-EDS spectral image of Figure 2(b) was collected from the boxed area indicated in Figure 2(a); the spectral image was collected using image pixels of 20×20 nm^2, each collected for 50 ms. Figure 2(d)–2(f) show the Ti, Co and O maps extracted from Figure 2(b). From these maps it becomes evident that the formation of hybrid TiO_2/Co_3O_4 is complete at 600 °C unlike catalyst synthesised at 350 °C (Figure S3, supplementary information, available online). XRD patterns of the pristine Co_3O_4, TiO_2 and TiO_2 supported Co_3O_4 prepared at 600 °C is presented in Figure S4 (supplementary information, available online). XRD patterns of the pristine Co_3O_4 and TiO_2 match the JCPDF card no: 42-1467 and 86-1156 respectively. Both pristine TiO_2 and Co_3O_4 peaks could be observed in the XRD patterns of the composite structure. This complements the observation from EDS and SAED as discussed above.

Figure 3 shows the EELS spectra of the Co_3O_4-TiO_2 structures annealed at 350 and 600 °C in comparison to that of pristine Co_3O_4 and TiO_2 at 350 °C. Figure 3(a) compares the energy loss near edge fine structure (ELNEFS) of the Co $L_{3,2}$ edges of a pristine Co_3O_4 specimen to that of the hybrid TiO_2/Co_3O_4 materials synthesised at 350 and 600 °C. The Co_3O_4 spectrum shows two sharp peaks at 789 eV (L_3) and 804 eV (L_2), which corresponds to the Co $2p_{3/2}$ and Co $2p_{1/2}$ spin–orbit peaks of the Co_3O_4 spinel structure (Hagelin-Weaver *et al.* 2004). The L_3/L_2 ratio is widely considered a fingerprint method for determining the Co valence state. From Figure 5(a) an L_3/L_2 ratio of 1.59, 1.46 and 1.49 is calculated for Co_3O_4, TiO_2/Co_3O_4 @350 °C and TiO_2/Co_3O_4 @600 °C, respectively. From these values it can be deduced that the pristine Co_3O_4 exists in the Co^{3+} oxidation state, typical of the Co_3O_4 electronic structure (Wang *et al.* 2000). Upon incorporation of the TiO_2 into the Co_3O_4 matrix it can be seen that the L_3/L_2 ratio decreases for the structures synthesised at both 350 °C and 600 °C.

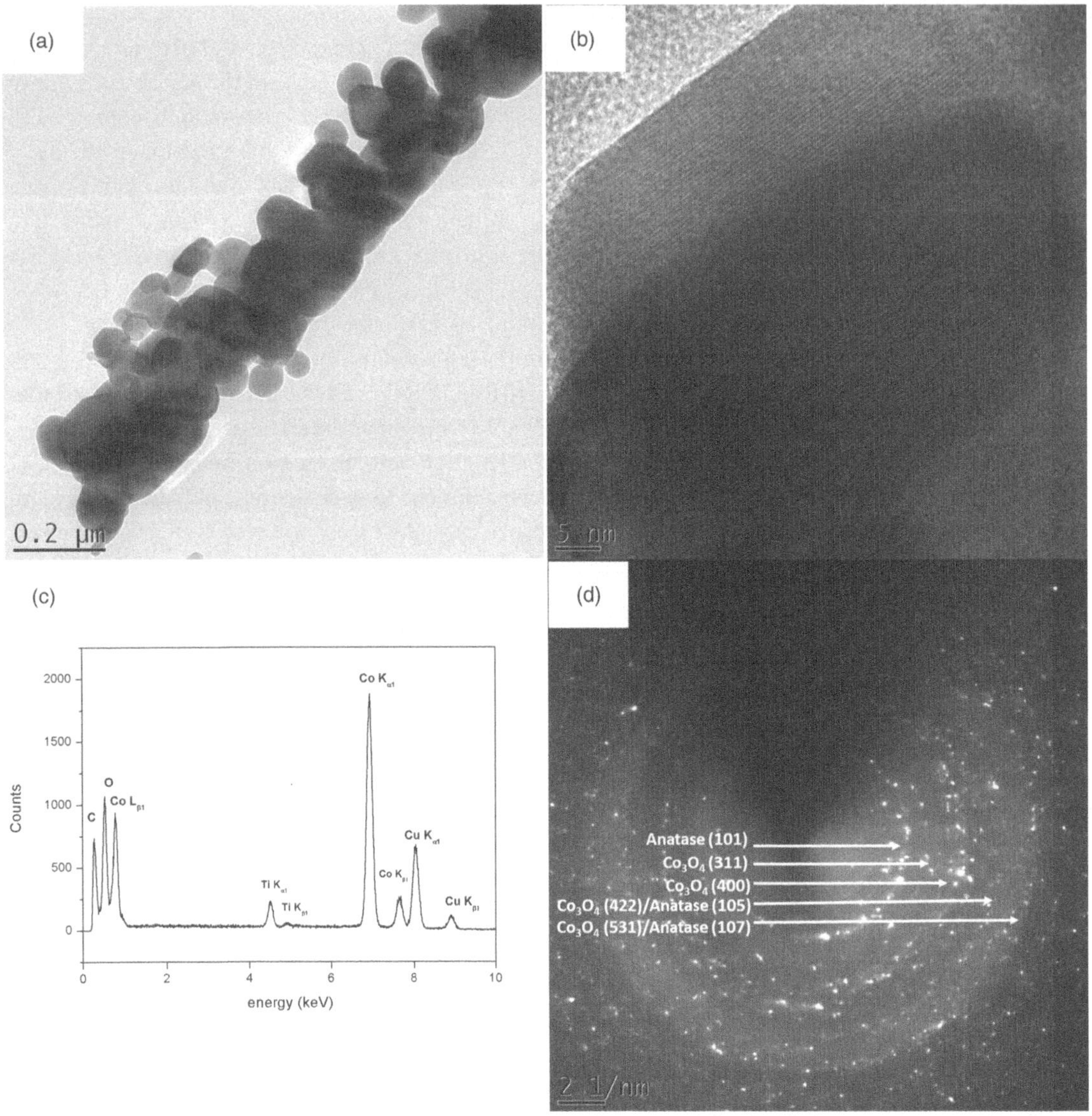

Figure 1 | (a) and (b) TEM bright field micrographs of sample, (c) EDS spectrum and (d) SAED pattern (sample prepared at 600 °C).

This suggest a re-coordination of the Co valency, caused by the TiO_2 diffusion into the Co_3O_4 matrix, resulting in the re-coordination of the Co valency due to the introduction of the Ti^{4+} ion. This successfully highlights the incorporation of TiO_2 into the Co_3O_4 lattice structure.

Figure 3(b) compares the ELNEFS of the Ti-$L_{3,2}$ edges of pure anatase TiO_2 to that of the TiO_2/Co_3O_4 synthesised at 350 and 600 °C. Unlike the case of the Co-$L_{3,2}$ ionisation edge, the L_3 and L_2 edges are separated by ~2.5 eV, with the L_2 edge more intense than L_3. At closer inspection it can be seen that both the L_2 and L_3 edges exhibit shoulders on the lower energy side of the peaks. This is typical of the anatase TiO_2 crystal structure caused by crystal-field splitting of the unoccupied, antibonding t_{2g} and e_g orbitals which contain mostly titanium d character (Brydson *et al.* 1989). From these line shapes it becomes obvious that the TiO_2 structure undergoes no discernible change in lattice at either 350 or 600 °C and remains anatase. From the SAED results of Figure 1(d) it was shown that the TiO_2 unit cell volume decreases, which makes it favourable for diffusion into the Co_3O_4 lattice. The oxygen K edge at 532 eV exhibits three peaks at approximately 546, 552 and 566 eV. The first peak at 538–546 eV is indicative of lattice oxides (Abu-Zied *et al.* 2015) and hence the reshaping of this peak indicates a re-coordination in the lattice of the oxides. From Figure 3(b) it can be seen that the oxygen K edge becomes more pronounced at 600 °C due to the presence of the TiO_2 in the Co_3O_4 lattice, unlike the case at 350 °C. This confirms,

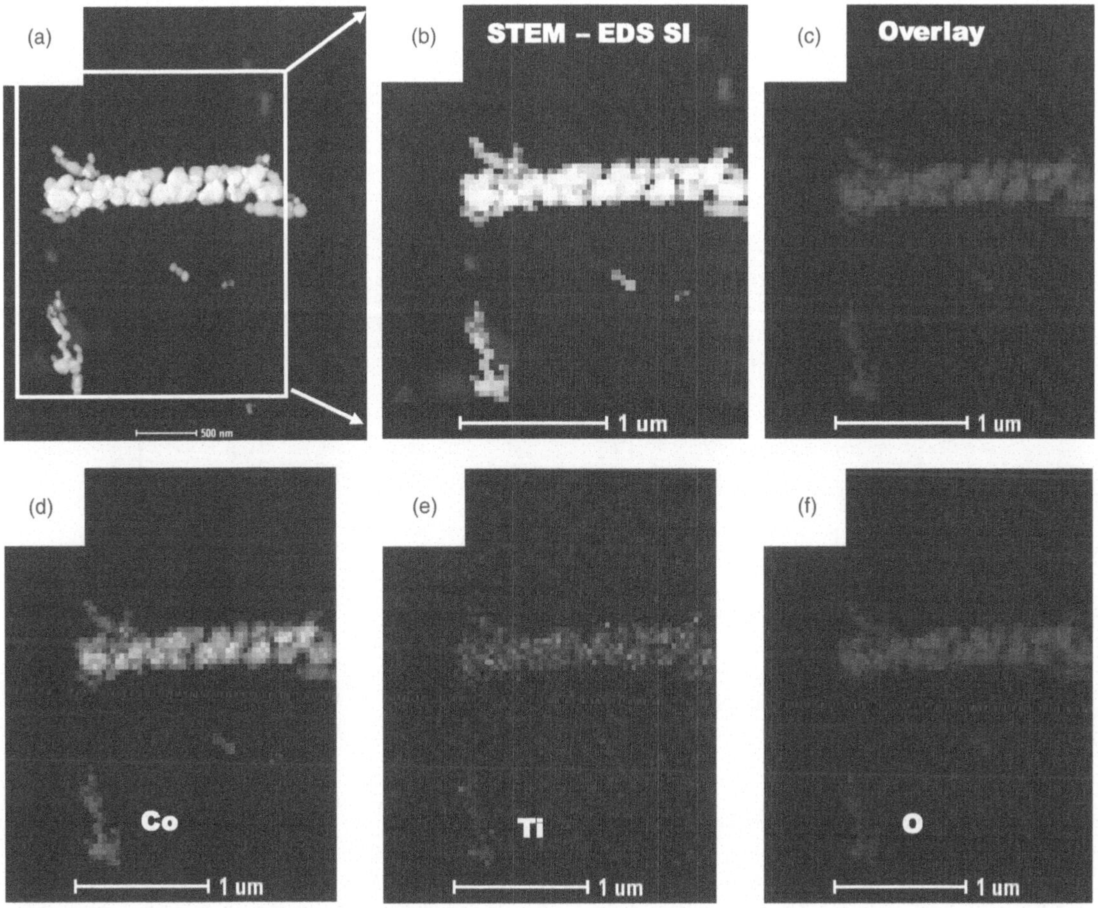

Figure 2 | (a) HAADF STEM micrograph of the composite Co_3O_4-TiO_2 material, (b) STEM-EDS spectral image of the area indicated in (a), (c) overlayed EDS map, (d)–(f) Co, Ti and oxygen maps extracted from the STEM-EDS spectral image of (b).

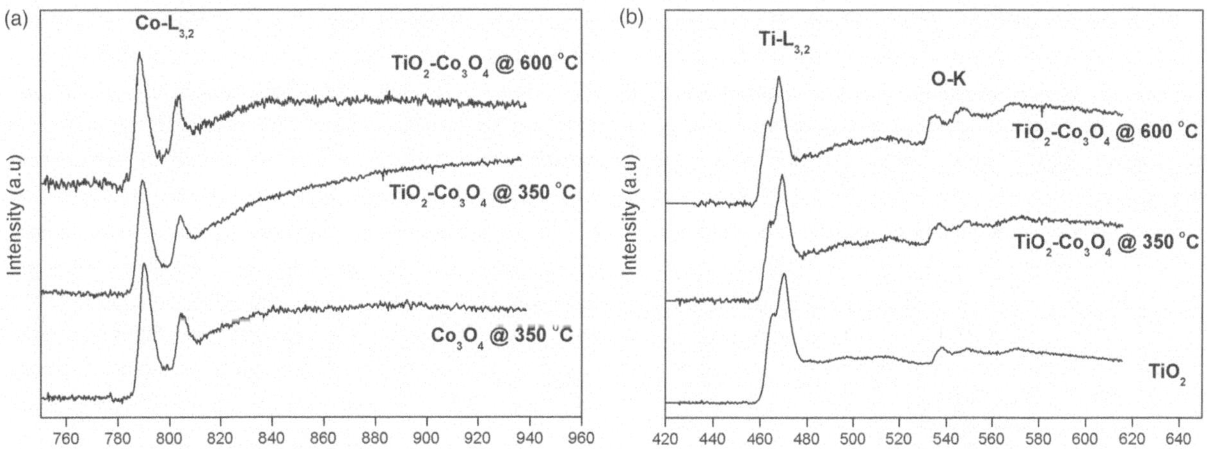

Figure 3 | Electron energy loss spectra comparing the energy loss near edge fine structures (ELNEFS) of the (a) Co-$L_{3,2}$ ionisation edge and (b) Ti-$L_{3,2}$ and O-K ionisation edges (ELNEFS) of the structures grown at 350 and 600 °C to that of pure Co_3O_4 and TiO_2 particles respectively.

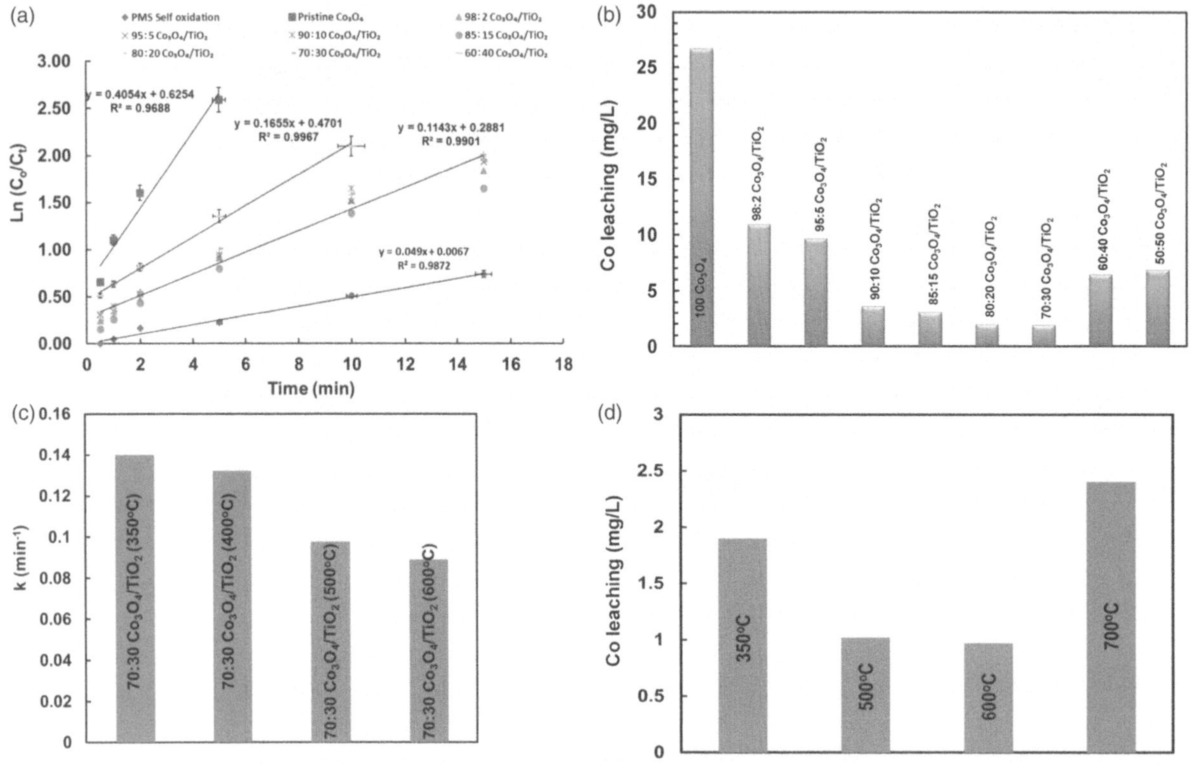

Figure 4 | Effect of Co_3O_4/TiO_2 ratio on (a) degradation kinetics of MO and (b) Co leaching; comparison of (c) degradation kinetics and (d) co-leaching for 70:30 Co_3O_4/TiO_2 catalyst synthesised at various temperatures (0.03 g/L catalyst load, 0.03 g/L oxone, 40 mg/L MO and 298 K)

once more, the analysis of the TEM and SAED results of Figures 1 and 2.

Catalytic activities and leaching of TiO_2 supported Co_3O_4 material

Figure 4(a) demonstrates the kinetic curves of heterogeneous PMS activation for the degradation of MO, while using pristine Co_3O_4 and various Co_3O_4 and TiO_2 ratios. Degradation of MO using PMS alone was also investigated to evaluate the role of PMS self-oxidation on the degradation of the dye. It can be seen that pristine Co_3O_4 exhibits the highest rate constant (k; 0.41 min^{-1}) while PMS on its own exhibits the lowest rate constant ($k = 0.049\ min^{-1}$). It is interesting to note that the rate constant for the composite materials, in the range of 98:2 Co_3O_4/TiO_2 to 70:30 Co_3O_4/TiO_2 remains almost similar ($k = {\sim}0.11\ min^{-1}$), however lower compared to pristine Co_3O_4. This suggests that by introducing the TiO_2 as a support the quantity of Co_3O_4 utilised could be diminished (cobalt leaching is undesirable and Co^{2+} is toxic) while the catalytic activity could be maintained. Further reduction of Co_3O_4 amount in the Co_3O_4/TiO_2 composite material (60:40 Co_3O_4/TiO_2) resulted in higher catalytic activity ($k = 0.17\ min^{-1}$) compared to other ratios studied in this study. Cobalt leaching is highlighted in Figure 4(b). It can be seen from Figure 4(b) that the cobalt leaching decreased with increasing TiO_2 amount in the composite material, up to a ratio of 70:30 Co_3O_4/TiO_2. A leaching amount of 1.9 and 26.7 mg/L was reported for 70:30 Co_3O_4/TiO_2 and pristine Co_3O_4. This highlights that by introducing TiO_2 as a support the amount of cobalt leaching can be reduced significantly. However, any further increase in TiO_2 amount in the composite structure increased the cobalt leaching. This is due to the fact that, when too much TiO_2 is used, a proper hetero-structure cannot form and leads to higher cobalt leaching. This results in a higher MO degradation rate constant as was observed in the case of 60:40 Co_3O_4/TiO_2 catalyst (Figure 4(b)). The 70:30 Co_3O_4/TiO_2 ratio was used as the optimum ratio as this catalyst exhibited minimum leaching. Removal of MO due to adsorption on the surface of Co_3O_4 and TiO_2 and interaction between TiO_2 and oxone was insignificant (Figure S5, supplementary information, available online; removal efficiency of MO in the presence of TiO_2 and oxone was lower than PMS self-oxidation). The effect of calcination temperature on the catalytic activity as well as on the cobalt leaching is shown in Figure 4(c) and 4(d). It can be seen that, with increasing calcination temperature catalytic activity decreases

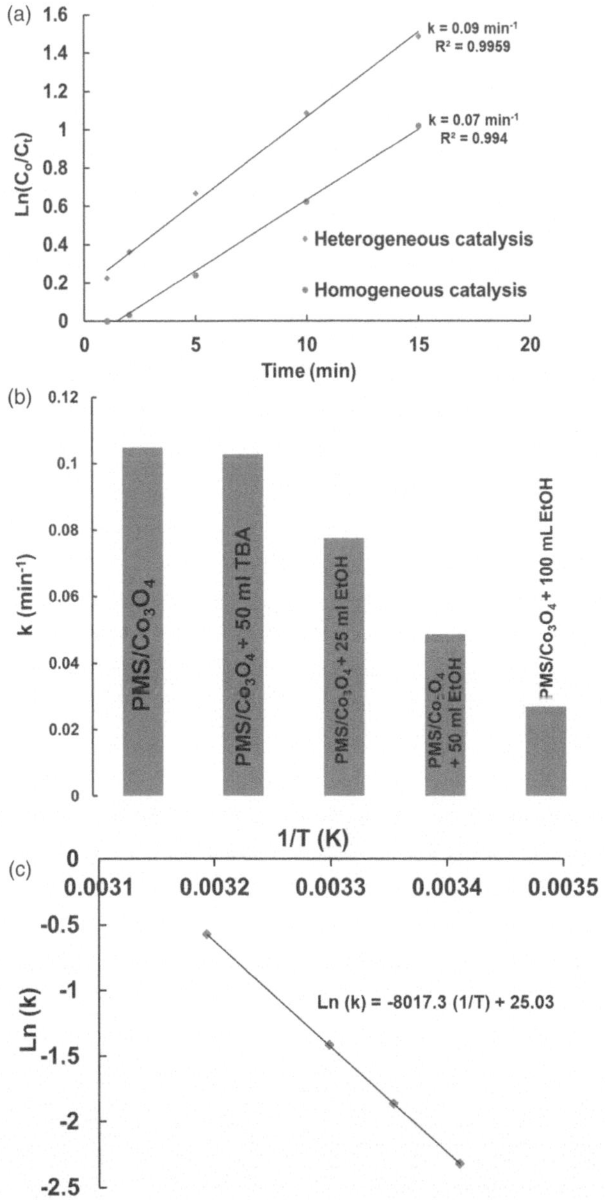

Figure 5 | (a) Comparison between homogeneous and heterogeneous PMS activation, (b) effect of quenching agent on MO degradation kinetic and (c) Arrhenius plot to determine the activation energy (0.03 g/L catalyst load, 0.03 g/L oxone, 40 mg/L MO and 298 K).

(Figure 4(c)), while the cobalt leaching first decreases with the increase of temperature (Figure 4(d)) in the range of 350–600 °C, but increases for catalyst calcined at 700 °C. This highlights a change in catalyst structure with temperature as was demonstrated earlier from the TEM analysis. Hence it can be postulated that there is a formation of a strong Co-O-Ti bond in the temperature range of 350–600 °C. This is due to the substitution of Ti^{4+} ions into Co_3O_4 host lattice as was previously confirmed from the ELNEFS study. Hence, a reduction of cobalt leaching was observed for that temperature range. However, cobalt oxide reduces to metallic cobalt at high temperature. It is well known that Co^0 can release Co^{2+} (Shi *et al.* 2015). Also, high calcination temperature might increase the Co_3O_4 grain size, thus reducing the contact area to the surface of the TiO_2 support. As a result, catalysts calcined at higher temperature are susceptible to leaching (Yang *et al.* 2009). Thus there was an increase of cobalt leaching when the catalyst was calcined at 700 °C. Catalyst (70:30 Co_3O_4/TiO_2) synthesised at 600 °C exhibited the lowest amount of leaching (0.9 mg/L) and was used further in this study unless otherwise stated.

Homogeneous experiments using dissolved cobalt concentrations (1.9 mg/L), higher than that leached (0.9 mg/L) from Co_3O_4/TiO_2 composite/hybrid, were performed to distinguish the role of homogeneous/heterogeneous catalysis in the degradation of MO. It can be seen from Figure 5(a) that the heterogeneous catalysis is more effective than the homogeneous catalysis. This implies that the Co_3O_4/TiO_2 catalyst calcined at 600 °C activated PMS via the heterogeneous route as a minimum of around 1 mg/L of dissolved Co is required to achieve the same extent of degradation via homogeneous PMS activation (Anipsitakis *et al.* 2006). The role of reactive species ($^{\bullet}OH$ and $SO_4^{\bullet -}$) in the degradation of MO was evaluated by a quenching test using EtOH and TBA as quenching agent. It has been established previously in literature that both $^{\bullet}OH$ and $SO_4^{\bullet -}$ could be generated from the PMS activation reaction (Anipsitakis & Dionysiou 2003; Saputra *et al.* 2013b). It can be seen from Figure 5(b) that in the presence of EtOH (various volumes added as can be seen from Figure 5(b)) notable decrease in reaction rate constant (k) occurred. Two different volumes of TBA (25 and 50 mL) were used to evaluate the role of $^{\bullet}OH$ in the MO degradation process. There was no significant decrease in rate constants in the presence of 25 and 50 mL TBA. However, a slight decrease in the reaction rate constant ($k = 0.08\ min^{-1}$) in the presence of 50 mL TBA compared to virgin solution ($k = 0.1\ min^{-1}$) was observed. This implies that $SO_4^{\bullet -}$ is the dominant reactive species that was generated by activating the PMS. The mechanism of PMS activation is given below (Wang *et al.* 2015):

$$Co^{2+} + HSO_5^- \rightarrow Co^{3+} + SO_4^{\bullet -} + HO^- \quad (2)$$

$$Co^{2+} + HSO_5^- \rightarrow Co^{3+} + SO_4^{2-} + {}^{\bullet}OH \quad (3)$$

$$^{\bullet}OH + SO_4^{\bullet -} + MO \rightarrow \text{several steps} \rightarrow CO_2 + H_2O + SO_4^{2-} \quad (4)$$

The effect of reaction temperature on the degradation kinetics of MO was investigated. Reaction rate constants increased significantly with increasing reaction temperature. Reaction rate constants of ~0.09, 0.16, 0.28 and 0.52 min^{-1} were obtained for temperatures of 20, 25, 30 and 40 °C respectively. This phenomenon is due to the rupture of the O-O bond and generation of $SO_4^{\bullet -}$ (Shi *et al.* 2012; Deng *et al.* 2017b). A higher reaction temperature also translates to a lower activation energy barrier. The activation energy (Ea), can be calculated with the classical Arrhenius equation as shown in Figure 5(c). The activation energy for 70:30 Co_3O_4/TiO_2 was found to be 66 kJ/mol. The Ea value is in the range (41–70 kJ/mol) of other reported Ea values for Co_3O_4 based systems (Saputra *et al.* 2013b; Deng *et al.* 2017a), and is higher than the activation energy required for diffusion-controlled reactions, which typically ranges from 10 to13 kJ/mol. This means that the MO removal was catalyst surface controlled.

Recyclability, effect of dye concentration and chemical oxygen demand removal efficiency of the catalyst

Recyclability or reusability of a heterogeneous catalyst is very important both from practical and economical points of view. It is always desirable to have a catalyst that can be reused without significant catalytic performance loss in industrial applications. The heterojunction catalyst was readily recycled just by centrifuging it from the solution and used in another batch for experiment without prior drying or calcination. The catalyst was recycled up to four times without any significant loss of catalytic activity as can be seen from Figure 6(a). In the first cycle 94% MO was removed after 30 min of reaction compared to 88% MO removal after 30 min in the fourth cycle. The effect of dye concentration on the catalytic activity of the catalyst was studied from 20 mg/L to 200 mg/L. It can be seen (Figure 6(b)) that the rate constant decreases with increases in initial dye concentration. Hence, it can be postulated that

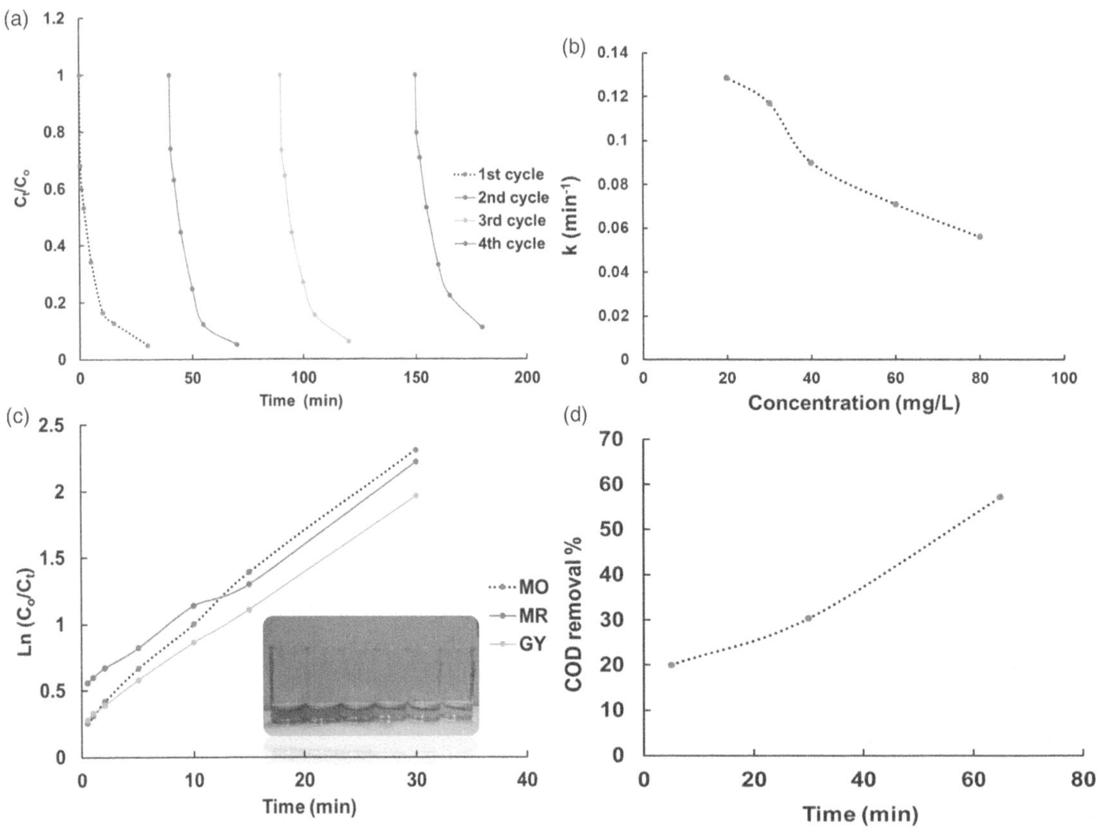

Figure 6 | (a) Recyclability of the as-prepared catalyst, (b) effect of initial dye concentration on MO degradation kinetics, (c) degradation kinetics of simulated dye bath water and (d) COD removal efficiency of the prepared catalyst.

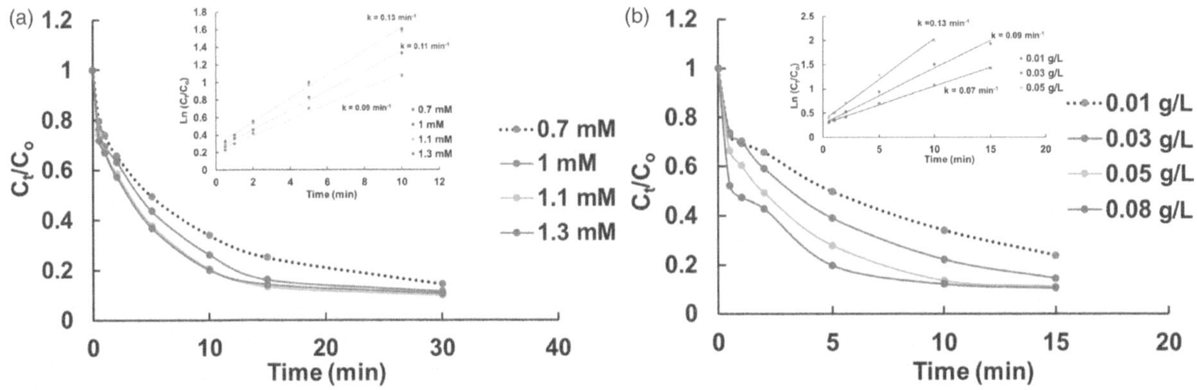

Figure 7 | Effect of PMS (a) and catalyst dosage (b) on the degradation of MO.

the dye concentration is a rate limiting step in the Co_3O_4/TiO_2-PMS oxidation reaction in this study. Synthetic dye bath effluent was prepared as an equal-amount mixture of three different dyes (60 mg/L total dye concentration), namely MO, golden yellow (GY) and methyl red (MR). This was done to evaluate the potential applications of the composite material in treating real industrial effluent. It can be seen from Figure 6(c) that the Co_3O_4/TiO_2 composite effectively decoloured the mixed dye. After 30 min of reaction a degradation of 95% was achieved for MO, 89% for MR and 86% for GY. Figure 6(c) shows the degradation kinetics of the mixed dye bath. Interestingly no significant difference in rate constants was observed for the three different dyes in the mixed bath. This implies that the catalyst has potential in degrading various dye types. The synthetic dye bath became colourless after 60 min of reaction (inset Figure 6(c)). From the environmental point of view, degradation of any organic pollutant, such as MO, is only the first step. It is well known that many intermediate compounds are formed during the degradation of MO. Chemical oxygen demand (COD) was measured to quantify the amount of organics present in the solution. It can be seen from Figure 6(d) that 57% of COD was removed after 60 minutes of catalytic reaction. This highlights the potential of the prepared material in practical wastewater treatment application.

Effect of PMS and catalyst dosage on MO degradation

Experiments were carried out to explore the effects of PMS and catalyst loading on the MO degradation in the oxidation of Co_3O_4/TiO_2. An increase in PMS loading from 0.7 to 1.1 mM and increase in catalyst loading from 0.01 to 0.05 g/L increased the MO degradation efficiency respectively (Figure 7(a) and 7(b)). However, a further increase in PMS and catalyst dosage to 1.3 mM and 0.08 g/L respectively did not enhance the MO degradation efficiency. When PMS and catalyst are present in excess dosage it might result in less availability of $SO_4^{\bullet-}$ due to the quenching mechanism. Huang and co-authors (Huang *et al.* 2014) showed, using EPR (electron paramagnetic resonance) technique, that when the PMS and catalyst are in excess it promotes the formation of nitroxide radicals and decreases the formation of $^{\bullet}OH$ and $SO4^{\bullet-}$ radicals. This results in lower degradation of the substrate.

CONCLUSION

This study investigated the heterogeneous PMS activation by TiO_2 supported Co_3O_4 catalyst. An organic binder mediated route was explored to create alloying/hybridisation between Co_3O_4 and TiO_2. Calcination temperature played an important role in controlling the catalyst morphology and Co-O-Ti bond. The pristine Co_3O_4 showed significantly higher removal efficiency of MO compared to TiO_2 supported Co_3O_4. However, TiO_2 supported Co_3O_4 resulted in significantly lower leaching than pristine Co_3O_4 material. This was due to the successful substitution of Ti^{4+} into the Co_3O_4 host lattice as was observed from the ELNEFS study, creating a strong Co-O-Ti bond. The 70:30 Co_3O_4/TiO_2 (calcined at 600 °C) exhibited the lowest amount of leaching. The as-prepared material also exhibited very good recyclability up to four cycles.

ACKNOWLEDGEMENTS

Mahabubur Chowdhury would like to thank the Cape Peninsula University of Technology for the research grant towards this project (URF 2017).

CONFLICTS OF INTEREST

The authors declare no conflicts of interest.

REFERENCES

Abu-Zied, B. M., Bawaked, S. M., Kosa, S. A. & Schwieger, W. 2015 Effect of microwave power on the thermal genesis of Co_3O_4 nanoparticles from cobalt oxalate micro-rods. *Applied Surface Science* **351**, 600–609.

Anipsitakis, G. P. & Dionysiou, D. D. 2003 Degradation of organic contaminants in water with sulfate radicals generated by the conjunction of peroxymonosulfate with cobalt. *Environmental Science & Technology* **37** (20), 4790–4797.

Anipsitakis, G. P. & Dionysiou, D. D. 2004 Radical generation by the interaction of transition metals with common oxidants. *Environmental Science & Technology* **38** (13), 3705–3712.

Anipsitakis, G. P., Dionysiou, D. D. & Gonzalez, M. A. 2006 Cobalt-mediated activation of peroxymonosulfate and sulfate radical attack on phenolic compounds. Implications of chloride ions. *Environmental Science & Technology* **40** (3), 1000–1007.

Brydson, R., Sauer, H., Engel, W., Thomass, J. M., Zeitler, E., Kosugi, N. & Kuroda, H. 1989 Electron energy loss and X-ray absorption spectroscopy of rutile and anatase: a test of structural sensitivity. *Journal of Physics: Condensed Matter* **1** (4), 797.

Chan, K. H. & Chu, W. 2009 Degradation of atrazine by cobalt-mediated activation of peroxymonosulfate: different cobalt counteranions in homogenous process and cobalt oxide catalysts in photolytic heterogeneous process. *Water Research* **43** (9), 2513–2521.

Chen, X., Chen, J., Qiao, X., Wang, D. & Cai, X. 2008 Performance of nano-Co_3O_4/peroxymonosulfate system: kinetics and mechanism study using Acid Orange 7 as a model compound. *Applied Catalysis B: Environmental* **80** (1–2), 116–121.

Chowdhury, M., Oputu, O., Kebede, M., Cummings, F., Cespedes, O., Maelsand, A. & Fester, V. 2015a Rapid and large-scale synthesis of Co_3O_4 octahedron particles with very high catalytic activity, good supercapacitance and unique magnetic properties. *RSC Advances* **5** (127), 104991–105002.

Chowdhury, M., Ntiribinyange, M., Nyamayaro, K. & Fester, V. 2015b Photocatalytic activities of ultra-small β-FeOOH and TiO_2 heterojunction structure under simulated solar irradiation. *Materials Research Bulletin* **68**, 133–141.

Chowdhury, M. R., Mpongwana, N., Cummings, F., Fester, V. & Ntwampe, S. 2016 β-FeOOH/TiO_2 Heterojunction for visible light driven photocatalytic inactivation of *E. coli*. In: *Photocatalysis* (W. Cao, ed.). Intech, The Netherlands.

Deng, J., Feng, S., Zhang, K., Li, J., Wang, H., Zhang, T. & Ma, X. 2017a Heterogeneous activation of peroxymonosulfate using ordered mesoporous Co_3O_4 for the degradation of chloramphenicol at neutral pH. *Chemical Engineering Journal* **308**, 505–515.

Deng, J., Ge, Y., Tan, C., Wang, H., Li, Q., Zhou, S. & Zhang, K. 2017b Degradation of ciprofloxacin using α-MnO_2 activated peroxymonosulfate process: effect of water constituents, degradation intermediates and toxicity evaluation. *Chemical Engineering Journal* **330**, 1390–1400.

Fleeger, J. W., Carman, K. R. & Nisbet, R. M. 2003 Indirect effects of contaminants in aquatic ecosystems. *Science of the Total Environment* **317** (1), 207–233.

Hagelin-Weaver, H. A. E., Hoflund, G. B., Minahan, D. M. & Salaita, G. N. 2004 Electron energy loss spectroscopic investigation of Co metal, CoO, and Co_3O_4 before and after Ar^+ bombardment. *Applied Surface Science* **235** (4), 420–448.

Huang, Z., Bao, H., Yao, Y., Lu, W. & Chen, W. 2014 Novel green activation processes and mechanism of peroxymonosulfate based on supported cobalt phthalocyanine catalyst. *Applied Catalysis B: Environmental* **154–155**, 36–43.

Hui, Z., Jianbo, W., Chuanxin, Z., Xiangyang, M., Ning, D., Jiangping, T. & Deren, Y. 2008 From cobalt nitrate carbonate hydroxide hydrate nanowires to porous Co_3O_4 nanorods for high performance lithium-ion battery electrodes. *Nanotechnology* **19** (3), 035711.

Ibupoto, Z. H., Elhag, S., AlSalhi, M. S., Nur, O. & Willander, M. 2014 Effect of urea on the morphology of Co_3O_4 nanostructures and their application for potentiometric glucose biosensor. *Electroanalysis* **26** (8), 1773–1781.

Kim, Y. J., Gao, B., Han, S. Y., Jung, M. H., Chakraborty, A. K., Ko, T., Lee, C. & Lee, W. I. 2009 Heterojunction of $FeTiO_3$ nanodisc and TiO_2 nanoparticle for a novel visible light photocatalyst. *Journal of Physical Chemistry C* **113** (5), 19179–19184.

Liang, H., Sun, H., Patel, A., Shukla, P., Zhu, Z. H. & Wang, S. 2012 Excellent performance of mesoporous Co_3O_4/MnO_2 nanoparticles in heterogeneous activation of peroxymonosulfate for phenol degradation in aqueous solutions. *Applied Catalysis B: Environmental* **127**, 330–335.

Miyamura, H., Matsubara, R., Miyazaki, Y. & Kobayashi, S. 2007 Aerobic oxidation of alcohols at room temperature and atmospheric conditions catalyzed by reusable gold nanoclusters stabilized by the benzene rings of polystyrene derivatives. *Angewandte Chemie* **119** (22), 4229–4232.

Muhammad, S., Saputra, E., Sun, H., Ang, H.-M., Tadé, M. O. & Wang, S. 2012 Heterogeneous catalytic oxidation of aqueous phenol on red mud-supported cobalt catalysts. *Industrial & Engineering Chemistry Research* **51** (47), 15351–15359.

Oghenochuko, O., Chowdhury, M., Nyamayaro, K. P. U. o., Cummings, F., Fester, V. & Fatoki, O. S. 2015 Novel β-FeOOH/NiO composite material as a potential catalyst for catalytic ozonation degradation of 4-chlorophenol. *RSC Advances* **5**, 59513–59521.

Saputra, E., Muhammad, S., Sun, H., Ang, H. M., Tadé, M. O. & Wang, S. 2012 Red mud and fly ash supported Co catalysts for phenol oxidation. *Catalysis Today* **190** (1), 68–72.

Saputra, E., Muhammad, S., Sun, H., Ang, H.-M., Tadé, M. O. & Wang, S. 2013a A comparative study of spinel structured Mn_3O_4, Co_3O_4 and Fe_3O_4 nanoparticles in catalytic oxidation of phenolic contaminants in aqueous solutions. *Journal of Colloid and Interface Science* **407**, 467–473.

Saputra, E., Muhammad, S., Sun, H., Ang, H.-M., Tadé, M. O. & Wang, S. 2013b Manganese oxides at different oxidation states for heterogeneous activation of peroxymonosulfate for phenol degradation in aqueous solutions. *Applied Catalysis B: Environmental* **142–143**, 729–735.

Shi, P., Su, R., Zhu, S., Zhu, M., Li, D. & Xu, S. 2012 Supported cobalt oxide on graphene oxide: highly efficient catalysts for the removal of Orange II from water. *Journal of Hazardous Materials* **229–230**, 331–339.

Shi, P., Wang, X., Zhou, X., Min, Y., Fan, J. & Yao, W. 2015 Influence of calcination temperature on the catalytic performance of Co_3O_4/GO nanocomposites for Orange II degradation. *RSC Advances* **5** (43), 34125–34133.

Shukla, P., Wang, S., Singh, K., Ang, H. M. & Tadé, M. O. 2010a Cobalt exchanged zeolites for heterogeneous catalytic oxidation of phenol in the presence of peroxymonosulphate. *Applied Catalysis B: Environmental* **99** (1–2), 163–169.

Shukla, P. R., Wang, S., Sun, H., Ang, H. M. & Tadé, M. 2010b Activated carbon supported cobalt catalysts for advanced oxidation of organic contaminants in aqueous solution. *Applied Catalysis B: Environmental* **100** (3–4), 529–534.

Tan, Y., Gao, Q., Li, Z., Tian, W., Qian, W., Yang, C. & Zhang, H. 2016 Unique 1D Co_3O_4 crystallized nanofibers with (220) oriented facets as high-performance lithium ion battery anode material. *Scientific Reports* **6**, 26460.

Wang, Z. L., Yin, J. S. & Jiang, Y. D. 2000 EELS analysis of cation valence states and oxygen vacancies in magnetic oxides. *Micron* **31** (5), 571–580.

Wang, G., Shen, X., Horvat, J., Wang, B., Liu, H., Wexler, D. & Yao, J. 2009 Hydrothermal synthesis and optical, magnetic, and supercapacitance properties of nanoporous cobalt oxide nanorods. *The Journal of Physical Chemistry C* **113** (11), 4357–4361.

Wang, Y., Zhou, L., Duan, X., Sun, H., Tin, E. L., Jin, W. & Wang, S. 2015 Photochemical degradation of phenol solutions on Co_3O_4 nanorods with sulfate radicals. *Catalysis Today* **258**, 576–584.

Wei, G., Liang, X., He, Z., Liao, Y., Xie, Z., Liu, P., Ji, S., He, H., Li, D. & Zhang, J. 2015 Heterogeneous activation of oxone by substituted magnetites $Fe_{3-x}M_xO_4$ (Cr, Mn, Co, Ni) for degradation of Acid Orange II at neutral pH. *Journal of Molecular Catalysis A: Chemical* **398** (Supplement C), 86–94.

Yang, Q., Choi, H. & Dionysiou, D. D. 2007 Nanocrystalline cobalt oxide immobilized on titanium dioxide nanoparticles for the heterogeneous activation of peroxymonosulfate. *Applied Catalysis B: Environmental* **74** (1–2), 170–178.

Yang, Q., Choi, H., Chen, Y. & Dionysiou, D. D. 2008 Heterogeneous activation of peroxymonosulfate by supported cobalt catalysts for the degradation of 2,4-dichlorophenol in water: the effect of support, cobalt precursor, and UV radiation. *Applied Catalysis B: Environmental* **77** (3), 300–307.

Yang, Q., Choi, H., Al-Abed, S. R. & Dionysiou, D. D. 2009 Iron–cobalt mixed oxide nanocatalysts: heterogeneous peroxymonosulfate activation, cobalt leaching, and ferromagnetic properties for environmental applications. *Applied Catalysis B: Environmental* **88** (3–4), 462–469.

Zhang, W., Tay, H. L., Lim, S. S., Wang, Y., Zhong, Z. & Xu, R. 2010 Supported cobalt oxide on MgO: highly efficient catalysts for degradation of organic dyes in dilute solutions. *Applied Catalysis B: Environmental* **95** (1–2), 93–99.

Zhu, Y., Chen, S., Quan, X. & Zhang, Y. 2013 Cobalt implanted TiO_2 nanocatalyst for heterogeneous activation of peroxymonosulfate. *RSC Advances* **3** (2), 520–525.

First received 27 April 2018; accepted in revised form 20 August 2018. Available online 30 August 2018

TiO_2 coated on PVDF membrane for dye wastewater treatment by a photocatalytic membrane

L. Penboon, A. Khrueakham and S. Sairiam

ABSTRACT

This work investigated the membrane modification to increase hydrophilicity aiming for application in a photocatalytic membrane. The modified poly(vinylidene fluoride) (PVDF) flat sheet membrane was coated with titanium dioxide (TiO_2). The chemical and physical changes of membranes were characterized by contact angle, scanning electron microscopy and porosity. The results found that TiO_2 successfully modified the PVDF membrane by decreasing the contact angle from 127.1° to 114.3° on increasing the TiO_2 concentration from 0.1 to 1.0 g/L. The contact angles were decreased with the increase of coating time. The water permeation after PVDF membrane modification decreased compared with the original membrane, whereas water fluxes increased with the increase of TiO_2 concentration. TiO_2 particles were observed on the membrane surface while the porosity of the membrane coated with TiO_2 decreased. The optimum condition of membrane coating was obtained at 1.0 g/L of TiO_2 and 60 min of coating time. The decolorization of RB 19 by photolysis and photolysis followed by filtration was less than 10%; in contrast, the decolorization of RB 19 by photocatalytic oxidation was as good as a photocatalytic membrane for 60 min. The TiO_2 photocatalytic membrane is a feasible technology for RB 19 decolorization in wastewater.

Key words | dye, membrane modification, photocatalytic membrane, PVDF, wastewater treatment

L. Penboon
S. Sairiam (corresponding author)
Department of Environmental Science, Faculty of Science,
Chulalongkorn University,
Bangkok 10330,
Thailand
E-mail: *sermpong.s@chula.ac.th*

A. Khrueakham
Division of Research Administration and Academic Service,
Kasetsart University Chalermphrakiat Sakon Nakhon Province Campus,
Sakon Nakhon 47000,
Thailand

S. Sairiam
Center of Excellence on Hazardous Substance Management (HSM),
Chulalongkorn University,
Bangkok 10330,
Thailand

INTRODUCTION

The high strength of colored effluents from textile wastewater may become visual eyesores and causes aesthetic pollution, eutrophication and perturbations in aquatic life. In addition, color-containing dyestuffs have been found to be toxic and carcinogenic to aquatic environments (Allah *et al.* 2009). Membrane technology has attracted research work in the last few decades. Membranes offer certain advantages, such as low energy, chemical consumption and steady operation, and they are widely used for separation applications (Iglesias *et al.* 2016). Poly(vinylidene fluoride) (PVDF) membranes are extensively used in industries for application due to their excellent chemical and thermal stability, and resistance to radiation. The major drawback of membrane application is the flux decline, leading to a shortened membrane life (Méricq *et al.* 2015). To address this challenge, membrane modification by TiO_2 has suggested an alternative method to improve the membrane performance by increasing the hydrophilicity due to the intrinsic hydrophobic property of the PVDF membrane. TiO_2 received attention due to its high catalytic activity, stability and low cost. The PVDF membrane with TiO_2 needed to be resistant toward the radicals produced by TiO_2-induced photocatalysis. The generation of hydroxyl radicals (•OH) by UV light can cause damage to organic compounds (Suresh *et al.* 2015). Unlike the conventional methods of using particulates of TiO_2, the separation step is required to remove TiO_2 particles after wastewater treatment, as it is difficult to remove them from the treated water (Aguedach *et al.* 2005). Therefore it is allowed to combine the photocatalytic mechanism for organic degradation with the filtration process into a single reactor (Zhang *et al.* 2014). The PVDF membrane could be functionalized with TiO_2 but the strong binding of TiO_2 on the membrane surface by a coating that is needed to avoid the release of TiO_2 leads to a decrease of reactivity of TiO_2. TiO_2 is used for the hydrophilic modification of the membrane because of its antibacterial, self-cleaning and hydrophilic properties on exposure to UV irradiation (Cruz *et al.* 2014). TiO_2 is inexpensive; therefore, membranes are expected to be cost-effective. This work is aiming for a feasible approach for

doi: 10.2166/wst.2019.023

the modification of PVDF membrane with TiO_2 by a dipping-coating method and application of the modified PVDF membrane for dye wastewater treatment. Reactive dyes have reactive groups in their structure and thus have the ability to bind covalently to organic fibers. In addition, Reactive Blue 19 (RB 19) is easily and commonly used in the textile industry (Taheria *et al.* 2013; Khan *et al.* 2015).

MATERIALS AND METHODS

Materials

PVDF flat sheet membrane was purchased from Durapore®. The specifications of the membrane were reported by the manufacturer as listed in Table 1. A commercially available titanium dioxide photocatalyst with the crystalline structure of anatase was supplied from Brenntag Ingredients (Thailand) Public Co., Ltd. The particles were analyzed with a laser particle sizer (Malvern Mastersizer 3000, UK). Particle sizes from 0.188 to 14.5 μm, with an average of 1.62 μm (D [4,3]) were obtained and the zeta potential was +1.30 mV by zeta sizer (Malvern Model ZSP, UK). The specific surface area of 7.39 m^2/g was obtained. RB 19 was kindly supplied from Dystar Thai Ltd. The properties of RB 19 are presented in Table 2. Hydrochloric acid (HCl) and sodium hydroxide (NaOH) of analytical grade were purchased from Merck. Deionized water (DI) was used for preparing all solutions.

Experimental setup

The feed solution was supplied using a peristaltic pump (Masterflex, Model 77201-62) through the pressure gauge before entering the membrane module. The permeability of membranes was determined by using the membrane module, which was a rectangular cross-flow type that was equipped with a rectangular flat sheet membrane with an effective area of 59.5 cm^2. The permeate water was collected to measure the pure water flux by using expression (1).

$$J_w = \frac{V}{A \times t} \quad (1)$$

where J is pure water flux (L/m^2 h), V is permeate volume (L), A is the effective membrane area (m^2), t is filtration time (h).

For the microfiltration experiments, the permeate was collected and returned to the feed tank. The feed flow rate was fixed at 0.850 L/min and applied pressure was 1.5 bar for all experiments.

Table 1 | Specification of the PVDF flat sheet membrane

Membrane pore size (μm)	0.22
Membrane porosity (%)	75
Contact angle (°)	128
Thickness (μm)	125
Surface area (cm^2)	59.5

Table 2 | Chemical structure and properties of RB 19 (Banaei *et al.* 2017)

Dye	Reactive Blue 19 (RB 19)
Chemical formula	$C_{22}H_{16}N_2Na_2O_{11}S_3$
Commercial name	Remazol Brilliant Blue R
Color index no.	61200
Class	Anthraquinone
Molecular weight (g/mol)	626.53
λ_{max} (nm)	592
Molecular structure	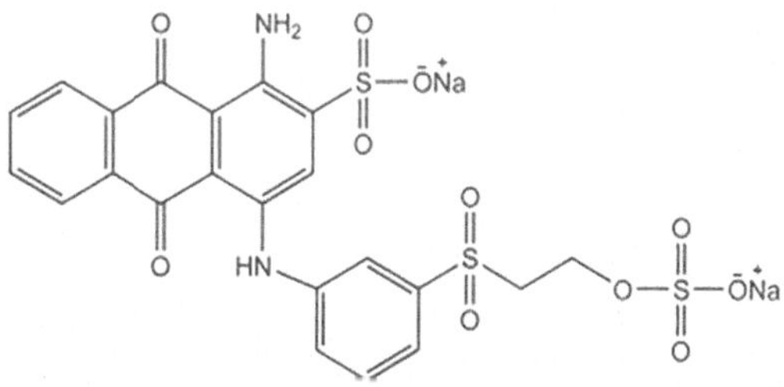

Membrane modification by TiO_2 coating

Membrane modification

PVDF flat sheet membranes were submerged in the TiO_2 solution (0.1–1.0 g/L) under magnetic stirrer for 5–60 min and then TiO_2 was attached on the membrane surface. The modified membranes were rinsed with DI water. Then, the modified membranes were dried and kept at room temperature before physical and chemical characterizations.

Membrane characterization

Membrane hydrophilicity was studied by measuring the contact angle with a goniometer (Model 100-00-220,

Ramehart, USA) of the membrane at 25 °C. The DI water was dropped on the membrane sample surface at five different spots, and the average of the measured contact angles was reported as the water contact angle. The morphology and chemical composition at the top surface of the membranes was observed by a scanning electron microscope and energy dispersive X-ray spectrophotometer (SEM-EDS, JSM-6610LV (JEOL)). All samples were dried at room temperature and were sputtered with gold. The porosity of the membranes was measured by a gravimetric method (Li *et al.* 2009) as the following expression:

$$\varepsilon = \frac{m_1 - m_2}{\rho_w \times A \times l} \tag{2}$$

where m_1 is the weight of the wet membrane, m_2 is the weight of the dry membrane, ρ_W is the water density (0.998 g cm^{-3}), A is the effective area of the membrane (m^2), l is the membrane thickness (m).

Decolorization of RB 19 by photolysis and photocatalytic oxidation

Photocatalytic oxidation for RB 19 decolorization with different TiO_2 concentrations (0.1–1.0 g/L) was studied at initial RB 19 concentration of 25 mg/L. The suspension was subjected to irradiation under UV light (UV-C lamp, 9 watts). The reactor used was a 1-L Pyrex beaker and the aqueous suspension was magnetically stirred for 30 min in the dark to ensure full equilibration of adsorption/desorption of the dye molecule on the catalyst surface. The photocatalytic activities were performed at room temperature. The pH of the suspension was controlled by using a pH meter (pH UB-10, Denver, USA) and adjusted to the desired pH by HCl (37%, Qrec) and NaOH (98%, Carlo). After that, the samples were collected at different times and were centrifuged at 5,000 rpm for 5 min. Then, samples were filtered by syringe filter 0.22 μm to remove the catalyst particles completely. The concentrations of RB 19 were determined by a spectrophotometer (Spectro SC, Labomed Inc., USA) at 592 nm. The decolorization efficiency is defined by the following expression:

$$\text{Removal efficiency (\%)} = \left(1 - \frac{C}{C_0}\right) \times 100 \tag{3}$$

where C_0 and C are the concentrations at time 0 and t, respectively.

For the photolysis of RB 19 decolorization, the RB 19 wastewater was studied as the same procedure with photocatalytic oxidation without the addition of TiO_2 particles.

Decolorization of Reactive Blue 19 dye wastewater by photocatalytic PVDF membrane using modified PVDF membrane coated with TiO_2

The modified PVDF membranes were applied in the photocatalytic membrane process for RB 19 dye wastewater treatment. The membrane used was 17 cm in length and 3.5 cm in width and was fixed tightly into the membrane module. The permeability of membranes was conducted with a dead-end filtration process. The UV-C light (9 Watts) was installed above the membrane module. The RB 19 wastewater was fed into the module by a peristaltic pump with a flow rate of 0.877 L/min and the pressure of 1.5 bar for 60 min and returned to the feed tank. The permeate was collected at different time intervals. The removal efficiency of RB 19 was determined by a spectrophotometer and is defined by expression (3).

RESULTS AND DISCUSSION

Characterizations of modified PVDF membrane

Contact angle

The contact angle of the modified PVDF membrane decreased with increased concentration of TiO_2. The contact angle of the original membrane was 128.1°. For the membrane modification, increasing the concentration of TiO_2 from 0.1 g/L to 1.0 g/L, the contact angle of membranes decreased from 127.3° to 114.3°, respectively, as shown in Figure 1(a). It was observed that increasing the coating time of TiO_2, the contact angles of the modified membrane had a lower contact angle than the original PVDF membrane, as shown in Figure 1(b). This was because the TiO_2 formed a hydrophilic layer on the top of the PVDF membrane surface.

Morphology

Figure 2 displays the morphology observations of the top surface of the original and modified PVDF flat sheet membranes. The results found that the top surface of the original PVDF membrane had no TiO_2 particles, as shown in Figure 2(a), while the modified PVDF

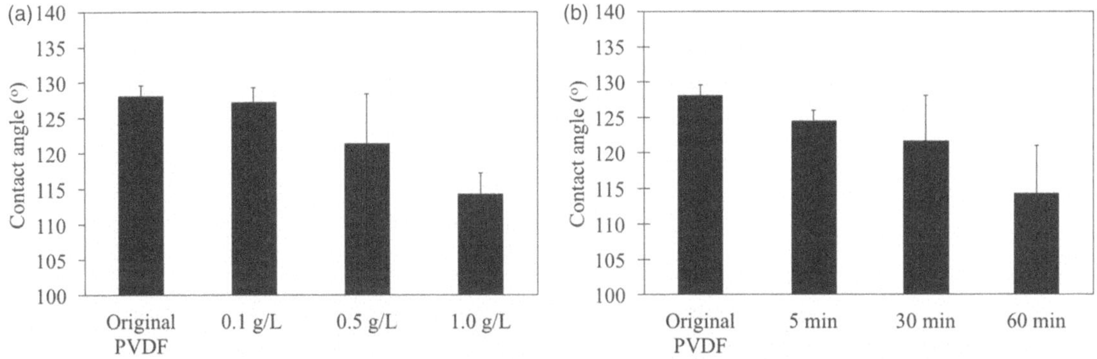

Figure 1 | Contact angle of original and modified PVDF membrane: (a) effect of TiO_2 concentration for 60 min; (b) effect of coating time under 1.0 g/L TiO_2.

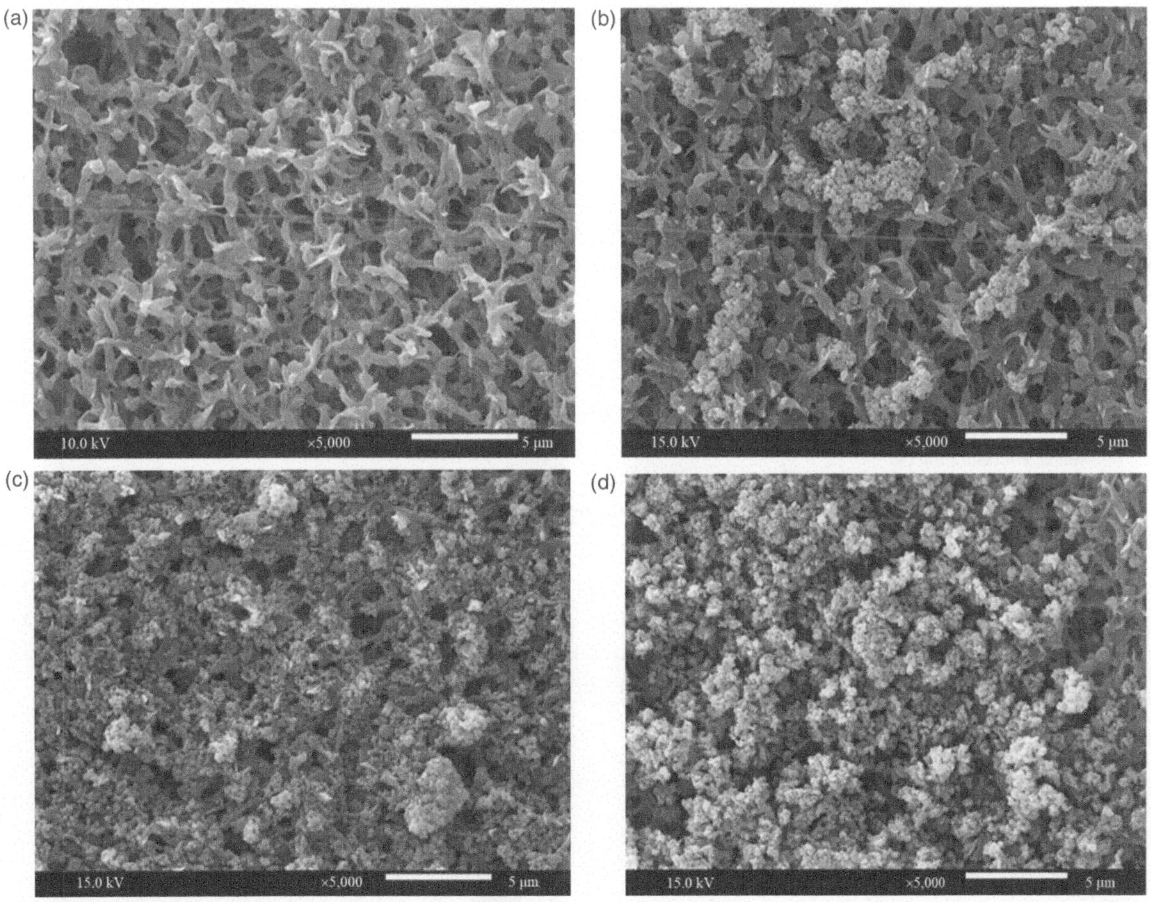

Figure 2 | SEM images of the top surface morphology of the PVDF membranes with 1 g/L TiO_2 concentration at different times: (a) PVDF membrane, (b) 5 min, (c) 30 min and (d) 60 min (magnification of 5,000×).

membranes with TiO_2 presented the deposition of TiO_2 particles on the membrane surface as presented in Figure 2(b)–2(e). In addition, TiO_2 particles increased on the modified membrane surface with an increased duration time of coating TiO_2. They were distributed as random groups that formed agglomerate clusters on the surface and in the pores of the PVDF membrane. In addition, the elemental analysis of TiO_2 particles on the surface membrane confirmed the successful TiO_2 coating by energy dispersive X-ray microanalysis (EDX) and the presence of three peaks relevant of TiO_2 at 0.2, 4.2 and 4.5 keV were found on the top of modified PVDF membranes. The titanium (Ti) content on the surface of the PVDF membrane coated with TiO_2 increased compared with that of the original membrane as presented in Table 3.

Porosity

The porosities of the original and modified PVDF membranes are shown in Figure 3. The PVDF membranes were modified by TiO_2 with different TiO_2 concentrations (0.1–1.0 g/L) and duration times (5–60 min). There were fluctuations in the results; however, it may be concluded that the changes in porosity after the TiO_2 coating were not significant. Overall, the porosities of TiO_2-coated PVDF were decreased. In addition, the porosities of modified PVDF membranes decreased with the addition of TiO_2 compared with the original PVDF membrane. The porosity of the original PVDF was 0.698 while the porosities of the modified membrane with 0.1 g/L TiO_2 were 0.626, 0.641, 0.571 and 0.340 with increased duration time from 5, 15, 30 and 60 min. Moreover, the porosity of the modified membrane with 0.5 g/L TiO_2 was 0.463, 0.352, 0.323 and 0.387, which was less than those of the original PVDF and the modified membrane with 1.0 g/L TiO_2 was 0.462, 0.771, 0.502 and 0.398. Furthermore, the TiO_2 particles were possibly aggregated with both the addition of TiO_2 and increased TiO_2 concentration that could block the membrane pores, subsequently reducing the membrane pores. The porosity of the membrane decreased due to the deposition of nanoparticles in the membrane (Fischer *et al.* 2015).

Table 3 | Change of chemical structure of membrane surface

	Atomic concentration (%)			
Surface	**C**	**F**	**O**	**Ti**
Original PVDF	60.32	37.76	2.42	–
1 g/L TiO_2 coated on PVDF				
15 min	59.81	38.54	1.33	0.32
30 min	60.87	36.48	2.08	0.58
60 min	53.80	36.62	7.13	2.54

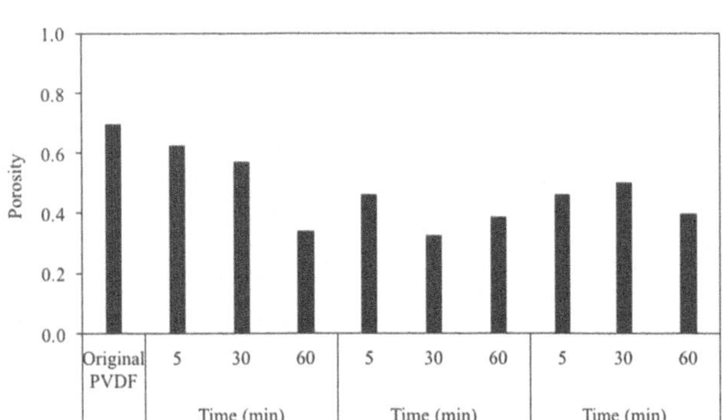

Figure 3 | The porosity of the original PVDF membrane and TiO_2 coated on PVDF membranes.

Water flux

The results of water permeation flux conducted with microfiltration are presented in Figure 4. The membranes were filtered in water under 1.5 bar pressure for 15 min to make the permeate reach a steady flux. The results revealed that the water flux of all modified membranes significantly decreased compared with the original PVDF membrane (165.63 L/m^2 h). The water fluxes of the modified membrane coated with 0.1, 0.5 and 1.0 g/L of TiO_2 for 60 min were 64.86, 104.55 and 143.24 L/m^2 h, respectively. The improvement of water flux may be due to the TiO_2, which made the membrane more hydrophilic, and enhanced the water molecules to pass through the membrane. According to the results of the contact angle measurements in Figure 1, the water contact angles of the membranes decreased with increased concentration of TiO_2 and coating time due to the hydrophilic groups on its surface. It was mentioned that the water flux decreased compared with the original PVDF membrane due to the decrease of porosity. Peng *et al.* (2018) reported that the membrane structure has an important role in determining the water permeability performance and membrane hydrophilicity.

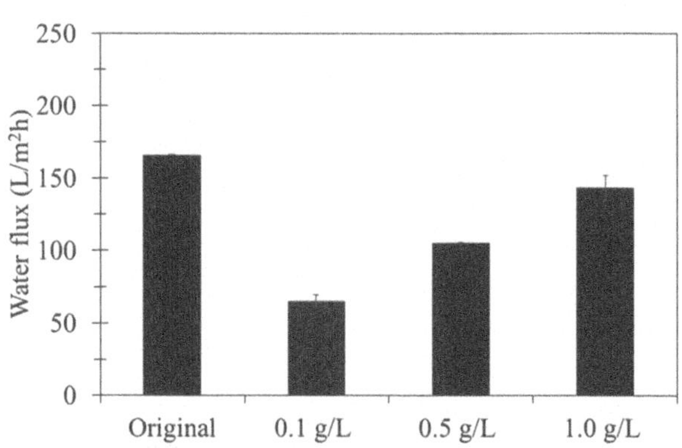

Figure 4 | Water flux of original and modified PVDF membranes coated with TiO_2 for 60 min.

Parameters affecting the decolorization of RB 19 wastewater by photocatalytic oxidation

Effect of initial pH

In this study, the decolorization of RB 19 was studied at different initial pH of RB 19 wastewater which was at pH 3, 5, 11 for 0.1 g/L of TiO_2 and the initial dye concentration of 25 mg/L at room temperature for 120 min with 9 W of UV light. The results found that the decolorization of RB 19 at pH 3 was higher than that at pH 5 and pH 11 by the photocatalytic process as illustrated in Figure 5(a). The decolorization of RB 19 increased from 46.6% to 57.9% when pH decreased from 5 to 3. This was because a pH change would affect the surface-charge properties of the TiO_2, leading to the absorption behavior of RB 19. The pH_{pzc} (point of zero charge) for TiO_2 is 6.3. In acidic condition (pH < 6.3), the TiO_2 surface remained positively charged and could adsorb the negatively charged (reaction (4)) RB 19 facilitating the degradation, resulting in the high photocatalytic efficiency obtained. Meanwhile, the decolorization of RB 19 was 48.2% at pH 11. In contrast, the decolorization of RB 19 decreased in alkaline condition (pH > 6.3) because the TiO_2 surface and dye anion were negatively charged (reaction (5)), leading to a reduction in photocatalytic efficiency of decolorization (Gaya *et al.* 2008). Therefore, initial pH of RB 19 wastewater, which is pH 3, was selected to study further.

$$\text{When pH} < \text{pzc: TiOH} + H^+ \rightarrow TiOH_2^+ \tag{4}$$

$$\text{When pH} > \text{pzc: TiOH} + OH^- \rightarrow TiO^- + H_2O \tag{5}$$

Effect of initial catalyst dosage

In this study, initial TiO_2 was investigated from 0.1 to 1 g/L with an initial dye concentration of 25 mg/L at pH 3 and room temperature for 120 min. The decolorization efficiency of RB 19 is illustrated in Figure 5(b). The results showed that the decolorization efficiency of RB 19 increased from 57.9% to 79.5% when the photocatalyst concentration increased from 0.1 to 0.5 g/L. The decolorization efficiency slightly increased from 79.5% to 86.1% after an increase in the catalyst concentration from 0.5 to 1.0 g/L. This was because the catalyst loading increased; as a result, there was increasing of the number of active sites and light penetration on the photocatalyst surface. Therefore, the number of hydroxyl radicals (•OH) increased (Nishio *et al.* 2006), resulting in the degradation of RB 19 by •OH. Daneshvar *et al.* (2004) reported that •OH is a powerful oxidizing agent that attacks organic compounds and intermediates, leading to CO_2 and H_2O. This result was likely in the study by Tambat *et al.* (2016) that the efficiency of organic compounds' degradation was dependent on the concentration of the photocatalyst.

The RB 19 decolorization by photolysis and photocatalytic oxidation

The decolorization of RB 19 by photolysis at initial pH 3 of RB 19 wastewater, and an initial dye concentration of 25 mg/L at room temperature for 120 min was investigated. The filtration process with photolysis was also studied to compare with photocatalytic oxidation. The results found that the decolorizations of RB 19 were 6.9% and 8.9% by direct photolysis and photolysis followed by filtration after

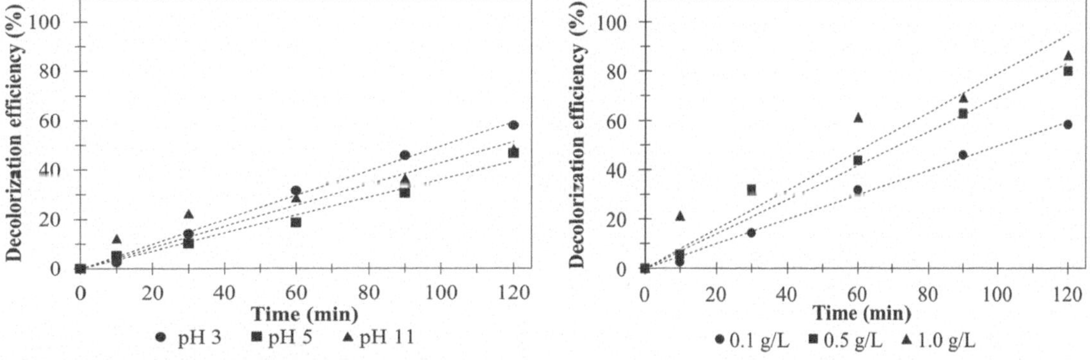

Figure 5 | Parameters affecting the RB 19 decolorization by photocatalytic oxidation: (a) effect of initial pH under 0.1 g/L TiO_2 and 25 mg/L RB 19; and (b) effect of initial TiO_2 concentration under 25 mg/L RB 19 and at pH 3.

60 min, respectively as presented in Figure 6. Therefore, under these irradiations, it was observed that the decolorization efficiency of RB 19 was low by photolysis and photolysis followed by filtration due to degradation of RB 19 on the TiO_2 surface absorption. The decolorization efficiency of RB 19 by photocatalytic oxidation was 61.1%, which was higher than that of photolysis and photolysis followed by filtration. This was because the RB 19 was oxidized by •OH. When the TiO_2 is illuminated with photons of energy $h\nu$ higher than (or equal to) their band-gap, electrons and holes are generated. TiO_2 was stimulated to excite an electron to produce •OH and superoxide radical anions ($O_2^{\cdot-}$). Then, organic compounds were degraded by •OH to carbon dioxide, water and inorganic anions (Leong *et al.* 2014; Ribeiro *et al.* 2015) as presented in the following reactions:

$$TiO_2 + h\nu \rightarrow e^- + h^+ \quad (6)$$

$$h^+ + H_2O \rightarrow \bullet OH + H^+ \quad (7)$$

$$e^- + O_2 \rightarrow O_2^{\cdot-} \quad (8)$$

$$\bullet OH + \text{dye} \rightarrow \text{intermediates} \rightarrow CO_2 + H_2O \quad (9)$$

Decolorization of Reactive Blue 19 dye wastewater by photocatalytic PVDF membrane

According to the results, the highest decolorization of RB 19 was by photocatalytic oxidation. However, the separation of TiO_2 after treatment is needed. Therefore, it is necessary to include the membrane in the process to prevent TiO_2 release to the environment. The membrane properties played an important role in the filtration process. Therefore, this part will study the decolorization of RB 19 on the membrane modification by TiO_2 coating on the membrane surface for the feasibility of the modified membrane in the photocatalytic membrane.

The TiO_2 coated on a PVDF flat sheet membrane is a material that contains both organic (polymer) and inorganic (TiO_2) components. After modifying the PVDF membrane, the membrane modified for 60 min of coating time was selected to study the decolorization of RB 19. The results found that the decolorization performance of RB 19 increased with the increase in the TiO_2 concentration coated on the membrane surface. The RB 19 decolorization of 1.0 g/L of TiO_2 coating on the membrane was highest (35%) compared with 0.1 g/L (19%) and 0.5 g/L (26%) due to the photocatalytic oxidation (reactions (6)–(9)) as presented in Figure 7. Then, RB 19 was oxidized by •OH to small molecules, carbon dioxide and water. The decolorization efficiency of RB 19 by a photocatalytic membrane coated with TiO_2 was lower than that of photocatalytic oxidation for 60 min. This was because the TiO_2 particles were attached on the membrane so that the light intensity could not have irradiated the TiO_2, resulting in a decrease in the efficiency.

Figure 8 illustrates a representative time profile of RB 19 solution at 25 mg/L. RB 19 is represented mainly by three peaks: (1) 592 nm, (2) 285 nm, (3) 230 nm. The peak at 592 nm corresponds to the chromophore group (blue color) (Guimaraes *et al.* 2012) and the peak located at 230 nm corresponds to the benzoic ring (Yuan *et al.* 2011). It was observed that as the photocatalytic oxidation and photocatalytic membrane proceeded, the peaks at 230 nm and 250 nm decreased as well as the peak at 592 nm. Therefore, the RB 19 decolorizations were achieved after treatment. In contrast, the peaks of RB 19 with photolysis

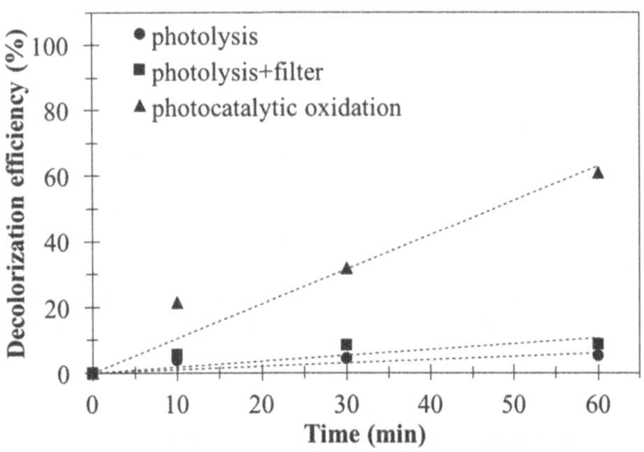

Figure 6 | Decolorization efficiency of RB 19 of photolysis, filtration and photocatalytic oxidation (25 mg/L RB 19 at pH 3 for 60 min).

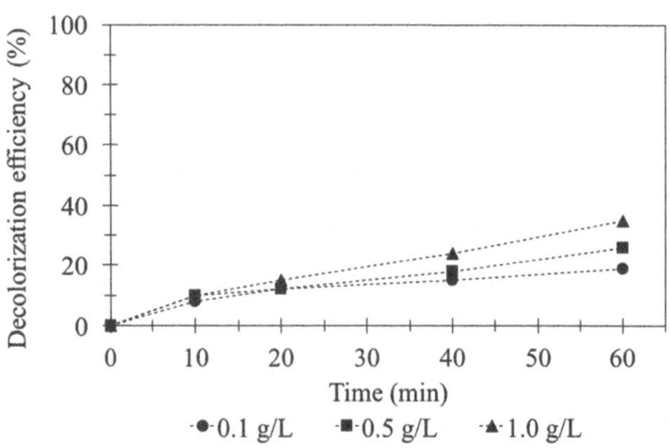

Figure 7 | RB 19 removal using a modified PVDF membrane with different TiO_2 concentrations under 25 mg/L RB 19 at pH 3.

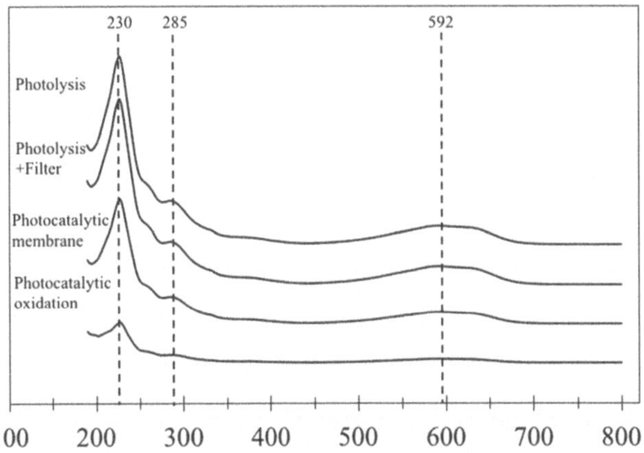

Figure 8 | UV-vis spectra of RB 19 solution at different treatments under 25 mg/L RB 19 at pH 3.

and photolysis followed by filtration did not significantly change.

CONCLUSIONS

The conditions of this study of membrane coating with TiO_2 were investigated by varying the experimental parameters, including TiO_2 concentration and coating time. The original and modified membranes were characterized for contact angle, SEM images and porosity. The results showed that the contact angle of the original PVDF membrane (128.1°) decreased after TiO_2 coating from 127.3° to 114.3°, with increase of the TiO_2 concentration from 0.1 to 1.0 g/L. The TiO_2 particles were observed on the membrane surface after coating, while the porosity of the PVDF membrane decreased. The water fluxes of the PVDF membrane modified by TiO_2 coating increased with the increase in the TiO_2 concentration. The decolorization efficiency of RB 19 by photolysis and photolysis followed by filtration was less than 10%, while the efficiency of photocatalytic oxidation of RB 19 presented 86.1% after 120 min. However, the decolorization performance of RB 19 by a photocatalytic membrane with TiO_2 coating was lower than that of the photocatalytic oxidation. Therefore, the photocatalytic membrane is a challenging and feasible method that can combine two processes in one unit for wastewater treatment.

ACKNOWLEDGEMENTS

This research was financially supported by Center of Excellence on Hazardous Substance Management (HSM), Chulalongkorn University, Thailand. The authors gratefully thank the Department of Environmental Science, Faculty of Science, Chulalongkorn University for the experimental facilities provided.

REFERENCES

Aguedach, A., Brosillon, S., Morvan, J. & Lhadi, E. K. 2005 Photocatalytic degradation of azo-dyes reactive black 5 and reactive yellow 145 in water over a newly deposited titanium dioxide. *Appl. Catal., B.* **57**, 55–62.

Allah, T. A. G., Kato, S., Satokawa, S. & Kojima, T. 2009 Treatment of synthetic dyes wastewater utilizing a magnetically separable photocatalyst ($TiO_2/SiO_2/Fe_3O_4$): parametric and kinetic studies. *Desalination* **244**, 1–11.

Banaei, A., Samadi, S., Karimi, S., Vojoudi, H., Pourbasheer, E. & Badiei, A. 2017 Synthesis of silica gel modified with 2,2′-(hexane-1,6-diylbis(oxy))dibenzaldehyde as a new adsorbent for the removal of Reactive Yellow 84 and Reactive Blue 19 dyes from aqueous solutions: equilibrium and thermodynamic studies. *Powder Technol.* **319**, 60–70.

Cruz, N. K. O., Semblante, G. U., Senoro, D. B., You, S. J. & Lu, S. C. 2014 Dye degradation and antifouling properties of polyvinylidene fluoride/titanium oxide membrane prepared by sol–gel method. *J. Taiwan Inst. Chem. Eng.* **45**, 192–201.

Daneshvar, N., Rabbani, M., Modirshahla, N. & Behnalady, M. A. 2004 Kinetic modeling of photocatalytic degradation of Acid Red 27 in UV/TiO_2 process. *J. Photochem. Photobiol. A* **168**, 39–45.

Fischer, K., Grimm, M., Meyers, J., Dietrich, C., Gläser, R. & Schulze, A. 2015 Photoactive microfiltration membranes via directed synthesis of TiO_2 nanoparticles on the polymer surface for removal of drugs from water. *J. Membr. Sci.* **478**, 49–57.

Gaya, U. L. & Abdullah, A. H. 2008 Heterogeneous photocatalytic degradation of organic contaminants over titanium dioxide: a review of fundamentals, progress and problems. *J. Photochem. Photobiol. C* **9**, 1–12.

Guimaraes, J. R., Maniero, M. G. & Nogueirade Araujo, R. 2012 A comparative study on the degradation of RB-19 dye in an aqueous medium by advanced oxidation processes. *J. Environ. Manage.* **110**, 33–39.

Iglesias, O., Rivero, M. J., Urtiaga, A. M. & Ortiz, I. 2016 Membrane-based photocatalytic systems for process intensification. *Chem. Eng. J.* **305**, 136–148.

Khan, M. A. N., Siddique, M., Wahid, F. & Khan, R. 2015 Removal of reactive blue 19 dye by sono, photo and sonophotocatalytic oxidation using visible light. *Ultrason. Sono.* **26**, 370–377.

Leong, S., Razmjou, A., Wang, K., Hapgood, K., Zhang, X. & Wang, H. 2014 TiO_2 based photocatalytic membranes: a review. *J. Membr. Sci.* **472**, 167–184.

Li, J. F., Xu, Z. L., Yang, H., Yu, L. Y. & Liu, M. 2009 Effect of TiO_2 nanoparticles on the surface morphology and performance of microporous PES membrane. *Appl. Surf. Sci.* **255**, 4725–4732.

Méricq, J. P., Mendret, J., Brosillon, S. & Faur, C. 2015 High performance PVDF-TiO_2 membranes for water treatment. *Chem. Eng. Sci.* **123**, 283–291.

Nishio, J., Tokumura, M., Znad, H. T. & Kawase, Y. 2006 Photocatalytic decolorization of azo-dye with zinc oxide powder in an external UV light irradiation slurry photoreactor. *J. Hazard. Mater.* **138**, 106–115.

Peng, T., Yu, Z., Pan, Y. & Zeng, G. 2018 Antibacterial photocatalytic self-cleaning poly(vinylidene fluoride) membrane for dye wastewater treatment. *Polym. Adv. Technol.* **29**, 254–262.

Ribeiro, A. R., Nines, O., Pereira, M. F. R. & Silva, A. M. T. 2015 An overview on the advanced oxidation processes applied for the treatment of water pollutants defined in the recently launched Directive 2013/39/EU. *Environ. Int.* **75**, 33–51.

Suresh, P., Vijaya, J. J. & Kennedy, L. J. 2015 Photocatalytic degradation of textile dyeing wastewater through microwave synthesized Zr-ZC, Ni-AC and Zn-AC. *Trans. Nonferrous Met. Soc. China* **25**, 4216–4225.

Taheria, M., Moghaddam, M. R. A. & Arami, M. 2013 Techno-economical optimization of Reactive Blue 19 removal by combined electrocoagulation/coagulation process through MOPSO using RSM and ANFIS models. *J. Environ. Manage.* **128**, 798–806.

Tambat, S., Umale, S. & Sontakke, S. 2016 Photocatalytic degradation of Milling Yellow dye using sol–gel synthesized CeO_2. *Mater. Res. Bull.* **76**, 466–472.

Yuan, R. X., Ramjaun, S. N., Wang, Z. H. & Liu, J. S. 2011 Effects of chloride ion on degradation of Acid Orange 7 by sulfate radical-based advanced oxidation process: implications for formation of chlorinated aromatic compounds. *J. Hazard. Mater.* **196**, 173–179.

Zhang, X., Wang, D., Lopez, D. R. S. & Costa, J. C. D. 2014 Fabrication of nanostructured TiO_2 hollow fiber photocatalytic membrane and application for wastewater treatment. *Chem. Eng. J.* **236**, 314–322.

First received 10 August 2018; accepted in revised form 7 January 2019. Available online 17 January 2019

Enhanced degradation of methylene blue by a solution plasma process catalyzed by incidentally co-generated copper nanoparticles

Isarawut Prasertsung, Sasin Kaewcharoen, Kornnika Kunpinit, Wattanachai Yaowarat, Nagahiro Saito and Tanapon Phenrat

ABSTRACT

This study presents a catalytic organic pollution treatment using the solution plasma process (SPP) with incidentally co-generated copper (Cu) nanoparticles via Cu electrode erosion. Methylene blue (MB) was used as a model organic contaminant. The treatment time was from 0 to 60 minutes at the plasma frequencies of 15 and 30 kHz. The treatment efficacy using the Cu electrode was compared with that of the tungsten (W) electrode. The high erosion-resistant W electrode provided no W nanoparticles, while the low erosion-resistant Cu electrode yielded incidental nanoparticles (10–20 nm), hypothesized to catalyze the MB degradation during the SPP. The percentage of MB degradation and the hydrogen peroxide (H_2O_2) generation were determined by an ultraviolet-visible spectrophotometer. The results showed that, after the SPP by the Cu electrode for 60 minutes, the MB was degraded up to 96%. Using the Cu electrode at a high plasma frequency strongly accelerated the Cu nanoparticle generation and MB treatment, although the amount of H_2O_2 generated during the SPP using the Cu electrode was less than that of the W electrode. The Cu nanoparticles were hypothesized to enhance MB degradation via both homogeneous (release of dissolved Cu ions) and heterogeneous (on the surface of the particles) catalytic processes.

Key words | catalytic degradation, copper nanoparticles, methylene blue, solution plasma

Isarawut Prasertsung
Sasin Kaewcharoen
Kornnika Kunpinit
Wattanachai Yaowarat
Chemical Engineering Program, Department of Industrial Engineering, Faculty of Engineering, Naresuan University, Phitsanulok 65000, Thailand

Isarawut Prasertsung
Department of Biotechnology, Faculty of Technology, Mahasarakham University, Mahasarakham 44150, Thailand

Nagahiro Saito
Reaction Kinetics and Dynamics Research Group, Department of Materials, Physics and Energy Engineering, Graduate School of Engineering, Nagoya University, Furo-cho, Chikusa-ku, Nagoya 464-8603, Japan

Tanapon Phenrat (corresponding author)
Research Unit for Integrated Natural Resources Remediation and Reclamation (IN3 R), Department of Civil Engineering, Faculty of Engineering, Naresuan University, Phitsanulok 65000, Thailand;
Center of Excellence for Sustainability of Health, Environment and Industry (SHEI), Faculty of Engineering, Naresuan University, Phitsanulok, 65000, Thailand;
and
Research Program of Toxic Substance Management in the Mining Industry, Center of Excellence on Hazardous Substance Management (HSM), Chulalongkorn University, Bangkok 10330, Thailand
E-mail: *pomphenrat@gmail.com*

INTRODUCTION

With a constant increase in population, water demand for domestic, industrial, and agricultural applications is also increasing. Due to the excessive usage of chemicals in industry, wastewater from various industries contains harmful compounds. For example, the wastewater from the paper and textile industries is highly contaminated by dyes and hazardous compounds (Ong *et al.* 2011; Manoj Kumar Reddy & Subrahmanyam 2012; Manoj Kumar Reddy *et al.* 2013a). To treat contaminants from these industries, various decontamination methods, including biological treatment, membrane filtration, ion exchange (Thamaraiselvan & Noel 2015), and adsorption (Bharathi & Ramesh 2013), have been extensively examined. However, some of these decontamination techniques may just transfer

doi: 10.2166/wst.2019.035

contaminants from one phase to another instead of desirably mineralizing them.

Advanced oxidation processes (AOPs) are a promising contaminant mineralization alternative via the generation and reaction of hydroxyl radicals. These radicals could oxidize organic pollutants present in the wastewater and contaminated groundwater (Crimi & Taylor 2007; Siegrist *et al.* 2011; Manoj Kumar Reddy *et al.* 2013b; Sacco *et al.* 2015; Shen *et al.* 2016; Pankaj *et al.* 2017; Sarangapani *et al.* 2017). The oxidation potential of hydroxyl radicals ($E_0 = 2.8$ V) is relatively high in comparison to that of conventional oxidants, such as ozone ($E_0 = 2.0$ V) and hydrogen peroxide ($E_0 = 1.7$ V), making the hydroxyl radical capable of promptly mineralizing various pollutants, including dyes and hazardous volatile and semi-volatile organic compounds. Various AOP techniques, such as photocatalysis, photo-Fenton, and O_3/ultraviolet (UV), have been extensively studied and utilized (Wang & Xu 2012; Bokare & Choi 2014). Nevertheless, the limitation of these techniques is the insufficient quantity of the oxidants, especially hydroxyl radical, to sustain and speed up contaminant degradation. To overcome this problem, a catalytic AOP is preferable (Khan *et al.* 2016; Thao *et al.* 2017).

Electrical discharge in the liquid phase, also known as the solution plasma process (SPP), has been widely utilized for nanomaterial synthesis, surface modification, water treatment, sterilization, and decomposition of organic contaminants and molecules (Locke *et al.* 2006; Takai 2008; Manoj Kumar Reddy & Subrahmanyam 2012; Prasertsung *et al.* 2012; Pootawang *et al.* 2013; Prasertsung *et al.* 2013). This process produces highly active species, such as hydroxyl radicals (OH•), hydroperoxyl radicals (HO_2•), free electrons (e^-), superoxide anions (O_2^-), and atomic oxygen anions (O^-) (Yuan *et al.* 2010; Wong & Mongkolnavin 2016). Since the molecular density of the liquid phase is much higher than that of the gas phase, it is reasonable to expect a higher reaction rate of active species generated by the SPP in comparison to that of other gas phase generations (Pocostales *et al.* 2010; Tichonovas *et al.* 2013; Pankaj *et al.* 2017; Sarangapani *et al.* 2017).

Solution plasma conditions, such as electrode materials, appear to influence the degradation of organic molecules and the synthesis of nanomaterials. For example, recent research found that among three metals (tungsten (W), copper (Cu), and iron (Fe)), at high pulse frequency, the Fe electrodes strongly promoted the degradation of chitosan, making chitosan more water soluble (Tantiplapol *et al.* 2015). Similarly, Fe and Cu electrodes could grow carbon nanotubes from propane via AC arc and dielectric barrier discharge plasma, while the W electrode could not (Sun *et al.* 2016). Nevertheless, to our knowledge, only a few studies investigated the effects of electrode materials on the degradation rate of organic contaminants using the SPP. A recent study reported that, in comparison to the titanium dioxide (TiO_2) electrode, using W as a high-voltage electrode in pulse corona discharge with needle-plate electrode geometry caused the erosion of W and released dissolved WO_4^- ions, which are involved in the catalytic degradation of dimethyl sulfoxide using hydrogen peroxide (H_2O_2) (Lukes *et al.* 2011). For instance, platinum-tungsten alloy as a high-voltage electrode decreased the yield of H_2O_2 produced by a pulsed corona discharge in water. At the same time, when used as high-voltage electrodes, these materials increased the efficiency of the plasma chemical removal of organic compounds from water due to their catalytic effects. When the electrode is coated with a thin ceramic layer, the electric field strength on the electrode surface may be enhanced by the concentration of the pre-discharge current in small open pores. As a result of the inhomogeneous nature of the ceramic layer, the electric field may be higher, and it may be easier than expected to initiate streamers from some spots. Such a system can be promoted with its catalytic activity or be designed as a support for a suitable catalyst of the plasma chemical reactions (Lukes *et al.* 2011).

In the current study, we hypothesized that low erosion-resistant electrode materials, such as Cu, could be sputtered during plasma treatment, yielding Cu nanoparticles, which are involved in pollution abatement. The Cu nanoparticles were intentionally engineered for use as a solid catalyst to enhance the degradation rate of organic contaminants. Nevertheless, to our knowledge, only a few researchers have investigated the role of incidental nanoparticles on pollution abatement (Glover *et al.* 2011; Srivastava *et al.* 2013), and none have investigated the SPP treatment with incidentally co-generated Cu nanoparticles. For this purpose, methylene blue (MB) was used as a model of organic contaminants to evaluate the catalytic treatment in a plasma reactor with different types of electrodes and applied pulse frequencies. Two types of electrodes, one with high erosion resistance (W) and one with low erosion resistance (Cu), were used. The sputtered solid metal particles were characterized in terms of the concentration, shape, and size. The MB degradation extent and rate as well as the formation of H_2O_2 in the electrical discharge in the solution were also quantified.

EXPERIMENTAL SECTION

Solution plasma setup

The setup of the SPP, modified from our previous study (Tantiplapol *et al.* 2015), is shown in Figure 1. The needle electrodes, which are 1 mm in diameter and covered by a ceramic tube, were fixed by silicone in a glass reactor and used to generate a pulsed electric discharge. The high-frequency bipolar pulsed DC power supply was connected to the needle electrodes. The frequencies, voltages, and pulse widths used in the experiment are described next.

Solution plasma treatment of methylene blue

The MB powder was dissolved in distilled water to obtain 125 mg/L of the MB solution, which was placed in the glass reactor. The plasma in the solution was produced at the fixed voltage of 1.6 kV and pulse width of 3 μs. Two types of electrodes, which were different in erosion resistance – 0.5% thoriated W and Cu, were used. The frequencies of power supply varied in the range of 15–30 kHz, and the treatment time of the plasma varied at 0–60 minutes. To enhance the uniformity of the MB solution to contact the plasma, the solution was continuously stirred by a magnetic stirrer. To characterize the species in plasma-treated MB solutions, optical emission spectroscopy (OES) was used to monitor the light emitted from the plasma in the wavelength range of 200–1,100 nm. The emission was detected at the plasma treatment time of 1 minute, through the quartz glass window, with an optical fiber placed 1 mm in front of the glass chamber. Data were acquired with Avantes software.

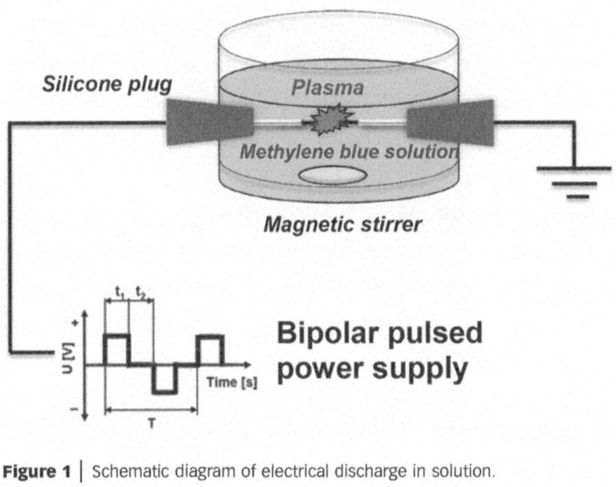

Figure 1 | Schematic diagram of electrical discharge in solution.

The degradation of MB was monitored for every 10-minute interval of plasma treatment using a double beam UV spectrophotometer (Shimadzu) at the wavelength of 668 nm. The degradation percentage was calculated using Equation (1).

$$\% \text{ degradation} = \frac{C_0 - C_t}{C_0} \times 100 \tag{1}$$

where C_0 and C_t are the initial concentration and the concentration of the MB solution at time t, respectively. All experiments were conducted in duplicate.

H_2O_2 quantification

In SPP treatment, the H_2O_2 can be produced by recombination reaction of OH• radicals. Therefore, the H_2O_2 generated in the solution was determined colorimetrically using the reaction of H_2O_2 with titanyl ions (Ti^{4+}), giving the yellow-color complex of pertitanic acid ($TiO_2H_2O_2$) (Chandana *et al.* 2015).

$$Ti^{4+} + H_2O_2 + 2H_2O \rightarrow TiO_2H_2O_2 + 4H^+ \tag{2}$$

In brief, 1.00 mL of the solution and 2.0 mL of titanium sulfate (12 g/L in 40% sulfuric acid) were mixed together and diluted to 10.00 mL with distilled water in a volumetric flask. The absorbance of the mixture was measured at 410 nm with a 5-mm colorimetric cuvette. The standard curve was obtained using the standard H_2O_2 solution and is shown in Figure 2.

Microscopic characterization of electrodes and incidentally co-generated Cu nanoparticles

The morphology of the Cu electrode surface was examined by scanning electron microscopy (SEM; JEOL, JSM-6400). To identify the shape and size of the sputtered solid particles obtained from the plasma treatment of MB using a Cu electrode, a transmission electron microscope (TEM; Philips, EM208) was employed. After plasma treatment for 60 minutes, 50 μL of the solution was collected and dispersed in 2 mL of ethanol. A small drop of solution was placed onto a grid prior to evaporation at room temperature overnight. The TEM microscope was operated at an accelerating voltage of 120 kV.

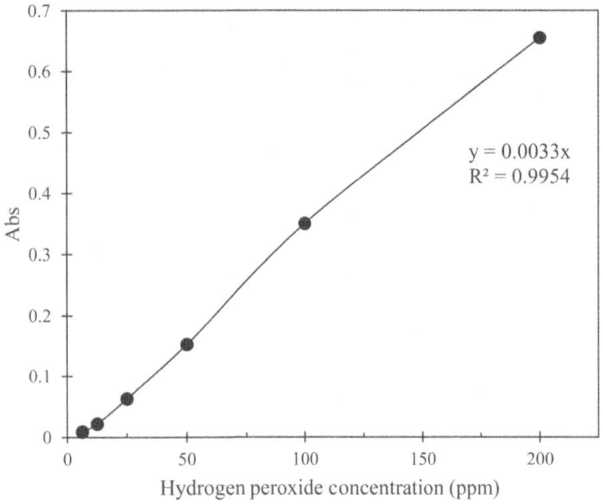

Figure 2 | Standard curve of H_2O_2.

RESULTS AND DISCUSSION

Species in the emission spectra of plasma-treated MB solution

The emission spectra of the plasma-treated MB solution using the two different types of electrodes, W (high erosion-resistant) and Cu (low erosion-resistant), which were observed at a treatment time of 1 minute, are presented in Figure 3. The OES spectra of the plasma-treated MB solution using the W electrode exhibited strong peaks at the wavelengths of 486.0, 656.5, 777.3, and 844.0 nm, corresponding to Hα, Hα, O^- radicals, and O^- radicals, respectively (Baroch *et al.* 2008). This was possibly due to the ionic current generated in the MB solution during the solution plasma treatment. As a result, the reactive species could be produced by the decomposition of water molecules (Watthanaphanit *et al.* 2014). On the other hand, for the Cu electrode, the additional spectrum peaks were observed at the wavelengths of 309.0 and 521.8 nm, corresponding to OH radicals and Cu I, respectively (Tepe *et al.* 1997). The Cu I species could be generated from the erosion of electrode material during plasma treatment, resulting in Cu particles or ions. This was because the erosion resistance of the Cu electrodes is lower than that of W. The volumetric erosion rate of Cu and W was measured to be 13.4×10^{-8} and 6.4×10^{-8} cm^3/s, respectively (Holzer & Locke 2008). Therefore, the Cu particle, which was generated by the electrode release of material, was easily formed by electrolysis (Potocký *et al.* 2009). This hypothesis will be verified later in this study.

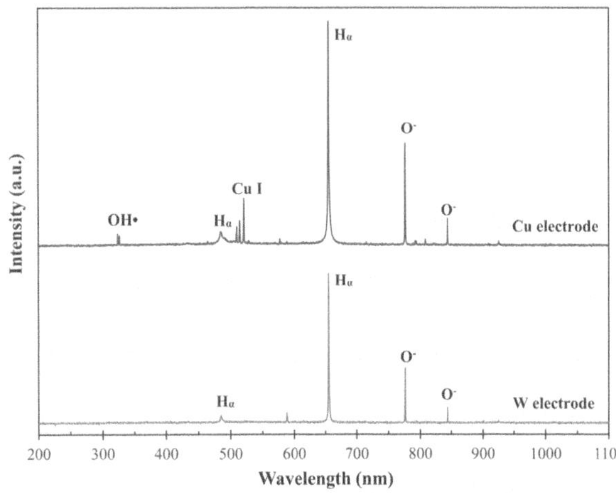

Figure 3 | OES spectrum of plasma-treated MB solution at the treatment time of 1 minute.

Kinetics of MB degradation

Figure 4(a) shows the percentage of MB degradation as a function of treatment time for the two types of electrodes at 15 kHz. The results showed that, after plasma treatment using the W electrode for 10 minutes, the percentage of MB degradation rapidly increased to 40%. After that, a slight increase in the percentage of MB degradation was noticed until the treatment time reached 60 minutes. The maximum percentage of MB degradation treated by plasma using the W electrode was 79%. The degradation of MB was predominantly caused by free radicals, such as hydroxyl that is generated during plasma treatment. The mechanism of MB degradation by hydroxyl radicals has been discussed in detail elsewhere (Manoj Kumar Reddy *et al.* 2013a; Chandana *et al.* 2015).

On the other hand, for the plasma treatment using the Cu electrode at 15 kHz, the MB degradation reaction was dramatically enhanced. The percentage of MB degradation by plasma using the Cu electrode was much higher than that of W. The maximum percentage of MB degradation by plasma using the Cu electrode was 92%. A stronger degradation effect from the use of the Cu electrode could be attributed to copper particles or ions, which were generated by the erosion of the Cu electrode via electrolysis during plasma treatment.

Furthermore, the applied pulse frequency played a substantial role in the percentage of MB degradation, as shown in Figure 4(b). Noticeably, the degradation rate of MB was increased with the increase in applied pulse frequency from 15 to 30 kHz. The greater improvement on the degradation rate of MB at a high pulse frequency could be attributed to the energy input increase when the

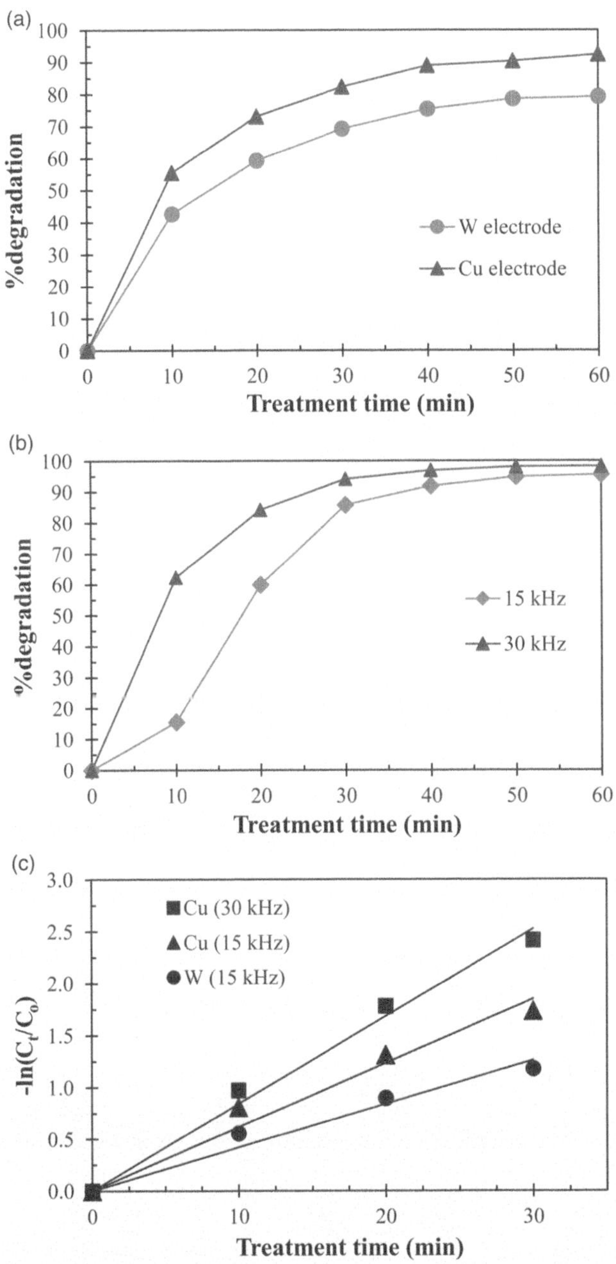

Figure 4 | (a) Percentage degradation of methylene blue as a function of treatment time and types of electrodes at 15 kHz, (b) percentage degradation of methylene blue by Cu electrodes as a function of treatment time at the applied pulse frequency of 15 and 30 kHz, and (c) first order kinetic plots for the MB degradation.

pulse frequency increased, as previously reported (Kang *et al.* 2013). The increase of energy input could enhance the number of hydroxyl radicals during plasma treatment and thus promote the MB degradation process (Chandana *et al.* 2015). In addition, an increase of energy could strongly affect the concentration of Cu nanoparticles. This could be described by an accelerated erosion of Cu electrodes during the plasma treatment, which provided a high number of solid particles. As the result, the MB degradation process could be catalytically enhanced, resulting in a high degradation rate.

According to the result of the MB degradation, the initial rate constants were calculated by fitting the data to the first order rate equation:

$$-\ln(C_t/C_0) = kt \tag{3}$$

where C_t is the concentration of the MB solution at reaction time (t), C_0 is the initial concentration of the MB solution, k (min^{-1}) is the degradation rate constant, and t is the reaction time. A linear relationship between $-\ln(C_t/C_0)$ and the reaction time for the degradation of MB by plasma (reaction time from 0 to 30 min) is shown in Figure 4(c). The k value of the degradation of MB by plasma using the W electrode was 4.2×10^{-2} min^{-1} at 15 kHz. Interestingly, the k values of degradation of the MB were enhanced and measured as 6.1×10^{-2} min^{-1} at 15 kHz when the Cu electrode was introduced. The obtained k value for the plasma treatment using Cu was higher than that of the W electrode.

The pulse frequency clearly affected the MB degradation rate. Obviously, the MB degradation was increased with the increasing pulse frequency. The k value of the MB degradation treated by plasma at the pulse frequency of 30 kHz was 8.4×10^{-2} min^{-1}. According to the calculation of k, the degradation reaction of MB by plasma was evaluated to be in the following order: Cu at 30 kHz (8.4×10^{-2} min^{-1}) > Cu at 15 kHz (6.1×10^{-2} min^{-1}) > W at 15 kHz (4.2×10^{-2} min^{-1}). Therefore, the finding in this study proved that the SPP treatment using the Cu electrode with high applied pulse frequency (30 kHz) is effective for MB degradation.

Microscopic change of Cu electrodes and incidentally co-generated Cu nanoparticles

Figure 5 shows a SEM micrograph of Cu electrodes (a) before and (b) after plasma treatment. The results showed a smooth surface of the Cu electrode before the plasma was applied. However, the electrode surface became rough when the plasma was applied for 60 minutes. This ensured that the erosion of the Cu electrode via material sputtering existed during plasma treatment. Figure 6 shows a TEM micrograph of Cu solid particles generated during plasma treatment of the MB solution by electrode material sputtering. Noticeably, the Cu nanoparticles are nearly spherical with a random orientation. The sizes of the nanoparticles are in the range of 5–10 nm.

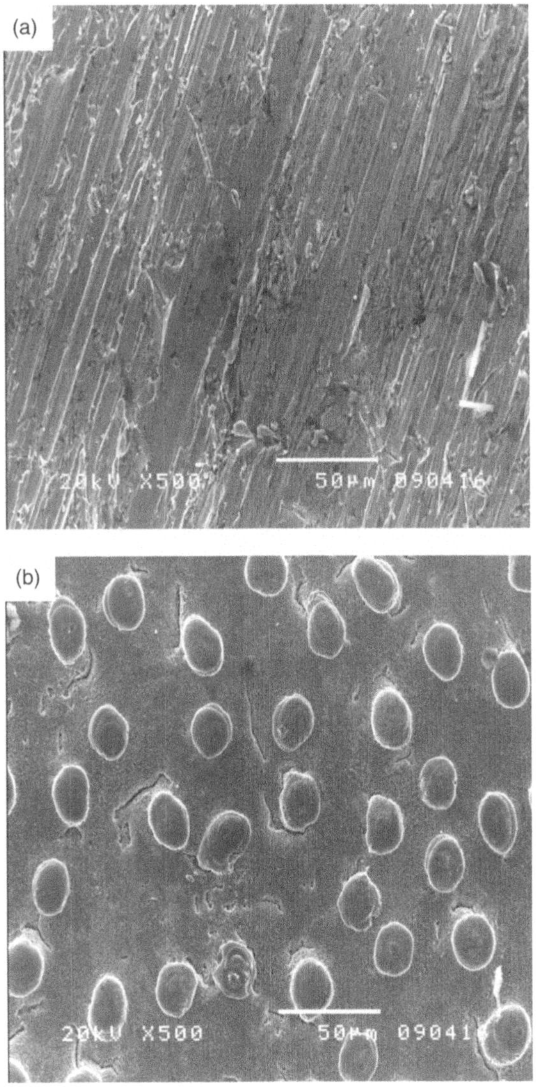

Figure 5 | SEM micrographs of copper electrode surface: (a) before plasma treatment and (b) after plasma treatment.

Mechanisms of enhanced MB degradation by Cu nanoparticles

The MB degradation is predominantly affected by H_2O_2, which is generated during plasma treatment by the recombination reaction of hydroxyl radicals as follows (Tantiplapol *et al.* 2015).

$$OH\bullet + OH\bullet \rightarrow H_2O_2 \quad (4)$$

A couple of possible mechanisms of enhanced MB degradation by Cu nanoparticles formed during the SPP can be described as follows. First, Cu nanoparticles could act as a solid catalyst in MB degradation, as previously

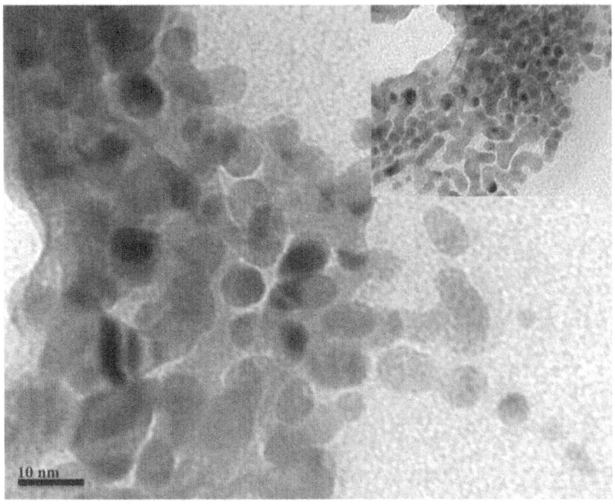

Figure 6 | Transmission electron micrograph of copper nanoparticles generated by the solution plasma treatment.

described in a recent study (Ndolomingo & Meijboom 2015). These authors proposed the MB degradation mechanism shown in Equation (5).

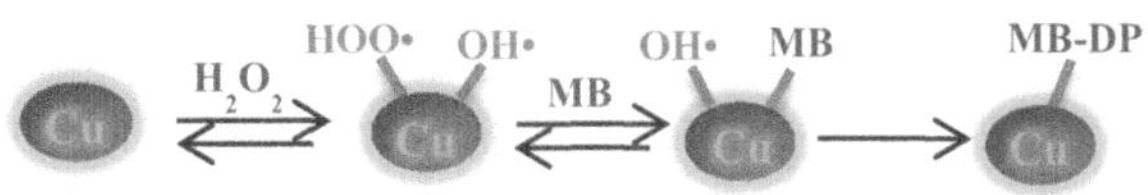

(5)

The peroxide can react on the surface of the Cu nanoparticles leading to the absorption of the reactive species, especially hydroxyl radicals. The MB reacts with hydroxyl radicals, resulting in the decomposition of MB on the surface of the nanoparticles followed by the desorption of by-products from the nanoparticle surface.

Second, these nanoparticles could be transformed into Cu ions through oxidation reaction at the electrode during plasma treatment (Sarahney *et al.* 2012). As reported in a recent study (Friedrich *et al.* 2012), Cu ions can also act as a catalyst, which could enhance the Fenton reaction (cupro-Fenton reaction). Therefore, the decomposition of the H_2O_2, which is generated in the system during plasma treatment by the recombination reaction of hydroxyl radicals, could occur and provide a high number of hydroxyl radicals (as noted in Figure 3) as follows (Friedrich *et al.* 2012).

$$H_2O_2 + Cu^{+} + H^{+} \rightarrow OH \cdot + Cu^{2+} + H_2O \quad (6)$$

This result corresponds with the generation of H_2O_2 during plasma treatment as illustrated in Figure 7. Notably, the amount of H_2O_2 produced during plasma treatment using the W electrode was dramatically increased with the

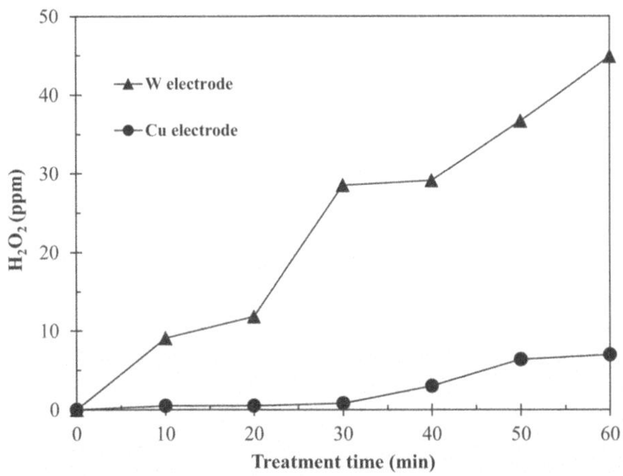

Figure 7 | The production of hydrogen peroxide (H_2O_2) as a function of treatment time and types of electrodes.

increasing treatment time, while the amount of H_2O_2 produced during plasma treatment using the Cu electrode seemed to be relatively constant or just slightly increased. The increase of H_2O_2 during plasma treatment using W could be attributed to the recombination reaction of hydroxyl radicals, but no further transformation of H_2O_2 to hydroxyl radicals took place, limiting the MB degradation kinetics. In contrast, in the case of the Cu electrode, the generated H_2O_2 during plasma treatment could be transformed to hydroxyl radicals again, as described above, participating in enhanced MB degradation. As a result, the amount of H_2O_2 was constant. Therefore, the decomposition reaction of H_2O_2 by Fenton reaction may also be responsible for the enhancement of MB degradation (Wang & Jiang 2009).

CONCLUSION

In this study, solution plasma was introduced to treat the MB solution. The percentage of plasma-treated MB solution was increased with the increasing plasma treatment time, compared to that of an untreated sample. The degradation rate was greatly affected by the types of electrodes and the applied pulse frequency. The plasma treatment of the MB solution using the Cu electrode and a high pulse frequency strongly promoted the degradation rate of MB. We identified the incidentally co-generated Cu nanoparticles via the Cu electrode erosion during the solution plasma treatment. We also found that these nanoparticles strongly promoted the degradation of the MB solution during the SPP by heterogeneous and/or homogeneous catalytic degradation.

ACKNOWLEDGEMENTS

The electrical discharge in the solution apparatus support from EcoTopia Science Institute and the Technology and Department of Materials Engineering, Nagoya University, and the financial support from Thailand Research Fund (TRF) (grant number TRG5880028) are gratefully acknowledged. T. Phenrat thanks Office of Higher Education Commission (OHEC) and S&T Postgraduate Education and Research Development Office (PERDO) for the financial support of the research program.

REFERENCES

Baroch, P., Anita, V., Saito, N. & Takai, O. 2008 Bipolar pulsed electrical discharge for decomposition of organic compounds in water. *J. Electrostat.* **66**, 294–299.

Bharathi, K. S. & Ramesh, S. T. 2013 Removal of dyes using agricultural waste as low-cost adsorbents: a review. *Appl. Water Sci.* **3** (4), 773–790.

Bokare, A. D. & Choi, W. 2014 Review of iron-free Fenton-like systems for activating H_2O_2 in advanced oxidation processes. *J. Hazard. Mater.* **275**, 121–135.

Chandana, L., Manoj Kumar Reddy, P. & Subrahmanyam, C. 2015 Atmospheric pressure non-thermal plasma jet for the degradation of methylene blue in aqueous medium. *Chem. Eng. J.* **282**, 116–122.

Crimi, M. & Taylor, J. 2007 Experimental evaluation of catalyzed hydrogen peroxide and sodium persulfate for destruction of BTEX contaminants. *Soil Sediment Contam.* **16**, 29–45.

Friedrich, L. C., Silva Zanta, C. L., Machulek, A., Silva, V. O. & Quina, F. H. 2012 Interference of inorganic ions on phenol degradation by the Fenton reaction. *Sci. Agric.* **69**, 347–351.

Glover, R. D., Miller, J. M. & Hutchison, J. E. 2011 Generation of metal nanoparticles from silver and copper objects: nanoparticle dynamics on surfaces and potential sources of nanoparticles in the environment. *ACS Nano* **5** (11), 8950–8957.

Holzer, F. & Locke, B. R. 2008 Influence of high voltage needle electrode material on hydrogen peroxide formation and electrode erosion in a hybrid gas–liquid series electrical discharge reactor. *Plasma Chem. Plasma Process.* **28** (1), 1–13.

Kang, J., Li, O. L. & Saito, N. 2013 Synthesis of structure-controlled carbon nano spheres by solution plasma process. *Carbon* **60**, 293–298.

Khan, S., He, X., Khan, H. M., Boccelli, D. & Dionysiou, D. D. 2016 Efficient degradation of lindane in aqueous solution by iron (II) and/or UV activated peroxymonosulfate. *J. Photochem. Photobiol. A* **316**, 37–43.

Locke, B. R., Sato, M., Sunka, P., Hoffmann, M. R. & Chang, J. S. 2006 Electrohydraulic discharge and nonthermal plasma for water treatment. *Ind. Eng. Chem. Res.* **45** (3), 882–905.

Lukes, P., Clupek, M., Babicky, V., Sisrova, I. & Janda, V. 2011 The catalytic role of tungsten electrode material in the plasmachemical activity of a pulsed corona discharge in water. *Plasma Sources Sci. Technol.* **20**, 034011.

Manoj Kumar Reddy, P. & Subrahmanyam, C. 2012 Green approach for wastewater treatment – degradation and mineralization of aqueous organic pollutants by discharge plasma. *Ind. Eng. Chem. Res.* **51** (34), 11097–11103.

Manoj Kumar Reddy, P., Rama Raju, B., Karuppiah, J., Linga Reddy, E. & Subrahmanyam, C. 2013a Degradation and mineralization of methylene blue by dielectric barrier discharge non-thermal plasma reactor. *Chem. Eng. J.* **217**, 41–47.

Manoj Kumar Reddy, P., Ramaraju, B. & Subrahmanyam, C. 2013b Degradation of malachite green by dielectric barrier discharge plasma. *Water Sci. Technol.* **67** (5), 1097–1104.

Ndolomingo, M. J. & Meijboom, R. 2015 Kinetic analysis of catalytic oxidation of methylene blue over γ-Al_2O_3 supported copper nanoparticles. *Appl. Catal. A.* **506**, 33–43.

Ong, S.-T., Keng, P.-S., Lee, W.-N., Ha, S.-T. & Hung, Y.-T. 2011 Dye waste treatment. *Water* **3** (4), 157–176.

Pankaj, S. K., Wan, Z., Colonna, W. & Keener, K. M. 2017 Degradation kinetics of organic dyes in water by high voltage atmospheric air and modified air cold plasma. *Water Sci. Technol.* **76** (3–4), 567–574.

Pocostales, J. P., Sein, M. M., Knolle, W., von Sonntag, C. & Schmidt, T. C. 2010 Degradation of ozone-refractory organic phosphates in wastewater by ozone and ozone/hydrogen peroxide (peroxone): the role of ozone consumption by dissolved organic matter. *Environ. Sci. Technol.* **44** (21), 8248–8253.

Pootawang, P., Saito, N. & Lee, S. Y. 2013 Discharge time dependence of a solution plasma process for colloidal copper nanoparticle synthesis and particle characteristics. *Nanotechnology* **24** (5), 055604.

Potocký, Š., Saito, N. & Takai, O. 2009 Needle electrode erosion in water plasma discharge. *Thin Solid Films* **518**, 918–923.

Prasertsung, I., Damrongsakkul, S., Terashima, C., Saito, N. & Takai, O. 2012 Preparation of low molecular weight chitosan using solution plasma system. *Carbohydr. Polym.* **87** (4), 2745–2749.

Prasertsung, I., Damrongsakkul, S. & Saito, N. 2013 Crosslinking of a gelatin solutions induced by pulsed electrical discharges in solutions. *Plasma Process. Polym.* **10** (9), 792–797.

Sacco, O., Vaiano, V., Han, C., Sannino, D. & Dionysiou, D. D. 2015 Photocatalytic removal of atrazine using N-doped TiO_2 supported on phosphors. *Appl. Catal. B: Environ.* **164**, 462–474.

Sarahney, H., Mao, X. & Alshawabkeha, A. N. 2012 Role of iron anode oxidation on transformation of chromium by electrolysis. *Electrochim. Acta* **86**, 96–101.

Sarangapani, C., Dixit, Y., Milosavljevic, V., Bourke, P., Sullivan, C. & Cullen, P. J. 2017 Optimization of atmospheric air plasma for degradation of organic dyes in wastewater. *Water Sci. Technol.* **75** (1–2), 207–219.

Shen, C., Wu, S., Chen, H., Rashid, S. & Wen, Y. 2016 Phthalate degradation by glow discharge plasma enhanced with pyrite in aqueous solution. *Water Sci. Technol.* **74** (6), 1365–1375.

Siegrist, R. L., Crimi, M. & Simpkin, T. J. 2011 *In Situ Chemical Oxidation for Groundwater Remediation*. Springer Science + Business Media, New York.

Srivastava, S. K., Yamada, R., Ogino, C. & Kondo, A. 2013 Biogenic synthesis and characterization of gold nanoparticles by *Escherichia coli* K12 and its heterogeneous catalysis in degradation of 4-nitrophenol. *Nanoscale Res. Lett.* **8** (1), 70.

Sun, D. L., Hong, R. Y., Wang, F., Liu, J. Y. & Rajesh Kumar, M. 2016 Synthesis and modification of carbon nanomaterials via AC arc and dielectric barrier discharge plasma. *Chem. Eng. J.* **283**, 9–20.

Takai, O. 2008 Solution plasma processing (SPP). *Pure Appl. Chem.* **80**, 2003–2011.

Tantiplapol, T., Singsawat, Y., Narongsil, N., Damrongsakkul, S., Saito, N. & Prasertsung, I. 2015 Influences of solution plasma conditions on degradation rate and properties of chitosan. *Innov. Food Sci. Emerging Technol.* **32**, 116–120.

Tepe, R. K., Jacksier, T. & Barnes, R. M. 1997 Atomic emission of anhydrous hydrogen bromide: spectra from 200 to 900 nm by sealed inductively coupled plasma-atomic emission spectroscopy. *Spectrochim. Acta B* **52**, 2103–2113.

Thamaraiselvan, C. & Noel, M. 2015 Membrane processes for dye wastewater treatment: recent progress in fouling control. *Crit. Rev. Environ. Sci. Technol.* **45** (10), 1007–1040.

Thao, L. T. S., Dang, T. T. T., Khanitchaidecha, W., Channei, D. & Nakaruk, A. 2017 Photocatalytic degradation of organic dye under UV-A irradiation using TiO_2-vetiver multifunctional nano particles. *Materials* **10** (2), 122.

Tichonovas, M., Krugly, E., Racys, V., Hippler, R., Kauneliene, V., Stasiulaitiene, I. & Martuzevicius, D. 2013 Degradation of various textile dyes as wastewater pollutants under dielectric barrier discharge plasma treatment. *Chem. Eng. J.* **229**, 9–19.

Wang, J. L. & Xu, L. J. 2012 Advanced oxidation processes for wastewater treatment: formation of hydroxyl radical and application. *Crit. Rev. Environ. Sci. Technol.* **42** (3), 251–325.

Watthanaphanit, A., Heo, Y. K. & Saito, N. 2014 Influence of the discharge time of solution plasma process on the formation of gold nanoparticles in alginate matrix. *J. Taiwan Inst. Chem. Eng.* **45**, 3099–3103.

Wang, L. & Jiang, X. 2009 Unusual catalytic effects of iron salts on phenol degradation by glow discharge plasma in aqueous solution. *J. Hazard. Mater.* **161**, 926–932.

Wong, C. S. & Mongkolnavin, R. 2016 *Elements of Plasma Technology*. Springer, New York.

Yuan, M.-H., Narengerile, Watanabe, T. & Chang, C.-Y. 2010 DC water plasma at atmospheric pressure for the treatment of aqueous phenol. *Environ. Sci. Technol.* **44** (12), 4710–4715.

First received 29 June 2018; accepted in revised form 11 January 2019. Available online 23 January 2019

Coupled anaerobic ammonium oxidation and hydrogenotrophic denitrification for simultaneous NH_4-N and NO_3-N removal

Tatsuru Kamei, Rawintra Eamrat, Kenta Shinoda, Yasuhiro Tanaka and Futaba Kazama

ABSTRACT

Nitrate removal during anaerobic ammonium oxidation (anammox) treatment is a concern for optimization of the anammox process. This study demonstrated the applicability and long-term stability of the coupled anammox and hydrogenotrophic denitrification (CAHD) process as an alternative method for nitrate removal. Laboratory-scale fixed bed anammox reactors (FBR) supplied with H_2 to support denitrification were operated under two types of synthetic water. The FBRs showed simultaneous NH_4-N and NO_3-N removal, indicating that the CAHD process can support NO_3-N removal during the anammox process. Intermittent H_2 supply (e.g. 5 mL/min for a 1-L reactor, 14/6-min on/off cycle) helped maintain the CAHD process without deteriorating its performance under long-term operation and resulted in a nitrogen removal rate of 0.21 kg-N/m^3/d and ammonium, nitrate, and dissolved inorganic nitrogen removal efficiencies of 73.4%, 80.4%, and 77%, respectively. The microbial community structure related to the CAHD process was not influenced by changes in influent water quality, and included the anammox bacteria '*Candidatus* Jettenia' and a *Sulfuritalea hydrogenivorans*-like species as the dominant bacteria even after long-term reactor operation, suggesting that these bacteria are key to the CAHD process. These results indicate that the CAHD process is a promising method for enhancing the efficiency of anammox process.

Key words | anammox, hydrogen gas, hydrogenotrophic denitrification, nitrate

Tatsuru Kamei
Futaba Kazama (corresponding author)
Interdisciplinary Research Centre for River Basin Environment,
University of Yamanashi 4-3-11,
Kofu, Yamanashi,
Japan
E-mail: *kfutaba@yamanashi.ac.jp*

Rawintra Eamrat
Interdisciplinary Graduate School of Medicine and Engineering,
University of Yamanashi 4-3-11,
Takeda, Kofu, Yamanashi,
Japan

Kenta Shinoda
Yasuhiro Tanaka
Futaba Kazama
Department of Environmental Science, Faculty of Life and Environment Science,
University of Yamanashi 4-4-37,
Kofu, Yamanashi,
Japan

INTRODUCTION

The anaerobic ammonia oxidation (anammox) process relies on the activity of chemoautotrophic bacteria (so called anammox bacteria), which converts ammonium nitrogen (NH_4-N) and nitrite nitrogen (NO_2-N) into nitrogen gas (N_2) and some nitrate nitrogen (NO_3-N) (Strous *et al.* 1998). The anammox process promotes aeration and electron donor addition cost savings of 50% and 100%, respectively, compared to the current nitrification and denitrification processes for NH_4-N removal. To exploit this high cost effectiveness, numerous studies have attempted to apply the anammox process to treat wastewater containing high NH_4-N concentrations, such as landfill leachate, digested wastewater, and industrial wastewater (e.g. Zhang *et al.* 2011; Isaka *et al.* 2017). Based on recent efforts, a commercial model of an anammox plant has been developed and widely applied at the plant-scale at over 100 sites (Lackner *et al.* 2014).

However, the anammox process cannot readily remove NO_3-N due to the physiological characteristics of anammox bacteria. Moreover, NO_3-N, an industrial byproduct found in effluents, is a major concern in the anammox wastewater treatment process design, more so in the case of high nitrogen loadings. This is because, as revealed in the stoichiometry of the anammox process (Equation (1)), approximately 26% of NH_4-N is converted into NO_3-N (Sliekers *et al.* 2002), and this could lead to the concentration of NO_3-N exceeding the allowed limit in wastewater if the concentration of NH_4-N is high in the untreated wastewater. Thus, residual NO_3-N removal should be conducted in the effluent from anammox process. Further, the effluent from anammox process usually contains NO_3-N and residual NH_4-N, resulting in fluctuations in the NO_2-N generation efficiency in partial nitrification

doi: 10.2166/wst.2018.459

units, which are commonly installed in anammox treatment facilities (Xie *et al.* 2018). Therefore, secondary treatment is required after the anammox process to simultaneously remove residual NH_4-N and NO_3-N. As such, both NO_3-N removal by the anammox process and simultaneous NH_4-N and NO_3-N removal are necessary to enhance the treatment efficiency of the anammox process.

$$NH_4^+ + 1.32NO_2^- \rightarrow 1.02N_2 + 0.26NO_3^- + 2.03H_2O \quad (1)$$

Coupling the anammox process with heterotrophic denitrification is an option to enable NO_3-N removal and simultaneous NH_4-N and NO_3-N removal. This has been developed for the treatment of municipal wastewater containing organic carbon (Sumino *et al.* 2006; Takekawa *et al.* 2014). However, the influent of the anammox treatment system generally do not contain organic carbon, because the anammox process is inhibited by the presence of organic carbon (Ni *et al.* 2012). Activation of heterotrophic denitrification requires the addition of organic carbon, and applying this technology to mainstream anammox treatments could cause further water quality deterioration. Therefore, coupled anammox and denitrification that can be operated under inorganic conditions must be developed.

We previously evaluated the concept of simultaneous NH_4-N and NO_3-N removal in inorganic wastewater via anammox and autotrophic hydrogenotrophic denitrification (HD), which drives the denitrification process via NO_3-N and NO_2-N reduction (Lee & Rittmann 2002). This combination of the anammox and HD processes (hereafter referred to as the CAHD process) uses hydrogen gas (H_2) as an electron donor in the NO_3-N reduction step (Figure 1). The estimated equation for the simultaneous removal of NH_4-N and NO_3-N via the CAHD process was calculated

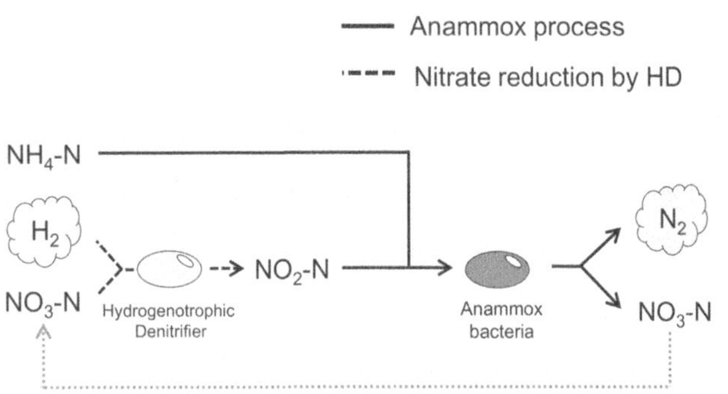

Figure 1 | Conceptual diagram of the coupled anammox and hydrogenotrophic denitrification (CAHD).

by combining the anammox and HD processes from a previous study (Kamei *et al.* 2015) (Equation (2)).

$$NH_4^+ + 1.06NO_3^- + 1.32H_2 \rightarrow 1.02N_2 + 3.32H_2O \quad (2)$$

This equation reveals that almost the same amount of NH_4-N and NO_3-N could be removed via the CAHD process. Since both microbial processes are autotrophic reactions (Smith *et al.* 1994; van de Graaf *et al.* 1996), excess sludge production could likely be minimized. H_2, unlike other electron donors (e.g. methanol, ethanol), does not inhibit the anammox process, and can be utilized in the coupling anammox and denitrification process for the simultaneous removal of NH_4-N and NO_3-N (Waki *et al.* 2013). Supplying H_2 can remove residual oxygen in influent water, which supports and stabilizes the anammox and HD processes in the CAHD process. Including this, the deterioration of the anammox process induced by electron donor addition may be neglected in the CAHD process.

Furthermore, H_2 utilization as an electron donor in this process can provide several advantages. Since HD activity can be controlled by changing the dose of H_2, the denitrification performance of CAHD process can be controlled and changed easily. On the other hand, a secondary treatment system for residual electron donor can be eliminated in the CAHD process, since H_2 is naturally released from treated water to the atmosphere.

Recently, complete oxidation of NH_4-N to NO_3-N by nitrification and subsequent coupling denitrification and anammox is focused as a new concept of anammox treatment, since complete nitrification is not required, sensitive system control compared to partial nitrification (Si *et al.* 2018). According to current research work, the CAHD process can also be applied to the treatment of inorganic wastewater containing only NH_4-N, by combining nitrification for oxidizing NH_4-N to NO_3-N. The CAHD process could be a promising technology for wastewater treatment of anammox process based on the several advantages mentioned above. To develop the CAHD process, the key factors for combining the two processes, the performance stability, and the microbial structure, should be evaluated comprehensively. However, limited information is available on the nitrogen removal efficiency of the CAHD process, although many studies have been conducted on the anammox and HD processes independently. Therefore, in this study, we evaluated (1) the performance stability of a synthetic inorganic wastewater treatment using CAHD process to estimate the long-term performance of the system and (2) the major microbial species related to the CAHD process to help clarify the mechanism.

METHODS

Experimental design

We conducted two independent experiments to evaluate the performance of the CAHD process using synthetic inorganic wastewater (Experiment I), and the long-term stability of the nitrogen removal performance (Experiment II) in fixed-bed reactors (FBRs). The FBRs were inoculated with anammox sludge immobilized on non-woven fabric as a microbial seed. The anammox sludge used in this research was obtained via long-term cultivation of activated sludge. The initial biomass concentration was set to ~12.5 g wet weight (w.w)/L in each experiment. Beside this, seed sludge for HD was not inoculated in these experiments, since we had found the concurrent occurrence of anammox and HD processes after cultivation of anammox sludge under H_2 supplying condition as reported in previous research (Kamei *et al.* 2015). The FBRs were operated at a water temperature of 35 °C. H_2 (more than 99.99% purity) produced from a water electrolytic H_2 generator (HG-260; GL Science, Tokyo, Japan) was supplied through a commercial air stone diffuser with a 15 × 30-mm diameter. The H_2 flow rates were set at 2.5 to 5 mL/min for a reactor volume of 1 L, and were changed based on the experimental set up. The dissolved oxygen (DO) concentration of the influent in each experiment was maintained below 0.3 mg/L to minimize the inhibition of anaerobic processes in the presence of oxygen.

Details of the experimental conditions are summarized in Table 1. In Experiment I, two sets of FBRs with 4.2-L working volumes were prepared and operated under both pH-controlled and pH-uncontrolled conditions to compare their nitrogen removal efficiencies. Figure 2 summarizes the overall experimental set up. The pH inside the reactor was maintained at 7.8–8.3 via the periodic supply of carbon dioxide gas. The FBRs were operated under four operational conditions (Runs 1–4) to evaluate the CAHD process through sequencing operations (Run 1) and batch-mode operations to elucidate the presence of the anammox process (Run 2), the HD process (Run 3), and the effect of reduced H_2 on the CAHD process under the sequencing reactor operation (Run 4). The batch experiments of Run 2 and 3 were conducted only to pH controlled reactor after Run 1. Subsequently, the reactor was operated again in the sequencing mode setting for Run 4. In Experiment I, synthetic inorganic wastewater was prepared using tap water and adding reagents (Table 2). $(NH_4)_2SO_4$, $NaNO_2$, and $NaNO_3$ were added as sources of NH_4-N, NO_2-N and NO_3-N. The dose pattern was varied based on each run with the initial concentrations set at 40 mg-N/L for each of the nitrogen sources (Table 1).

In Experiment II, the long-term performance stability was evaluated under the conditions of Run 5. The FBRs were restarted and operated with intermittent H_2 gas supply with the pH controlled at 7.8–8.3 via periodic carbon dioxide supply. A synthetic medium for anammox bacterial enrichment (Strous *et al.* 1998) was used as the influent water to assess the effect of intermittent H_2 supply on the CAHD process (Table S1, available with the online version of this paper). In this experiment, $(NH_4)_2SO_4$ and $NaNO_3$ were added to obtain NH_4-N and NO_3-N concentrations of 40 mg-N/L each. The H_2 was supplied in the same volume as that in Experiment I, although the reactor volume was changed to 2 L. To compare the performance changes and intermittent gas patterns, the H_2 supply rate was calculated using Equation (3):

$$H_2 \text{ supply ratio } (\%) = 100 \times (T_{H2}/T_A) \tag{3}$$

Table 1 | Details of the experimental set up

Experiment	Reactor volume (L)	HRT[a] (hour)	Initial sludge (g w.w/L)	Runs	Operational mode	NH_4-N (mg-N/L)	NO_3-N (mg-N/L)	NO_2-N (mg-N/L)	H_2 flow rate (mL/min/$L_{reactor}$)
I	4.2	7	12.5	Run 1	Continuous	40	40	–	5
				Run 2	Batch	40	–	40	–
				Run 3	Batch	40	40	–	5
				Run 4	Continuous	40	40	–	2.5, 3.8, 5
II	2	7	12.5	Run 5	Continuous	40	40	–	5 (Intermittent supply)

[a]HRT: Hydraulic retention time.

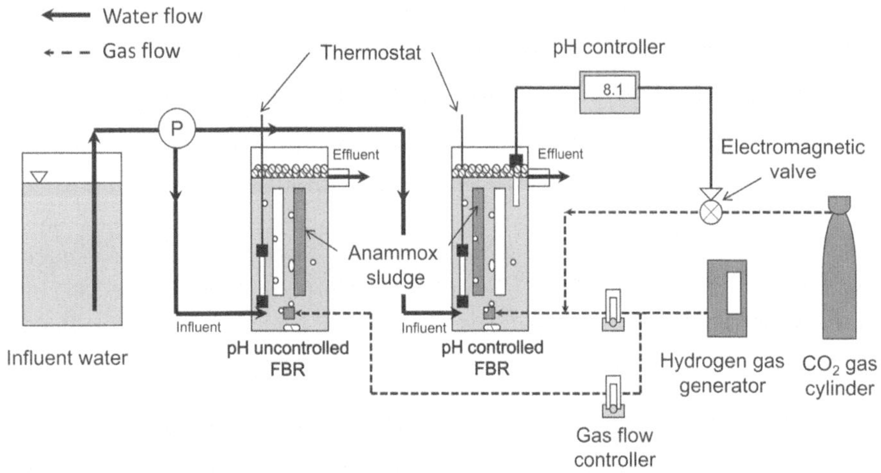

Figure 2 | Experimental set up of the fixed-bed reactors (FBRs) in Experiment I.

Table 2 | Components of the synthetic inorganic wastewater used in Experiment I

	(g/L)
$Na_2HPO_4\ 12H_2O$	0.04
KCl	0.01
$CaCl_2\ 2H_2O$	0.08
$MgSO_4\ 7H_2O$	0.07
$NaHCO_3$	1.2

where T_{H2} represents the H_2 supply time during one cycle and T_A is the duration of one cycle (20 min).

Water sampling and analysis

The DO concentration was measured with a portable DO meter (Multi 3410; WTW, Weilheim, Germany). Influent and effluent water samples were collected and filtered through 0.45-μm membrane filters (Merck Millipore, Darmstadt, Germany). The samples were frozen and stored until the water quality analysis. To measure inorganic nitrogen concentrations, we used the phenate method to determine the NH_4-N concentration (JWWA 1993) and the colorimetric and ultraviolet methods to determine the NO_2-N and NO_3-N concentrations (APHA 1998). Based on the results of the water quality analysis, the nitrogen loading rate (NLR) and nitrogen removal rate (NRR) were calculated using Equations (4) and (5).

$$\mathrm{NLR}\,(\text{kg-N/m}^3/\text{d}) = C_{in} \times Q/V \quad (4)$$

$$\mathrm{NRR}\,(\text{kg-N/m}^3/\text{d}) = \Delta C \times Q/V \quad (5)$$

where C_{in} represents the dissolved inorganic nitrogen (DIN) concentration as the sum of NH_4-N, NO_2-N, and NO_3-N in influent water and ΔC denotes the difference in DIN difference between influent and effluent water from the reactor. Meanwhile, Q and V represent the water flow rate and volume of the FBRs in each experimental condition, respectively.

Microbial community analysis using bacterial 16S rRNA amplicon sequences

Samples for the microbial community analysis were collected from the FBRs of both Experiments I and II, and were collected from the surface of the microbial carriers (day 0) and suspended solids (on each operational day). Total nucleic acids were extracted from 0.1 g w.w of sample using a FastDNA SPIN Kit for Soil (MP-Biomedicals, Santa Ana, CA, USA). A partial region (the V4 hypervariable region) of the bacterial 16S rRNA gene was amplified by polymerase chain reaction (PCR) using the primers Eub-515f (5′-GTGCCAGCMGCCGCGGTAA-3′) and Eub-806r (5′-GGACTACSVGGGTATCTAAT-3′) (Caporaso *et al.* 2011). The total bacterial concentration in each sample collected from the FBRs ranged from 1.8×10^8 to 2.4×10^9 copies/g w.w biomass. To minimize the PCR biomass, the PCR reaction comprised 20 cycles. The amplified products were subjected to next-generation sequencing using an Illumina MiSeq gene sequencer by a commercial sequencing service (FASMAC, Atsugi, Japan). In total, 48,105–87,786 sequence reads were obtained and classified into operational taxonomical units (OTUs) based on a 3%

divergence (i.e. 97% similarity). In total, 116–155 OTUs in each biomass sample were obtained after classification.

RESULTS AND DISCUSSION

CAHD process in synthetic inorganic wastewater treatment

Figure 3 presents the NH_4-N and NO_3-N removal efficiencies and NO_2-N concentration in the reactors. Under pH controlled conditions NH_4-N and NO_3-N concentrations decreased and NO_2-N concentrations slightly increased after supplying H_2. The pH-controlled reactor maintained simultaneous NH_4-N and NO_3-N removal for more 30 days continuously, with NH_4-N and NO_3-N removal efficiencies of 46% and 80%, respectively, at the end of the experiment. The NO_2-N concentration was maintained at around 5 mg-N/L throughout the experiment. By contrast, although NO_3-N removal and NO_2-N accumulation were observed in the FBR with uncontrolled pH, no significant NH_4-N removal was observed. Since simultaneous removal of NH_4-N and NO_3-N were detected, enhanced NRR and stability were only observed under the pH-controlled condition (Figure 4). The NRR in this reactor reached 0.18 kg-N/m^3/d for an NLR of 0.32 kg-N/m^3/d and achieved a DIN removal rate of 56% after 30 days of operation. By contrast, the reactor with uncontrolled pH showed a two-fold lower NRR, reaching 0.09 kg-N/m^3/d by day 30. In addition, the pH inside the reactor reached 8.7 after 5 days of operation, and then remained at around 8.6–8.9 until end of the operation. Since the reactor pH exceeded the suitable growth range for anammox bacteria (6.5–8.5) (Ali *et al.* 2015), the anammox process was inhibited in the pH uncontrolled reactor. Further, the NH_4-N and NO_2-N removal via the anammox process was suppressed, resulting in NO_2-N accumulation after NO_3-N reduction by the HD process in this reactor. In the CAHD process, denitrification mainly occurs via the anammox reaction. Thus, the pH uncontrolled reactor showed a lower NRR than the pH controlled reactor.

The results of the batch experiments in Runs 2 and 3 showed the simultaneous occurrence of the anammox and HD processes under pH-controlled conditions (Figure 5). Simultaneous removal of NH_4-N and NO_2-N with slight increments in NO_3-N were detected under conditions conducive to the anammox process (Run 2), where NO_2-N was added instead of NO_3-N (Figure 5(a)). The NO_2-N to NH_4-N removal ratio throughout the experimental period was 1.21, which was similar to the theoretical value based on the anammox process. In addition, NO_3-N removal with NO_2-N accumulation occurred only under H_2 supply in the batch experiment of Run 3 (Figure 5(b)). The occurrence of NO_3-N reduction and NO_2-N accumulation depended on the supply of H_2, which is suggestive of the concurrent occurrence of the HD process in the reactor and might support simultaneous NH_4-N and NO_3-N removal along with the anammox process. Further, pH control should be maintained to avoid anammox process deterioration and achieve simultaneous NH_4-N and NO_3-N removal via the CAHD process.

However, NH_4-N degradation was insignificant under H_2 supply in Run 3. Additionally, the NO_3-N removal was dominant under continuous operation in Run 1, as shown in the

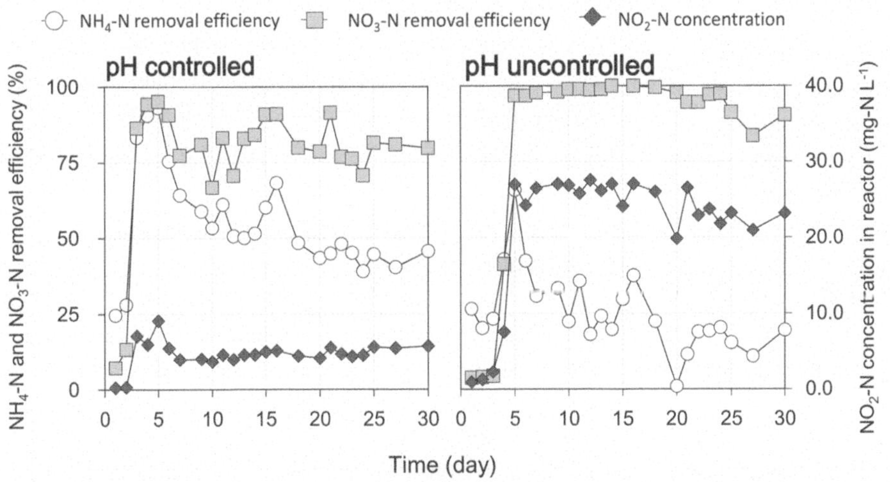

Figure 3 | Removal efficiency of NH_4-N and NO_3-N, and NO_2-N concentration changes for the pH-controlled and pH-uncontrolled conditions in Run 1 of Experiment I.

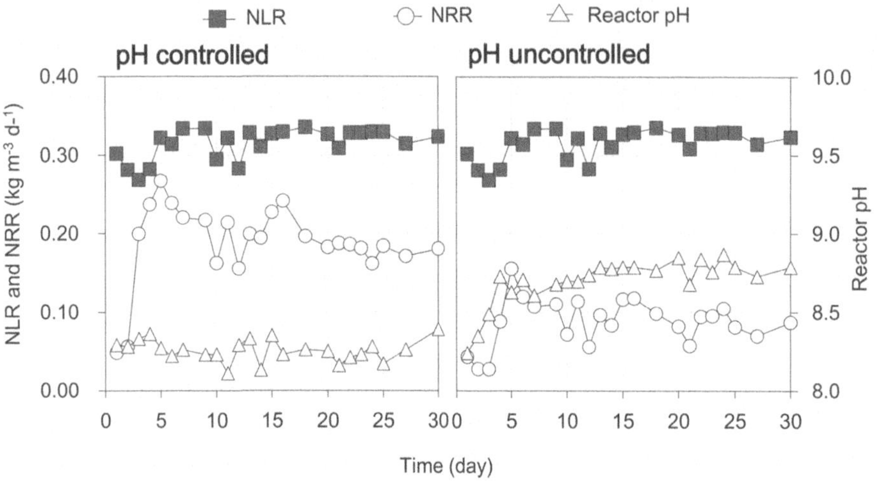

Figure 4 | NLR, NRR, and reactor pH during the CAHD process using synthetic wastewater in Run 1 of Experiment I.

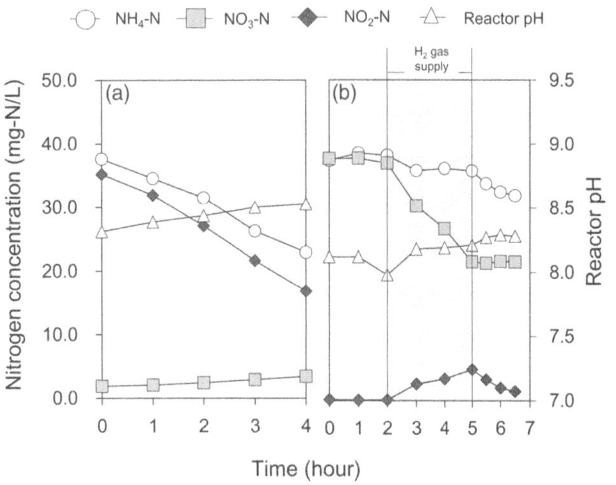

Figure 5 | Changes in the nitrogen concentration in the batch mode operations of the pH controlled FBR to clarify the presence of (a) the anammox process in Run 2 and (b) the HD process in Run 3 of Experiment I.

performance results detected at day 30 (NH_4-N removal efficiency: 48%, NO_3-N removal efficiency: 80%) and the higher NO_3-N to NH-N removal ratio (NO_3-N/NH_4-N = 1.3–2.1), which was greater than the theoretical ratio for the CAHD process (NO_3-N/NH_4-N = 1.06), after achieving stable denitrification (Figure S1, available with the online version of this paper). These results suggest the predominance of NO_2-N reduction via the HD process even after achieving simultaneous NH_4-N and NO_3-N removals.

Insufficient NO_2-N supply for the anammox process would limit the denitrification efficiency of the CAHD process. Adequate H_2 supply induced NO_2-N reduction in the HD process, as reported by Lee & Rittmann (2002) (e.g. 0.18 g H_2 per 1 g NO_3-N reduction and 0.21 g of H_2 per 1 g NO_2-N reduction). Thus, the pH-controlled FBR was operated with reduced H_2 flow rates in Run 4 to minimize NO_2-N reduction and achieve NO_2-N accumulation in the HD process. However, the results revealed suppressed NO_3-N reduction in the HD process due to a lack of H_2, resulting in less NO_2-N accumulation and a decrease in the NH_4-N removal via the anammox process (Table S2, available online). To minimize NO_2-N reduction in the HD process and maintain the long-term operation of the CAHD process, we considered decreasing the contact time between H_2 and HD bacteria as an alternative method according to the results of Run 3 (Figure 5(b)). In Run 3, NO_3-N reduction and NO_2-N accumulation occurred under H_2 supply. Moreover, NO_2-N might have been consumed in the anammox process without H_2 supply where NH_4-N was removed; this is indicative of the suppression of NO_2-N reduction in the HD process without H_2 supply. Thus, intermittent H_2 supply can control the suppression of NO_2-N reduction in the HD process. Hence, the FBR was restarted and successfully operated with intermittent H_2 supply for another 180 days continuously in Experiment II, supporting the long-term maintenance of the CAHD process.

Long-term nitrogen removal performance under intermittent H_2 supply

After 30 days of operation of the CAHD process under continuous H_2 supply, intermittent H_2 supply was started under the conditions for Run 5 in Experiment II. The denitrification performance of the FBR was maintained, and simultaneous removal of NH_4-N and NO_3-N was detected continuously for more than 180 days (Figure 6). FBR

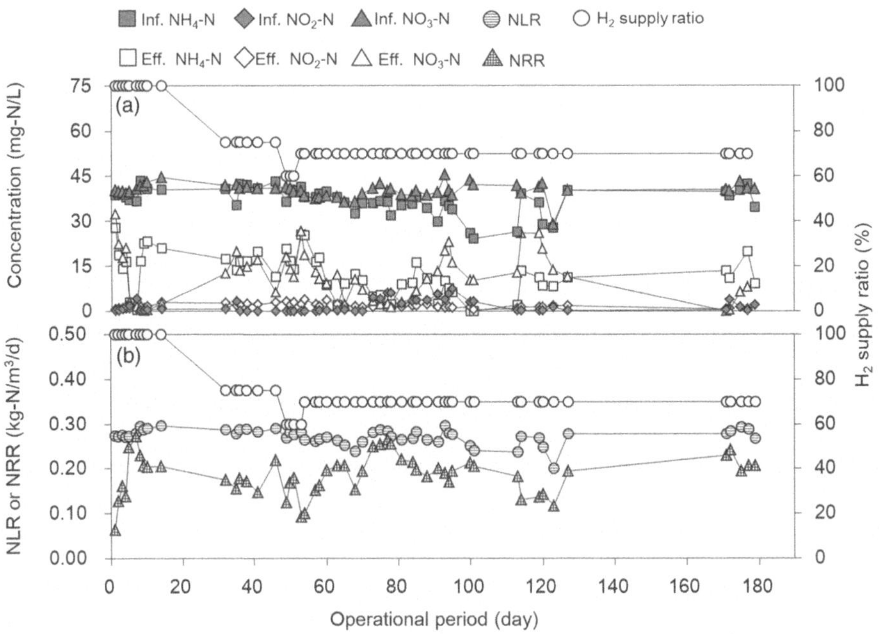

Figure 6 | Changes in (a) nitrogen concentration in effluent and influent and (b) NLR and NRR during the long-term operation of the FBRs under Run 5 of Experiment II.

operation under a H_2 supply rate of 70% (14/6 min on/off cycle) seemed to be much more stable than the other conditions (75%: 15/5 min on/off cycle; 60%: 12/8 min on/off cycle). Although the performance was occasionally fractured under these conditions, the effluent concentrations of NH_4-N and NO_3-N were below 15 and 10 mg-N/L in most samples, respectively (Figure 6(a)). At the end of the run, the NH_4-N, NO_3-N, and DIN removal rates reached 73.4%, 80.4%, and 77%, respectively, while achieving an NRR of 0.21 kg-N/m^3/d under an NLR of 0.27 kg-N/m^3/d (Figure 6(b)). Compared to the results of Experiment I, the NO_3-N to NH_4-N removal ratio was less than 1.6 after beginning intermittent H_2 supply, which was closer to the theoretical value of the HD process than the results of Experiment I (Figure S2, available online). Therefore, intermittent H_2 supply could avoid NO_2-N reduction via the HD process while maintaining NO_3-N reduction, resulting in good performance of the anammox process for NH_4-N removal without a lack of NO_2-N.

High gas production costs and lower gas use efficiencies are major bottlenecks of H_2 use in microbiological nitrogen removal systems, including the CAHD process, since H_2 has a low solubility in water (i.e. standard-state saturation concentration of 1.6 mg/L). Recent studies have attempted to overcome these bottlenecks with newly developed applications, for instance the development of closed and pressurized reactor operations (Epsztein *et al.* 2017), special gas diffusers such as hollow fiber membranes, and microbubble diffuser systems (Eamrat *et al.* 2016). Moreover, Epsztein *et al.* (2017) developed a sealed and pressurized reactor that yielded an H_2 use efficiency of 100% during the denitrification process. Meanwhile, Eamrat *et al.* (2016) reported that the use of a microbubble diffuser with intermittent H_2 supply enhanced the gas bubble retention time, resulting in H_2 savings of 75% compared to continuous gas supply during NO_3-N removal (Eamrat *et al.* 2016). Although further optimization of this process is necessary, the CAHD process could be maintained by controlling the reduction of NO_3-N and accumulation of NO_2-N via intermittent H_2 supply. The H_2 volume required to support NO_3-N reduction and NO_2-N accumulation in the CAHD process may be further minimized by applying this newly developed technology, making the CAHD process more feasible for application to the simultaneous removal of NH_4-N and NO_3-N from inorganic wastewaters, such as secondary effluent from anammox treatment and wastewater containing NH_4-N after oxidizing of NH_4-N to NO_3-N by nitrification process.

Analysis of the microbial community in the CAHD process

We analyzed the changes in the microbial community structure of pH-controlled FBRs during Run 1 of Experiment I and Run 5 of Experiment II, and summarized these at the phylum level although a Proteobacteria abundance was described based on subclass classification (Figure 7(a)). The major taxa in each sample included Planctomycetes, β-Proteobacteria, Chlorobi, and Armatimonadetes. However, the relative abundances of Planctomycetes and β-Proteobacteria changed markedly after beginning H_2 supply. Consequently, β-Proteobacteria became the predominant taxon in each condition, accounting for 40.6% (day 30, Run 1, Experiment I) and 70.6% (day 171, Run 5, Experiment II) of the total bacteria. Moreover, it occurred in the presence of NO_3-N reduction in both experiments. By contrast, although the relative abundance decreased in the presence of H_2, Planctomycetes was the second-most dominant taxon over the course of the operational period in both experiments.

The major bacterial genera were '*Candidatus* Jettenia' in Planctomycetes and *Dok59* in β-Proteobacteria in both Experiments I and II (Figure 7(b)), indicative of similar microbial communities in the inoculum and suspended sludge in both FBRs, even though the wastewater types and operational periods differed. The genus '*Candidatus* Jettenia' is known for its activity in the anammox process, including several proposed species (Ali *et al.* 2015). In Experiment II, this genus still accounted for ~15.2% of the total bacteria after 180 days of operation. In addition, another anammox genus, '*Candidatus* Brocadia', appeared and accounted for 1.5% of the total bacteria on day 171, indicating that the reactor conditions were still conducive to anammox bacterial growth, although the overall relative abundance of anammox bacteria decreased in the suspended sludge.

The changes in sludge appearance also suggested the existence of the anammox process in the reactor during long-term reactor operation in Experiment II (Figure S3, available online). Brownish-colored sludge accumulated in the reactor, and most of the sludge became brownish over the course of Experiment II. However, the microbial carrier as the inoculum maintained a reddish color with tiny anammox granules in the suspended sludge. Since the water quality analysis results revealed NH_4-N removal, these results indicated that the anammox process, and thus nitrogen removal activity, continued in the reactor even after long-term operation with intermittent H_2 supply in Experiment II.

After starting H_2 supply, the genus *Dok59* increased and became the dominant genus, ahead of the genera *Thauera* and *Hydrogenophaga*, which are hydrogen-dependent denitrifiers (Kämpfer *et al.* 2005; Mao *et al.* 2013). After long-term system operation in Experiment II, the relative abundance of this genus reached 65%, representing the dominant taxon. *Dok59* is an unclassified genus in the family Rhodocyclaceae. Detailed information of *Dok59* is very limited, although several studies have reported its

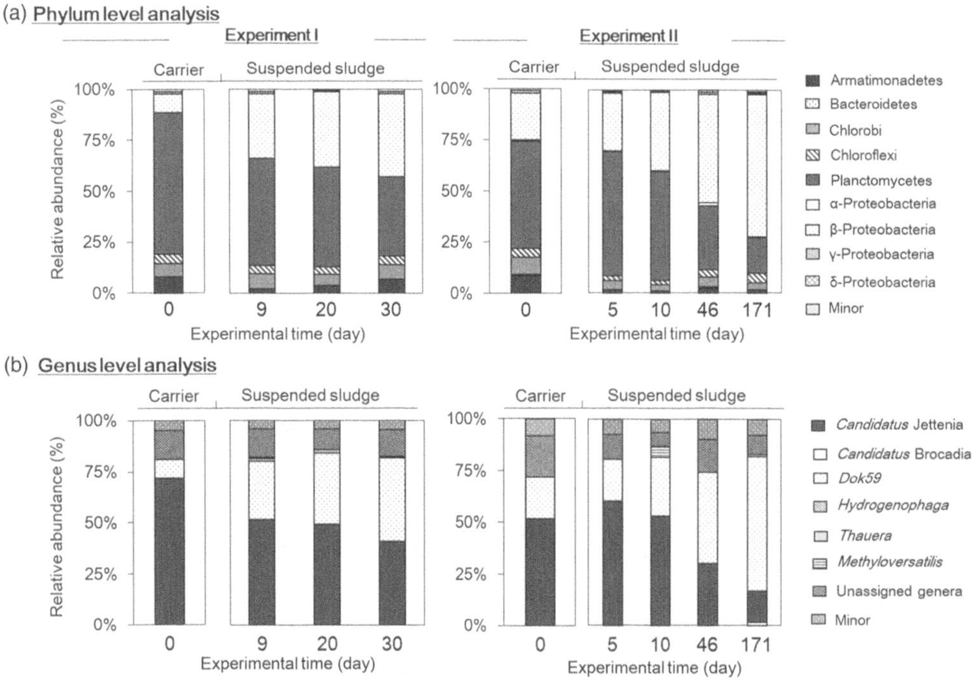

Figure 7 | Microbial community analysis of pH controlled FBRs in Run 1 of Experiment I and Run 5 of Experiment II. (a) Phylum-level analysis. (b) Genus level analysis. Relative abundances <0.5% are classified as 'Minor'.

detection in biomass from anammox reactors (Chang *et al.* 2011). The homology of the *Dok59* sequence from the next-generation sequencing analysis was compared to that in the BLAST database provided by the National Center for Biotechnology Information (NCBI) (https://blast.ncbi.nlm.nih.gov/Blast.cgi), and showed 96% similarity to the *Sulfuritalea hydrogenivorans* strain sk43H^T (family: Rhodocyclaceae). This species is an autotrophic bacterium isolated from fresh lake sediment, and can perform nitrate reduction using thiosulfate, elemental sulfur, and H_2 as electron donors (Kojima & Fukui 2011). The *Sulfuritalea hydrogenivorans*-like species could exist as members of the microbial community of enriched anammox sludge by using NO_3-N produced via the anammox process and alternative electron donors, and may have become highly enriched after beginning the experimental run by using H_2 as the major electron donor. These results indicate that anammox bacteria and the *Sulfuritalea hydrogenivorans*-like species might perform simultaneous NH_4-N and NO_3-N reduction in the CAHD process.

CONCLUSIONS

Coupled denitrification via anammox and HD removed NH_4-N and NO_3-N simultaneously from synthetic inorganic wastewater, indicating that this coupled denitrification process could be applied to inorganic wastewater treatment. Intermittent H_2 supply (e.g. 14/6 min on/off cycle; H_2 flow rate: 5 mL/min/$L_{reactor}$) might be a critical factor in maintaining the CAHD process for stable long-term performance, resulting in NH_4-N, NO_3-N, and DIN removal efficiencies of 73.4%, 80.4%, and 77%, respectively, after 180 days of operation. The *Sulfuritalea hydrogenivorans*-like species and the anammox bacteria '*Candidatus* Jettenia' coexisted in the FBRs, indicating that simultaneous NH_4-N and NO_3-N removal was achieved by these two microbes. Further studies should be conducted to optimize the H_2 volume for NO_3-N reduction and NO_2-N accumulation in H_2-dependent denitrification and evaluate the treatment performance and microbial community changes through experiments using actual wastewater.

ACKNOWLEDGEMENTS

This research was performed with partial financial assistance from the 'Project for Hydro-microbiological Approach for Water Security in Kathmandu Valley, Nepal' under the Science and Technology Research Partnership for the Sustainable Development program of Japan Science and Technology Agency and Japan International Cooperation Agency, as well as a Japan Society for the Promotion of Science KAKENHI Grant (grant number 26340059).

REFERENCES

Ali, M., Oshiki, M., Awata, T., Isobe, K., Kimura, Z., Yoshikawa, H., Hira, D., Kindaichi, T., Satoh, H., Fujii, T. & Okabe, S. 2015 Physiological characterization of anaerobic ammonium oxidizing bacterium '*Candidatus* Jettenia caeni'. *Environ. Microbiol.* **17**, 2172–2189.

American Public Health Association/American Water Works Association/Water Environment Federation (APHA) 1998 *Standard Methods for the Examination of Water and Wastewater*, 20th edn. American Public Health Association/American Water Works Association/Water Environment Federation (APHA), Washington, DC, USA.

Caporaso, J. G., Lauber, C. L., Walters, W. A., Berg Lyons, D., Lozupone, C. A., Turnbaugh, P. J., Fierer, N. & Knight, R. 2011 Global patterns of 16S rRNA diversity at a depth of millions of sequences per sample. *Proceedings of the National Academy of Sciences of the United States of America* **108**, 4516–4522.

Chang, C. Y., Tanong, K., Xu, J. & Shon, H. 2011 Microbial community analysis of an aerobic nitrifying-denitrifying MBR treating ABS resin wastewater. *Bioresource Technology* **102**, 5337–5344.

Eamrat, R., Mochizuki, T., Kamei, T. & Kazama, F. 2016 Hydrogenotrophic denitrification activity under intermittent hydrogen supply using micro-bubble system hydrogenotrophic denitrification activity under intermittent hydrogen supply using micro-bubble system. *Naresuan University Engineering Journal* **11**, 47–51.

Epsztein, R., Beliavski, M., Tarre, S. & Green, M. 2017 Pressurized hydrogenotrophic denitrification reactor for small water systems. *Journal of Environmental Management* **216**, 315–319.

Isaka, K., Kimura, Y., Matsuura, M., Osaka, T. & Tsuneda, S. 2017 First full-scale nitritation-anammox plant using gel entrapment technology for ammonia plant effluent. *Biochemical Engineering Journal* **122**, 115–122.

JWWA 1993 *Standard Method for Examination of Water*. Japan Water Works Association (JWWA), Tokyo, Japan.

Kamei, T., Naitoh, D., Khanitchaidecha, W. & Kazama, F. 2015 Simultaneous removal of ammonium and nitrate by a combination of ANAMMOX and hydrogenotrophic denitrification. *Journal of Water and Environment Technology* **13**, 167–178. doi:10.2965/jwet.2015.167.

Kämpfer, P., Schulze, R., Jäckel, U., Malik, K. a., Amann, R. & Spring, S. 2005 *Hydrogenophaga defluvii* sp. nov. and *Hydrogenophaga atypica* sp. nov., isolated from activated sludge. *International Journal of Systemic and Evolutionary Microbiology* **55**, 341–344.

Kojima, H. & Fukui, M. 2011 *Sulfuritalea hydrogenivorans* gen. nov., sp. nov., a facultative autotroph isolated from a freshwater lake. *International Journal of Systemic and Evolutionary Microbiology* **61**, 1651–1655.

Lackner, S., Gilbert, E. M., Vlaeminck, S. E., Joss, A., Horn, H. & van Loosdrecht, M. C. M. 2014 Full-scale partial nitritation/anammox experiences – an application survey. *Water Research* **55**, 292–303.

Lee, K.-C. & Rittmann, B. E. 2002 Applying a novel autohydrogenotrophic hollow-fiber membrane biofilm reactor for denitrification of drinking water. *Water Research* **36**, 2040–2052.

Mao, Y., Xia, Y. & Zhang, T. 2013 Characterization of *Thauera*-dominated hydrogen-oxidizing autotrophic denitrifying microbial communities by using high-throughput sequencing. *Bioresource Technology* **128**, 703–710.

Ni, S. Q., Ni, J. Y., Hu, D. L. & Sung, S. 2012 Effect of organic matter on the performance of granular anammox process. *Bioresource Technology* **110**, 701–705.

Si, Z., Peng, Y., Yang, A., Zhang, S., Li, B., Wang, B. & Wang, S. 2018 Rapid nitrite production via partial denitrification: pilot-scale operation and microbial community analysis. *Environmental Science: Water Research & Technology* **4**, 80–86.

Sliekers, A. O., Derwort, N., Gomez, J. L. C., Strous, M., Kuenen, J. G. & Jetten, M. S. M. 2002 Completely autotrophic nitrogen removal over nitrite in one single reactor. *Water Research* **36**, 2475–2482.

Smith, R. L., Ceazan, M. L. & Brooks, M. H. 1994 Autotrophic, hydrogen-oxidizing, denitrifying bacteria in ground water, potential agents for bioremediation of nitrate contamination. *Applied and Environmental Microbiology* **60**, 1949–1955.

Strous, M., Heijnen, J. J., Kuenen, J. G. & Jetten, M. S. M. 1998 The sequencing batch reactor as a powerful tool for the study of slowly growing anaerobic ammonium-oxidizing microorganisms. *Applied and Environmental Microbiology* **50**, 589–596.

Sumino, T., Isaka, K., Ikuta, H., Saiki, Y. & Yokota, T. 2006 Nitrogen removal from wastewater using simultaneous nitrate reduction and anaerobic ammonium oxidation in single reactor. *Journal of Bioscience and Bioengineering* **102**, 346–351.

Takekawa, M., Park, G., Soda, S. & Ike, M. 2014 Simultaneous anammox and denitrification (SAD) process in sequencing batch reactors. *Bioresource Technology* **174**, 159–166.

van de Graaf, A. A., de Bruijn, P., Robertson, L. A., Jetten, M. S. M. & Kuenen, J. G. 1996 Autotrophic growth of anaerobic ammonium-oxidizing micro-organisms in a fluidized bed reactor. *Microbiology* **142**, 2187–2196.

Waki, M., Yasuda, T., Fukumoto, Y., Kuroda, K. & Suzuki, K. 2013 Effect of electron donors on anammox coupling with nitrate reduction for removing nitrogen from nitrate and ammonium. *Bioresource Technology* **130**, 592–598.

Xie, G.-J., Liu, T., Cai, C., Hu, S. & Yuan, Z. 2018 Achieving high-level nitrogen removal in mainstream by coupling anammox with denitrifying anaerobic methane oxidation in a membrane biofilm reactor. *Water Research* **131**, 196–204.

Zhang, L., Yang, J., Hira, D., Fujii, T., Zhang, W. & Furukawa, K. 2011 High-rate nitrogen removal from anaerobic digester liquor using an up-flow anammox reactor with polyethylene sponge as a biomass carrier. *Journal of Bioscience and Bioengineering* **111**, 306–311.

First received 31 July 2018; accepted in revised form 20 October 2018. Available online 30 October 2018

Enzymatic hydrolysis of floatable fatty wastes from dairy and meat food-processing industries and further anaerobic digestion

N. C. Pascale, J. J. Chastinet, D. M. Bila, G. L. Sant`Anna Jr, S. L. Quitério and S. M. R. Vendramel

ABSTRACT

The wastewater from food-processing industries is generally heavily charged with lipids and proteins. Flotation process is commonly applied to separate the hydrophobic material phase, producing flotation froth, a waste that has high levels of fats and proteins. Enzymatic hydrolysis may be used to overcome the difficulty of fat biotransformation in a subsequent anaerobic digestion. In the present work, wastes from the flotation process of two industries (dairy and poultry slaughterhouse) were hydrolyzed with a commercial lipase and without enzyme addition (control). The effect of adjusting the pH at the beginning of the hydrolytic assays was also investigated. The long chain free fatty acids (LCFAs) released were identified and quantified and 5-d digestion assays were conducted with the hydrolyzed material. The results indicated that the hydrolysis assays conducted with initial pH adjusted to 7.0 and the utilization of a commercial enzyme promoted a higher increase in amounts of LCFAs, particularly of unsaturated acids. In most anaerobic digestion assays, the specific methane production showed a decreasing trend with the increase of unsaturated fatty acids in the medium. In general, the utilization of a commercial enzyme (lipase) in the hydrolysis process did not contribute to enhancing methane production in 5-d anaerobic digestion assays.

Key words | anaerobic digestion, fatty wastes, hydrolysis, lipase, lipids

N. C. Pascale
J. J. Chastinet
S. L. Quitério
S. M. R. Vendramel (corresponding author)
Laboratory of Environmental Sciences,
Federal Institute of Education,
Science and Technology of Rio de Janeiro,
121 Senador Furtado Street,121, 20270-021 Rio de Janeiro, RJ,
Brazil
E-mail: *simone.vendramel@ifrj.edu.br*

D. M. Bila
G. L. Sant`Anna Jr
Department of Sanitary and Environmental Engineering,
State University of Rio de Janeiro,
524 São Francisco Xavier Street, room 5029-F,
20550-900 Rio de Janeiro, RJ,
Brazil

INTRODUCTION

The wastewater from several food-processing industries such as dairy and meat contain high concentrations of organic matter, mainly lipids and proteins. To remove particulate organic matter, the flotation process is commonly applied to remove hydrophobic materials from the aqueous phase, resulting in the recovery of large amounts of fats and greases (fatty waste) from the wastewater. Thus, a generally pasty waste (flotation froth) is generated, whose environmentally safe disposal or treatment is required. Anaerobic digestion could be an appropriate technology to reduce the pollution potential of this kind of fatty waste besides producing biomethane, which is a renewable source of energy. According to Neves *et al.* (2009), potentially, fats present high theoretical conversion yields to methane (0.99 L g^{-1}) when compared with carbohydrates (0.42 L g^{-1}) and proteins (0.63 L g^{-1}). However, the hydrolysis of lipids is frequently a limiting step for the anaerobic process. In general, fats are removed by several techniques to avoid operation problems in biological treatment systems. Problems caused by excessive oil and grease include the reduction in cell-aqueous phase transfer rates, sedimentation hindrance due to the development of filamentous microorganisms, development and flotation of sludge with poor activity, clogging and the emergence of unpleasant odors (Valladão *et al.* 2007).

The simplest lipids containing fatty acids are triacylglycerols, which are composed of three fatty acids, each bonded to a glycerol molecule by an ester bond. When these substances are hydrolyzed, glycerol and free fatty acids are liberated. Although the splitting of these molecules by hydrolysis seems to be beneficial for the subsequent biodegradation process, some triacylglycerols release long chain fatty acids (LCFAs) and these substances may hinder, particularly, anaerobic digestion. The LCFAs have already been reported, some decades ago, as inhibitors of

doi: 10.2166/wst.2018.508

the anaerobic process (Hanaki *et al.* 1981; Koster & Cramer 1987). Meanwhile, the cleavage of triacylglycerols by hydrolysis can increase the biodisponibility of organic molecules in the medium, improving the biodegradation process (Battimelli *et al.* 2010).

So, it seems that a kind of dilemma exists: is hydrolysis (mainly enzymatic hydrolysis) of fats really beneficial for the subsequent anaerobic digestion? Or does hydrolysis generate LCFAs that have an inhibitory effect on anaerobic digestion?

A relative abundant literature evidenced that, for industrial wastewaters containing oils and grease (O&G) contents up to 1,500 mg L^{-1}, the previous enzymatic hydrolysis enhanced methane production (Leal *et al.* 2006; Sousa *et al.* 2013; Duarte *et al.* 2015; Meng *et al.* 2017). The successful results obtained with wastewaters were a source of motivation to investigate whether enzymatic hydrolysis would be able to improve the anaerobic biodegradation of froth wastes, a very different matrix and, consequently, enhance methane production. Thus, the hydrolysis of the floatable fatty wastes from dairy and meat food-processing industries was investigated.

METHODS

The fatty wastes were collected from flotation units of local industries (dairy and poultry slaughterhouse), and stored at −4.0 °C until use. Working with flotation froths presents challenges because their composition and physical structure can change during storage, even when maintained at low temperatures. In the present investigation, we have made the choice of working with only one sample of each type of flotation froth, and we performed the experiments in parallel or in a short period of time to avoid significant changes in the waste properties. To investigate hydrolysis and anaerobic digestion, duplicate assays were made. The wastes' characteristics were determined according to the standard procedures of APHA (2005). Lipids were determined according to the adapted method of Postma & Stroes (1968). Proteins were determined by the classical method of Lowry *et al.* (1951).

Hydrolysis assays

The enzyme lipase commercialized by the company Biocatalysts was used in the hydrolysis assays. Lipomod MD is a solid commercial lipase preparation, which presents a nominal activity of 11,500 U g^{-1}. The assays were conducted in stirred flasks containing the fatty waste and the enzyme preparation (0.10% w/w). The ion calcium (lipase activator) was added to the reaction medium (0.15% w/w). For a set of experiments, the fatty waste pH was adjusted to 7.0 before performing the hydrolysis. For the other set of assays, the initial pH was not adjusted. The hydrolysis was conducted for 12 or 24 hours under agitation at the temperature of 30 °C. The control assays were always performed without enzyme and with calcium addition under the same conditions as described above. The parameter used to establish the best conditions of hydrolysis was the acidity index (IA), which was determined according to the procedure named Free Fatty Acids Method Ca 5a-40 (AOCS 2009).

Samples were withdrawn from the flasks at the end of the reaction and the determination of the liberated long chain fatty acids was made by high resolution gas chromatography–mass spectrometry (GC-MS) using the following equipment: HRGC-MS Agilent Technologies, model 7890A-5975C, with a CTC PAL automatic injector. The column used was a DB-FFAP (15 m × 0.10 mm, 0.10 μm). The detector temperature was maintained at 240 °C. The sample injection volume was 1.0 μL and the injection flow rate was 0.5 mL min^{-1}, with a split ratio of 1:100. Helium was used as the entrainment gas at a flow rate of 1.0 mL min^{-1}. The selective monitoring of the specific ions of each fatty acid in the mass spectrometer detector was performed in scanning mode with a mass range of 40–400 m/z. The performance evaluation of the method followed the DOQ-CGCRE-008 (INMETRO 2016). The free fatty acid fractions from the liquid-liquid extraction phase were transesterified before chromatographic analysis. The chromatographic technique was validated for 12 organic acids, including 8 LCFAs (C12:0; C14:0; C14:1; C16:0; C16:1; C18:0; C18:1; C18:2).

Anaerobic digestion assays

The hydrolyzed wastes were named hydrolyzed 0.1%, when lipase was used, and control when no commercial enzyme was used. The anaerobic digestion of the hydrolyzed wastes was carried out using sealed 100-mL penicillin-type flasks and 60 mL working volume of a blend of anaerobic sludge and waste. The proportion of sludge and waste added in each anaerobic digestion assay was 1.4:1 (w/w). The flasks were sealed with rubber stoppers and aluminum seals, connected to plastic syringes to collect the produced biogas. This procedure was based on the work published elsewhere (Duarte *et al.* 2015).

Anaerobic sludge from an industrial digester treating poultry processing wastewater was used as inoculum, and contained a concentration of volatile suspended solids of

57 g mL^{-1} and specific methanogenic activity of 2.72 mL_{CH4} $gVSS^{-1}$ h^{-1}. The pH of the digestion medium was adjusted to 7.0 before feeding the medium to the flasks. Incubation was conducted at 35 °C, without agitation, for 5 days.

The volume of biogas produced in the anaerobic biodegradation assays was measured through the dislocation of the embolus of plastic syringes connected to the penicillin flasks. The methane content of the biogas was determined by gas chromatography (Agilent 7890A with flame ionization detector).

RESULTS AND DISCUSSION

The waste from the dairy industry (FW-D) is a flotation froth with a *mousse* aspect, with low capacity for phase separation. The waste from the poultry slaughterhouse (FW-P) do not have such an aspect, it was more aqueous and presented sedimentable solids. Both wastes have a characteristic unpleasant odor. The main physical and chemical characteristics of the wastes are presented in Table 1.

The two wastes have different properties. FW-D has a high level of fats, almost twofold higher than that found in the FW-P waste. The levels of chemical oxygen demand (COD) and O&G are higher in the flotation froth from the dairy industry (FW-D), which also has a lower humidity. These differences will certainly affect the hydrolysis and the anaerobic digestion of these wastes.

Hydrolysis assays

The increase in the acidity index in relation to the initial index (IA/IAo) for assays that lasted 12 and 24 hours with and without initial pH adjustment and with 0.1% w/w of commercial enzyme addition and control (no enzyme addition) is reported in Table 2. The increase of IA was moderate, 2.35 fold was the maximum result, and the best results were observed for the poultry fatty waste (FW-P). A slight increase in the ratio (IA/IAo) was observed when the initial pH was adjusted to 7.0 as well, when the reaction time was higher (as expected). Although the acidity increase in the control assays was significant, in most assays the values obtained were lower than those achieved in the corresponding assays with enzyme addition. The acidity increase in the control assays can be attributed to the autochthonous microorganisms adapted to the substrates (mainly fats and proteins) that are able to produce hydrolytic enzymes, including lipases.

Table 1 | The main physical and chemical characteristics of FW-P (poultry fatty waste) and FW-D (dairy fatty waste)

Characteristics	FW-P	FW-D
Density, kg L^{-1}	0.990 ± 0.008	0.939 ± 0.004
Humidity, %	90.27 ± 0.82	84.54 ± 0.41
pH	5.7 ± 0.1	5.3 ± 0.1
Alkalinity, $gCaCO_3$ L^{-1}	0.79 ± 0.01	1.27 ± 0.02
Proteins, g L^{-1}	13.2 ± 0.4	10.2 ± 0.3
COD, g L^{-1}	41 ± 5	73 ± 8
O&G, %	7.1 ± 0.5	12.5 ± 0.8
Lipids, g L^{-1}	15.1 ± 0.6	54.8 ± 0.8
Total LCFAs, g L^{-1}	0.76 ± 0.13	5.85 ± 0.45
IA, mg L^{-1}	4.99 ± 0.001	10.22 ± 0.03

In the present work, the term LCFA corresponds to free fatty acids. Only 8 LCFAs were found at detectable levels at the end of the hydrolysis reactions, which lasted 12 and 24 h. The relative amount of each LCFA and the total concentration of saturated and unsaturated fatty acids are shown in Tables 3 and 4, to FW-P and FW-D respectively.

Oleic acid was the predominant LCFA (unsaturated) in the reaction media of both fatty wastes (FW-D and FW-P). It was found in percentages comprised between 55 and 72% of the LCFAs detected. The original content of this LCFA in the raw waste was 61.3 and 54.7% for FW-P and FW-P, respectively. Palmitoleic acid was the second unsaturated acid found in hydrolysis experiments, but it was detected at low percentages (below 9%) for both wastes. Linoleic acid (unsaturated) was also found in low percentages, below 2% for FW-D and below 7% for FW-P.

The concentration of unsaturated LCFAs increased during the hydrolysis process (from 12 to 24 h) in 50% of the experiments with FW-D. For the poultry waste (FW-P) that increase was observed in 100% of the experiments (control and with enzyme addition, with and without pH adjustment). A slight increase in the percentage of unsaturated acids was observed in most experiments. They were

Table 2 | Relative increase in the acidity index (IA/IAo) observed in the hydrolysis assays conducted during 12 and 24 h, employing enzyme contents of 0.1% (w/w) or not (average results of duplicate assays)

	pH no adjustment				pH adjusted			
	Control		Hydrolyzed 0.1%		Control		Hydrolyzed 0.1%	
Wastes	12 h	24 h	12 h	24 h	12 h	24 h	12 h	24 h
FW-P	1.44	1.54	1.94	1.93	1.78	1.92	2.27	2.35
FW-D	1.09	1.12	1.10	1.14	1.13	1.22	1.14	1.25

Table 3 | LCFAs composition of the raw waste (FW-P) and as determined at the end of the hydrolysis reaction (12 and 24 h). Experiments with or without initial pH adjustment

		Control				Hydrolyzed (0.1%)			
		No pH adjustment		pH adjusted		No pH adjustment		pH adjusted	
%LCFA	Raw	12 h	24 h	12 h	24 h	12 h	24 h	12 h	24 h
Lauric [C12:0]	0.4	0.3	0.2	0.4	0.2	0.6	0.2	0.4	0.3
Myristic [C14:0]	3.1	7.5	3.9	1.9	1.1	1.7	3.6	2.0	1.0
Myristoleic [C14:1]	n.d.	n.d.	n.d.	n.d.	n.d.	n.d.	n.d.	n.d.	n.d.
Palmitic [C16:0]	19.3	21.8	17.0	12.1	12.2	16.3	18.3	15.5	8.3
Palmitoleic [C16:1]	4.5	0.2	3.0	5.8	7.6	5.7	3.7	4.5	8.4
Stearic [C18:0]	9.0	10.9	7.7	5.7	5.2	6.2	8.2	6.9	3.7
Oleic [C18:1]	61.3	59.0	66.4	70.6	68.3	63.7	62.4	67.1	71.8
Linoleic [C18:2]	2.0	0.2	1.9	3.4	5.5	5.8	3.5	3.6	6.5
Saturated LCFA (mg L^{-1})	240	450	410	665	615	340	550	380	840
Unsaturated LCFA (mg L^{-1})	515	660	1,015	1,890	2,680	1,040	1,270	1,580	5,480
Total LCFA (mg L^{-1})	755	1,110	1,425	2,555	3,295	1,380	1,820	1,960	6,320

n.d not detected – below the detection limit of the method.

Table 4 | LCFAs composition of the raw waste (FW-D) and determined at the end of the hydrolysis reaction (12 and 24 h). Experiments with or without initial pH adjustment

		Control				Hydrolyzed (0.1%)			
		No pH adjustment		pH adjusted		No pH adjustment		pH adjusted	
%LCFA	Raw	12 h	24 h	12 h	24 h	12 h	24 h	12 h	24 h
Lauric [C12:0]	1.5	1.3	1.4	0.9	0.7	1.1	1.4	0.7	0.6
Myristic [C14:0]	7.8	7.1	7.1	5.1	4.2	6.0	7.1	4.0	3.5
Myristoleic [C14:1]	0.1	0.7	1.0	0.7	0.9	0.5	0.9	0.8	0.7
Palmitic [C16:0]	21.9	21.3	20.1	15.4	14.4	18.4	20.2	13.0	11.8
Palmitoleic [C16:1]	4.3	4.4	4.2	5.2	6.4	4.9	4.7	5.4	5.4
Stearic [C18:0]	9.6	9.7	8.7	7.2	7.2	8.5	9.0	6.5	5.8
Oleic [C18:1]	54.7	56.0	58.2	65.8	65.5	60.5	57.0	69.8	71.3
Linoleic [C18:2]	0.1	0.2	0.2	0.4	1.6	0.6	0.6	0.7	1.6
Saturated LCFA (mg L^{-1})	2,390	2,630	2,790	2,920	2,400	2,650	2,690	2,510	2,620
Unsaturated LCFA (mg L^{-1})	3,455	4,040	4,670	7,270	6,660	5,150	4,440	7,880	9,480
Total LCFA (mg L^{-1})	5,845	6,670	7,460	10,190	9,060	7,800	7,130	10,390	12,100

predominant in the reaction media, corresponding to 61 to 79% of the LCFAs in the experiments with the dairy fatty waste (FW-D) and 59 to 87% for the poultry fatty waste (FW-P).

The most abundant saturated LCFAs were palmitic, stearic and myristic in hydrolysis media of both wastes (FW-D and FW-P). The percentages of those acids did not change appreciably during hydrolysis (from 12 to 24 h); the same occurred, in general, with the concentration of the saturated LCFAs. The most prominent increase of saturated LCFAs concentration occurred in the first 12 h of hydrolysis, as can be observed from raw waste data in Tables 3 and 4. Saturated acids corresponded to 32% and 41% of the total LCFA concentration for FW-P and FW-D, respectively. The distribution of LCFA had a small change when the commercial lipase was used. For FW-P there was a relative increase of the oleic acid content and a reduction of the palmitic acid content. For FW-D, this same trend was observed.

Hydrolysis was conducted under non-sterile conditions and for long time (24 h) at the temperature of 30 °C. Thus, hydrolysis of triglycerides was occurring, catalyzed by enzymes produced by the autochthonous microorganisms as well as by the commercial enzyme (in the experiments with enzyme addition). The microorganisms may also be able to promote the degradation of the fatty acids liberated from the triglycerides. Thus, there is a balance between LCFAs liberation by hydrolysis and occasional transformation of these substances by autochthonous microorganisms.

The pH adjustment, carried out with a concentrated solution of NaOH before the beginning of the assays, promoted an increase in the total LCFA concentration caused by basic hydrolysis. As hydrolysis proceeded, the total LCFAs content always increased for the waste FW-P. An increase in this parameter was also observed for the waste FW-D until 12 h of reaction. After that, a small decrease was observed in some experiments.

The pH adjustment at the beginning of the reaction was an important factor affecting the hydrolysis process. Figure 1 illustrates this effect for both wastes, comparing data obtained after 24 h of reaction for the control experiments and the assays conducted with commercial enzyme addition (0.1% w/w).

The utilization of commercial enzyme (lipase 0.1%) effectively promoted an increase in LCFAs concentrations, in comparison with control assays, when the initial pH was adjusted. In such a case, the augmentation of total LCFAs concentration was much more pronounced for unsaturated acids. In a study with wastewater from a poultry slaughterhouse presenting 800 mg L^{-1} of O&G, during enzymatic hydrolysis using lipase, the LCFAs concentration decreased over a reaction period of 4 to 8 h. This decline continued until 24 h and was attributed to the consumption of LCFAs by microorganisms (Valladão *et al.* 2011). Thus, despite the differences between wastewater and flotation froth properties, the consumption of LCFAs by microorganisms seems to have occurred in some experiments with both wastes (FW-P and FW-D).

Biodigestion assays

The specific methane production (mLg^{-1} waste) was calculated based on the volume of methane produced in 5 days and is represented as SMP-5d. Data are shown in Table 5. In general, this parameter was higher for the waste from the poultry slaughterhouse (FW-P). Production of methane was lower for the dairy waste (FW-D). The specific methane

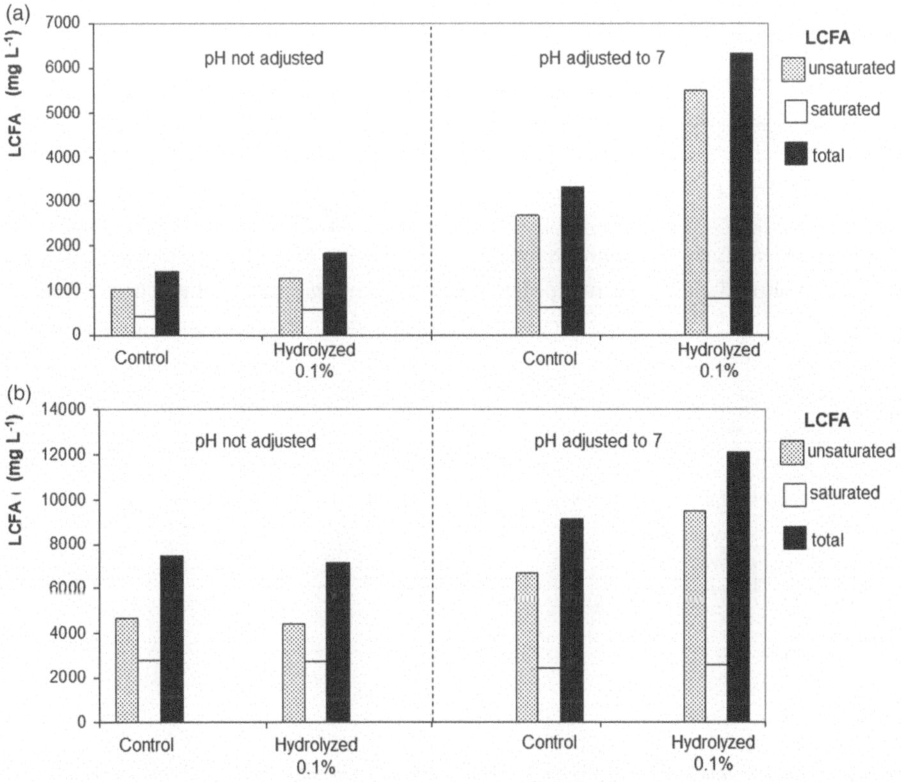

Figure 1 | Variation of LCFA concentration in experiments with and without initial pH adjustment. (a) FW-P, (b) FW-D. Time of hydrolysis considered: 24 h.

Table 5 | Specific methane production (SMP-5d) determined for all digestion assays (average results of duplicate assays)

	Control				Hydrolyzed (0.1%)			
	No pH adjustment		pH adjusted		No pH adjustment		pH adjusted	
	12 h	24 h	12 h	24 h	12 h	24 h	12 h	24 h
FW-D specific methane production (mL g^{-1} waste)	1.32	0.40	1.29	1.34	0.28	1.15	1.30	2.46
FW-P specific methane production (mL g^{-1} waste)	2.28	2.30	3.10	2.20	2.19	1.77	2.64	1.90

production results obtained with the hydrolyzed wastes (lipase addition) were lower than those achieved in the control assays for the poultry slaughterhouse (FW-P). Thus, the utilization of the commercial enzyme was not effective in this case. For the fatty waste from the dairy industry (FW-D) the beneficial effect of enzymatic hydrolysis was only observed when the hydrolysis was carried out for 24 h with initial pH adjustment. The effect of adjusting the waste pH before hydrolysis on SMP-5d was slightly positive for almost all digestion assays.

Figure 2 illustrates the methane production during the biodigestion assays conducted with the two wastes previously submitted to hydrolysis under the following conditions: waste pH adjusted to 7.0 just before the hydrolysis and time of reaction of 24 h. For the fatty waste (FW-P) the assays named control and hydrolyzed had almost the same profile of CH_4 production. For the waste (FW-D) the profiles of CH_4 production showed different patterns. The assay named control presented a lag time and the volume produced increased very little in the interval 50–120 h. It is worth noting that three of the methane production profiles in that figure show an increasing trend at the end of the assay (120 h or 5 d). Certainly, to have a more accurate evaluation, long term assays should be performed in future experiments.

The profiles of methane production from complex fatty-wastes may present an interesting behavior. In long term experiments (63-d) of anaerobic digestion of the flotation froth from a dairy industry conducted in an automatic system (AMPTS) by members of our research group, a significant increase in methane production was observed after 16 days of incubation. In that work, for the material hydrolyzed without addition of commercial lipase (control), the specific methane production increased from 0.56 to 33.4 mL g^{-1} waste over the period of 8 to 63 days. For the material hydrolyzed with lipase (0.1% w/w) the methane production raised from 0.81 to 34.4 mL g^{-1} waste in the same period. Thus, it is possible that high yields of methane can be achieved with the two tested wastes in long term digestion assays, since in 5 days expressive specific methane production values were attained in comparison with those reported in the cited work for 8 days (Bila *et al.* 2016).

The inhibition of anaerobic digestion by LCFAs, mainly unsaturated LCFAs, was reported in many works with wastewaters and synthetic solutions containing triglycerides (Lalman & Bagley 2000; Cirne *et al.* 2007). However, the inhibition phenomenon is complex, because the adaptation of microorganisms may occur and the inhibition intensity seems to dependent on the anaerobic sludge characteristics, in particular its biological diversity, since some methanogens

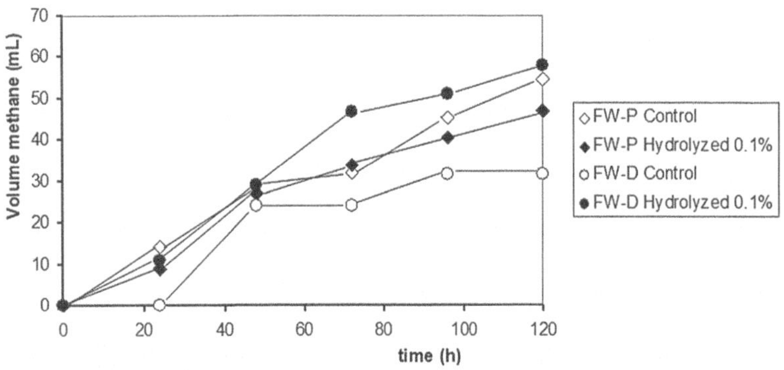

Figure 2 | Production of methane during anaerobic digestion of hydrolyzed wastes: control (without enzyme addition) and hydrolyzed (with commercial enzyme addition 0.1% w/w). Hydrolysis was conducted for 24 h and the initial pH of the wastes was adjusted to 7.0 at the beginning of the reaction.

are more sensitive to LCFAs than others (Silva *et al.* 2015). A severe inhibition by unsaturated LCFAs was reported (Templer *et al.* 2006), in particular, by linoleic and oleic acids.

In Figure 3, the concentration of unsaturated LCFAs was plotted against the specific methane production (SME-5d), using all data obtained. A different symbol was used for each experimental condition investigated. For a given set of experimental conditions (the same waste, the same pH management, without or with commercial enzyme addition, 12 or 24 hours of hydrolysis), lines in Figure 3 indicate an increasing trend of the specific methane production with the decrease of unsaturated LCFAs concentration. However, for one experimental condition (FW-D, pH adjusted, 24-h hydrolysis), represented by a closed triangle, the result did not fit to the general trend observed for the other assays. For that experiment, SMP-5d was high, even with a relatively high concentration of unsaturated LCFAs in the medium. We could not find a reasonable explanation for that result, which deviates from the trend observed for all other experiments.

Comparing the two wastes, it is clear that the one from the dairy industry showed worse results in terms of SME-5d. However, in anaerobic degradation assays lasting long times, several tens of days, adaptation of microorganisms can overcome the inhibitory effects caused by LCFAs, as already observed (Templer *et al.* 2006; Cirne *et al.* 2007). For wastes with very high content of organic material, including fats and proteins and presenting a pasty consistency, like FW-D, long times of digestion will be necessary to reach high methane production yields.

The hydrolysis step, with or without the addition of commercial lipase, promoted an increase in the concentration of LCFAs. For the waste richer in fats (FW-D), the larger amount of LCFAS liberated (mainly unsaturated acids) led to smaller methane yields, at least, in 5-d incubation assays. For the waste FW-P, presenting lower fat content, lower LCFAs concentrations were attained at the end of the hydrolytic step and higher methane production yields were observed.

Certainly, the results obtained in anaerobic digestion assays can be considered as preliminary ones, and our effort will be oriented to perform experiments in a pilot scale system installed in a dairy industry and/or in a slaughterhouse. This system will face the inherent variability of composition and physical structure of flotation froths and will produce very reliable results, allowing determination of the economic benefit of methane production in the context of energy consumption by the industry.

CONCLUSION

The step of waste hydrolysis (with or without addition of a commercial enzyme) promoted increase of the acidity index in most experiments and enhanced the formation of LCFAs. In comparison with the raw wastes, the concentration of LCFAs increased in the hydrolysis step and more intensely when the initial pH of the reaction was adjusted to 7.0. When this occurred, the utilization of the commercial enzyme led to higher LCFAs concentrations for the two wastes studied. The free fatty acids composition (in percentage) did not change appreciably during hydrolysis. Oleic acid was the preponderant LCFA in the reaction media of both wastes. Unsaturated LCFAs were always found in higher concentrations in all hydrolysis experiments.

The specific methane production, determined in a 5-d period, was higher for the waste FW-P, for which the amount of liberated unsaturated fatty acids was smaller, as a function of its lower fat content. The amount of

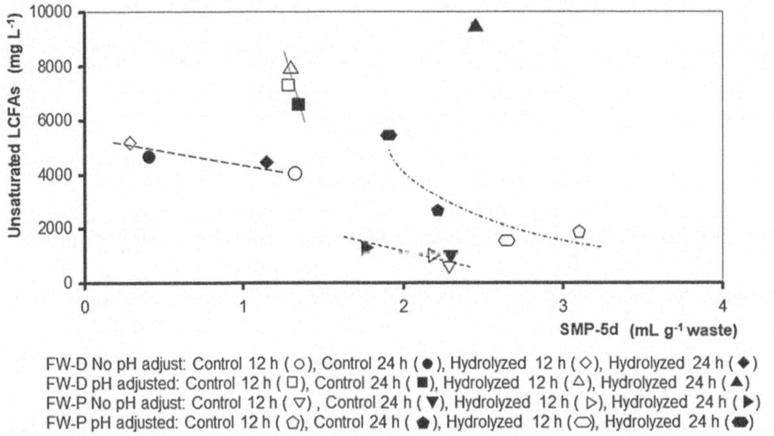

Figure 3 | Effect of unsaturated LCFAs in the hydrolyzed waste (with and without enzyme addition) on the specific methane production (SMP-5d) (average results of duplicate assays).

unsaturated LCFAs seems to promote some inhibition of methane production. A trend of reduction of the specific methane production with unsaturated LCFAs concentration was observed. Finally, concerning methane production, the addition of commercial lipase in the hydrolysis step was only beneficial for the waste from the poultry industry (FW-P). The potential of the autochthonous microorganisms cannot be neglected, because adequately stimulated they can produce enzymes, making unnecessary the utilization of commercial enzymes. With respect for the dilemma mentioned before, it is possible to state that the positive effect of hydrolysis on methane production is limited when the waste presents a high fat content. In this case, the inhibition effect on anaerobic digestion seems to surpass the occasional benefits of fat hydrolysis.

ACKNOWLEDGEMENTS

This study was supported by funds from 'Conselho Nacional de Desenvolvimento Científico e Tecnológico' (CNPq).

REFERENCES

American Oil Chemists Society 2009 *Official Method Ca 5a-40. Free Fatty Acids*. Applicable to all crude and refined vegetables oils, marine oils and animal fats. https://pt.scribd.com/document/270517200/Ca-5a-40-FFA (accessed 15 January 2017).

APHA/AWWA/WEF 2005 *Standard Methods for the Examination of Water and Wastewater*, 21st edn. American Public Health Association/American Water Works Association/Water Environment Federation, Washington, DC, USA.

Battimelli, A., Torrijos, M., Moletta, R. & Delgenès, J. P. 2010 Slaughterhouse fatty waste saponification to increase biogas yield. *Bioresource Technology* **101**, 3388–3393.

Bila, D. M., Vendramel, S. M. R. & Sant'Anna Jr, G. L. 2016 Traitement enzymatique et digestion anaérobie d'une mousse de flottation. (Enzymatic treatment and anaerobic digestion of a flotation foam.) Annales 5èmes Journées de la Methanisation, Chambéry, France, 4 pp., édition électronique.

Cirne, D. G., Paloumet, X., Bjornsson, L., Alves, M. M. & Mattiasson, B. 2007 Anaerobic digestion of lipid-rich waste – effect of lipid concentration. *Renewable Energy*. **32**, 965–975.

Duarte, J. G., Silva, L. L. S., Freire, D. M. G., Cammarota, M. C. & Gutarra, M. L. E. 2015 Enzymatic hydrolysis and anaerobic biological treatment of fish industry effluent: evaluation of the mesophilic and thermophilic conditions. *Renewable Energy* **83**, 455–462.

Hanaki, K., Matsuo, T. & Nagase, M. 1981 Mechanism of inhibition caused by long-chain fatty acids in anaerobic digestion process. *Biotechnology and Bioengineering* **23**, 1591–1610.

Instituto Nacional de Metrologia, Qualidade e Tecnologia – INMETRO 2016 *DOQ-CGCRE-008: Orientação Sobre Validação de Métodos Analíticos. CoordenaçãoGeral de Acreditação*. (Guidance on validation of analytical methods. General Coordination of Creditation.) http://www.inmetro.gov.br/Sidoq/Arquivos/CGCRE/DOQ/DOQ-CGCRE-8_05.pdf (accessed 20 January 2017).

Koster, I. W. & Cramer, A. 1987 Inhibition of methanogenesis from acetate in granular sludge by long-chain fatty acids. *Applied and Environmental Microbiology* **53**, 403–409.

Lalman, J. A. & Bagley, D. M. 2000 Anaerobic degradation and inhibitory effects of linoleic acid. *Water Research* **34** (17), 4220–4228.

Leal, M. C. M. R., Freire, D. M. G., Cammarota, M. C. & Sant'Anna Jr, G. L. 2006 Effect of enzymatic hydrolysis on the anaerobic treatment of dairy wastewater. *Process Biochemistry* **41**, 1173–1178.

Lowry, O. H., Rosenbrough, N. J., Farr, R. L. & Randall, R. J. 1951 Protein measurement with the Folin phenol reagent. *Journal of Biology and Chemistry* **193**, 265–275.

Meng, Y., Luan, F., Yuan, H., Chen, X. & Li, X. 2017 Enhancing anaerobic digestion performance of crude lipid in food waste by enzymatic pretreatment. *Bioresource Technology* **224**, 48–55.

Neves, L., Pereira, M. A., Mota, M. & Alves, M. M. 2009 Detection and quantification of long chain fatty acids in liquid and solid samples and its relevance to understand anaerobic digestion of lipids. *Bioresource Technology* **100**, 91–96.

Postma, T. & Stroes, J. A. P. 1968 Lipid screening in clinical chemistry. *Clinica Chimica Acta* **22**, 569–578.

Silva, S. A., Salvador, A. F., Cavaleiro, A. J., Pereira, M. A., Stams, A. J. M., Alves, M. M. & Sousa, D. Z. 2015 Toxicity of unsaturated and saturates long chain fatty acids towards Methanosaeta concilii. In: *Proceedings 14th World Conference on Anaerobic Digestion, Viña del Mar, Chile*, p. 4, electronic edition.

Sousa, D. Z., Salvador, A. F., Ramos, J., Guedes, A. P., Barbosa, S., Stams, A. J. M., Alves, M. M. & Pereira, M. A. 2013 Activity and viability of methanogens in anaerobic digestion of unsaturated and saturated long-chain fatty acids. *Applied and Environmental Microbiology* **79** (14), 4239–4245.

Templer, J., Lalman, J. A., Jing, N. & Ndegwa, P. M. 2006 Influence of long chain fatty acids on hydrogen metabolism. *Biotechnology Progress* **22**, 199–207.

Valladão, A. B. G., Freire, D. M. G. & Cammarota, M. C. 2007 Enzymatic pre-hydrolysis applied to the anaerobic treatment of effluents from poultry slaughterhouses. *International Biodeterioration & Biodegradation* **60**, 219–225.

Valladão, A. B. G., Torres, A. G., Freire, D. M. G. & Cammarota, M. C. 2011 Profiles of fatty acids and triacylglycerols and their influence on the anaerobic biodegradability of effluents from poultry slaughterhouse. *Bioresource Technology* **102**, 7043–7050.

First received 1 August 2018; accepted in revised form 2 December 2018. Available online 11 December 2018

Increasing activated sludge aggregation by magnetite nanoparticles addition

Dayane Gonzaga Domingos, Rosana Oliveira Henriques, Jéssica Antunes Xavier, Nelson Libardi Junior and Rejane Helena Ribeiro da Costa

ABSTRACT

The behavior of mixed samples (activated sludge and wastewater) was investigated after the addition of magnetite nanoparticles (NP-magnetite) in jar tests, with objective to increase the aggregation of sludge and improve its settleability, aiming for future application in granular sequencing batch reactors. The NP-magnetite was synthesized by the method of ion co-precipitation of Fe^{2+} and Fe^{3+} and characterized. The process of sludge sedimentation was evaluated for two different experimental strategies, with NP-magnetite concentrations between 25 and 150 mg L^{-1}. The concentration of 75 mg L^{-1} was the most favorable to process, as evidenced by sludge volumetric index and density. The results presented indicate that the addition of nanoparticles has the potential to improve aerobic granular systems, increasing the settleability of seed sludge.

Key words | activated sludge, aerobic granular sludge, jar test, magnetite nanoparticles, sequencing batch reactor

Dayane Gonzaga Domingos
Jéssica Antunes Xavier
Nelson Libardi Junior
Rejane Helena Ribeiro da Costa (corresponding author)
Department of Sanitary and Environmental Engineering,
Federal University of Santa Catarina, Campus Universitário UFSC/CTC,
88040-900, Florianópolis,
Brazil
E-mail: *rejane.costa@ufsc.br*

Rosana Oliveira Henriques
Department of Chemical Engineering,
Federal University of Santa Catarina,
Çampus Universitário UFSC/CTC, 88040-900,
Florianópolis,
Brazil

INTRODUCTION

Aerobic granular sludge technology has been developed for approximately 20 years as an evolution of the conventional activated sludge process. This technology is characterized by the development of microbial granules under aerobic conditions, applying selective pressure to retain fast settling sludge, allowing easy solid–liquid separation and higher reactor biomass concentrations (8–15 g L^{-1}). In addition, specific features of the aerobic granular biomass are the presence of slow growing phosphate-accumulating bacteria, the production of high amounts of extracellular polysaccharides (EPS), the ability to perform simultaneous nitrification and denitrification due to the aerobic, anoxic and anaerobic granule layers, achieving good nutrient removal in a single tank and, consequently, reducing the required space and treatment costs (Pronk *et al.* 2017).

Aerobic granules formation is still an unpredictable phenomenon, whereas one of the current assets for the success of this process is the reduced granulation time and maintenance of its stability during organic shocks and long process periods. According to Guimarães *et al.* (2018) the treatment of real sewage with aerobic granular technology has the drawbacks of long granulation times, unpredictable granule morphology, granule disintegrations and inefficient nutrient removal. Addition of compounds to enhance granule formation is an alternative to overcome the formation and stability problems related to this wastewater treatment technology (Liang *et al.* 2017; Sun *et al.* 2017).

The use of particles at nanoscale has been proposed as a nanotechnology strategy in the biological wastewater treatment field as well as in other environmental applications such as the remediation of groundwater (Phenrat *et al.* 2015). There is an increasing number of scientific publications regarding the applications of nanomaterials in water and wastewater treatment like disinfection and microbial control, catalysis/oxidation, adsorption, membrane technologies, sensing and monitoring (Qu *et al.* 2013). In addition, the toxicological effects of nanomaterials over microbial populations has been evaluated (Perazzoli *et al.* 2017). The removal of organic pollutants such as nitrogen has been already tested (Ni *et al.* 2013). However, there is no scientific publication regarding the application of magnetite nanoparticles (NPs) in a sequencing batch reactor (SBR) treating real domestic wastewater to stimulate the aerobic granulation process.

NP systems can be obtained from materials with different chemical natures, such as metals, metal oxides,

doi: 10.2166/wst.2019.055

silicates, ceramics, polymers and biomolecules, showing different morphologies. Among many NP metal oxides, magnetite (Fe_3O_4) has been attracting the attention of researchers (Abu-Dief & Abdel-Fatah 2017).

Many applications of magnetic NPs are based on use of magnetic fields to manipulate their properties, which is directly dependent on the effectiveness of the particle magnetic moment and field gradient. Ferromagnetic and antiferromagnetic states resemble each other, with difference in the alignment of spins; the individual moments are aligned randomly without an external magnetic field. Ferromagnetic materials contain multiple domains, in which the alignment of the magnetic dipole moment causes high spontaneous magnetization, not needing an external magnetic field. The application of a magnetic field could improve biological treatment of domestic wastewater (Silva *et al.* 2012).

The undesirable effects of various NPs present in many domestic and personal care products, entering sewage systems and wastewater treatment plants have been reported (Sundaram & Kumar 2017). In wastewater treatment plants, NPs such as TiO_2, ZnO_2, CeO_2 and Ag have the potential to impact the microbial community present in the microbial flocs, responsible for the chemical oxygen demand, nitrogen and phosphorus removal. Perazzoli *et al.* (2017) found that only high NP-magnetite concentrations ($>200\ mg\ L^{-1}$) would impact ammonium-oxidizing bacteria in an enriched activated sludge community. However, Ni *et al.* (2013) showed that long-term exposure of activated sludge to low concentrations of NP-magnetite ($50\ mg\ L^{-1}$) improved nitrogen removal due to a magnetic-induced effect and maintained a good floc structure, by the increase in EPS from 46.9 to $91.4\ mg\ g^{-1}$ of volatile suspended solids (VSS). The addition of magnetic NPs to wastewater treatment processes has been of interest since they have been considered ideal candidates for application in biological processes due to their high surface area, biocompatibility, low toxicity, stability in physiological conditions and tendency to be adsorbed by the activated sludge flocs (Liang *et al.* 2017).

NPs are able to aggregate the sludge, acting as a core for the microbial attachment. This characteristic is interesting for aerobic granulation processes. The magnetic field resulting from NPs is able to promote the accumulation of iron in biological sludge biomass, reducing the granule formation time by stimulating the production of EPS and improving microbial activity in the middle (Wang *et al.* 2012). In this context, Liang *et al.* (2017) proposed the addition of magnetic NPs in a process of aerobic granular sludge for the treatment of synthetic wastewater. The authors observed the reduction of the granule formation from 45 to 20 days, as well as better morphological characteristics such as its smooth and compact surface with less filamentous structures.

In this context, the aim of this work was to evaluate the effects of addition of NP-magnetite on the aggregation of the activated sludge flocs, aiming for the application of NP-magnetite in aerobic granular SBRs treating domestic wastewater, in order to accelerate the aerobic granulation and overcome granule disintegration processes.

METHODS

Synthesis and characterization of NP-magnetite

The synthesis of NPs was performed according to Lee *et al.* (2009), by means of co-precipitation of Fe^{2+} and Fe^{3+} ions. Eleven grams of iron(II) sulfate and 5.65 g of ammonium oxalate were added to 100 mL of distilled water, with vigorous shaking until total dissolution of salts, resulting in an FeC_2O_5 complex. After, 16 g of iron(III) sulfate was added, followed by the pH adjustment to 9.0 by the addition of ammonium hydroxide 32% (v/v), resulting in the precipitation of magnetite particles. The solution was stirred (1,200 rpm) and heated (75 °C) for 1 h, when the NPs were separated and recovered from the medium using a magnetic bar, followed by a washing with distilled water and drying by the use of a freeze-dryer.

The morphological and structural analysis of NP-magnetite samples was performed using scanning electronic microscopy (SEM) and transmission electronic microscopy (TEM). The scanning electron microscope (JSM-6390LV – JEOL) was operated at 15 kV, using samples in a metallic stub with carbon tape covered with a thin gold layer. TEM analysis (JEM-1011-JEOL) was performed at 100 kV, diluting the 100 mg samples in 10 mL ethanol and maintaining in ultrasound for 5 min at 25 °C. A portion was deposited in a grid of copper coated with carbon film Formvar® 200 mesh and 3.05 mm in diameter and kept at room temperature for 24 h for the evaporation of the solvent. The crystalline phase and particle size were determined by an X-ray diffractometer (XRD) (Cad-4 of Enraf-Nonious®) with an angular variation of 2θ between 0 and 90°, with step of 0.02°, 1 s step-time, CuKα radiation (40 kV/30 mA) and wavelength 1,5406 nm, using the equations of Scherrer and Warren (Cruz-Izquierdo *et al.* 2014).

The determination of chemical composition was conducted by infrared spectroscopy with Fourier transform attenuated total reflectance (Agilent, Cary 660). The average of 20 scans in the range 4,000–500 cm^{-1} and resolution of 4 cm^{-1} were taken. The magnetic properties of particles were observed through a vibrating sample magnetometer (VSM) (3473-70 Electromagnet), operating with the magnetic fields between −20 and 20 kOe at room temperature. Zeta potential was determined by a Nanosizer Zetasizer (Malvern). The 100 mg sample was diluted in distilled water and kept in an ultrasound bath for 2 min at room temperature (25 ± 1 °C).

NP-magnetite assays

The experiments were performed in a micro-controlled jar-test equipment (Milan 2036), containing six jars of 2 L each. The tests were conducted in order to determine the concentration of NP-magnetite that resulted in the best settleability. Each jar was filled with 1 L of mixed sample and NP-magnetite in the concentrations of 25, 50, 75, 100 and 150 mg L^{-1}. The samples used in the tests comprised 20% of activated sludge, obtained from a domestic wastewater treatment plant, and 80% raw sewage collected from a municipal sewage network. The samples contained total suspended solids and sludge volumetric index (SVI) of 1 ± 0.09 g L^{-1} and 234.12 ± 15.87 mL g^{-1}, respectively. The pH was near neutral (7.05) and the temperature was approximately 24 °C.

The experiments were conducted in two strategies. The first strategy was a usual batch flocculation assay. Strategy 2 aimed to simulate the reactions of agitation and settling occurring in an SBR. Strategy 1 comprised an intense agitation of 120 rpm for 5 min to completely disperse the additive and promote coagulation, then a reduction of agitation to initiate flocculation for 15 min at 40 rpm, followed by the sedimentation for the time of 30 min (Devrimci *et al.* 2012). Strategy 2 consisted of agitation at 120 rpm for 4 h followed by the sludge sedimentation for 30 min. The temperature of the experiments was maintained between 22 and 24 °C. All the tests were conducted in triplicate.

The pH and temperature were monitored using a multiparameter probe (Hanna HI-98191). The turbidity was measured after 30 min of sedimentation (Hach 2100N). The settleability process was evaluated by SVI after 30 min of sedimentation. The biomass density was determined with dextran blue according to Beun *et al.* (2002).

RESULTS AND DISCUSSION

NP-magnetite characterization

The SEM images of NP-magnetite are shown in Figure 1, where several clusters of particles can be observed. The agglomeration of particles is possibly caused by the drying process. According to Figure 2, the particles presented spherical structures in the nanometer scale. The average size of NP-magnetite was 5.5 ± 0.3 nm, calculated by the data obtained from XRD. This was confirmed in Figure 2 where the TEM images show particles smaller than 10 nm. The same particle behavior was reported by Henriques *et al.* (2018), obtaining spherical particles with average size of 5.5 nm.

A crystalline structure of the NP-magnetite was identified by XRD. The diffraction peaks ($2\theta = 30$, 35, 43, 53, 57, 63°) correspond to a face-centered cubic structure, corresponding to the product obtained by reaction of co-precipitation as being mainly magnetite NPs.

The spectrogram of the particles presented in Figure 3 shows the vibration between asymmetrical and symmetrical stretches of dicarboxylic anions (-CO_2-) present in oxalates that can be seen in the wavelengths of 1,656 and 1,432 cm^{-1}, respectively. At wavelengths of 1,299 and 1,118 cm^{-1} are the peaks related to the links between C-stretch of oxalate. The 729 cm^{-1} peak refers to a double bond between carbons (-C=C-). The broad band at 3,364 cm^{-1} refers to the presence of hydroxyls, which can be attributed to moisture-adsorbed samples (Silverstein & Webster 1998).

Figure 1 | SEM micrographs of the NP-magnetite with magnifications of 100× (A), 750× (B) and 5000× (C).

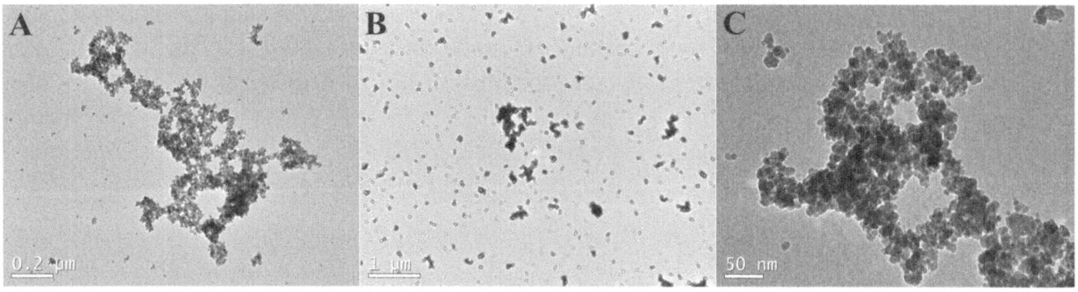

Figure 2 | TEM micrographs of the NP-magnetite with scales of 0.2 μm (A), 1 μm (B) and 50 nm (C).

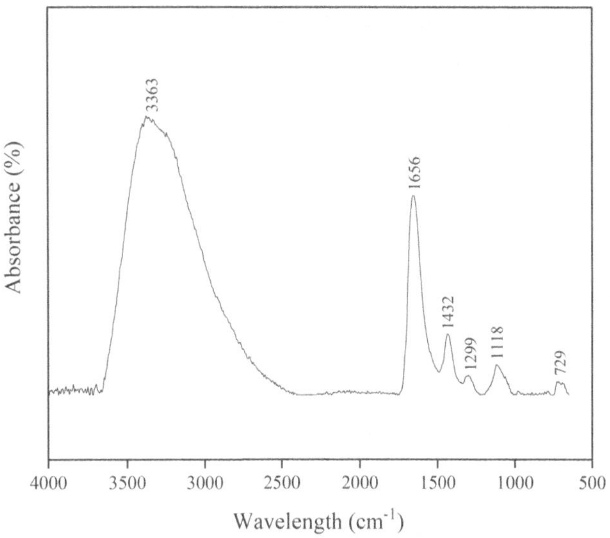

Figure 3 | NP-magnetite spectrogram with the peaks obtained in different wavelengths.

The magnetic properties for NP-magnetite were determined by magnetization curves obtained with the VSM. From the data obtained, values of saturation magnetization (Ms), coercivity (Hc) and remaining magnetization (Mr) were determined.

The Ms value was 50.33 emu g^{-1}, considered a low value, which may be associated with the co-precipitation procedure, since magnetite usually presents Ms values between 60 and 90 emu g^{-1}, varying according to its size (Henriques *et al.* 2018). The values found for Hc and Mr were null for both indexes, proving that the particles present superparamagnetic behavior. The absence of hysteresis in the magnetization curve can be related to absence of strong dipole–dipole forces between the particles, characteristic of superparamagnetic NPs (Mamani *et al.* 2014).

Through measures of zeta potential of NP-magnetite, information about the stability of NPs in an aqueous solution was obtained. According to Henriques *et al.* (2018), when the zeta potential presents values between −30 and +30 mV, forces of repulsion from the surface of the NP loads overcome the forces of attraction, preventing the particles from coagulating. For the NP-magnetite samples, the value obtained was -17.9 ± 5.6 mV, indicating that NPs showed a good electrophoretic mobility, as well as a low agglomeration tendency.

Influence of NP-magnetite in aerobic sludge settleability

The jar tests were conducted simultaneously with control samples (without addition of NP-magnetite) and the tested NP-magnetite concentrations. The two tested strategies were compared in terms of pH and turbidity, as presented in the Table 1.

There was no significant influence of NP-magnetite in the pH values, which remained in range of 7.1–7.4. This is interesting for the application of NP-magnetite in a granular SBR, which will not have its biomass affected by pH changes. Moreover pH and temperature are basic physicochemical parameters of control, which are essential for the formation of granular biomass in granular SBRs, impacting on granule structure, microbial growth, phosphorus removal and other parameters (Halim *et al.* 2016).

Table 1 | Influence of the NP-magnetite concentration over the pH and turbidity for the tested strategies

	Strategy 1		Strategy 2	
NP-magnetite (mg L^{-1})	**pH**	**Turbidity (NTU)**	**pH**	**Turbidity (NTU)**
Control	7.1 ± 0.2	21.8 ± 13.1	7.3 ± 0.3	10.4 ± 3.3
25	7.3 ± 0.3	20.9 ± 14.3	7.3 ± 0.4	10.2 ± 2.3
50	7.3 ± 0.2	21.4 ± 12.2	7.2 ± 0.3	10.9 ± 3.4
75	7.3 ± 0.3	21.7 ± 15.4	7.2 ± 0.3	10.9 ± 3.4
100	7.3 ± 0.2	22.3 ± 13.8	7.1 ± 0.1	12.3 ± 3.8
150	7.4 ± 0.3	23.1 ± 14.9	7.3 ± 0.3	9.9 ± 2.3

According to Table 1, there was a low turbidity removal in both strategies, with the largest removal efficiencies of 4 and 5% obtained with use of 25 mg L^{-1} for strategy 1 and 150 mg L^{-1} for strategy 2, respectively. However, there was no significance difference between the values for the turbidity. The extended agitation time in strategy 2 would require higher NP-magnetite concentration to achieve similar removal efficiency to that of strategy 1. Hwang *et al.* (2011) reported that the use of iron oxide NPs coated with surfactant for the treatment of sanitary wastewater in SBRs was inefficient in turbidity removal, increasing the concentration of soluble solids. Li (2005) reported that the addition of iron in an SBR with activated sludge was able to remove turbidity using concentrations lower than 23 mg L^{-1} whereas higher concentrations reduced the turbidity removal efficiency.

Figure 4 presents the sludge behavior in terms of SVI for the tested NP-magnetite concentrations for the strategies 1 and 2.

According to Figure 4, the addition of NP-magnetite at a concentration of 75 mg L^{-1} resulted in SVI reductions of 18% and 28%, for strategies 1 and 2, respectively. The concentration of 150 mg L^{-1} resulted in SVI reductions of 21% and 22%, for strategies 1 and 2, respectively. The results showed that addition of NP-magnetite improved sludge sedimentation and the sludge aggregation in both tests. The large standard deviations presented in Figure 4 are related to the variability of activated sludge samples used in this test. In addition, the SVI method is highly susceptible to analysis errors. However, it is clear that the addition of NP-magnetite induced the sedimentation, compared to samples without NP-magnetite.

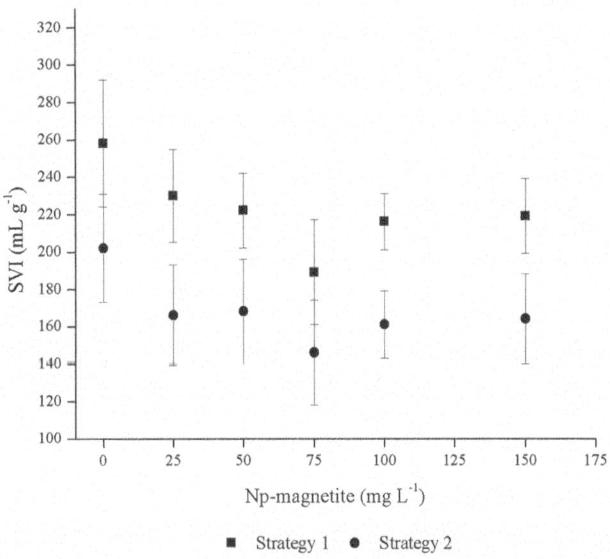

Figure 4 | SVI of the sludge with the addition of NP-magnetite in different concentrations.

You *et al.* (2016) showed that NP added over a period of 8 hours in an SBR showed efficiency in sedimentation, corroborating the results obtained in strategy 2. Li (2005) showed that addition of iron in an activated sludge SBR reduced the presence of particulate materials. However, the authors reported that a high concentration of iron (>35 mg L^{-1}) can cause the formation of small flocs, which can contribute to the formation of soluble solids and increased turbidity.

The NP-magnetite has the property of rapid sedimentation due to magnetic dipole–dipole attraction between the particles; because of its magnetic attraction, it is able to attract not only the particles, but also bacteria and may result in flocs with larger diameters. Not only are the magnetic properties of NP-magnetite able to influence the sedimentation, its polar feature and negative surface charge cause hydrogen bonds to occur between the hydroxyl and oxalate groups in biomass through the dipole–dipole interaction, favoring the sedimentation. However, the magnetization exerted by NP-magnetite was low, resulting in little influence on the attraction between flakes. If magnetization exerted by NP-magnetite was higher (values between 60 and 90 emu g^{-1}), it could have a greater effect on the sludge handling, and the settling ability could be better.

The volumetric index of sludge is closely related to its density. The influence of the NP-magnetite on the biomass density is presented in the Figure 5.

The highest flocs density was achieved by the use of 75 mg L^{-1} of NP-magnetite, which resulted in an increase of density values of 21% and 74% when compared to the control samples, for strategies 1 and 2, respectively. In fact, higher density structure provides flocs or aerobic granules with a high specific gravity, resulting in an enhanced settling ability, as indicated by SVI measurements. The aggregation behavior of sludge is highly important for solid–liquid separation, the effluent turbidity and dewatering properties in biological wastewater treatment systems (Ni *et al.* 2013). The biomass density was higher and SVI values were lower with the addition of NP-magnetite. Another possible effect of NP-magnetite could be that cations in magnetic NPs were absorbed by activated sludge essentially to form bridges between negatively charged bacterial cells by attraction of electrostatic charge, which caused the agglomeration of sludge (Liang *et al.* 2017).

According to Barr *et al.* (2010) one of the factors contributing to development of granules from activated sludge

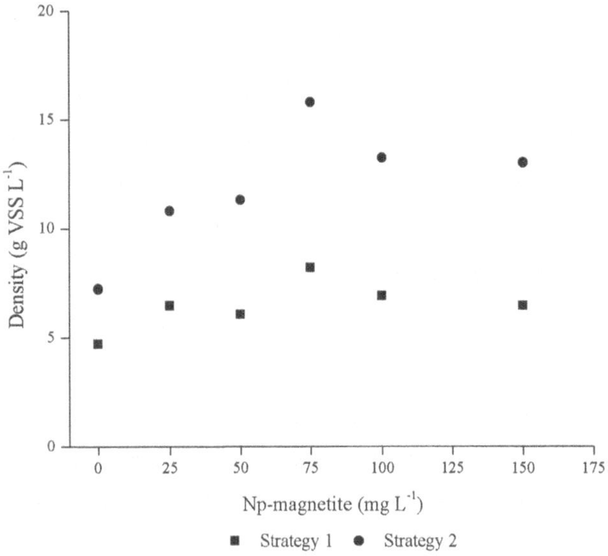

Figure 5 | Density of the sludge with the addition of NP-magnetite in different concentrations.

was presence of cores for microbial bio-fixation. NP-magnetite can generate a magnetic field and provide adequate force to propel bacteria into small particles. This type of cell attachment to these particles has been proposed as the initiation step for the granulation. Moreover, once the initial aggregates were formed, subsequent granulation could be considered as the phenomenon of an increase of these aggregate thicknesses (Liang *et al.* 2017). The tests performed in this work were short (1 and 4.5 hours); probably with application in reactors there will be biological interactions of biomass with the NP-magnetite, which were not possible to evaluate in the jar test. Previous literature stated that the magnetic field in the NPs decreased the time of total granulation, stimulating the excretion of EPS (Wang *et al.* 2012). Yan *et al.* (2015) showed that the existence of metallic ions reduced the EPS zeta potential, which contributed to adhesion among the aggregates. More than that, addition of NP may have increased hydrophobicity of the cell surface; according to Donlan & Costerton (2002), adhesion on microbial cell surface becomes better with increasing hydrophobicity.

CONCLUSIONS

The addition of NP-magnetite can increase the aggregation of biomass into flocs and increase its ability to retain particles and colloidal material, favoring the sedimentation, according to the SVI results. For both tested strategies, the concentration of 75 mg L^{-1} of NP-magnetite resulted in the best values, showing that the longer mixing applied in strategy 2 does not have great impact on the results, suggesting that the shear forces applied in SBRs would not impact the aggregation effect of the flocs by the addition of NP-magnetite. Since the selective pressure in aerobic granular reactors washes out the flocs without good settleability features, the addition of the NP-magnetite would contribute to the reduction of granulation time, by means of maintaining dense flocs in the reactor, stimulating a faster granule formation.

ACKNOWLEDGEMENTS

The authors thank Santa Catarina State Sanitation Company, Chemical Engineering Analytical Centre, Electronic Microscopy Laboratory, Fine Films and Surfaces Laboratory, Interdisciplinary Laboratory for Nanostructures Development and the Biochemical Engineering Laboratory of the Federal University of Santa Catarina. The authors acknowledge the financial support from CAPES and CNPQ.

REFERENCES

Abu-Dief, A. M. & Abdel-Fatah, S. M. 2017 Development and functionalization of magnetic nanoparticles as powerful and green catalysts for organic synthesis. *Journal of Basic and Applied Sciences* **7**, 55–67. DOI:10.1016/j.bjbas.2017.05.008.

Barr, J. J., Cook, A. E. & Bond, P. L. 2010 Granule formation mechanisms within an aerobic wastewater system for phosphorus removal. *Applied Environmental Microbiology* **76** (22), 7588–7597. DOI:10.1128/AEM.00864-10.

Beun, J. J., Loosdrecht, M. C. & Heijnen, J. J. 2002 Aerobic granulation in a sequencing batch airlift reactor. *Water Research* **63** (3), 702–712. DOI:10.1016/S0043-1354(01)00250-0.

Cruz-Izquierdo, A., Picó, E. A., López, C., Serra, J. L. & Llama, M. J. 2014 Magnetic cross-linked enzyme aggregates (mCLEAs) of *Candida antarctica* lipase: an efficient and stable biocatalyst for biodiesel synthesis. *PLoS One* **9** (12), e115202. DOI:10.1371/journal.pone.0115202.

Devrimci, H. A., Yuksel, A. M. & Sanin, F. D. 2012 Algal alginate: a potential coagulant for drinking water treatment. *Desalination* **7** (299), 16–21. DOI:10.1016/j.desal.2012.05.004.

Donlan, R. M. & Costerton, J. M. 2002 Biofilms: survival mechanisms of clinically relevant microorganisms. *Clinical Microbiology Reviews* **15** (2), 167–193. DOI:10.1128/CMR.15.2.167-193.2002.

Guimarães, B. L., Wagner, J., Akaboci, T. R. V., Daudt, G. C., Nielsen, P. H., Loosdrecht, M. C. M., Weissbrodt, D. G. &

Costa, R. H. R. 2018 Elucidating performance failures in use of granular sludge for nutrient removal from domestic wastewater in warm coastal climate region. *Environmental Technology* DOI:10.1080/09593330.2018.1551938.

Halim, M. H., Anuar, A. N., Jamal, N. S. A., Azmi, S. I., Ujang, Z. & Bob, M. M. 2016 Influence of high temperature on the performance of aerobic granular sludge in biological treatment of wastewater. *Journal of Environmental Management* **184**, 71–280. DOI:10.1016/j.jenvman.2016.09.079.

Henriques, R. O., Bork, J. A., Lorente-Fernandez, G., Guisan, J. M., Furigo Jr, A., Oliveira, D. & Pessela, B. C. 2018 Co-immobilization of lipases and β-d-galactosidase onto magnetic nanoparticle supports: biochemical characterization. *Molecular Catalysis* **453**, 12–21. DOI:10.1016/j.mcat.2018.04.022.

Hwang, S., Martinez, D., Perez, P. & Rinaldi, C. 2011 Effect of surfactant-coated iron oxide nanoparticles on the effluent water quality from a simulated sequencing batch reactor treating domestic wastewater. *Environmental Pollution* **159** (12), 3411–3415. DOI:10.1016/j.envpol.2011.08.032.

Lee, D., Ponvel, K. M., Kim, M., Hwang, S., Ahn, I. & Lee, C. 2009 Immobilization of lipase on hydrophobic nano-sized magnetite particles. *Journal of Molecular Catalysis B: Enzymatic* **57** (1–4), 62–66. DOI:10.1016/j.molcatb.2008.06.017.

Li, J. 2005 Effects of Fe(III) on floc characteristics of activated sludge. *Journal of Chemical Technology and Biotechnology* **80** (3), 313–319. DOI:10.1002/jctb.1169.

Liang, X. Y., Gao, B. Y. & Ni, S. Q. 2017 Effects of magnetic nanoparticles on aerobic granulation process. *Bioresource Technology* **227**, 44–49. DOI:10.1016/j.biortech.2016.12.038.

Mamani, J. B., Gamarra, L. F. & Brito, G. E. S. 2014 Synthesis and characterization of Fe_3O_4 nanoparticles with perspectives in biomedical applications. *Materials Research* **17** (3), 542–549. DOI:10.1590/S1516-14392014005000050.

Ni, S., Ni, J., Yang, N. & Wang, J. 2013 Effect of magnetic nanoparticles on the performance of activated sludge treatment system. *Bioresource Technology* **143**, 555–561. DOI:10.1016/j.biortech.2013.06.041.

Perazzoli, S., Michels, C. & Soares, H. M. 2017 Magnetite nanoparticles influence the ammonium-oxidizing bacteria activity during nitritation process. *Water Science and Technology* **75** (1), 165–172. DOI:10.2166/wst.2016.497.

Phenrat, T., Thongboot, T. & Lowry, G. V. 2015 Electromagnetic induction of zerovalent iron (ZVI) powder and nanoscale zerovalent iron (NZVI) particles enhances dechlorination of trichloroethylene in contaminated groundwater and soil: proof of concept. *Environmental Science and Technology* **50**, 872–880. DOI:10.1021/acs.est.5b04485.

Pronk, M., Giesen, A., Thompson, A., Robertson, S. & Loosdrecht, L. C. M. 2017 Aerobic granular biomass technology: advances in design, applications and further developments. *Water Practice and Technology* **12** (4), 987–996. DOI:10.2166/wpt.2017.101.

Qu, X., Alvarez, P. J. J. & Li, Q. 2013 Applications of nanotechnology in water and wastewater treatment. *Water Research* **47**, 3931–3946. DOI:10.1016/j.watres.2012.09.058.

Silva, A. K. A., Di Corato, R., Gazeau, F., Pellegrino, T. & Wilhelm, C. 2012 Magnetophoresis at the nanoscale: tracking the magnetic targeting efficiency of nanovectors. *Nanomedicine* **7** (11), 1713–1727. DOI:10.2217/nnm.12.40.

Silverstein, R. & Webster, F. 1998 *Spectrometric Identification of Organic Compounds*. Wiley, New York, p. 550.

Sun, C., Zhan, H. & Wang, L. 2017 Enhanced aerobic sludge granulation by seeding concentrated activated sludge with Ca-alginate gel. *Water Environment Research* **89** (12), 2078–2087. DOI:10.2175/106143017X14839994522786.

Sundaram, B. & Kumar, A. 2017 Long-term effect of metal oxide nanoparticles on activated sludge. *Water Science and Technology* **75** (2), 462–473. DOI:10.2166/wst.2016.541.

Wang, X. H., Dia, M., Yang, Y., Shi, Y., Gao, M. & Wang, S. 2012 Enhanced aerobic nitrifying granulation by static magnetic field. *Bioresource Technology* **110**, 105–110. DOI:10.1016/j.biortech.2012.01.108.

Yan, L., Liu, Y., Wen, Y., Ren, Y., Hao, G. & Zhang, Y. 2015 Role and significance of extracellular polymeric substances from granular sludge for simultaneous removal of organic matter and ammonia nitrogen. *Bioresource Technology* **179**, 460–466.

You, G., Hou, J., Wang, P., Xu, Y., Miao, L., Yang, Y. & Luo, H. 2016 Effects of CeO_2 nanoparticles on sludge aggregation and the role of extracellular polymeric substances – explanation based on extended DLVO. *Environmental Research* **151**, 698–705. DOI:10.1016/j.envres.2016.08.023.

First received 1 August 2018; accepted in revised form 28 January 2019. Available online 11 February 2019

Nutrient concentration in sediments accumulated in pre-treatment basins of urban LID technologies

F. K. F. Geronimo, M. C. Maniquiz-Redillas, J. S. Hong and L. H. Kim

ABSTRACT

In this study, the contribution of pre-treatment basins of low impact development (LID) technologies to nutrient reduction performance was evaluated by understanding the distribution of nutrient in sediments accumulated in each system. The captured sediments were mostly silt to medium sand ranging from 9% to 92% of the sediments collected. Greater average N and P concentrations were found in silt particles amounting to 345 mg/kg and 696 mg/kg, respectively compared to sand and gravel. Although, N concentrations in accumulated sediments were found to be highly variable at different particle sizes (CV: 0.24 to 0.77) compared to P concentration (CV: 0.08 to 0.36) attributed to effective P treatment mechanism through deposition compared to complex nitrogen removal mechanisms. In addition, the difference between N and P concentrations of sediments collected in the pre-treatment basins of LID technologies and *in-situ* soil was attributed to the continuous pollutant input to the LID technologies during storm events. The study proved that pre-treatment basins of stormwater LID technologies reinforced the nutrient removal performances through sediment retention. The findings of this research may be used to design pre-treatment basins of LID technologies considering nutrients as a limiting factor.

Key words | low impact development, nutrients, pre-treatment basins, sediments

F. K. F. Geronimo
J. S. Hong
L. H. Kim (corresponding author)
Department of Civil and Environmental Engineering,
Kongju National University,
1223-24 Cheonandaero, Seobukgu, Cheonan City, Chungnamdo, 31080,
Republic of Korea
E-mail: *leehyung@kongju.ac.kr*

M. C. Maniquiz-Redillas
Civil Engineering Department, Gokongwei College of Engineering,
De La Salle University-Manila,
2401 Taft Avenue, Malate Manila,
Philippines

INTRODUCTION

The transition of inherent landscape to an extremely developed environment increases the impervious surfaces covered by building rooftops, landscaping and transportation land uses such as roads, highways, parking lots and sidewalks. Development of urban surfaces prevents infiltration, thereby increasing the runoff directed to stormwater network and deteriorating the water resources in highly urbanized areas (Qiu 2013). Substantial amount of nutrients in urban storm runoff contributes to the eutrophication of receiving water bodies leading to algal bloom (Taylor *et al.* 2005). Excess phosphorus can lead to decreased biological productivity while excess nitrogen is harmful to marine ecology. These pollutants were usually transported as sediment-bound contaminants (Vaze & Chiew 2004).

An innovative way of controlling nutrients in stormwater runoff was through low impact development (LID) technologies an example of a nature-based solution to address the negative effects of climate change and urbanisation to natural urban water cycle (Zölch *et al.* 2017). LID technologies mimic the predeveloped state of an area, thereby preventing the water cycle disruption by employing different physical, chemical and biological mechanisms. Ahiablame *et al.* 2012 have already summarized and reported the performance of the LID technologies in controlling several pollutants including nutrients in urban stormwater runoff from different studies. TN and TP were removed by 7% to 99% and −3% to 99%, respectively. High variability of nutrient removal in LID technologies was due to the complexities of the chemistry of these pollutants making it difficult to attain relatively similar removal efficiencies, especially when applied to different site and environmental conditions (Geronimo *et al.* 2015). Designs of LID differ depending on target pollutants. Pre-treatment units were usually employed to LID technologies to reduce the flow of incoming runoff or receive the initial highly polluted runoff commonly known as first flush (Maniquiz-Redillas *et al.* 2014). Since the pre-treatment units receive the first flush of pollutants, sediments, in which particulate-bound pollutants were attached, settle, resulting to minimized clogging in the filter media unit, thereby reducing maintenance cost.

doi: 10.2166/wst.2019.033

Currently, limited studies about sediment accumulation, its corresponding nutrient concentration and its implication to the performance of LID technologies has been studied. As such, hydrologic and hydraulic factors affecting nutrient transport in urban catchments and inside LID technologies were identified in this study. The contribution of pre-treatment basins of LID technologies to nutrient reduction performance was also evaluated by understanding the distribution of nutrient in sediments accumulated in each system. In order to apply similar design of LID with pre-treatment units to other urban catchments, design criteria considering nutrient load accumulation in pre-treatment basin of LID technologies were derived.

MATERIAL AND METHODS

Description of LID technologies

Three LID technologies were developed and monitored during selected storm events from May 2009 to October 2016 with design characteristics summarized in Table 1. An infiltration trench (IT) and two hybrid constructed wetlands (HCW1 and HCW2) were installed inside Kongju National University, Cheonan City, South Korea to treat urban stormwater runoff from 100% impervious road and parking areas. The three LID facilities were similarly designed with pre-treatment basin. However, IT was designed with infiltration capability while there were no infiltration capabilities employed in the design of HCW1 and HCW2. The schematic diagram of each LID technologies including the sampling points and flow path were exhibited in Figure 1. The site location and media configuration of the three LID technologies were already previously reported by Maniquiz-Redillas *et al.* 2014. The pretreatment basins of these LID technologies comprised almost 23% to 38% of the total facility storage volume.

Table 1 | Design characteristics of LID technologies

	LID technology		
Parameter	**Infiltration trench (IT)**	**Small hybrid wetland (HCW1)**	**Hybrid sub-surface flow wetland (HCW2)**
Runoff source	Road	Road and parking lot	Road and parking lot
Catchment area (m^2)	520	323	425
Infiltration capability	Yes	None	None
Filter media	Woodchip, sand and gravel	Woodchip, sand and gravel	Bioceramic, sand and gravel
Facility aspect ratio (L:W:H)	1:0.2:0.26	1:0.15:0.1	1:0.14:0.1
SV/TV[a] (%)	45.4	33.9	30.6
PV/SV[b] (%)	22.9	37.8	26.2

[a]Facility storage volume to total volume.
[b]Pre-treatment basin storage volume to facility storage volume.

Storm event monitoring, pre-treatment basin maintenance, and analytical analysis

Selected storm events were monitored from the year of construction of LID technologies, 2009 for IT and 2010 both for HCW1 and HCW2, up to 2016 to evaluate the efficiency in urban storm water runoff management. Following the typical sampling scheme in South Korea, first grab sample was collected as soon as the runoff entered the inflow ports of the LID technologies or when discharge was produced by the LID technologies. This was followed by grab sample collection at a time interval of 5, 10, 15, 30 and 60 min, respectively. After the first hour of runoff and discharge, one grab sample was collected for each succeeding hour until a maximum of 12 samples were collected or until the storm event ended. The collected water samples were analytically tested for total suspended solids (TSS), total nitrogen (TN) and total phosphorus (TP) concentration according to *Standard Methods for the Examination of Water and Wastewater* (APHA *et al.* 1992). In addition to water samples, hydrologic and hydraulic parameters including antecedent dry days (ADD), rainfall depth and intensity, rainfall and runoff duration, runoff and discharged volumes and inflow and outflow rates were measured and recorded in the location of LID during storm events.

Part of the maintenance routine conducted for the LID technologies was sediment collection in the pre-treatment basin. Sediment samples accumulated in the pre-treatment basins of the three LID technologies were collected once every one to two years from 2011 to 2015. Sediment accumulation rate was calculated by dividing the weight of sediments collected in the pre-treatment basins to the product of the catchment area and time between sediment collections. Particle size analysis was conducted to determine the size distribution of the collected sediments. Chemical characteristics of the sediments including nitrogen (N) and phosphorus (P) were analytically tested based on the method reported by Carter & Gregorich (2006).

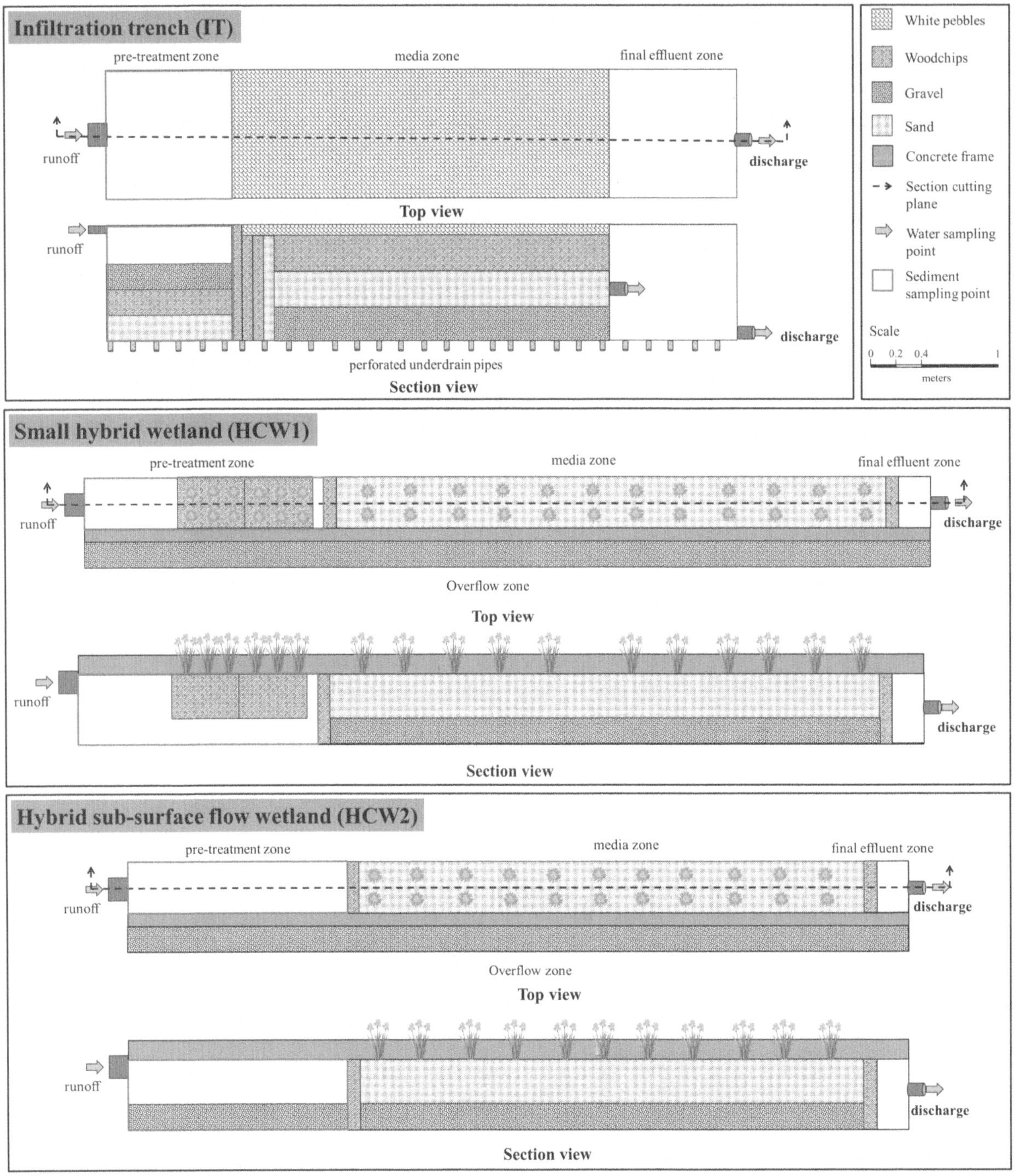

Figure 1 | Schematic diagram of each LID technology with the corresponding sampling points.

In addition, phosphorus types including organic phosphorus (organic P), apatite phosphorus (apatite P), non-apatite inorganic phosphorus (NAIP) and labile phosphorus (labile P) which comprises the total phosphorus in sediments were tested using the method reported by Hieltjes & Lijklema (1980). Nutrient load accumulation rate was calculated by multiplying the sediment accumulation ratio with the corresponding nutrient concentration of the sediments.

Event mean concentration (EMC) was calculated by dividing the total inflow or total outflow pollutant load with the runoff or discharged volume, respectively. In addition, pollutant removal efficiency of the system was calculated by subtracting the fraction of outflow load to inflow load to 1 multiplied to 100. Results were statistically analyzed using SYSTAT 9.0 and OriginPro 8 software package including analysis of variance (one-way ANOVA) and Pearson correlation analysis. Significant differences were accepted at the $p = 0.05$, signifying 95% confidence level.

RESULTS AND DISCUSSION

Stormwater characteristics and hydrologic factors affecting its transport in stormwater runoff

The mean hydrologic and hydraulic characteristics of monitored storm events were summarized in Table 2. Storm events with rainfall depth ranging from 1 to 91 mm were monitored in the LID technologies. 76% of the 96 monitored rainfall events were less than 20 mm which was equal to the percentage of small rainfall occurring in Cheonan City, South Korea reported by Maniquiz *et al.* 2010. Antecedent dry days (ADD) in between monitored storm events were ranging from 0.2 to 34.21 days with the shortest usually occurring in the summer season and the longest between winter to spring season. Hydrologic parameters including ADD, rainfall depth, average rainfall intensity, rainfall duration and runoff duration of the three LID technologies were not significantly different ($p < 0.05$). This finding was attributed to the proximity of each LID technology subjecting them to the same environmental conditions (Maniquiz-Redillas & Kim 2016). Among the three LID facilities, IT attained the highest mean stormwater runoff volume reduction efficiency amounting to 62% which was significantly different from HCW1 and HCW2 which only has 41% and 36%, respectively ($p < 0.05$). The infiltration mechanism employed in IT was found to be the affecting factor in enabling more storm water volume to be reduced since infiltration processes smear peak flows and emphasizes groundwater recharge (Davis *et al.* 2006).

In order to effectively analyze the nutrient concentration in the accumulated sediment of the pre-treatment basins of LID technologies, it is important to determine the characteristics of the stormwater which is the source of these nutrients. Exhibited in Figure 2 are the ranges of EMC from stormwater runoff and discharged by the LID technologies. IT, HCW1 and HCW2 significantly reduced the inflow TSS, TN and TP EMC with an average removal efficiency of 65%, 29% and 30%, respectively ($p < 0.05$). TN and TP load were also found to be highly correlated with TSS load both in runoff, and discharge of the LID technologies ($r = 0.86$ to 0.97). Linear regression analysis, however, revealed that only TP load removal efficiency by the LID technologies was a function of TSS load removal efficiency ($R^2 = 0.63$, $p < 0.001$). These findings implied that the retained TN and TP will be stored in the system until physicochemical or biological treatment will take place in the LID technologies or when maintenance procedures such as accumulated sediment collection will be performed. Hydrologic and hydraulic parameters including ADD, rainfall depth, rainfall duration and runoff duration was found to affect the inflow TSS, TN and TP load with Pearson correlation coefficient (r) ranging from 0.76 to 0.99. These findings implied that higher inflow TSS, TN and TP load can be measured from the stormwater runoff from the urban catchment areas with higher ADD and rainfall depth or longer rainfall and runoff duration. In addition, TSS, TN and TP removal were found to be inversely proportional to the several hydrologic and hydraulic factors including ADD, rainfall depth, rainfall duration, runoff duration, runoff volume and average inflow rate ($r = 0.63$ to 0.99). The factors identified in these results may be used to design LID that targets nutrients in urban stormwater runoff. Hydrologic and hydraulic factors including volume, average flow, peak flow, HRT and runoff duration were also identified to have significantly affected the performance of a tree box filter, an infiltration type LID technology (Geronimo *et al.* 2016).

Table 2 | Hydrologic and hydraulic characteristics of monitored storm events

		LID technology		
Parameter	**Unit**	**IT**	**HCW1**	**HCW2**
N storm events		38	32	26
Antecedent dry days[a]	day	5.51	6.45	6.73
Rainfall depth[a]	mm	10.14	6.94	8.25
Average rainfall intensity[a]	mm/hr	3.19	2.94	4.59
Rainfall duration[a]	hr	3.82	3.40	2.98
Runoff duration[a]	hr	2.81	2.07	2.19
Runoff volume[a]	m^3	4.28	1.57	1.94
Discharge volume[a]	m^3	2.47	1.26	1.97
Average inflow rate[a]	m^3/hr	1.33	11.77	5.78
Average outflow rate[a]	m^3/hr	1.00	0.76	1.86
Peak inflow rate[a]	m^3/hr	8.10	3.74	2.73
Peak outflow rate[a]	m^3/hr	3.75	3.03	2.35

[a]Mean.

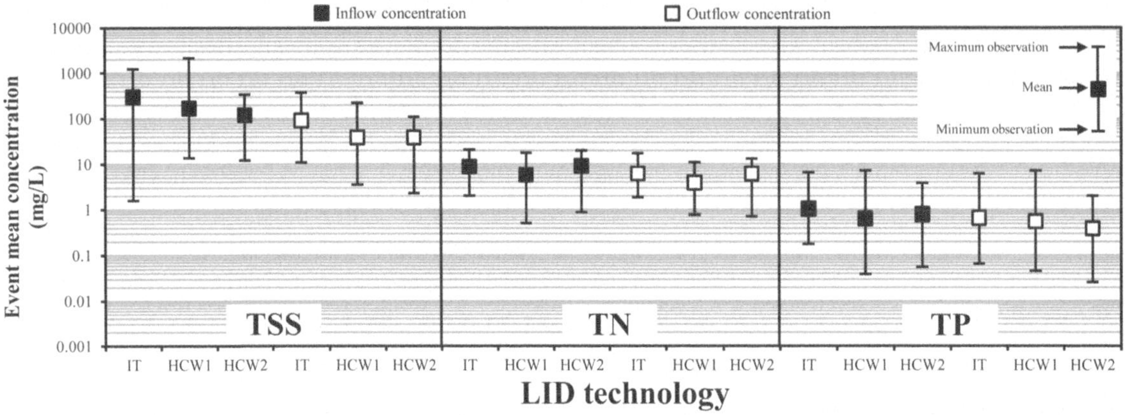

Figure 2 | Ranges of pollutant event mean concentration in each LID technology.

Distribution of nutrient concentration in sediments

The log-normal particle size distribution of sediments collected in the pre-treatment basins was exhibited in Figure 3. Apparently, the captured sediments were mostly silt to medium sand ranging from 9% to 92% of the sediments collected. Silt comprises less than 1% to 14% of the sediments collected in the LID facilities. Usually, the pre-treatment facilities of stormwater LID were designed to treat first flush during storm events. These findings complement the results of Egodawatta & Goonetilleke (2008) and Muthusamy *et al.* (2018) that most of the sediments washed off during the first flush were either <200 μm or 0 to 1,000 μm, respectively. As shown in Figure 4, greater average N and P concentrations were found in silt particles amounting to 345 mg/kg and 696 mg/kg, respectively, compared to sand and gravel. However, silt was found to be the lowest contributor of N and P loads in the sediment of about 5% to 7% only due to its relatively low fraction of the total weight of the sediments collected compared to sand and gravel. Sand contributed the highest nutrient load in the sediments collected from the pre-treatment basins of the LID technologies containing 80% and 79% of the present N and P loads, respectively. Although, N concentrations in accumulated sediments were found to be highly variable at different particle sizes with coefficient of variation (CV) ranging from 0.24 to 0.77 compared to P concentration (CV: 0.08 to 0.36). This finding was attributed to the effective P treatment mechanism through deposition compared to complex nitrogen removal mechanisms (Boers *et al.* 1998). By enhancing the silt deposition or settling in the sedimentation basins of the LID technologies, the overall nutrient treatment performance of the LID technologies may be expected. It was found that the P concentrations found in the sediments of the different stormwater LID technologies of this study were almost in the same range as that of the study conducted by Dalu *et al.* 2018. It was also found in the same study that P concentrations in urban sediments may be as high as 2.5 times that of agricultural sediments (Dalu *et al.* 2018).

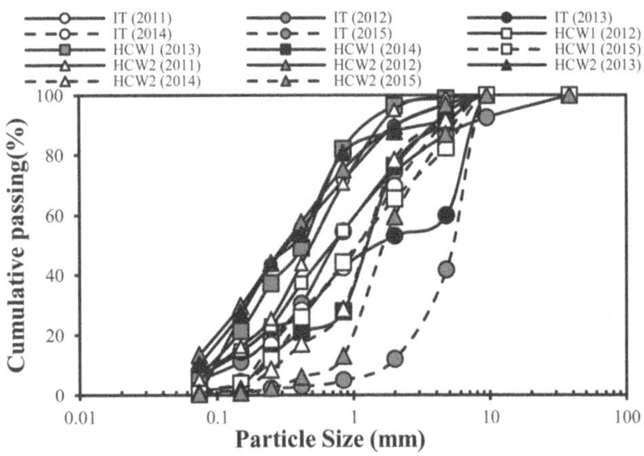

Figure 3 | Particle size distribution of sediment captured by pre-treatment basins of LID technologies.

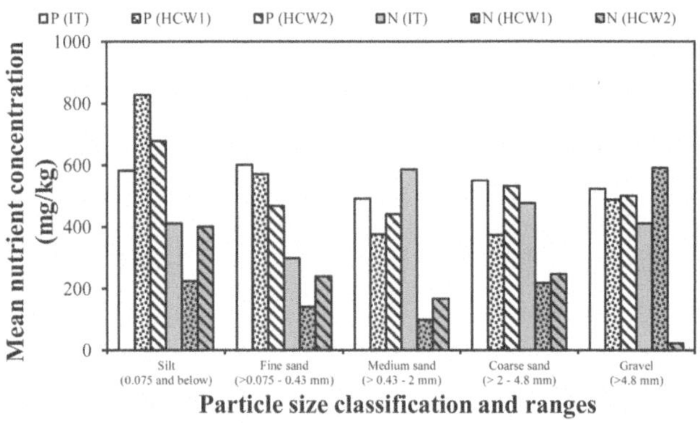

Figure 4 | Nutrient concentration at varying sediment particle size ranges captured by pre-treatment basins of LID technologies.

On the other hand, the N concentration in sediments collected from an urban river receiving runoff from agricultural areas were found to be ten folds higher than the N concentration in this study (Huang *et al.* 2018). Sediment accumulation rate was greater in IT by almost three to four times compared to HCW1 and HCW2 due to the recurring wet and dry state of the pre-treatment basin of IT. This occurrence was the result of employing infiltration mechanism in IT which was not present both in HCW1 and HCW2 causing the pre-treatment basins to be ponded even during the days without storm events. The comparison between N and P concentrations of collected sediments and *in-situ* soil is shown in Figure 5. Sediment N concentrations were 10% to 66% greater than the *in-situ* soil. On the other hand, P concentrations of the sediments collected in LID pre-treatment basins were almost two to three times greater than the P concentration of *in-situ* soil. The difference between the concentrations was attributed to the continuous pollutant input to the LID technologies during storm events which were initially received by the pre-treatment basins. It was identified that the dominant P form in sediment was organic P followed by NAIP. The difference in P concentration in collected sediment and soil was mainly due to the loss of organic P in soil. The dynamics of soil organic P forms were greatly influenced by the interactions of biological, chemical and physical properties of soil (Stewart & Tiessen 1987).

Design of pre-treatment basin of LID technologies considering nutrient load accumulation rate

Design criteria for LID technologies such as bioretention focus on rainfall, flood and water-quality control which were estimated based on the surface to catchment area ratio (Chin 2017). However, in Korea, designers of stormwater treatment technologies considered first flush criteria

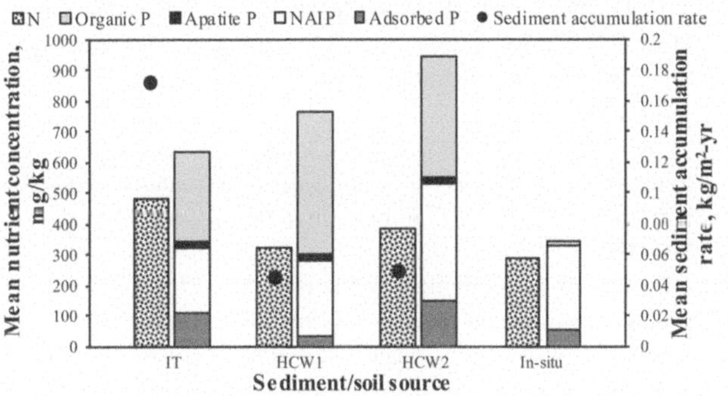

Figure 5 | Comparison of mean N and P concentrations in the sediments captured by pre-treatment basins of LID technologies and *in-situ* soil and sediment accumulation rates per LID technology.

for sizing (Lee *et al.* 2010). In order to effectively design similar LID technologies considering the nutrient load accumulation rate, regression plots shown in Figure 6 were generated. The relationship of both pre-treatment basin volume to catchment area ratio (PV/CA) and the storage volume to catchment area ratio (SV/CA) to the nutrient load accumulation rate in the pre-treatment basins of LID technologies were considered in the development of this design criteria. Apparently, an increase in the PV/CA or SV/CA corresponded to increased nutrient load accumulation capacity of the LID technologies. Considering an LID facility with N or P load accumulation rate of 50 mg/m^2-yr, PV/CA and SV/CA ratio ranging from 0.19 to 0.21 and 0.57 to 0.82, respectively, must be considered.

CONCLUSIONS

Understanding the nutrient concentration variability in sediments accumulated in LID technologies proved the importance of pre-treatment basins in reinforcing the nutrient removal performances of stormwater LID technologies through sediment retention. Based on the results of this study, it was found that hydrologic factors including ADD, rainfall depth, rainfall duration and runoff duration affected the transport of TSS, TN and TP in urban catchments. Transport of TSS in urban catchments and inside LID facilities was also identified as an affecting factor for nutrient transport. However, linear regression analysis revealed that only TP load removal efficiency was found to be highly affected by TSS load removal efficiency ($R^2 = 0.63$, $p < 0.001$). The analysis of nutrient concentration in varying sediment particle sizes demonstrated that greater N and P

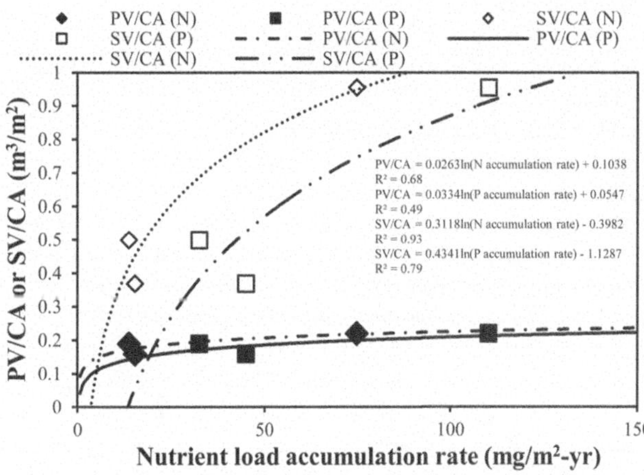

Figure 6 | Relationship between nutrient load accumulation rate through accumulated sediments in pre-treatment basins with PV/CA and SV/CA.

concentrations were found in silt particles compared to sand and gravel. In addition, the difference between N and P concentrations of sediments collected in the pre-treatment basins of LID technologies and *in-situ* soil was attributed to the continuous pollutant input to the LID technologies during storm events. Depending on the target nutrient load accumulation rate, estimation of PV/CA and SV/CA may be conducted based on the regression model developed in this study. It is recommended that nutrient concentrations of accumulated sediments in the pre-treatment basins of LID technologies be used to classify the treatment mechanisms occurring inside each unit in the system. The findings of this research may be used to design pre-treatment basins of LID technologies considering nutrients as a limiting factor.

ACKNOWLEDGEMENT

This research is supported by the Korean Ministry of Environment (MOE) through the 'Public Technology Program based on Environmental Policy' with grant number: 2016000200002. The authors are grateful for their support.

REFERENCES

Ahiablame, L. M., Engel, B. A. & Chaubey, I. 2012 Effectiveness of low impact development practices: literature review and suggestions for future research. *Water, Air, & Soil Pollution* **223** (7), 4253–4273.

APHA (American Public Health Association), AWWA (American Water Works Association) & WEF (Water Environment Federation) 1992 *Standard Methods for the Examination of Water and Wastewater*, 18th edn. APHA (American Public Health Association), Washington, DC, USA.

Boers, P. C. M., Van Raaphorst, W. & Van der Molen, D. T. 1998 Phosphorus retention in sediments. *Water Science and Technology* **37** (3), 31–39.

Carter, M. R. & Gregorich, E. G. 2006 *Soil Sampling and Methods of Analysis*, 2nd edn. Canadian Society of Soil Science, Boca Raton, FL, USA.

Chin, D. A. 2017 Designing bioretention areas for stormwater management. *Environmental Processes* **4** (1), 1–13.

Dalu, T., Wasserman, R. J., Wu, Q., Froneman, W. P. & Weyl, O. L. 2018 River sediment metal and nutrient variations along an urban–agriculture gradient in an arid austral landscape: implications for environmental health. *Environmental Science and Pollution Research* **25** (3), 2842–2852.

Davis, A. P., Shokouhian, M., Sharma, H. & Minami, C. 2006 Water quality improvement through bioretention media: nitrogen and phosphorus removal. *Water Environment Research* **78** (3), 284–293.

Egodawatta, P. & Goonetilleke, A. 2008 Understanding road surface pollutant wash-off and underlying physical processes using simulated rainfall. *Water Science and Technology* **57** (8), 1241–1246.

Geronimo, F. K. F., Maniquiz-Redillas, M. C. & Kim, L. H. 2015 Fate and removal of nutrients in bioretention systems. *Desalination and Water Treatment* **53** (11), 3072–3079.

Geronimo, F. K. F., Hong, J. S. & Kim, L. H. 2016 Hydrologic and hydraulic factors affecting the long-term treatment performance of an urban stormwater tree box filter. *Journal of Korean Society of Water Environment* **33** (6), 715–721.

Hieltjes, A. H. & Lijklema, L. 1980 Fractionation of inorganic phosphates in calcareous sediments. *Journal of Environmental Quality* **9** (3), 405–407.

Huang, S., Chen, C. & Jaffé, P. R. 2018 Seasonal distribution of nitrifiers and denitrifiers in urban river sediments affected by agricultural activities. *Science of The Total Environment* **642**, 1282–1291.

Lee, E. J., Maniquiz, M. C., Gorme, J. B. & Kim, L. H. 2010 Determination of cost-effective first flush criteria for BMP sizing. *Desalination and Water Treatment* **19** (1–3), 157–163.

Maniquiz, M. C., Choi, J., Lee, S., Cho, H. J. & Kim, L. H. 2010 Appropriate methods in determining the event mean concentration and pollutant removal efficiency of a best management practice. *Environmental Engineering Research* **15** (4), 215–223.

Maniquiz-Redillas, M. C. & Kim, L. H. 2016 Evaluation of the capability of low-impact development practices for the removal of heavy metal from urban stormwater runoff. *Environmental Technology* **37** (18), 2265–2272.

Maniquiz-Redillas, M. C., Geronimo, F. K. F. & Kim, L. H. 2014 Investigation on the effectiveness of pretreatment in stormwater management technologies. *Journal of Environmental Sciences* **26** (9), 1824–1830.

Muthusamy, M., Tait, S., Schellart, A., Beg, M. N. A., Carvalho, R. F. & de Lima, J. L. 2018 Improving understanding of the underlying physical process of sediment wash-off from urban road surfaces. *Journal of Hydrology* **557**, 426–433.

Qiu, Z. 2013 Comparative assessment of stormwater and nonpoint source pollution best management practices in suburban watershed management. *Water* **5** (1), 280–291.

Stewart, J. W. B. & Tiessen, H. 1987 Dynamics of soil organic phosphorus. *Biogeochemistry* **4** (1), 41–60.

Taylor, G. D., Fletcher, T. D., Wong, T. H., Breen, P. F. & Duncan, H. P. 2005 Nitrogen composition in urban runoff – implications for stormwater management. *Water Research* **39** (10), 1982–1989.

Vaze, J. & Chiew, F. H. 2004 Nutrient loads associated with different sediment sizes in urban stormwater and surface pollutants. *Journal of Environmental Engineering* **130** (4), 391–396.

Zölch, T., Henze, L., Keilholz, P. & Pauleit, S. 2017 Regulating urban surface runoff through nature-based solutions–an assessment at the micro-scale. *Environmental Research* **157**, 135–144.

First received 10 August 2018; accepted in revised form 11 January 2019. Available online 23 January 2019

Reduction of silver nanoparticle toxicity affecting ammonia oxidation using cell entrapment technique

Nguyen Thanh Giao, Tawan Limpiyakorn, Pumis Thuptimdang, Thunyalux Ratpukdi and Sumana Siripattanakul-Ratpukdi

ABSTRACT

Occurrence of silver nanoparticles (AgNPs) in wastewater treatment systems could impact the ammonia oxidation (AO). This study investigated the reduction of AgNPs and dissociated silver ion (Ag^+) toxicity on nitrifying sludge using cell entrapment technique. Three entrapment materials, including barium alginate (BA), polyvinyl alcohol (PVA), and a mixture of polyvinyl alcohol and barium alginate (PVA-BA), were applied. The BA beads provided the highest reduction of silver toxicity (up to 90%) and durability. Live/dead assays showed fatality of entrapped cells after exposure to AgNPs and Ag^+. The maximum AO rate of the BA-entrapped cells was 5.6 mg-N/g-MLSS/h. The AO kinetics under the presence of silver followed an uncompetitive inhibition kinetic model. The experiments with AgNPs and Ag^+ gave the apparent maximum AO rates of 4.2 and 4.8 mg-N/g-MLSS/h, respectively. The apparent half-saturation constants of the BA-entrapped cells under the presence of silver were 10.5 to 13.4 mg/L. Scanning electron microscopic observation coupled with energy-dispersive X-ray spectroscopy indicated no silver inside the beads. This elucidates that the silver toxicity can be reduced by preventing silver penetration through the porous material, leading to less microbial cell damage. This study revealed the potential of the entrapment technology for mitigating the effect of silver species on nitrification.

Key words | entrapped cells, nitrifying activated sludge, silver nanoparticles

Nguyen Thanh Giao
International Program in Hazardous Substance and Environmental Management, Graduate School, Chulalongkorn University,
Bangkok 10330,
Thailand
and
Department of Environmental Management, College of Environment and Natural Resources, Can Tho University,
Can Tho 90000, Viet Nam

Nguyen Thanh Giao
Tawan Limpiyakorn
Thunyalux Ratpukdi
Sumana Siripattanakul-Ratpukdi (corresponding author)
Center of Excellence on Hazardous Substance Management,
Bangkok 10330, Thailand
E-mail: *sumana.r@kku.ac.th*; *jeans_sumana@yahoo.com*

Tawan Limpiyakorn
Department of Environmental Engineering, Faculty of Engineering and Research Unit Control of Emerging Micropollutants in Environment, Chulalongkorn University,
Bangkok 10330, Thailand
and
Research Network of Chulalongkorn University and National Nanotechnology Center (RNN),
Bangkok, Thailand

Pumis Thuptimdang
Department of Chemistry, Faculty of Science and Environmental Science Research Center (ESRC), University of Chiang Mai,
Chiang Mai 56000, Thailand

Thunyalux Ratpukdi
Sumana Siripattanakul-Ratpukdi
Department of Environmental Engineering, Faculty of Engineering and Research Center for Environmental and Hazardous Substance Management,
Khon Kaen University,
Khon Kaen 40002, Thailand

INTRODUCTION

Silver nanoparticles (AgNPs) have been used in many applications, such as textile, medical, and plastic products (Pantic 2014). The large amount of AgNPs used resulted in residue being deposited in wastewater treatment plants (WWTPs) (Mitrano *et al.* 2011; Hoque *et al.* 2012; Li *et al.* 2013). With the superior antimicrobial ability of the particles, microorganisms in WWTPs, including nitrifying microorganisms, might be inhibited (Chopra 2007). Previous works reported

doi: 10.2166/wst.2019.075

that ammonia oxidation (AO) activity of nitrifying sludge could be seriously inhibited by silver species, including AgNPs and liberated Ag^+ (Choi *et al.* 2008; Jeong *et al.* 2012; Giao *et al.* 2017). The silver species that partially inhibited the AO process followed uncompetitive inhibition-like kinetics with the AgNPs and liberated Ag^+ inhibitory constants of 73.5 and 0.29 mg/L, respectively (Giao *et al.* 2017).

The potential of the cell entrapment technique to alleviate the negative impact of toxic substances on microbial activity has been studied. Cell entrapment materials reduce cell exposure to toxic chemicals leading to lower microbial inhibition (Saez *et al.* 2012; Siripattanakul-Ratpukdi 2012; Siripattanakul-Ratpukdi *et al.* 2014; Lucchesi *et al.* 2015). Prior studies have also indicated that the entrapment materials assist living microorganisms to resist adverse environmental conditions (e.g. pH and temperature changes) and to retain higher cell numbers in the system without reducing the microbial activity (Siripattanakul & Khan 2010; Hu & Yang 2015). Siripattanakul-Ratpukdi *et al.* (2014) found much lower inhibition of nitrification in the entrapped cell system influenced by AgNPs. The previous work preliminarily proved the feasibility of the entrapped cell application. However, the treatment performance and the durability of entrapped cells depended on the entrapment materials (Paul & Vignais 1980; Liu *et al.* 2012). For example, changing salt solutions (from calcium chloride to barium chloride) during the preparation of gel beads could significantly improve the physical and chemical properties. This, in turn, could lead to higher yield and activity of the entrapped cells (Paul & Vignais 1980). The addition of sodium alginate to polyvinyl alcohol (PVA) could also enhance the stability of the produced beads (Liu *et al.* 2012).

This study investigated the reduction of AgNP toxicity affecting AO using cell entrapment techniques. The screening of the entrapment materials (barium alginate (BA), polyvinyl alcohol (PVA), and mixture of polyvinyl alcohol and barium alginate (PVA-BA)) was based on nitrification performance and bead durability. The influence of silver species (AgNPs and Ag^+) on the inhibition of AO was evaluated. The AO kinetics of selected entrapped cells were also investigated. Microbial cell viability and morphology were observed by confocal laser scanning microscopy (CLSM) and scanning electron microscopy coupled with an energy dispersive X-ray (SEM-EDX). This primarily elucidated the mechanism of silver toxicity reduction. The results of this study could be used as fundamental information for AgNP contaminated wastewater treatment in the future.

MATERIALS AND METHODS

AgNPs and Ag^+

Silver nanoparticles (5,000 mg/L) were obtained from PrimeNANO technology (Prime Nanotechnology Co. Ltd, Chulalongkorn University, Thailand). Transmission electron microscopy showed spherical AgNP sizes in the range of 2–12 nm (Giao *et al.* 2017). AgNP suspensions prepared in inorganic synthetic wastewater (SWW) peaked at a wavelength of 400 nm, which indicated that AgNPs were well dispersed in the experimental condition. Additionally, the residue of Ag^+ from AgNPs of 10 mg/L was found to be less than the method detection limit (3 μg/L). Silver nitrate ($AgNO_3$, Merck, Darmstadt, Germany) was used as a source of Ag^+.

Nitrifying sludge acclimation

The seed sludge was taken from a continuous-flow enriching nitrifier reactor. The continuous reactor had been operating for more than five years with equal hydraulic (HRT) and solid (SRT) retention times of four days. During nitrifying sludge acclimation, SWW was comprised of 1.9821 g/L $(NH_4)_2SO_4$, 0.2 g/L NaCl, 0.2 g/L K_2HPO_4, 0.4 g/L $MgCl_2.6H_2O$, 0.1 g/L $CaCl_2.2H_2O$, 0.5 g/L KCl, and 1.0 g/L $NaHCO_3$ with an addition of 1 mL of inorganic salt solution. The inorganic salt solution contained 40 g/L $MgSO_4.7H_2O$, 40 g/L $CaCl_2.2H_2O$, 200 g/L KH_2PO_4, 1 g/L $FeSO_4.7H_2O$, 0.1 g/L Na_2MoO_4, 0.2 g/L $MnCl_2.4H_2O$, 0.02 g/L $CuSO_4.5H_2O$, 0.1 g/L $ZnSO_4.7H_2O$, and 0.002 g/L $CoCl_2.6H_2O$. The NH_4^+-N concentration of 429 ± 62 mg-N/L was maintained. For the silver exposure experiment, wastewater was comprised of 1.9821 g/L $(NH_4)_2SO_4$, 1 g/L $NaHCO_3$, 0.1 g/L K_2HPO_4, 0.025 g/L $MgSO_4$, and 0.025 g/L NaCl with AgNPs and Ag^+ at 10 mg/L and 0.1 mg/L, respectively.

Entrapped cell preparation

The enriched nitrifying sludge was harvested and centrifuged at 5,000 rpm for 20 min and the clear supernatant was removed. The concentrated cells were re-suspended, well-mixed, and centrifuged in SWW as a washing process five times. The washed cells were used for preparing the entrapped cells. The washed cells were added to the entrapment matrices to reach a final cell concentration in the matrix of 2,000 mg-MLSS/L.

Preparation of the BA and PVA-entrapped cells followed previous works (Siripattanakul-Ratpukdi & Tongkliang 2012; Siripattanakul-Ratpukdi *et al.* 2014). To prepare BA-entrapped cells, 2% (w/v) alginic acid sodium salt and 5% (w/v) barium chloride were used. The alginic acid sodium salt solution and the nitrifying sludge were homogeneously mixed and then dropped into the barium chloride solution using a peristaltic pump at 20 rpm. The gel beads were hardened in the barium chloride solution for 1 h. BA gel beads with diameters of 3–6 mm were formed.

For the PVA-entrapped cells, 10% (w/v) polyvinyl alcohol (PVA, 99%, fully hydrolyzed) was applied. The mixture of the PVA solution and the nitrifying sludge was dropped into a saturated boric acid solution for 30 min before it was transferred to a 1 M sodium phosphate buffer for 2 h to harden the produced gel beads. PVA gel beads of 3–6 mm diameters were formed. The other entrapment material applied in this study was a PVA-BA matrix. The combination of PVA and alginate was developed and successfully applied for entrapping phenol-degrading bacteria (Wu & Wisecarver 1992). The PVA-BA-entrapped cells were prepared by using 10% (w/v) polyvinyl and 1% (w/v) alginic acid sodium salt. The combined gel solution was mixed with the sludge. The preparation protocol for PVA-BA-entrapped cells followed that of the PVA gel, as described above.

Experimental setup and procedure

AO batch experiments were performed. A glass reactor with 100 mL working volume was applied. The reactor was operated under room temperature and 2 mg/L dissolved oxygen (DO). The study was divided into two parts: (1) inhibited AO by entrapped cells and (2) kinetics of AO by selected entrapped cells. The first part emphasized on selection of appropriate entrapment material (among BA, PVA, and PVA-BA) for the situation. The selection was based on AO percentage, bacterial viability, and material durability. For the second part, AO kinetics by selected entrapped cells was performed. This part focused on AO modeling based on Monod equation. Also, bacterial cell and bead morphology after the experiment was investigated using SEM-EDX. Figure 1 is summarized overall experiments.

Inhibited AO by entrapped cells

AO experiments on free cells, entrapped cells, and only materials were performed under the absence and presence of silver species. NH_4^+-N, Ag^+, and AgNP concentrations of 70, 0.1, and 10 mg/L, respectively, were applied. The

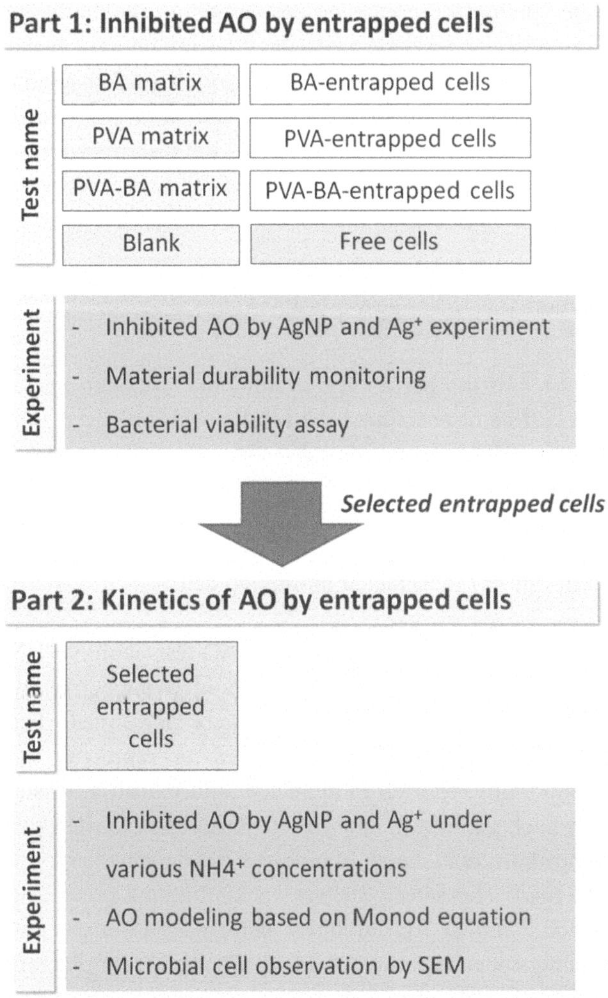

Figure 1 | Experimental procedure for AO by entrapped and free cells under presence of silver.

$NH4^+$-N concentration was selected following the concentrations reported in municipal wastewater (Bai *et al.* 2012). The Ag^+ and AgNP concentrations were chosen based on the inhibition information reported previously (Giao *et al.* 2017). In the previous work, AO with AgNP concentrations of 0, 1, 10, and 100 mg/L was tested. The tests with AgNPs of 10 and 100 mg/L gave similar inhibition (63 and 76% compared to the test without AgNP, respectively). Therefore the AgNP concentration at 10 mg/L was chosen for this study. The selected Ag^+ concentration (0.1 mg/L) was the released concentration from 10 mg/L AgNP suspension (Giao *et al.* 2017). A nitrifying sludge of 50 mg-MLSS/L was used. The AO experiments were conducted with cells entrapped in BA, PVA, and PVA-BA. 100 mL of the reactor with different entrapped cells was aerated to obtain DO at 2 mg/L. Water samples were taken after a 60-h test. The selected period (60 hour) was preliminarily tested and AO had completely reached a plateau (the preliminary result shown in Figure S1 in the supplementary material, available with the online version of this paper). Inhibition percentage of AO was estimated and compared to the reactors without silver. The inhibition percentage of AO activity was calculated based on Equation (1).

$$\text{Inhibition (\%)} = \frac{AR_C - AR_i}{AR_C} \times 100 \tag{1}$$

where AR_C is the percentage of ammonia reduction in the controls (free or entrapped cell reactor with no silver) and AR_i is percentage of ammonia reduction under the presence of AgNPs or Ag^+.

Kinetics of AO by selected entrapped cells

For the kinetic experiments, the 60-h AO tests with NH_4^+-N (10 to 180 mg-N/L) and silver (AgNPs of 10 mg/L, Ag^+ of 0.1 mg/L, and no silver addition) were performed. The NH_4^+-N concentrations were selected to represent the concentrations reported in municipal and industrial wastewater, such as coking and rubber production wastewater (Limpiyakorn *et al.* 2005; Chaiprapat & Sdoodee 2007; Bai *et al.* 2012). The selected silver concentrations previously reported a similar AO inhibition percentage from the free nitrifying sludge (Giao *et al.* 2017). 100 mL of the reactors and the selected entrapped and free cells were aerated for 60 h. The aqueous samples were collected at 0, 5, 17, 42, and 60 h to monitor AO rate. The concentration of NH_4^+-N was measured using the salicylate–hypochlorite method (Bower & Holm-Hansen 1980). The $NH4^+$-N measurement method was selected because it is more suitable when several ions, such as Ca^{2+}, Mg^{2+}, and Cl^-, are present in the sample (Le & Boyd 2012). This method is frequently used for the analysis of ammonia in seawater.

The NH_4^+-N concentration was used to calculate the AO rate and kinetics. Reduction of NH_4^+-N concentration was plotted versus time to calculate the AO rate in mg-N/L/h. The specific AO (q) was estimated as follows in Equation (2).

$$q = \frac{1}{X} \cdot \frac{dS}{dt} \tag{2}$$

where q is the specific AO rate (mg-N/g-MLSS · h), X is the initial biomass concentration applied in this study (mg-MLSS/L), S is the initial NH_4^+-N concentration (mg/L), and t is the time elapsed during the AO experiment (h).

The specific AO rate was then calculated for the kinetic parameters based on the Monod model. The traditional Monod equation depends on microbial growth. However, in this study, the microbial cells were entrapped in the matrices. Thus, the modified Monod equation based on substrate utilization was applied. Theoretically, the Monod model cannot directly be applied under inhibitory conditions. In previous work, an uncompetitive-like inhibition model was proposed (Giao *et al.* 2017). The q and the half-saturation constant (or Monod constant) for AO (K_s) were calculated based on the inhibitory constant (K_i) and the inhibitor concentration (I), as presented in Equation (3). To determine K_i, experiments under various inhibitor concentrations were required. In this study, the focus was only on the treatment by different entrapped cells. Therefore, complete uncompetitive-like kinetics of AO under the presence of silver species might be further investigated. In this study, the apparent specific AO rate (q') and half-saturation constant (K'_S) for AO were proposed (Equation (4)).

$$q = \frac{(q_{\max}/(1 + (I/K_i))) \cdot S}{K_s/(1 + (I/K_i)) + S} \tag{3}$$

$$q' = \frac{q'_{\max} S}{K'_s + S} \tag{4}$$

q' and $q'_{\max}$ are the apparent specific AO rate and the apparent maximum specific AO rate (mg-N/g-MLSS · h), respectively. K'_s is the apparent half-saturation constant (or Monod constant) for AO (mg/L). For estimation of the kinetic parameters, $q'_{\max}$ and K'_s were calculated using a linear plot that followed the Lineweaver-Burk equation (Equation (5)). Triplicate experimental data were used for kinetic parameter estimation. Experimental and model data were statistically confirmed by non-linear regression analysis using Sigma Plot software provided by Khon Kaen University, Thailand (R^2 of 0.86–0.96, $p < 0.0001$).

$$\frac{1}{q'} = \frac{K'_s}{q'_{\max}} \cdot \frac{1}{S} + \frac{1}{q'_{\max}} \tag{5}$$

Free and entrapped cell viability using live/dead assay

Free and entrapped cell viability was investigated. Ten entrapped cell beads after the experiment were collected. The bead sample was washed five times with de-ionized water and five times with a NaCl solution (0.85%). The washed beads were cut using a pair of tweezers. The tweezed beads were dipped in 10 mL of NaCl (0.85%) and then gently squeezed to let the cells out of the entrapment matrices. The suspended cells (free and de-entrapped cells) were centrifuged at 5,000 rpm for 5 min. The supernatant was discarded, and the pellets were collected. The pellets were re-suspended and concentrated to obtain the free and de-entrapped cell samples for the live/dead assay.

The staining procedure was performed using a staining kit (LIVE/DEAD® BacLight™ Bacterial Viability, Molecular Probes, Invitrogen). The samples were observed under CLSM (FluoView FV10i, Olympus, Japan). The following two samples were compared: (1) microbial cells directly taken from the nitrifying reactor and (2) killed cells (autoclaved at 121 °C, at 100 kPa for 15 min), stained for estimation of microbial viability. The percentage of live and dead cells was estimated based on 30 CLSM-images. The results were presented as % dead cells (±standard deviation).

Scanning electron microscopic observation

After the kinetic experiment, microbial cells and bead physiology were observed using SEM-EDX. The BA-entrapped cells (selected matrix) were prepared for SEM observations according to the procedure by Siripattanakul-Ratpukdi *et al.* (2014). Briefly, the entrapped cell beads from the experiment were collected and washed with de-ionized water. The collected gel beads were fixed using 1% osmium tetroxide for 1 h. The fixed gel beads were then rinsed three times with de-ionized water for 10 min. The beads were dehydrated with a series of ethanol that ranged from 30 to 95%. The absolute ethanol was applied as the last step of the dehydration. The dehydrated beads were dried using a critical point dryer (Balzers, CPD 020, Liechtenstein). The beads were divided into two parts using a razor blade in liquid nitrogen, attached to a stub, and coated with gold. The dried BA, PVA, and PVA-BA beads were observed using SEM with an energy-dispersive spectroscopy attachment (JEOL, JSM-5410LV, Tokyo, Japan). Elemental compositions of selected areas were analyzed using an energy dispersive X-ray (EDX, Oxford Instruments, Model X-MaxN, UK) to identify aggregates of silver.

Statistical analysis

The percentage of AO inhibition and membrane compromised cells caused by AgNPs and Ag^+ was presented as an averaged percent inhibition (%) ±standard deviation (SD), regardless of the initial NH_4^+-N concentrations. A statistical comparison was performed using IBM SPSS statistics for Windows, Version 19.0 (IBM Corp., Armonk, NY, USA). The significance of the differences ($p < 0.05$) was determined by a one-way analysis of variance (ANOVA) followed by Duncan's test (Ahrari *et al.* 2015).

RESULTS AND DISCUSSION

Inhibited AO by entrapped cells

Percentage of AO inhibition

An AO experiment under various conditions was performed. Blank AO test and AO tests by only entrapment matrices (no cells) were conducted (results presented in Figure S2 in the supplementary material, available with the online version of this paper). It was found that natural ammonia removal (based on blank experiment) and ammonia adsorption on entrapment matrices did not obviously occur. For the tests with nitrifying sludge, under the absence of silver contamination, the nitrifying sludge reduced ammonia. Percentages of AO by the free (94.5%) and entrapped (90.2–95.5%) nitrifying sludge were similar (Figure 2). The entrapped cells did not run into limited substrate diffusion problem (Siripattanakul-Ratpukdi *et al.* 2014). This showed potential of entrapped cell utilization in the future.

During the experiments with silver, AO performance was significantly reduced. Under the presence of AgNP (10 mg/L), the free and entrapped cells reduced ammonia by 31.3 and 55.0–70.2%, respectively. With regards to the influence of Ag^+ (0.1 mg/L), the free and entrapped cells oxidized ammonia by 19.4 and 54.2–75.5%, respectively. Previously, Giao *et al.* (2017) reported that the AO inhibition by silver caused by microbial cell damage. In this study, the entrapment matrix led to tortuosity of silver to reach to

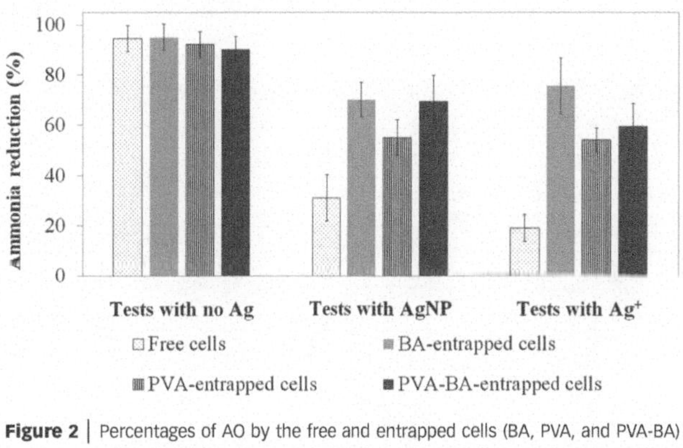

Figure 2 | Percentages of AO by the free and entrapped cells (BA, PVA, and PVA-BA) under the absence (Ag 0 mg/L) and presence of silver species (AgNPs 10 mg/L and Ag^+ 0.1 mg/L) after the 60-h experiment. The error bar indicates the standard deviation.

Table 1 | Inhibition percentage of AO caused by AgNPs and Ag^+

Microorganism	Silver species and concentration	Inhibition percentage of AO (%)	Reference
BA-entrapped nitrifying sludge	AgNPs at 10 mg/L	20.8	This study
BA-entrapped nitrifying sludge	Ag^+ at 0.1 mg/L	15.0	This study
Free nitrifying sludge	AgNPs at 10 mg/L	63.3	Giao *et al.* (2017)
Free nitrifying sludge	Ag^+ at 0.1 mg/L	75.3	Giao *et al.* (2017)
Calcium alginate-entrapped nitrifying sludge	AgNPs at 5 mg/L	19.0–35.0	Siripattanakul-Ratpukdi *et al.* (2014)
PVA-entrapped nitrifying sludge	AgNPs at 1 mg/L	18.0–95.0	Siripattanakul-Ratpukdi *et al.* (2014)
Free nitrifying sludge	Ag^+ at 1 mg/L	13.5	Liang *et al.* (2010)
Free nitrifying sludge	AgNPs at 1 mg/L	41.4	Liang *et al.* (2010)
Free nitrifying sludge	AgNPs at 1 mg/L	100.0	Choi *et al.* (2009)

microbial cells resulting better AO by the entrapped cells. Figure 2 clearly shows that AgNPs at 10 mg/L and Ag^+ at 0.10 mg/L (represented liberated dissolved silver from AgNPs) partially inhibited AO of the BA-entrapped cells by 25.4 and 19.4%, the PVA-entrapped cells by 37.0 and 38.2%, and the PVA-BA-entrapped cells by 20.7 and 30.4%, respectively. The result revealed that the BA and PVA-BA-entrapped cells performed better than the PVA-entrapped cells. The different entrapment materials produced different gel structures resulting in altered AO inhibition. When compared to the result of free cells reported previously, inhibition percentages of AO by AgNPs at 10 mg/L and Ag^+ at 0.10 mg/L were 63.3% and 75.3%, respectively (Table 1). The results clearly showed lower AO inhibition in the entrapped cell experiment.

We preliminarily monitored NO_2^- plus NO_3^--N concentration during the experiments as shown in Table S1 (in supplementary material, available online). The result stated that the entrapment material (BA) did not show the ammonia removal and nitrite plus nitrate production. Nitrite plus nitrate accumulated in all tests with microbial cells under the presence or the absence of silver. However, it was obvious that silver influenced AO and nitrite plus nitrate production. At AgNPs 10 mg/L and Ag+ 0.10 mg/L, inhibition of nitrite plus nitrate production was 41% and 56%, respectively.

Durability of entrapped cells

In this sub-section, the stability of the entrapped cell beads was the focus. Only the durability of the entrapped beads after 30 days of exposure to AgNPs and Ag^+ was observed. Figure 3 preliminarily indicates the durability of the gel beads. During the experiment, damaged beads were quantified based on bead swelling and abrasion compared to beads at the beginning. After 30 days, no damaged BA beads (0%) were observed while the PVA (56%) and PVA-BA (10%) beads were obviously swollen and damaged. This phenomenon was also reported in a previous study (Bach & Dinh 2014). The PVA beads deformed because of the hydrophilic nature. The cross-linking of PVA and boric acid could be reversed, resulting in bead swelling. The results revealed that the addition of

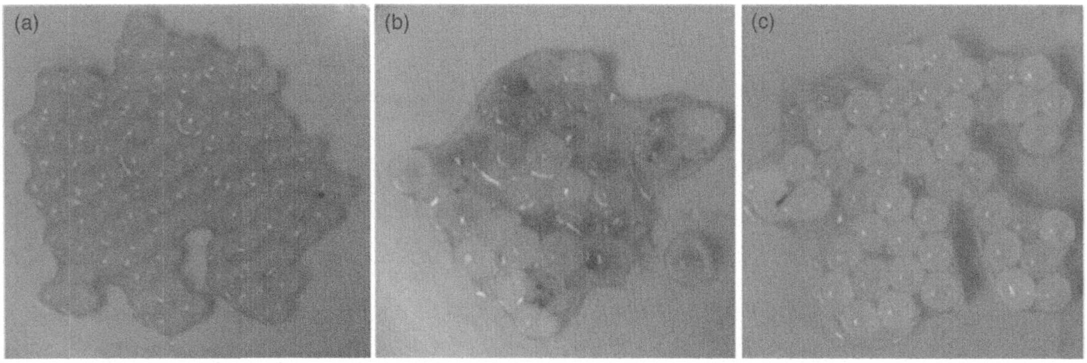

Figure 3 | Sludge entrapped in (a) BA, (b) PVA, and (c) PVA-BA after shaking for 30 days at 200 rpm with AgNPs at 10 mg/L.

BA into the PVA gel could improve stability in SWW. This result was in line with other studies that showed that a combination of PVA and BA led to higher stability and storage longevity of the produced gel beads (Paul & Vignais 1980; Liu *et al.* 2012). Among the cell entrapment techniques, the BA-entrapped cells were the most stable beads. Similarly, a previous study reported that BA had high gel strength and flexibility (Taweetanawanit *et al.* 2017). The BA-entrapped cells were stable during long-term operation and supported the growth of the entrapped microbes (Siripattanakul-Ratpukdi & Tongkliang 2012).

The cell viability assay (after the 60-h experiment), based on bacterial cell membrane damage is shown in Figure 4. The proportions of dead cells between the free cells (collected directly from the enriching reactor) and the controlled BA- and PVA-entrapped cells (entrapped cells without AgNPs or Ag^+ exposure) were insignificantly different ($p > 0.05$). The result revealed that BA and PVA were not toxic to bacterial cells. Prior studies reported that BA and PVA were not harmful to microbes (Siripattanakul & Khan 2010; Wadhawan *et al.* 2011; Vecchiatini *et al.* 2015).

Conversely, the percentage of dead cells from the PVA-BA-entrapped cell test (no silver) was significantly higher than free cells ($p < 0.05$). As discussed earlier, BA and PVA are non-toxic to microbial cells, and likewise the mixture of PVA-BA should not cause toxicity to the entrapped cells. The damaged cells in the PVA-BA test might be due to the viscosity of the material (PVA-BA) resulting in less diffusion of oxygen and ammonia (substrate) into the beads. Among the entrapped cells, BA was the most promising entrapment material for mitigation of the toxicity of AgNPs and Ag^+. The BA matrix promoted high treatment performance (low inhibition) and cell viability.

Kinetics of AO by BA-entrapped cells

The kinetics of AO from the BA-entrapped cells are demonstrated in Figure 5. Table 2 is a summary of the AO kinetic parameters. Figure 5 shows that the experimental results well fitted Monod model. Under absence of silver, self-substrate inhibition did not occur, resulting in the steady specific AO rates for the tests with NH_4^+ higher than 100 mg/L. Based on free ammonia calculation, only 0.56 mg/L free ammonia may form (Anthonisen *et al.* 1976). Free ammonia of 10–150 mg/L could inhibit AO. At the concentration related to this study, no inhibition by free ammonia would take place. The result from this study revealed that the tested NH_4^+ concentrations (0–180 mg-N/L) did not show adverse effect from nitrogenous species.

Results from the tests with no silver indicated that the entrapped cell system encountered the substrate diffusion problem resulting in much lower q_{max} values compared to that of the free cells. Under the presence of silver, both K'_S and q'_{max} values of the free cell systems were much lower than those without silver. This result implied that inhibitors (AgNPs or Ag+) obviously influenced AO in the free cell system. The entrapped cell tests with silver (both species) had similar kinetic parameter values (Table 2). This highlights the success of the cell entrapment application for reducing AgNPs and liberated Ag^+ toxicity. Table 1 confirms that the entrapped cells reduced the influence of AgNPs and Ag^+ by three and five times, respectively, as compared to those from the suspended cells at the same conditions.

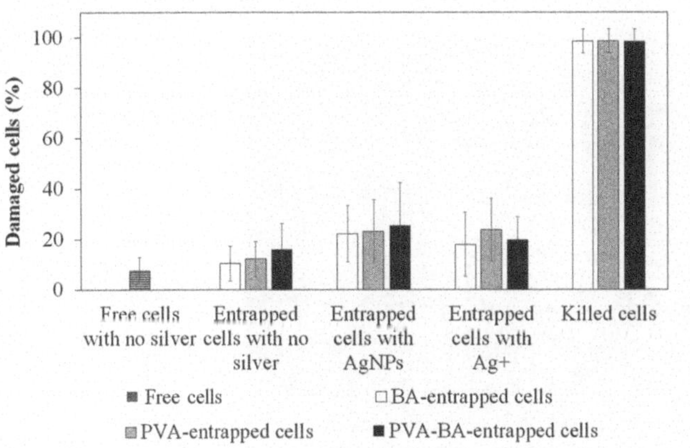

Figure 4 | Percentage of damaged cells from the tests with non-inhibited free cells (free cells with no Ag), non-inhibited entrapped cells (entrapped cells with no Ag), inhibited entrapped cells by 10 mg/L of AgNPs (entrapped cells with AgNPs) and 0.1 mg/L of Ag^+ (entrapped cells with Ag^+). Error bars indicate the standard deviation.

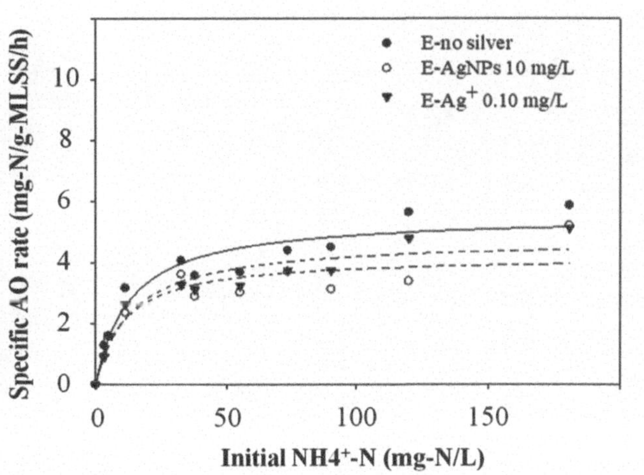

Figure 5 | Kinetics of AO by the nitrifying sludge entrapped in BA with no silver (•), AgNPs at 10 mg/L (○), and Ag^+ at 0.1 mg/L (▾).

Table 2 | AO kinetic parameters under the absence and presence of silver

Cell mode	Silver species and concentration	Kinetic parameters	
Test with no silver		K_s (mg/L)	q_{max} (mg-N/g-MLSS · h)
Free cells	No silver	17.8[a]	12.8[a]
BA-entrapped cells	No silver	12.9	5.6
Test with silver		K'_s (mg/L)	q'_{max} (mg-N/g-MLSS · h)
Free cells	AgNPs at 10 mg/L	3.3[a]	3.3[a]
	Ag^+ at 0.1 mg/L	3.7[a]	1.3[a]
BA-entrapped cells	AgNPs at 10 mg/L	10.5	4.2
	Ag^+ at 0.1 mg/L	13.4	4.8

[a]Results from Giao *et al.* (2017).

Microscopic observation of BA-entrapped cells

Figure 6 presents SEM images of the BA-entrapped cells. As shown in Figure 5(a) and 5(b), several micro-colonies laid on the BA-matrices. A previous study indicated that cross-linking between barium and alginate created a dense network with an abundance of fine pores (Siripattanakul-Ratpukdi & Tongkliang 2012). This study also found numerous small pores (Figure 6(c)). The pore sizes ranged from 0.1 to 0.3 μm, which were much smaller than the single cells (1–2 μm), resulting in no micro-colonies in the pores. The small porous structure was related to the low maximum

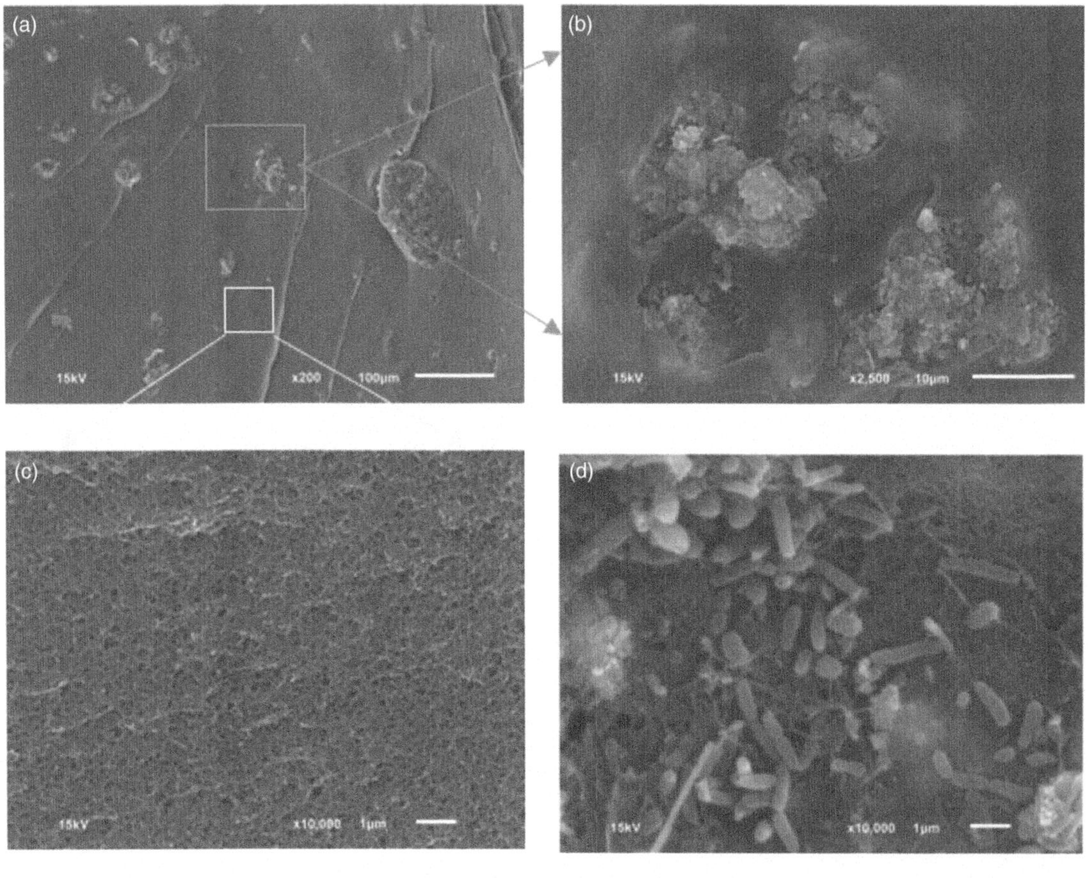

Figure 6 | SEM micrographs for observation of BA-entrapped cells after 60 h of exposure to AgNPs at 10 mg/L. (a) Cross-section of the BA-entrapped beads with several colonies of cells found in the BA matrices, observed at 200 ×. (b) Microbial cells were surrounded by BA-matrices, observed at 2,500 ×. (c) The microporous structures formed by ionic-cross linking between alginate and barium ions. The sizes of the pores were estimated as 0.1–0.3 μm, observed at 10,000 ×. (d) A cluster of cells with long- (majority) and short-rod (minority) shapes, observed at 10,000 × .

AO rates found in the kinetic experiment of the entrapped cells. This structure was torturous for the substrate to reach the microbial cells inside the BA beads. The dense structure protected the cells from silver species, leading to similar AO rates in the experiments by the entrapped cells under absence and presence of silver species.

It was noted by Giao *et al.* (2017) that the microbial cells changed from long rod-shaped to shorter rod-shaped in the presence of AgNPs. In this study, most of the cells remained long rod-shaped (Figure 6(d)). This could primarily indicate that the microbial cells were not stressed from the toxic substance; therefore, cell morphological adaptation was not observed. During SEM observation, element (silver) analysis using an EDX technique was also performed. In the entrapped cell experiment, silver clusters were not detected inside the beads. This confirmed that the entrapment matrices reduced AgNPs and liberated Ag^+ transporting into the beads.

CONCLUSION

The entrapment materials, including BA, PVA, and PVA-BA, were non-toxic to microbial cells. In the short-term AO test, BA was the most promising entrapment material for reduction of AgNPs and Ag^+ toxicity. The inhibitory kinetics results showed that the cells entrapped in BA maintained high microbial cell viability (approximately 20% dead cells) and AO up to 90% in the presence of AgNPs or Ag^+. The kinetic parameters confirmed low influence of AgNPs and Ag^+ on AO by the entrapped cells. Microscopic observation of the entrapped cells suggested that the porous structure of matrices limited penetration of silver, resulting in lower cell damage and AO inhibition. Future study on the long-term impact of AgNPs and Ag^+ on AO activity and microbial community change in entrapped cells should be investigated.

ACKNOWLEDGEMENTS

The authors acknowledge financial and equipment support from the Research Center for Environmental and Hazardous Substance Management (Khon Kaen University) and Research Career Development Grant RSA6080054 (Thailand Research Fund (TRF) and Khon Kaen University). This work was supported by the National Nanotechnology Center (NANOTEC), NSTDA, Ministry of Science and Technology, Thailand, through its program of Research Network NANOTEC (RNN). Any opinions, findings, and conclusions or recommendations expressed in this material are those of the authors and do not necessarily reflect the views of the grant agencies.

REFERENCES

Ahrari, F., Eslami, N., Rajabi, O., Ghazvini, K. & Barati, S. 2015 The antimicrobial sensitivity of *Streptococcus mutans* and *Streptococcus sangius* to colloidal solutions of different nanoparticles applied as mouthwashes. *Dental Research Journal* **12** (1), 44–49.

Anthonisen, A. C., Loehr, R. C., Prakasam, T. B. S. & Srinath, E. G. 1976 Inhibition of nitrification by ammonia and nitrous acid. *Journal of the Water Pollution Control Federation* **48** (5), 835–852.

Bach, L. T. & Dinh, P. V. 2014 Immobilized bacteria by using PVA (polyvinyl alcohol) crosslinked with sodium sulfate. *International Journal of Science and Engineering (IJSE)* **7** (1), 41–47.

Bai, Y., Sun, Q., Wen, D. & Tang, X. 2012 Abundance of ammonia-oxidizing bacteria and archaea in industrial and domestic wastewater treatment systems. *FEMS Microbiology Ecology* **80** (2), 323–330.

Bower, C. E. & Holm-Hansen, T. 1980 A salicylate-hypochlorite method for determining ammonia in seawater. *Fishery and Aquaculture Science* **37**, 794–798.

Chaiprapat, S. & Sdoodee, S. 2007 Effects of wastewater recycling from natural rubber smoked sheet production on economic crops in southern Thailand. *Resources, Conservation and Recycling* **51**, 577–590.

Choi, O., Deng, K. K., Kim, N. J., Ross, J., Surampallie, R. Y. & Hua, Z. 2008 The inhibitory effects of silver nanoparticles, silver ions, and silver chloride colloids on microbial growth. *Water Research* **42**, 3066–3074.

Choi, O., Cleuenger, T., Deng, B., Surampalli, R., Ross, L. & Hu, Z. 2009 Role of sulfide and ligand strength in controlling nanosilver toxicity. *Water Research* **43**, 1879–1886.

Chopra, I. 2007 The increasing use of silver-based products as antimicrobial agents: a useful development or a cause for concern. *Journal of Antimicrobial Chemotherapy* **59**, 587–590.

Giao, N. T., Limpiyakorn, T., Kunapongkiti, P., Thuptimdang, P. & Siripattanakul-Ratpukdi, S. 2017 Influence of silver nanoparticles and liberated silver ions on nitrifying sludge: ammonia oxidation inhibitory kinetics and mechanism. *Environmental Science and Pollution Research* **24** (10), 9229–9240.

Hoque, M. E., Khosravi, K., Newman, K. & Metcalfe, C. D. 2012 Detection and characterization of silver nanoparticles in aqueous matrices using asymmetric-flow field flow fractionation with inductively coupled plasma mass spectrometry. *Journal of Chromatography A* **1233**, 109–115.

Hu, J. & Yang, Q. 2015 Microbial degradation of di-*N*-butylphthalate by *Micrococcus* sp. immobilized with polyvinyl alcohol. *Desalination and Water Treatment* **56**, 2457–2463.

Jeong, E., Chae, S. R., Kang, S. T. & Shin, H. S. 2012 Effects of silver nanoparticles on biological nitrogen removal processes. *Water Science and Technology* **65** (7), 1298–1303.

Le, P. T. T. & Boyd, C. E. 2012 Comparison of phenate and salicylate methods for determination of total ammonia nitrogen in freshwater and saline water. *Journal of the World Aquaculture Society* **43** (6), 885–889.

Li, L., Hartmann, G., Döblinger, M. & Schuster, M. 2013 Quantification of nanoscale silver particles removal and release from municipal wastewater treatment plants in Germany. *Environmental Science and Technology* **47**, 7317–7323.

Liang, Z., Das, A. & Hu, Z. 2010 Bacterial response to a shock load of nanosilver in an activated sludge treatment system. *Water Research* **44**, 5432–5438.

Limpiyakorn, T., Shinohara, Y., Kurisu, F. & Yagi, O. 2005 Communities of ammonia-oxidizing bacteria in activated sludge of various sewage treatment plants in Tokyo. *FEMS Microbiology Ecology* **54** (2), 205–217.

Liu, Z.-Q., Zhou, M., Zhang, X.-H., Xu, J.-M., Xue, Y.-P. & Zheng, Y.-G. 2012 Biosynthesis of iminodiacetic acid from iminodiacetonitrile by immobilized recombinant *Escherichia coli* harboring nitrilase. *Journal of Molecular Microbiology and Biotechnology* **22**, 35–47.

Lucchesi, G. I., Bergero, M. F., Boeris, P. S., López, G. A., Heredia, R. M. & Liffourrena, A. S. 2015 Immobilization of *P. putida* a (ATCC 12633) cells using Ca-alginate: environmental applications for the removal of cationic surfactants pollutants in industrial wastewater. *Alginic Acid: Chemical Structure, Uses and Health Benefits* **2015**, 105–118.

Mitrano, D. M., Lesher, E. K., Bednar, A., Monserud, J., Higgins, C. P. & Ranville, J. F. 2011 Detecting nanoparticulate silver using single-particle inductively coupled plasma-mass spectrometry. *Environmental Toxicology and Chemistry* **31**, 115–121.

Pantic, I. 2014 Application of silver nanoparticles in experimental physiology and clinical medicine: current status and future prospects. *Reviews on Advanced Materials Science* **2014** (37), 15–19.

Paul, F. & Vignais, P. M. 1980 Photophosphorylation in bacterial chromatophores entrapped in alginate gel: improvement of the physical and biochemical properties of gel beads with barium as gel-inducing agent. *Enzyme and Microbial Technology* **2** (4), 281–287.

Saez, J. M., Benimeli, C. S. & Amoroso, M. J. 2012 Lindane removal by pure and mixed cultures of immobilized actinobacteria. *Chemosphere* **89**, 982–987.

Siripattanakul, S. & Khan, E. 2010 Fundamentals and applications of entrapped cell bioaugmentation for contaminant removal. *Emerging Environmental Technologies* **2**, 147–170.

Siripattanakul-Ratpukdi, S. 2012 Entrapped cell system for decentralized hospital wastewater treatment: inhibitory effect of disinfectants. *Environmental Technology* **33** (20), 2319–2328.

Siripattanakul-Ratpukdi, S. & Tongkliang, T. 2012 Municipal wastewater treatment using barium alginate entrapped activated sludge: adjustment of utilization conditions. *International Journal of Chemical Engineering and Applications* **3** (5), 328–332.

Siripattanakul-Ratpukdi, S., Ploychankul, C., Limpiyakorn, T., Vangnai, A. S., Rongsayamanont, C. & Khan, E. 2014 Mitigation of nitrification inhibition by silver nanoparticles using cell entrapment technique. *Journal of Nanoparticle Research* **16**, 2218.

Taweetanawanit, P., Radpukdee, T., Giao, N. T. & Siripattanakul-Ratpukdi, S. 2017 Mechanical and chemical stabilities of barium alginate gel: influence of chemical concentrations. *Key Engineering Materials* **718**, 62–66.

Vecchiatini, R., Penolazzi, L., Lambertini, E., Angelozzi, M., Morganti, C., Mazzitelli, S., Trombelli, L., Nastruzzi, C. & Piva, R. 2015 Effect of dynamic three-dimensional culture on osteogenic potential of human periodontal ligament-derived mesenchymal stem cells entrapped in alginate microbeads. *Journal of Periodontal Research* **50** (4), 544–553.

Wadhawan, T., Maruska, Z. B., Siripattanakul, S., Hill, C. B., Gupta, A., Prüb, B. M., McEvoy, J. M. & Khan, E. 2011 A new method to determine initial viability of entrapped cells using fluorescent nucleic acid staining. *Bioresource Technology* **102**, 1622–1627.

Wu, K.-Y. A. & Wisecarver, K. D. 1992 Cell immobilization using PVA crosslinked with boric acid. *Biotechnology and Bioengineering* **39**, 447–449.

First received 27 July 2018; accepted in revised form 10 February 2019. Available online 21 February 2019

Community composition of marine and brackish water ammonia-oxidizing consortia developed for aquaculture application

P. G. Preena, Cini Achuthan, V. J. Rejish Kumar, R. Boobal, G. D. Deepa, Jayesh Puthumana, Sunitha Poulose, I. S. Surekhamol and I. S. Bright Singh

ABSTRACT

To mitigate the toxicity of ammonia in aquaculture systems, marine and brackish water ammonia-oxidizing bacterial consortia have been developed and are used for activation of nitrifying bioreactors integrated to recirculating aquaculture systems. To shed more light on to these biological entities, diversity of both the consortia were analyzed based on random cloning of 16S rRNA gene and ammonia-oxidizing bacterial specific *amo*A gene sequences. The dendrograms of representative clones on the basis of amplified ribosomal DNA restriction analysis generated 22 and 19 clusters for marine and brackish water nitrifying consortia, respectively. Phylogenetic analysis demonstrated the presence of various autotrophic nitrifiers belonging to α-, β- and γ-Proteobacteria, anaerobic ammonia oxidizers, heterotrophic denitrifiers, Bacteroidetes, and Actinobacteria. Distribution patterns of the organisms within the two consortia were determined using the software Geneious and diversity indices were investigated using Mega 5.0, VITCOMIC and Primer 7. The abundance of ammonia oxidizers was found in the order of $2.21 \pm 0.25 \times 10^9$ copies/g wet weight of marine consortium and $6.20 \pm 0.23 \times 10^7$ copies/g of brackish water consortium. Besides, marine ammonia-oxidizing consortium exhibited higher mean population diversity and Shannon Wiener diversity than the brackish water counterparts.

Key words | ammonia monooxygenase, ammonia-oxidizing consortia, diversity, quantification, VITCOMIC

P. G. Preena
Cini Achuthan
V. J. Rejish Kumar
R. Boobal
G. D. Deepa
Jayesh Puthumana
Sunitha Poulose
I. S. Surekhamol
I. S. Bright Singh (corresponding author)
National Centre for Aquatic Animal Health,
Cochin University of Science and Technology,
Fine Arts Avenue, Cochin 682016,
India
E-mail: *isbsingh@gmail.com*

V. J. Rejish Kumar
Present address: Department of Aquaculture,
Kerala University of Fisheries and Ocean Studies,
Panangad, Kochi 682506,
India

INTRODUCTION

Recirculating aquaculture systems (RAS) are one among the sustainable aquaculture production facilities which stand out as the promise for intensification. Improved waste management, effective nutrient cycling, and diminished water consumption could be made possible only through this technology. However, one of the major constraints is the accumulation of nitrogenous wastes such as ammonia and nitrite in the system leading to physiological stress to the cultured species. To facilitate removal of ammonia nitrogen from RAS by establishing instantaneous nitrification, two category nitrifying bacterial consortia such as ammonia-oxidizing consortium for penaeid culture (AMOPCU-I), and ammonia-oxidizing consortium for non penaeid culture (AMONPCU-I) were developed through enrichment technique at National Centre for Aquatic Animal Health (NCAAH) (Achuthan *et al.* 2006). Besides, two nitrite oxidizing consortia such as NIOPCU-1 and NIONPCU-1 could also be developed to address nitrite in marine and brackish water aquaculture systems (Achuthan *et al.* 2006). At laboratory scale, these consortia were found to have potential to activate nitrifying bioreactors such as Packed Bed Bioreactor and Stringed Bed Suspended Bioreactor (Patent no. 241648), which could be integrated to shrimp larval rearing systems and broodstock maintenance facilities (Kumar *et al.* 2010). The consortia were also made use of in enhancing nitrification in *P. monodon* grow out systems by immobilizing on to wood powder (Manju *et al.* 2009). Over a period, investigations have been made on these

doi: 10.2166/wst.2019.029

consortia and during the process, NIOPCU-1 and NIONPCU-1 could be characterized (Preena *et al.* 2017b). Under such circumstances, the present investigation was undertaken to analyze the diversity of ammonia-oxidizing consortia such as AMOPCU-1and AMONPCU-1 to facilitate their better utilization and preservation.

Chemolithotrophic ammonia-oxidizing bacteria (AOB) play a vital role in the nitrogen cycle by transforming ammonia to nitrite in the presence of oxygen. Ammonia-oxidizing bacteria are the inevitable entities for the turnover of inorganic ammonia in several ecosystems and for biological wastewater treatment, thereby resolving the problems of aquatic toxicity. Oxidation of ammonia to nitrite is the rate-limiting step of nitrification. The codependent enzymes, ammonia monooxygenase (AMO), the membrane-bound multiple subunit enzyme and hydroxylamine oxidoreductase (HAO), an unusual enzyme with multiple c-type hemes, are the key enzymes responsible for energy conservation during the rate-limiting step. In this study, gene encoding ammonia monooxygenase has been considered to determine the ammonia-oxidizing efficacy of the consortia. Until recently, no single autotrophic bacterium had been reported to have the capability to oxidize ammonia directly to nitrate. However, *Nitrospira* sp. (van Kessel *et al.* 2015) has now been reported to have the potential to accomplish complete nitrification (oxidation of ammonia to nitrate) (Comammox). The Planctomycetes group, anammox, is also responsible for N removal by means of ammonia oxidation coupled with nitrite reduction. Ammonia oxidizers are normally found in low numbers in the environment as they are slow growers and sensitive to several environmental factors, and they are difficult to get segregated as pure cultures (Kasmuri & Lovitt 2018). Hence, sensitive and precise methods of detection, isolation, and identification are essential to understand more of their ecological importance in natural systems. Among them, nucleic acid-based methods are preferred to determine the diversity of nitrifiers in the light of deficiency of a culture-dependent approach to bring out diverse AOB from the environment (Wang *et al.* 2014).

The extent of the diversity of AOB greatly influences the stability of biological filters (Vannecke *et al.* 2016). Hence, it is clear that it becomes essential to make the consortia diverse to achieve robustness of nitrification in an aquaculture setting. In order to achieve this requirement, a fair understanding of the conditions which improve species richness, evenness and abundance of nitrifiers are needed. Since molecular approaches allow better understanding of diversity and evenness of AOB in the natural environment than revealed by way of cultivation-dependent methods alone (Rowan *et al.* 2003), molecular tools have been employed in this study for characterizing the diversity of ammonia oxidizers in two consortia of nitrifiers such as AMOPCU-1 and AMONPCU-1 by way of constructing 16S rRNA libraries. Molecular characterization using amplified ribosomal DNA restriction analysis provided the phylogenetic framework to evaluate the diversity and distribution of nitrifying environmental populations. Genes encoding rate-limiting enzymes turn out to be better targeted for fine-scale resolution functionally and ecologically as they are under high selection pressures (Rotthauwe *et al.* 1997). This realization suggests that assessment of microbial community based on the abundance of functional *amo* gene is a better strategy for assessing the health of either a biological filter or nitrifying microbial consortium for their effectiveness in regulating nitrogen removal. The desired *amo* genes were quantified using the calibration curve which corresponded to threshold cycles to exact concentrations of template DNA. Salinity was implicated as a key regulator in determining species composition, diversity, distribution, and nitrification potential of ammonia-oxidizing consortia (Cui *et al.* 2016). Hence, distinction of species composition between marine and brackish water consortia provides a basis to evaluate the contribution of defined species to nitrification in different salinity environment. It is also reported that species richness and distribution vary among different ecosystems as established by Rowan *et al.* (2003) in wastewater plants.

MATERIALS AND METHODS

Enrichment of marine and brackish water ammonia-oxidizing consortia

Ammonia-oxidizing bacterial consortia such as AMOPCU-1 and AMONPCU-1 (Achuthan *et al.* 2006), maintained as part of Microbial Culture Collection at NCAAH, was used for the present study and the consortia were amplified in the 2 L fermentor (Bioflo 2000, New Brunswick Scientific, USA) following Preena *et al.* (2017b).

DNA extraction and 16S rRNA gene amplification

DNA was extracted metagenomically from AMOPCU-1 and AMONPCU-1 following the protocol described by Preena *et al.* (2017b). The concentration of DNA was quantified using spectrophotometer (Hitachi U-2800, Hitachi Corp., Japan) at Abs_{260}. For the amplification of 16S rRNA gene,

16S –1F (5′-GAGTTTGATCCTGGCTCA-3′) and 16S-1R (5′-ACGGCTACCTTGTTACGACTT-3′) primer systems were used (Lane 1991). The separated polymerase chain reaction (PCR) products were observed on 1% agarose gel.

Amplification of the functional gene *amo*A

For amplification of the *amo*A gene, *amo*A-1F (5′-GGGGTTTCTACTGGTGGT-3′) and *amo*A-2R (5′-CCCCTCKGSAAAGCCTTCTTC-3′) (K = G or T; S = G or C) primers systems (Rotthauwe *et al.* 1997) were used. Standard thermal profile used for the amplification of *amo*A target sequence was slightly modified and are as follows: 5 min at 94 °C; pause at 80 °C to add polymerase; 42 cycles consisting of 90 s at 56.8 °C (annealing), 90 s at 72 °C (elongation), and 60 s at 94 °C (denaturation); and a final cycle consisting of 90 s at 60 °C and 10 min at 72 °C.

Cloning of universal 16S rRNA and *amo*A genes

PCR products of 16SrRNA and *amo*A genes from AMOPCU-1 and AMONPCU-1 were used for cloning into the pGEM-T Easy vector (Promega). The plasmids carrying positive inserts were extracted using plasmid mini-preparation kit (Sigma Aldrich, USA), where a 50 μL increased volume of reagent for elution has been used for the better recovery of plasmid.

Amplified ribosomal DNA restriction analysis (ARDRA) of 16S rRNA gene positive clones

ARDRA was performed to assess the diversity of AMOPCU-1 and AMONPCU-1. Diluted (50 times) plasmids containing inserts were re-amplified using the universal primers, 16S-1F (5′-GAGTTTGATCCTGGCTCA-3′) and 16S-1R (5′-ACGGCTACCTTGTTACGACTT-3′) at same PCR conditions for amplification of 16S rRNA gene. Each of the PCR products was subjected to digestion separately with *Hpa*II, *Alu*I and *Hae*III (New England BioLabs, USA). These tetra cutter restriction enzymes were found to be the useful tools for screening environmental clone libraries and the digestion reaction was carried out as performed by Preena *et al.* (2017b).

Cluster identification using unweighted average linkage

Quantity One software (Bio-Rad, USA) detected the individual loci of the banding pattern and NTSYSpc software was used for further analysis (Version 2.02i, Applied Biostatistics Inc., USA). NT edit programme transferred the data matrix from MS Excel spreadsheets to a proprietary matrix file and the programme NTSYSpc assessed the rectangular data matrix generated. Similarities were determined by a statistical module, sequential agglomerative hierarchical nested cluster method (SAHN) and clustering was performed by un-weighted pair group method with arithmetic mean (UPGMA).

Phylogenetic characterisation of universal 16SrRNA gene and functional *amo*A gene

The sequences of 16S rRNA clones based on ARDRA and randomly selected *amo*A gene inserted clones from AMOPCU-1 and AMONPCU-1 were detected using ABI Prism 3700 Big Dye sequencer (SciGenom, Cochin). The vector regions in the sequences were removed using VecScreen analysis tool of NCBI database and then matched with the GenBank database using Basic Local Alignment Search Tool (BLAST) algorithm and ClustalW (Altschul *et al.* 1990). Mega 5.0 software has been explored for the calculation of distance matrices of aligned data sets of the marine and brackish water clone library (Tamura *et al.* 2011). Thus, the 16S rRNA and *amo*A phylogenetic trees were created by means of UPGMA statistical method and Kimura 2-parameter substitution model. The frequency of occurrence of different organisms and bacterial community composition of AMOPCU-1 and AMONPCU-1 was evaluated by employing Geneious software and rapid visualization tool VITCOMIC (Mori *et al.* 2010). The diversity index included Shannon Wiener diversity for species richness, evenness, and distribution in both the consortia as determined using Primer 7 (Clarke & Gorley 2015). In addition, principal component analysis (PCA) has been performed for the OTU estimation using the Primer 7.

Determination of *amo*A gene titer using quantitative polymerase chain reaction (qPCR)

Standard curves of plasmids carrying *amo*A gene were constructed for quantifying the functional *amo*A gene present in both AMOPCU-1 and AMONPCU-1. Plasmids containing the *amo*A gene insert were extracted using a plasmid purification kit (Sigma Aldrich, USA) and kept as the standard in tenfold serial dilutions ranging from 3×10^5 to 30 copies. Initially, marine and brackish water consortia DNA concentration was quantified employing cubic meter. Marine and brackish DNA samples having uniform concentrations were used for quantifying the *amo*A gene. The PCR mix

(10 μL) comprised 5 μL Power SYBR Green master mix (Applied Biosystems, California, USA), 1 μL each of forward and reverse primers targeting *amo*A gene (*amo*A 1F and *amo*A 2R), 1 μL template DNA, and 2 μL MilliQ water. A final concentration of 600 nM primers was used. The PCR mix without template DNA was kept as the negative control and triplicates were maintained for all reactions. The reaction conditions were followed by an initial denaturation at 95 °C for 10 min, 40 cycles of 95 °C for 30 sec, annealing at 58 °C for 30 s and extension at 68 °C for 45 s and the data were analyzed using the StepOne™ Software v2.2.2 (Applied Biosystems). The DNA was quantified based on standard curves and compared by one-way analysis of variance (ANOVA).

Accession numbers

The obtained 16S rRNA and *amo*A gene sequences were deposited to GenBank under accession numbers KM382225- KM382245, KR537268- KR537286, FJ652050- FJ652051, FJ665506- FJ665508, FJ711765- FJ711766, GQ221068, KT429246- KT429269.

RESULTS AND DISCUSSION

DNA extraction of marine and brackish water ammonia-oxidizing consortia

Genomic DNA (>10 kb in size) was extracted from both AMOPCU-1 and AMONPCU-1. DNA extraction from the consortia required special treatment to disrupt bacterial cells probably because of the dense glycocalyx. Cell lysis using lysozyme and SDS, heat shock along with phenol-chloroform extraction made it a suitable extraction procedure for nitrifying bacterial consortia.

PCR amplification of the phylogenetic and functional genes

In this process, 1,500 bp 16S rRNA gene could be amplified from AMOPCU-1 and AMONPCU-1 using universal bacterial-specific primers and cloned to produce 16S rRNA gene representative library of each consortium. The functional gene ammonia monooxygenase (*amo*A) could be amplified from the consortia having product size of 491 bp, resulting in the construction of the *amo*A gene library.

Cloning of 16S rRNA and *amo*A genes

On cloning of *amo*A and 16S rRNA genes to determine the bacterial diversity of the consortia, one hundred and two 16S rRNA gene positive clones were obtained from AMOPCU-1 and one hundred and three from AMONPCU-1. This resulted in the generation of 16S rRNA and *amo*A gene libraries consisting 100 clones each. *Amo*A clones from AMOPCU-1 and AMONPCU-1 were randomly selected for sequencing, while the 16S rRNA clones were subjected to ARDRA.

Amplified ribosomal DNA restriction analysis and sequencing

The restricted fragment patterns, operational taxonomic units (OTUs), obtained from ARDRA helped to distinguish the taxonomic groups. The ARDRA profile grouped the clones into different clusters. ARDRA pattern of 100 clones from AMOPCU-1 showed dendrogram of 22 clusters (Figure S1). However, the digestion of an equal number of clones from AMONPCU-1 generated 19 clusters (Figure S2) exhibiting a different pattern. The dendrogram of the clones from AMOPCU-1 showed distance coefficient of 0.74 to 1.00 while in AMONPCU-1, it was 0.79 to 1.00. Sequencing of representative clones from each cluster of both consortia was performed. (Figures S1 and S2 are available with the online version of this paper.)

Phylogenetic analyses and diversity determination

Through culture-dependent techniques, less information could be generated on the diversity of complex environments such as aquaculture systems. In this context, culture-independent molecular methods were taken in to account for determining the universal and functional gene diversity of the consortia by employing four bio-informatics tools such as Mega 5.0, Geneious, VITCOMIC, and Primer 7. Ecological significance and phylogenetic relatedness were clearly understood through these studies. The phylogenetic tree constructed using 16S rRNA gene sequences revealed that majority of the hits were with alpha-Proteobacteria within both the consortia. The other clones from AMOPCU-1 clustered with Bacteroidetes, Actinobacteria, and beta-Proteobacteria, while AMONPCU-1 had the same hits along with gamma-Proteobacteria.

The phylogenetic tree constructed using the aligned sequences of AMOPCU-1 with 99% similarity with GenBank database showed major clustering within the taxonomic composition of the marine counterparts. The phylogenetic tree created with the assembled sequences of 22 OTUs generated a dendrogram comprised mainly of two groups separating themselves into seven clusters (Figure 1). The two groups were separated at around 2.5r. The clones were separated within the clusters at around 0.2r. The first group included few clones which clustered with alpha-Proteobacteria such as *Nitrobacter* sp., *Ochrobactrum anthropi, Mesorhizobium* sp., *Tistrella mobilis* and an uncultured bacterium clone. While various uncultured bacterium clones along with Actinobacteria named *Microbacterium, Gaiella*, and *Salinibacterium* grouped under another cluster in the first group along with a single existing sequence of an uncultured soil bacterium. The first group also consisted of Actinobacteria named *Microbacterium, Gaiella* and *Salinibacterium*, uncultured soil bacterium and various uncultured bacterium clones including *Vitellibacter*, as separate clusters. The second group comprised mainly three clusters with Bacteroidetes such as *Muricauda* and *Formosa* as one, *Rhodococcus* sp. and actinobacterium as the second cluster. Besides, beta-Proteobacteria such as *Nitrosomonas* sp. and *Alcaligenes* sp., along with separately branched alpha-Proteobacteria called *Caulobacter* sp. with various uncultured Phyllobacteria and hyphomicrobes formed the final cluster. Using the diversity estimation of Mega 5.0, the mean diversity in the entire population of AMOPCU-1 was found to be 2.584 ± 0.439.

Another bioinformatic tool, Geneious was also employed in the study for analyzing the diversity pattern of both the ammonia-oxidizing consortia. For the data analyses, representative 16S rRNA gene sequences were introduced into the Geneious platform for compiling the information. Ribosomal Database Project Database (RDP) was used for comparing the 16SrRNA sequences in Geneious 16S Biodiversity Tool. Thus, a red-green color gradient displaying krona chart, with red being the highest observed e-value (least significant) and green being the lowest (most significant), was generated based on the hierarchical graph of the microbial diversity in the sample. In the Geneious map (Figure 2), AMOPCU-1 was found mainly composed of Proteobacteria and Bacteroidetes followed by Actinobacteria and Chloroflexi. Among the Proteobacteria, an α group represented the major one which had 9% Phyllobacteraceae, 5% Hyphomicrobiaceae, 5% Brucellaceae, 5% Xanthobacteraceae, 5% Bradyrhizobiaceae, 5% Rhodospirillaceae and 5% Caulobacteraceae encompassing seven families. Beta-Proteobacteria included mainly 5% Nitrosomonadaceae and 5% Alcaligenaceae. Bacteroidetes included two main groups such as Sphinghobacteria, with 5% Sphinghobacteraceae and 5% Saprospiraceae and Flavobacteria, with 14% Flavobacteraceae making the population thick. Actinobacteria were

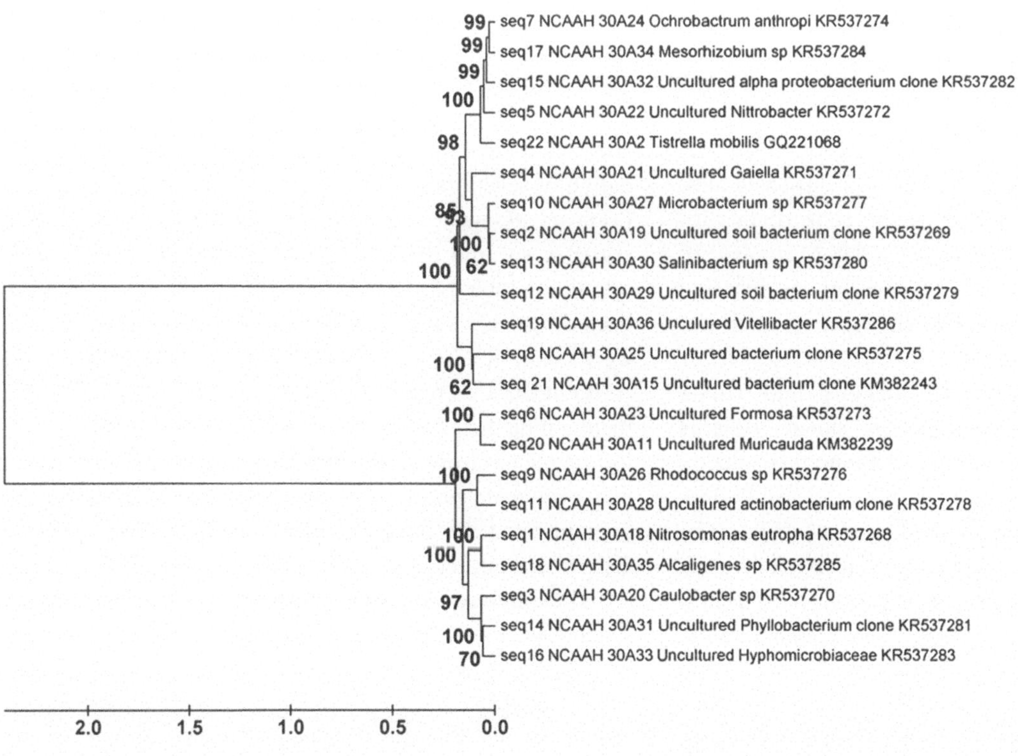

Figure 1 | Phylogenetic analysis of the 16S rRNA gene sequences from the marine ammonia-oxidizing consortium (AMOPCU-1) by using Mega 5.0 software.

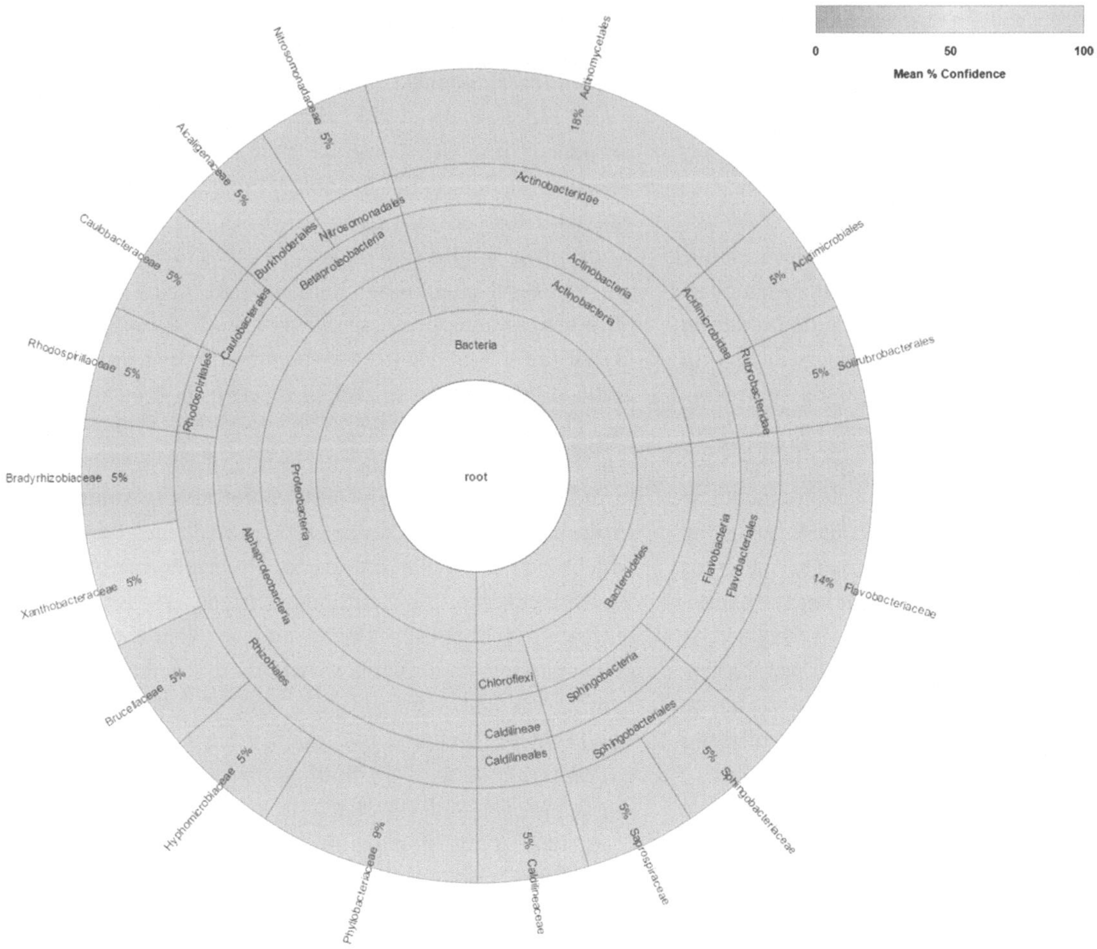

Figure 2 | Geneious map showing the distribution pattern of microbial groups within marine ammonia-oxidizing consortium (AMOPCU-1).

also present in abundant numbers having 18% Actinomycetales, 5% Acidimicrobiales and 5% Solirubrobacterales. Chloroflexi represented the minor part in consortium with 5% Caldilineaceae. The presence of a large number of Proteobacteria confirmed the existence of nitrifier-like organisms in the consortium. Different types of microbes made the consortium species-rich.

Comparative *amo*A sequence analysis and rRNA approach of nitrification-denitrification sludge systems detected a similar kind of coexistence of ammonia-oxidizing and nitrite oxidizing bacteria such as Bacteroidetes, Proteobacteria, and Actinobacteria (McLellan *et al.* 2010). The abundance of various alpha and beta-Proteobacteria such as *Nitrobacter*, *Nitrosomonas*, *Vitellibacter*, *Alcaligenes*, *Mesorhizobium*, *Tistrella*, *Caulobacter*, etc., highlighted the efficiency of the consortium to perform ammonia and nitrite oxidation and hydrocarbon degradation. As the AMO has broad substrate range, ammonia oxidizers such as *N. europaea* could also be utilized in the bioremediation of contaminated sediments and could also assist in the oxidation of several alternate substrates such as hydrocarbons (Yang *et al.* 1999). The presence of marine bacteria like *Formosa* sp. and *Salinibacterium* proved the influence of salinity on their existence in a marine environment. Actinobacteria groups such as *Microbacterium* and *Streptomyces* were already found resolved from AMOPCU-1 and AMONPCU-1 as reported by Preena *et al.* (2017a), thereby highlighting their dominant role within the consortia for performing heterotrophic denitrification as well. Actinobacteria like *Microbacterium*, *Rhodococcus* and *Gaiella* were demonstrated to have hydrocarbon degrading property also in wastewater (You *et al.* 2007).

The phylogenetic tree constructed using the aligned sequences of dendrogram generated 19 OTUs from AMONPCU-1 with 99% similar sequences from the GenBank database. This showed a major clustering of sequences within the taxonomic composition of the brackish water counterparts. This (Figure 3) consisted

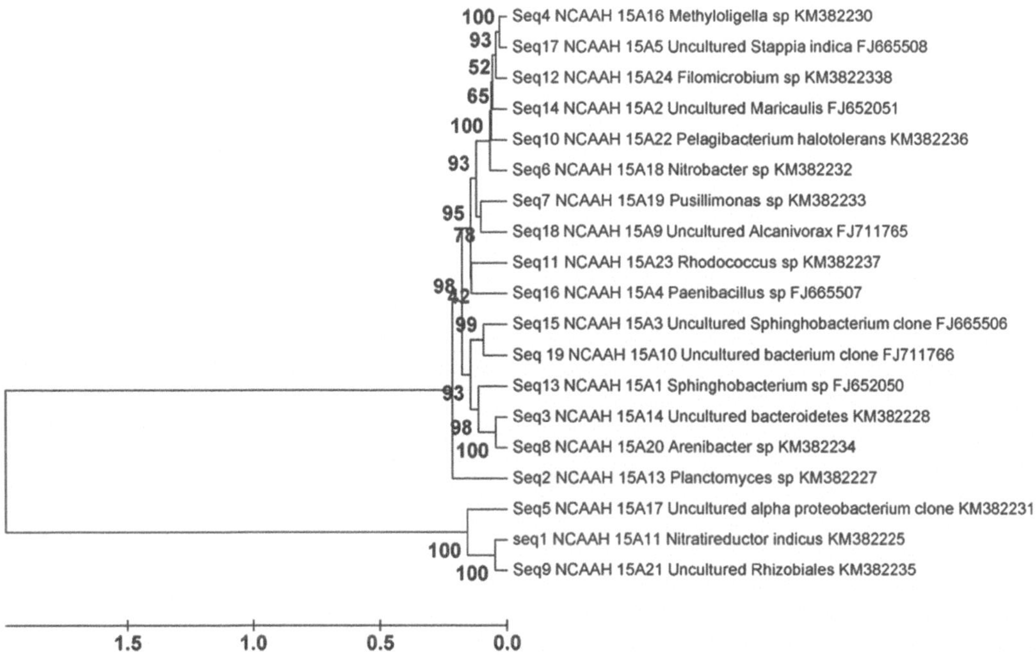

Figure 3 | Phylogenetic analysis of the 16S rRNA gene sequences from brackish water ammonia-oxidizing consortium (AMONPCU-1) by using Mega 5.0 software.

mainly of two groups with branches having altogether six clusters. The most dominant clone in AMONPCU-1, *Nitratireductor indicus,* responsible for denitrification, formed a separate group clustered with uncultured alpha-Proteobacterium and Rhizobiales at around 2.00r. This was the cluster found to be separated far away from the first group, which could be considered under group 2, and rest of the OTUs in the consortium was arranged under group 1. In group 1, *Planctomyces* sp., responsible for anaerobic ammonia oxidation, was found clustered separately at 0.2r. The remaining clusters of group 1 separated out at around 0.1r, which included *Alcanivorax* (gamma-Proteobacteria) and *Pusillimonas* (beta-Proteobacteria) as the one, and alpha-Proteobacteria such as *Methyloligella*, *Stappia indica*, *Filomicrobium* sp., *Maricaulis* sp., *Pelagibacterium* sp. and *Nitrobacter* sp. as another. Various Bacteroidetes named *Sphinghobacterium*, *Muricauda,* and *Arenibacter* got branched off differently along with *Paenibacillus* (Firmicutes) and *Rhodococcus* (Bacteroidetes) as another cluster. The diversity estimation using Mega 5.0, revealed the diversity of AMONPCU-1 as 1.328 ± 0.174 and the mean interpopulation diversity between AMOPCU-1 and AMONPCU-1 as 0.110 ± 0.023 and the overall mean distance between them as 1.717 ± 0.214.

In the Geneious map (Figure 4) of AMONPCU-1, the identified organisms were not found to be uniformly distributed; instead, some other groups were dominating, unlike the marine one. It consisted mainly Proteobacteria, Bacteroidetes, and Actinobacteria as in AMOPCU-1 followed by Planctomycetes and Firmicutes. Six types of alpha-Proteobacteria families such as Hyphomicrobiaceae (16%), Rhodobiaceae (5%), Bradyrhizobiaceae (5%), Phyllobacteraceae (5%), Hyphomonadaceae (5%) and Rhodobacteraceae (5%) were present along with 5% Alcanivoraceae belonging to β-Proteobacteria. In addition to α- and β-Proteobacteria, Alcanigenaceae of γ-Proteobacteria with 5% also was found to be present in brackish water one unlike marine consortium, where it was not observed. Only 5% Actinomycetales was observed in brackish water consortium, hence it was not the predominant one as in the case of marine consortium. A greater number of *Flavobacterium* with 16% and *Sphinghobacterium,* with 11% Chitinophagaceae and 5% Sphinghobacteraceae were found in the Bacteroidetes group. Accordingly, Bacteroidetes were found to be predominated in brackish water consortium than in the marine one. Unlike marine consortium, anaerobic ammonia oxidizers, Planctomycetes and Firmicutes also existed at 5% each. Even though different families were present in AMONPCU-1, they were not uniformly scattered as compared to those in AMOPCU-1. Hence, the species diversity appeared to be lower for AMONPCU-1 than AMOPCU-1 as observed in the case of nitrite-oxidizing consortia (NIOPCU-1 and NIONPCU-1) (Preena *et al.* 2017b).

Besides phylogenetic assessment, an overall taxonomic composition of AMOPCU-1 and AMONPCU-1 could be

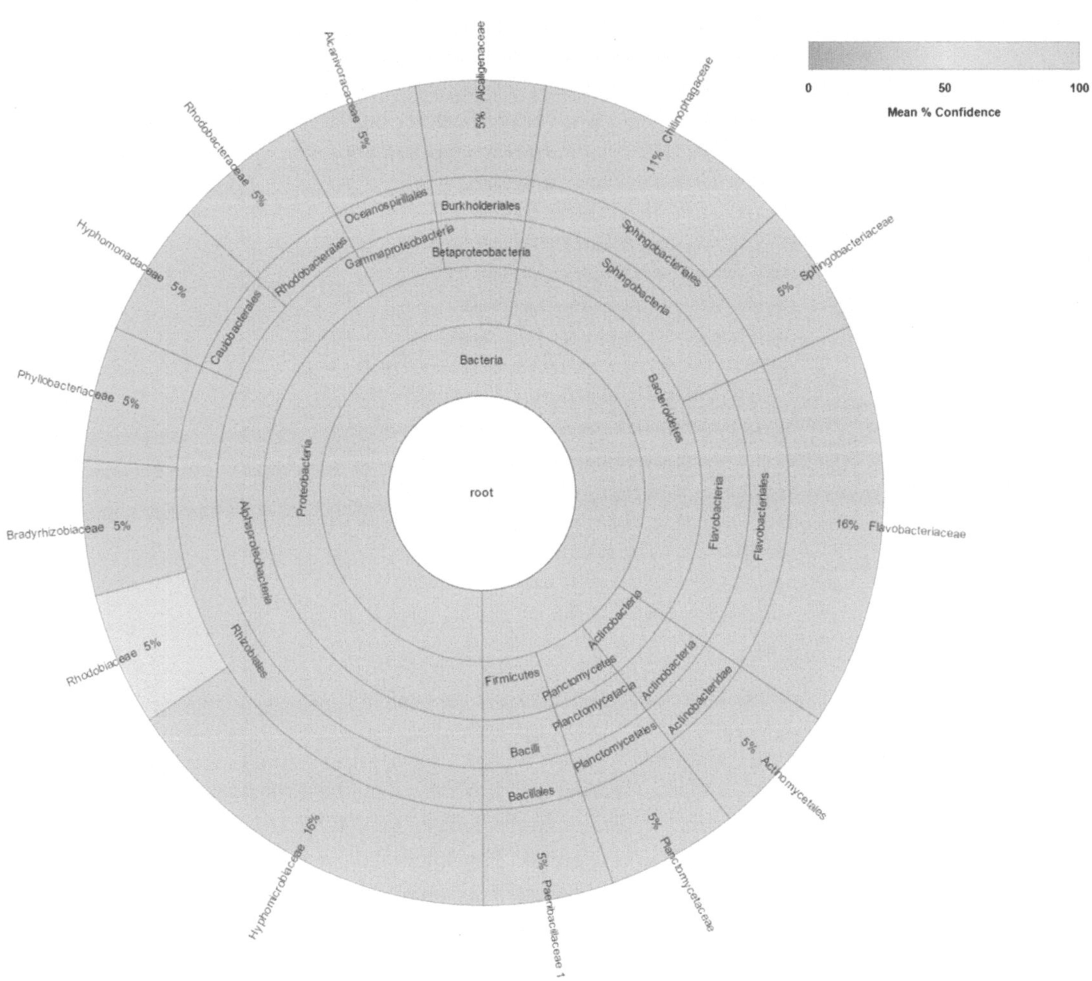

Figure 4 | Geneious map showing the distribution pattern of microbial groups within brackish water ammonia-oxidizing consortium (AMONPCU-1).

clearly determined using VITCOMIC software. In the VITCOMIC map (Figure 5), each phylum name was represented by various font colors. The percentage of similarities to database sequences were differentiated by size of circles, with an inner circle showing 80% followed by 85, 90, 95 and 100% similarity. The relative abundance of the sequences in the samples was denoted by the size of the dots. The red dots indicated the taxonomic group of AMOPCU-1, green of AMONPCU-1, grey the common taxonomic group belonging to AMOPCU-1 and AMONPCU-1. The VITCOMIC map showed that most of the taxonomic communities of both the consortia belonged to Proteobacteria and Bacteroidetes. The comparative taxonomic composition generated by VITCOMIC map revealed the relative dominance and species distribution found in the two consortia. The map showed that <10% of Bacteroidetes and <10% of Actinobacteria found commonly and equally in both the consortia. AMOPCU-1 showed the relative abundance of <5% Proteobacteria, Chlorobi, Gemmatimonodates, followed by <5% of Actinobacteria and Bacteroidetes. While AMONPCU-1 presented the relative abundance of <10% of Proteobacteria, Nitrospirae, Planctomycetes, Chlorobi, Firmicutes along with <10% of Bacteroidetes. Schreier *et al.* (2010) also reported the similar kind of coexistence of Actinobacteria, Bacterioidetes/Chlorobi, Firmicutes, *Nitrospirae*, Planctomycetales, and Proteobacteria in the microbial biofilter. Jacard index, Lenon index and Yue and Clayton Theta of 0.0789, 0.1578 and 0.0789, respectively, revealed the significant difference between both the communities.

Species diversity is mainly influenced by the species abundance, distribution, and frequency of occurrence. As the species richness and abundance increases, diversity also increases (Torsvik *et al.* 1996). Here, species richness (d = 5.693) seems to be high in AMOPCU-1 but evenness and dominance seem to be low (J' = 0.9455,

Figure 5 | VITCOMIC merged map results for marine (AMOPCU-1) and brackish water (AMONPCU-1) ammonia-oxidizing consortia.

1-lambda' = 0.9602) and vice versa for AMONPCU-1 (d = 5.114, J' = 0.9682, 1-lambda' = 0.9646). But as a whole, Shannon Wiener diversity index of AMOPCU-1 [H' (log2) = 4.216] was found to be higher than AMONPCU-1 [H' (log2) = 4.113]. On PCA the graph generated using Primer 7 software showed the distribution and clustering pattern of different OTUs recovered from both the marine and brackish water consortia (Figure S3, available online). A total of unique 19 OTUs were clustered in marine consortium and 16 OTUs in brackish one, while only three OTUs were found to be clustered common in both the consortia, thereby supporting the information obtained using other software tools that marine consortium was more diverse than the brackish one.

Even though 16S rRNA gene sequences provided basic information to evaluate the diversity and community composition of microbes in complex environments, phylogenetic analysis of 100 clones could reveal only a partial view of the phylogenetic breadth present in the consortia. Hence, in addition to 16S rRNA analyses, a functional gene analysis was also considered to unearth the diversity. Twelve clones were selected from the functional *amo*A gene library having 100 clones of each consortium for sequencing. Obtained sequences from each consortium were aligned with *amo*A proteins of nitrifiers as present in the GenBank and accordingly, phylogenetic trees were constructed (Figures S4 and S5, available online). The presence of the *amo*A gene within both the ammonia-oxidizing consortia confirmed the presence of autotrophic nitrifiers. All the cloned sequences from both the consortia have shown >98% similarity to species of uncultured *Nitrosomonas*, uncultured nitrifying bacterium, and uncultured ammonia-oxidizing beta-Proteobacteria. 16S rRNA analysis already hits with *Nitrosomonas* and uncultured beta-Proteobacteria, thereby confirming their predominant presence.

The studies carried out using different approaches indicated that various uncultured alpha-Proteobacteria, Actinobacteria, and Bacteroidetes were found common in both AMOPCU-1 and AMONPCU-1. But beta-Proteobacteria such as *Nitrosomonas* and alpha-Proteobacteria such as *Nitrobacter* were noticed only among AMOPCU-1 on analyzing 16S rRNA sequences. This might be due to the minority of these groups in AMONPCU-1. But even though diversity was found to be low in AMONPCU-1, it has been enriched with denitrifier, *Nitratireductor indicus*, anaerobic ammonia oxidizer, Planctomycetes and gamma-Proteobacteria which were not observed in AMOPCU-1. This observation is unlike in that of brackish water nitrite oxidizing consortium (NIONPCU-1), where those specific groups were not seen (Preena *et al.* 2017b). Accordingly, the unique combination of various uncultured autotrophs and heterotrophs in the consortia made them unique and efficient bioaugmentors.

Even though abundant and diverse archaea have been reported to decipher the role of ammonia oxidation (Francis *et al.* 2005), no archeaota was detected in AMOPCU-1 and AMONPCU-1. This might be due to the niche separation, influenced by the combined effect of oxygen concentration and ammonia availability, between the bacterial and archaeal ammonia oxidizers (Wang *et al.* 2013). Environmental 16S rRNA and *amo*A gene libraries significantly extended our knowledge on the natural diversity of AOB. It is suggested that the combined existence of *Nitrosomonas* and *Nitrobacter* could out-compete the other nitrifiers such as *Nitrospira* and *Nitrosospira* at high oxygen and substrate load which might be the reason of the absence of latter in the nitrifying bacterial consortia studied (Schramm *et al.* 2000). Diversity among the functional bacterial *amo*A gene in AMOPCU-1 and AMONPCU-1 was found to be too low as compared to 16S rRNA sequences. Recent studies mentioned the major role of *Nitrosomonas* in ammonia removal in saline wastewater systems even though they existed in a few populations (Sánchez *et al.* 2013). Thus, in the present study, the major occurrence of *Nitrosomonas* highlighted the nitrifying efficiency of the consortia even if highly diverse ammonia oxidizers were not observed on *amo*A gene analysis.

Quantification of *amo*A gene by quantitative PCR

The concentration of *amo*A gene in AMOPCU-1 was found to be 5.543×10^3 copies/60 ng of DNA and 6.20×10^2 copies in AMONPCU-1. As the Genomic DNA extracted from 1 g (wet weight) AMOPCU-1 was diluted to 4×10^5 and that of brackish water to 1×10^5, the copy number of *amo*A gene could be recorded as 2.21×10^9 and 6.20×10^7 per mL of genomic DNA, respectively. Thus, the total number of copies of *amo*A gene present in 1 g wet weight of AMOPCU-1 could be measured as $2.21 \pm 0.25 \times 10^9$/g and $6.20 \pm 0.23 \times 10^7$/g in AMONPCU-1. Accordingly, the copy number of *amo*A gene was found to be fairly high in both marine (AMOPCU-1) and brackish water (AMONPCU-1) consortium even though their diversity was low on qualitative analysis. The efficiency was inferred from the slope of the regression line as 94.499% and R^2 value could be observed as 0.929 (Figures S6 and S7, available online). Quantification was based on the threshold cycle, C_t. The marine nitrifying bacterial consortium, AMOPCU-1, was found to have the mean threshold cycle (ΔC_t) of 27.8, and the brackish water consortium, AMONPCU-1, 30.43. Thus, the threshold cycle itself revealed that *amo*A gene count was higher in AMOPCU-1 than in AMONPCU-1. The quantified *amo*A genes showed a significant difference between the two consortia statistically (ANOVA) (P value = 0.0029). Likewise, the abundance of various ammonia oxidizing and nitrite oxidizing bacteria was quantified by RT PCR in domestic wastewater treatment systems so as to determine the system performance (Bai *et al.* 2012).

The new information generated in this study was that the diverse organisms under the group γ Proteobacteria such as *Nitrosomonas* and α Proteobacteria such as *Nitrobacter* were enriched more in AMOPCU-1 compared to that in AMONPCU-1. Accordingly, it could be concluded that marine ammonia-oxidizing consortium was more potent in nitrification compared to the brackish water ammonia-oxidizing consortium. This indicates that sea water (marine environment) harbors more potent nitrifiers with more diversity in contrast to the observations in the previous studies where fresh waters have been demonstrated to have elevated nitrification potency with more diversity (Bernhard *et al.* 2005).

CONCLUSION

Molecular characterization by way of sequencing 16S rRNA revealed the coexistence of autotrophic nitrifiers, heterotrophic denitrifiers and anaerobic ammonia oxidizers in both marine and brackish water consortia. Functional *amo*A gene analysis corroborated with the presence of specific ammonia oxidizers and their quantitative analyses detected the abundance of ammonia oxidizers as $2.21 \pm 0.25 \times 10^9$ copies/g wet weight of marine ammonia-oxidizing consortium and $6.20 \pm 0.23 \times 10^7$ copies/g of that of the brackish water one. The phylogenetic analysis employing Mega 5.0, Geneious, VITCOMIC, and Primer 7 software suggested higher diversity in the marine ammonia-oxidizing consortium (AMOPCU-1) than that in the brackish water (AMONPCU-1) counterparts.

ACKNOWLEDGEMENTS

This work was carried out with the financial assistance from Department of Science and Technology (DST) under

INSPIRE Scheme with Grant number IF0193 and also acknowledges Cochin University of Science and Technology for the support.

CONFLICT OF INTEREST

The authors declare that we have no conflict of interest.

REFERENCES

Achuthan, C., Kumar, V. J. R., Manju, N. J., Philip, R. & Singh, I. S. B. 2006 Development of nitrifying bacterial consortia for immobilizing in nitrifying bioreactors designed for penaeid and non-penaeid larval rearing systems in the tropics. *Indian Journal of Geo-Marine Sciences* **35**, 240–248.

Altschul, S. F., Gish, W., Miller, W., Myers, E. W. & Lipman, D. J. 1990 Basic local alignment search tool. *Journal of Molecular Biology* **215**, 403–410.

Bai, Y., Sun, Q., Wen, D. & Tang, X. 2012 Abundance of ammonia-oxidizing bacteria and archaea in industrial and domestic wastewater treatment systems. *FEMS Microbial Ecology* **80**, 323–330.

Bernhard, A. E., Donn, T., Giblin, A. E. & Stahl, D. A. 2005 Loss of diversity of ammonia-oxidizing bacteria correlates with increasing salinity in an estuary system. *Environmental Microbiology* **7**, 1289–1297.

Clarke, K. R. & Gorley, R. N. 2015 PRIMER v7: User Manual/Tutorial, PRIMER-E, Plymouth, pp. 296.

Cui, Y.-W., Zhang, H.-Y., Ding, J.-R. & Peng, Y.-Z. 2016 The effects of salinity on nitrification using halophilic nitrifiers in a Sequencing Batch Reactor treating hypersaline waste water. *Scientific Reports* **6**, 24825.

Francis, C. A., Roberts, K. J., Beman, J. M., Santoro, A. E. & Oakley, B. B. 2005 Ubiquity and diversity of ammonia-oxidizing archaea in water columns and sediments of the ocean. *Proceedings of the National Academy of Sciences of the United States of America* **102**, 14683–14688.

Kasmuri, N. & Lovitt, R. W. 2018 Enriching the culture of ammonia oxidizing bacteria from soil and fishpond with bio-filters. *Journal of Engineering Science and Technology* **13**, 1562–1572.

Kumar, V. J. R., Joseph, V., Philip & Singh, I. S. B. 2010 Nitrification in brackishwater recirculating aquaculture system integrated with activated packed bed bioreactor. *Water Science and Technology* **61**, 797–805.

Lane, D. J. 1991 16S/23S rRNA sequencing. In: *Nucleic Acid Techniques in Bacterial Systematics* (E. Stackebrandt & M. Goodfellow, eds). Wiley, New York, NY, USA, pp. 115–147.

Manju, N. J., Deepesh, V., Achuthan, C., Rosamma, P. & Singh, I. S. B. 2009 Immobilization of nitrifying bacterial consortia on wood particles for bioaugmenting nitrification in shrimp culture systems. *Aquaculture* **29**, 65–75.

McLellan, S., Huse, S., Spitz Mueller, S., Andreishcheva, E. & Sogin, M. 2010 Diversity and population structure of sewage derived microorganisms in wastewater treatment plant influent. *Environmental Microbiology* **12**, 378–392.

Mori, H., Maruyama, F. & Kurokawa, K. 2010 VITCOMIC: visualization tool for taxonomic compositions of microbial communities based on 16S rRNA gene sequences. *BMC Bioinformatics* **11**, 332.

Preena, P. G., Manju, N. J., Deepesh, V., Thomas, A. & Singh, I. S. B. 2017a Genetic diversity of nitrate reducing bacteria in marine and brackishwater nitrifying bacterial consortia generated for activating nitrifying bioreactors in recirculating aquaculture systems. *Aquaculture Research* **48**, 5729–5740.

Preena, P. G., Kumar, V. J. R., Achuthan, C., George, R., Boobal, R., Nair, R. R. & Singh, I. S. B. 2017b Diversity of marine and brackishwater nitrite oxidizing consortia developed for activating nitrifying bioreactors in aquaculture. *International Journal of Environmental Science and Technology* (In press).

Rotthauwe, J. H., Witzel, K. P. & Liesack, W. 1997 The ammonia monooxygenase structural gene amoA as a functional marker: molecular fine-scale analysis of natural ammonia-oxidizing populations. *Applied and Environmental Microbiology* **63**, 4704–4712.

Rowan, A. K., Snapeb, J. R., Fearnsidec, D., Barerd, M. R., Curtise, T. P. & Head, I. M. 2003 Composition and diversity of ammonia-oxidizing bacterial communities in wastewater treatment reactors of different design treating identical wastewater. *FEMS Microbiology Ecology* **43**, 195–206.

Sánchez, O., Ferrera, I., González, J. M. & Mas, J. 2013 Assessing bacterial diversity in a sea water-processing waste water treatment plant by 454-pyrosequencing of the 16SrRNA and *amo*A genes. *Microbial Biotechnology* **4**, 435–442.

Schramm, A., de Beer, D., Gieseke, A. & Amann, R. 2000 Microenvironments and distribution of nitrifying bacteria in a membrane- bound biofilm. *Environmental Microbiology* **2**, 680–688.

Schreier, H. J., Mirzoyan, N. & Saito, K. 2010 Microbial diversity of biological filters in recirculating aquaculture systems. *Current Opinion in Biotechnology* **21**, 318–325.

Tamura, K., Peterson, D., Peterson, N., Stecher, G., Nei, M. & Kumar, S. 2011 MEGA5: molecular evolutionary genetics analysis using Maximum Likelihood, evolutionary distance, and Maximum Parsimony methods. *Molecular Biology and Evolution* **28**, 2731–2739.

Torsvik, V., Sorheim, R. & Gokseyr, J. 1996 Total bacterial diversity in soil and sediment communities – a review. *Journal of Industrial Microbiology* **17**, 170–178.

van Kessel, M. A. H. J., Speth, R., Albertsen, M., Nielsen, P. H., Op den Camp, H. J. M., Kartal, B., Jetten, M. S. M. & Lücker, S. 2015 Complete nitrification by a single microorganism. *Nature* **528**, 555–559

Vannecke, T. P. W., Bernet, N., Winkler, M. K. H., Santa-Catalina, G., Steyer, J. P. & Volcke, E. I. P. 2016 Influence of process dynamics on the microbial diversity in a nitrifying biofilm reactor: correlation analysis and simulation study. *Biotechnology and Bioengineering* **113**, 1962–1974.

Wang, Y. F., Feng, Y. Y., Ma, X. & Gu, J. D. 2013 Seasonal dynamics of ammonia/ammonium oxidizing prokaryotes in oxic and anoxic wetland sediments of subtropical coastal mangrove. *Applied Microbiology and Biotechnology* **97**, 7919–7934.

Wang, Z., Zhang, X.-X., Lu, X., Liu, B., Li, Y., Long, C. & Li, A. 2014 Abundance and diversity of bacterial nitrifiers and denitrifiers and their functional genes in tannery wastewater treatment plants revealed by high-throughput sequencing. *PLoS One* **9**, e113603.

Yang, L., Chang, Y. F. & Chou, M. S. 1999 Feasibility of bioremediation of trichloroethylene contaminated sites by nitrifying bacteria through cometabolism with ammonia. *Journal of Hazardous Materials* **69**, 111–126.

You, S. J., Tseng, D. H., Ou, S. H. & Chang, W. K. 2007 Performance and microbial diversity of a membrane bioreactor treating real textile dyeing wastewater. *Environmental Technology* **28**, 935–941.

First received 4 September 2018; accepted in revised form 9 January 2019. Available online 18 January 2019

IWA Publishing's authorised EU representative for General Product Safety Regulations is Diane D'Arras, 15 rue Duret, 75116 Paris, France, e-mail: safety@iwap.co.uk.

Printed and bound by CPI Group (UK) Ltd, Croydon, CR0 4YY
12/05/2026
02108868-0001